1. 2018 Science & You International Symposium and Summit on Culture of Science kicked off in Beijing on Sep. 15th, 2018

2. HAN Qide, Invited Speech, *Diversity and Inclusive Communication of Science Culture*
Honorary President of China Association for Science and Technology, Academician of Chinese Academy of Sciences

3. HUAI Jinpeng, Welcome Address
Secretary of the Leading Party Members' Group, Executive Vice President and Chief Executive Secretary of China Association for Science and Technology

4. Olivier RICHARD, Welcome Speech
Ministre Conseiller/Deputy Head of Mission

5. LI Hong, Chairmen of the Conference
Vice President of China Association for Science and Technology

6. Pierre MUTZENHARDT, Chairman of the Conference
President of University of Lorraine

7. Bernard SCHIELE, Co-Chair of Scientific Committee and Program Committee
Full Professor, Faculty of Communication of University of Quebec at Montreal

8. CHENG Donghong, Chair of Program Committee
President of Chinese Association of Natural Science Museums

9. Massimiano BUCCHI, Keynote Speech, *The Challenges of Science Communication 2.0: Credibility, Expertise and the Crisis Of Mediators*
Editor of Public Understanding of Science, Professor of Science and Technology and Society, University of Trento

10. Noyuri MIMA, Keynote Speech, *Potential of Local Science Festivals for a Sustainable Society of the Future*
Professor of Future University Hakodate, Deputy Director of the National Museum of Emerging Science and Innovation

11. LIAO Fangyu, Keynote Speech, *Scientific Culture Communication in Internet Environment*
Director of Computer Network Information Center, Chinese Academy of Sciences

12. GUO Yike, Plenary Session Speech, *Data Science and You*
Professor of Imperial College

13. LI Zhengfeng, Plenary Session Speech, *Thoughts on the Trend of Scientific Culture Development*
Vice Director of School of Social Science of Tsinghua University

14. Michel CLAESSENS, Plenary Session Speech, *Is There a Future for Science Communication?*
ITER Policy and Communication Officer

15. Jan RIISE, Plenary Session Speech, *Community and Societal Engagement in Science*
Manager for Engagement of City Future Research Center in Chalmers University of Technology

16. REN Fujun, Summary Speech, Director of Organizing Committee
President of National Academy of Innovation Strategy

17. WANG Yaping, Special Guest
The Four-level Astronaut of PLA Astronaut Brigade

18. ZHOU Wenbiao, Director of Organizing Committee
Secretary of the Leading Party Committee and Vice President of National Academy of Innovation Strategy

19. LUO Hui, Member of Scientific Committee
China Association for Science and Technology

20. ZHOU Daya, Secretary-general of Organizing Committee
Vice President of National Academy of Innovation Strategy

21. RUAN Cao, Vice Director of Organizing Committee
Vice President of National Academy of Innovation Strategy

22. Group Photo

SCIENCE & YOU

CREATE THE FUTURE

科学与你

2018“科学与你”国际研讨会暨科学文化高峰论坛论文集

Proceedings of the 2018 Science & You International Symposium and Summit on Culture of Science

刘萱 主编

赵勍 李响 马健铨 副主编

清華大學出版社

北 京

内 容 简 介

本书是2018“科学与你”国际研讨会暨科学文化高峰论坛的论文汇编，本次会议由中国科协支持，中国科协创新战略研究院与法国洛林大学共同主办。本次会议的主题是“理解、分享、参与：多元世界的新思考”，关注社会语境的多样性，展示公众、研究者、决策者的多元视角和新见解，希望打造有助于科学文化领域繁荣的学术平台。

图书在版编目(CIP)数据

科学与你：2018“科学与你”国际研讨会暨科学文化高峰论坛论文集：英文/刘萱主编.—北京：清华大学出版社，2019

ISBN 978-7-302-52486-1

Ⅰ. ①科…　Ⅱ. ①刘…　Ⅲ. ①科学技术－文化研究－国际学术会议－文集－英文　Ⅳ. G301-53

中国版本图书馆 CIP 数据核字(2019)第 043122 号

责任编辑： 魏贺佳
封面设计： 冯　帆
责任校对： 赵丽敏
责任印制： 宋　林

出版发行： 清华大学出版社
　　网　　址： http://www.tup.com.cn, http://www.wqbook.com
　　地　　址： 北京清华大学学研大厦 A 座　　**邮　　编：** 100084
　　社 总 机： 010-62770175　　**邮　　购：** 010-62786544
　　投稿与读者服务： 010-62776969, c-service@tup.tsinghua.edu.cn
　　质量反馈： 010-62772015, zhiliang@tup.tsinghua.edu.cn
印 装 者： 三河市铭诚印务有限公司
经　　销： 全国新华书店
开　　本： 185mm × 260mm　　**印　　张：** 21.25　　**插　　页：** 2　　**字　　数：** 693 千字
版　　次： 2019 年 4 月第 1 版　　**印　　次：** 2019 年 4 月第 1 次印刷
定　　价： 128.00 元

产品编号：081756-01

SCIENTIFIC COMMITTEE

CHAIR

Bernard SCHIELE（University of Quebec at Montreal, Canada）

WANG Chunfa（National Museum of China, China）

MEMBER

Martin BAUER（London School of Economics and Political Science, United Kingdom）

Franks BRADLEY（London School of Economics and Political Science, United Kingdom）

Massimiano BUCCHI（University of Trento, Italy）

CHEN Rui（National Academy of Innovation Strategy, China）

Michel CLAESSENS（ITER policy and communication officer, Belgium）

Denis ENTEMEYER（University of Lorraine, France）

Per HETLAND（University of Oslo, Norway）

KIM Hak Soo（Sogang University, South Korea）

Joëlle LE MAREC（CELSA Paris 4 Sorbonne, France）

LI Zhengfeng（Tsinghua University, China）

LIU Xuan（National Academy of Innovation Strategy, China）

Luisa MASSARANI（National Institute of Public Communication in Science and Technology, Brazil）

LUO Hui（National Academy of Innovation Strategy, China）

Gauhar RAZA（Council of Scientific & Industrial Research, India）

Gordon SAMMUT（University of Malta, Malta）

Watanabe MASATAKA（University of Tsukuba, Japan）

PROGRAM COMMITTEE

CHAIR

Bernard SCHIELE（University of Quebec at Montreal, Canada）

CHENG Donghong（Chinese Association of Natural Science Museums, China）

MEMBER

Julie ADAM（University of Lorraine, France）

Martin BAUER（London School of Economics and Political Science, United Kingdom）

Nicolas BECK（University of Lorraine, France）

Michel CLAESSENS（ITER policy and communication officer, Belgium）

Florence DAMOUR（University of Lorraine, France）

Denis ENTEMEYER（University of Lorraine, France）

HU Zhiqiang（University of Chinese Academy of Sciences, China）

LIU Xuan（National Academy of Innovation Strategy, China）

Gauhar RAZA（Council of Scientific & Industrial Research, India）

REN Fujun（National Academy of Innovation Strategy, China）

RUAN Cao（National Academy of Innovation Strategy, China）

Mariama TRAORÉ（University of Lorraine, France）

WANG Hongwei（National Academy of Innovation Strategy, China）

ZHAO Yandong（China Academy of Science and Technology Development Strategy, China）

ORGANIZER

PRESIDENTS

LI Hong（China Association for Science and Technology, China）
Pierre MUTZENHARDT（University of Lorraine, France）

DIRECTOR

REN Fujun（National Academy of Innovation Strategy, China）
ZHOU Wenbiao（National Academy of Innovation Strategy, China）

SECRETARY-GENERAL

ZHOU Daya（National Academy of Innovation Strategy, China）

VICE DIRECTOR

CHEN Rui（National Academy of Innovation Strategy, China）
RUAN Cao（National Academy of Innovation Strategy, China）

MEMBER

Julie ADAM（University of Lorraine, France）
Nicolas BECK（University of Lorraine, France）
HAO Qian（National Academy of Innovation Strategy, China）
LIU Xuan（National Academy of Innovation Strategy, China）
Pierre MUTZENHARDT（University of Lorraine, France）
SHI Yunyan（National Academy of Innovation Strategy, China）
Mariama TRAORÉ（University of Lorraine, France）
WANG Hongwei（National Academy of Innovation Strategy, China）

SECRETARIAT

LIU Xuan（National Academy of Innovation Strategy, China）
MA Jianquan（National Academy of Innovation Strategy, China）
LI Xiang（National Academy of Innovation Strategy, China）
ZHAO Ji（National Academy of Innovation Strategy, China）

SUPPORTERS OF CONFERENCE

Council of Scientific & Industrial Research, India
China Academy of Science and Technology Development Strategy, China
European Commission
International Fusion Reactor Program（ITER）, Belgium
Japanese Association of Science Communication, Japan
London School of Economics and Political Science, UK
National Institute of Public Communication in Science and Technology, Brazil
National Museum of China, China
National Museum of Emerging Science and Innovation, Japan
Sogang University, South Korea
Tsinghua University, China
University of Chinese Academy of Sciences, China
University of Quebec at Montreal, Canada
University of Malta, Malta
University of Oslo, Norway
University of Paris-Sorbonne, France
University of Trento, Italy
University of Tsukuba, Japan

FOREWORD

In September 15-17th 2018, the National Academy of Innovation Strategy (NAIS), affiliated to the China Association for Science and Technology (CAST) organized the Science & You International Conference in Beijing, in collaboration with the University of Lorraine. It is the first time that Science & You took place in China, in Asia and in non-French area. According to previous conferences and international standard of symposium, the organizer constructed international scientific committee and program committee in 2016, held committee meetings to discuss procedures and arrangements of the symposium, and finally held the whole conference successfully.

China has a long history and brilliant cultures of science. Simultaneously, with the rapid development of science and technology in modern context, our view on science has a bearing on the development of human beings and the prosperity of our society. CAST has been taking great efforts on promoting the culture of science, building related academic platform and improving international communication in this field. Thanks to the previous accumulation and working foundation, it is time to host a high level international academic conference, which helps to gather scholars in the field of culture of science for discussions in depth. Science & You originated from the University of Lorraine in France, one of most important centers of science in the world. Its first edition took place in Nancy (France) from 1st to 6th June 2015, and gathered 1000 attendees, 130 PhD students, and 10000 visitors of the general public. Science & You has been developing into an international brand in science culture field. Consequently, based on the common aim of internationalization, the platform was co-established.

Theme of this conference is Knowing, Sharing, Caring: New Insights for a Diverse World. The scientific committee of Science & You agrees that globalization rests upon diversity, including:

Science communication strategies and practices;

Contexts in which communication actions take place;

Publics at which is aimed science communication;

Questions, issues, responsibilities of science communicators as well as scientists;

Research in science of science communication, models, approaches, applications, impacts, reception, understanding etc.

In the proceeding, most of reported papers were collected in this book. The collection was divided into four sessions, which are:

1. Culture of Science in Public: Participation and Engagement;
2. Culture of Science in Media: Communication and Representation;
3. Culture of Science in Scientists and Their Community: Role and Ethos;
4. Culture of Science in Policy and Strategy: Governance and Legislation.

The aim of this collection is to provide a place of sharing, exchanges, discussions and debates, to acknowledge this diversity of contexts, publics, research, strategies and new insights, and to shape an environment conducive to decentering, and an opportunity for enrichment.

It is a splendid step but only the beginning of NAIS's contribution on constructing the high-level international communication platform, as well as collaboration with oversea institutions to promote the culture of science in China. NAIS will make contributions to researches and communications in the field of culture of science continuously and find its own distinctive role in the global academic community.

REN Fujun
President of National Academy of Innovation Strategy, CAST

CONTENTS

The Potential of Local Science Festivals for a Sustainable Society

——A case study of the hakodate international science festival

Noyuri Mima

Faculty of Systems Information Science, Future University Hakodate, Hakodateshi, Japan

Abstract: In this paper, I describe a local science festival that I began designing in 2008 and which has been held each year since 2009. I use this case to consider the significance of local science festivals as learning environments for citizens. It is worth mentioning that the design of the local science festival discussed here involved the application of learning theory and philosophy. Rather than 'knowledge transfer' or 'knowledge acquisition', science communication is contingent on inter-personal activity performed through conversations that are inseparable from the situations in which they occur. Science communication is a part of learning activities and should be defined as a process of interaction that transcends a single individual, emerging in the context of social relationships within a broadly inclusive community. Moreover, conviviality, the vernacular, and the commons will also be important when considering the significance of local science festivals. Many cities in Japan and around the world face situations like the one discussed here. It will be meaningful to share the lessons learned and design model used in this case as a 'cultural apparatus'. I believe that local science festivals will contribute to creating a sustainable and resilient society.

Keywords: Science Festival; Science Communication; Science Literacy; Learning; Local Context; Conviviality; Vernacular; Commons; Cultural Apparatus

1. Background

It has been 10 years since the inaugural science festival was held in the Japanese city of Hakodate. The festival is held annually at multiple venues in the Hakodate area for nine days every August. It offers a wide variety of science-related events for everyone from children to adults—for laypersons and experts alike. As a series of preliminary events, hands-on classes and experimental workshops for children are offered through the summer holidays beginning in mid-July. There are also events for adults in September.

The number of people involved in carrying out the operational tasks for the festival has been increasing each year; thus, it has gradually become an established event. People from diverse backgrounds bring their interests and ideas, forming multiple groups loosely joined with one another. Though funding has been a challenge each year, we have steadily gained the support of local businesses and leveraged the festival's strengths as a networked organization. Contributions of free drinks, product samples, and a free venue are appreciated. The festival is a place for participatory and collaborative practices generated by the mutual exercise of the knowledge, skills, and ideas brought by various individuals.

I was involved in establishing and drawing up the plans for the National Museum of Emerging Science and Innovation (Miraikan) in Tokyo, where I served as the deputy director from 2003 to 2006. At that time, the importance of science communication was beginning to gain recognition as a worldwide trend. After the completion of my term of office, when I returned from Tokyo to Hakodate, a regional city with a population of 260,000 people, it came to my attention that Hakodate did not have a science museum or science center. However, I knew that it would be difficult for such a small city to secure a budget for creating and operating a science museum/center.

Therefore, in 2008, I recruited some colleagues to establish Science Support Hakodate as an organization to promote science communication and develop citizens' science literacy. Since 2009,

116-2 Kamedanakanocho, Hakodateshi, Hokkaido 0418655 Japan. noyuri@fun.ac.jp.

we have held the Hakodate International Science Festival every summer. Through trial and error, we have built up our know-how and a record of achievement, resulting in the expansion of our circle of volunteers from industries, academia, and the government, such that our yearly calendar of activities is now more or less fixed.

2. Theory and Philosophy for Designing a Science Festival

To plan a science festival to be held in a local city, two perspectives were considered. One was a learning theory perspective regarding a festival as a place of learning and the other was a philosophical perspective regarding local communities.

1) Learning Theory Perspectives

The need to develop science literacy and promote science communication as a means of doing so has been discussed actively in recent years in science and technology policies and in the fields of science, technology, and social studies. However, these discussions have not yet—at least in the context of education and learning—touched on science festivals as a means of realizing this end (Mima and Watanabe, 2008).

A paradigm shift in learning psychology occurred with the transition that began in the 1980s from behaviorism to social constructivism to a situated model. Learning came to be reinterpreted from 'something passive' to 'something active' and from an 'individual enterprise' to a 'social enterprise' (Mima and Yamauchi, 2005). In other words, learners came to be seen as having the power to engage their own environment and seize knowledge, and learning came to be seen as a process of collaborative activity and discussion with others. It has been a shift from individual to collective learning focused on the importance of society, culture, and others. Moreover, this way of thinking has also demanded a shift in the nature of worldly things and knowledge. From something static and fixed, knowledge has come to be recognized as being socially constructed—something built communally in the context of communicative processes.

2) Philosophical Perspectives

When thinking about a science festival deeply rooted in a local community, it is useful to think about the concepts of conviviality, the vernacular, and the commons, as discussed by Ivan Illich, a philosopher who was active in the latter half of the 20th century. These important concepts are featured in Illich's most eminent works, *Deschooling Society* (Illich, 1971), *Tools for Conviviality* (Illich, 1973), and *Shadow Work* (Illich, 1981). These concepts have been reviewed again in the context of the digital society in recent years (Bollier, 2013).

To be 'convivial' means to live together happily in a state of mutual independence. Illich once lived in a village on the outskirts of Cuernavaca, Mexico, and is said to have taken this term from the Spanish word *conviviencial*, used to refer to the ties that linked the indigenous villagers to the commons and the festive interactions that occurred when a market was open (Kurihara, 2006). Local festivals are indeed convivial in this sense, representing an autonomous mechanism that makes life more vibrant and enjoyable.

'Vernacular' refers to the characteristic of being rooted in the realm of everyday life. The term describes something that is neither mass-produced elsewhere nor supplied by the government; rather, it is something of one's own. Vernacular space is something that emerges from the formation of our own space within the mutually beneficial commons in our communities. The vernacular is ultimately neither an acquisitive act realized in the form of an exchange of currency, nor is it an institutionalized service.

The term 'commons' originated from the idea of shared pasturage, and it signifies a communal environment accompanying a convivial life. It is a new social and political sphere in which people can create their own rules and solve problems tailored to local conditions in a grassroots fashion. In contrast to 'resources', commons are something 'shared by everyone', providing a space in which the subsistence activities of people take root.

Through these concepts, Illich argues, things are made rather than consumed; he regards human independence as something that builds up in the form of regional and autochthonous lifestyles in the company of others. Illich also points out the danger of relying on the products and care provided by groups of experts. He maintains that human happiness depends on our subjective understanding of the world; it is consistent with the recent significance of promoting science communication and developing science literacy.

3. Methods

To design and operate a science festival from learning theories and the philosophical viewpoints mentioned in section 2, the work by Mima and Watanabe (2008) was used as a model of community evolution. The following are descriptive of the situation in Hakodate City, where this practice, the model, and activities were implemented.

1) Background of the Initiative

- Hakodate, Hokkaido: the third largest city (260,000 people) in Hokkaido, one of the 47 prefectures of Japan
- Major industries: tourism; fisheries, especially for squid and kelp; and food product manufacturing (processing)
- Annual average income for a Hakodate citizen: $25,000 USD (2017)
- Number of higher education institutes: eight schools and approximately 4,000 students

2) The Evolution Model of a Learning Community

The Hakodate International Science Festival has been designed and operated according to the evolution model of the learning community in Mima and Watanabe (2008). The model has six phases from the phase in which activities are distributed through the phases of networking, organizing, integrating, synchronizing, and emerging as a convivial community (Fig. 1).

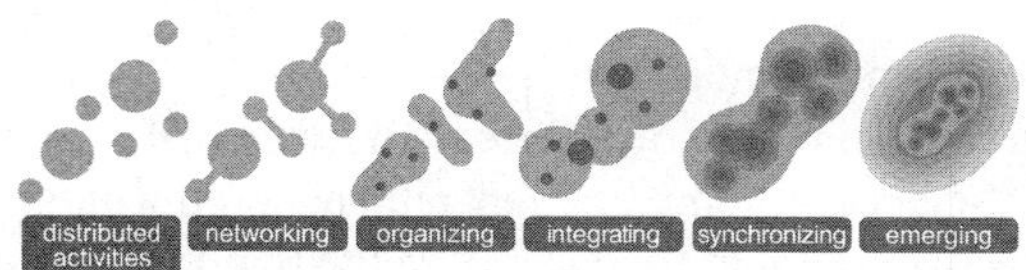

Fig. 1 The evolution model of a learning community. (modified based on Mima and Watanabe, 2008)

3) The Initiative and Its Activities

To operate the science festival, Science Support Hakodate (SSH) was established in 2008 in collaboration with the city administration, higher education institutions, research institutes, and a funding agency (Fig. 2). The mission of SSH is to promote science communication and improve science literacy in the Hakodate area through the Hakodate International Science Festival and other activities.

The tasks of SSH have been managed by 17 members of the steering committee appointed from each participating organization. Future University

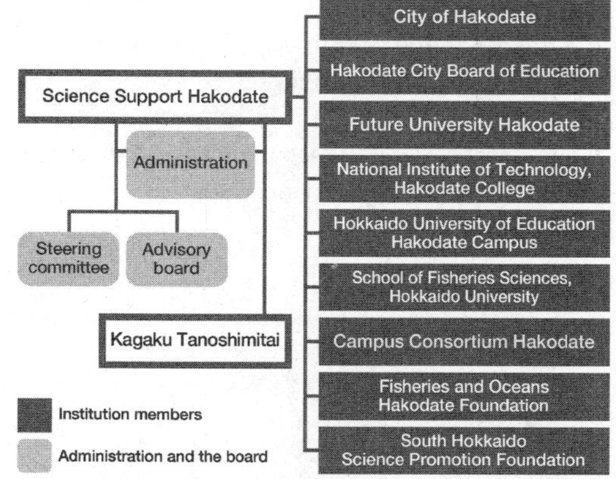

Fig. 2 Organizational chart for Science Support Hakodate.

Hakodate is in charge of the secretariat. The advisory board members from the heads of each participating institute evaluate the annual and medium-term plans. In addition, the voluntary group of citizens—Kagaku Tanoshimitai (Science Enjoyment Corps)—supports activities throughout the year.

The steering committee conducts monthly meetings, not only for practical administration but also for the active sharing of ideas. The coordinator makes various adjustments, and each member is in charge of the coordination and internal planning of individual projects within and outside of the affiliated institute. SSH oversees the publicity and design of various products to increase their visibility for citizens.

SSH is also conducting a training course—Hakodate Science Terakoya—as well as the science festival. This intensive course has been designed to develop science communicators and regional coordinators among college students and citizens.

4. Results

The results of our efforts over 10 years are as follows.

1) Hakodate International Science Festival

- Inauguration: August 2009
- Duration: nine days ending on the final Sunday in August
- Organizer: Science Support Hakodate and a steering committee of 17 people
- Annual revenue: includes monies from the social contribution budget of Future University Hakodate, corporate sponsors, etc., totaling $36,000 USD
- Expenditures: event-related expenses, advertising expenses, gratuities, etc.

Trends in numbers of participants and partners and a breakdown of exhibits and joint displays at the 2017 Hakodate International Science Festival are shown in Fig. 3 and Table 1.

There are several ways to contribute to the festival. The method and process are shown in Table 2 and Fig. 4.

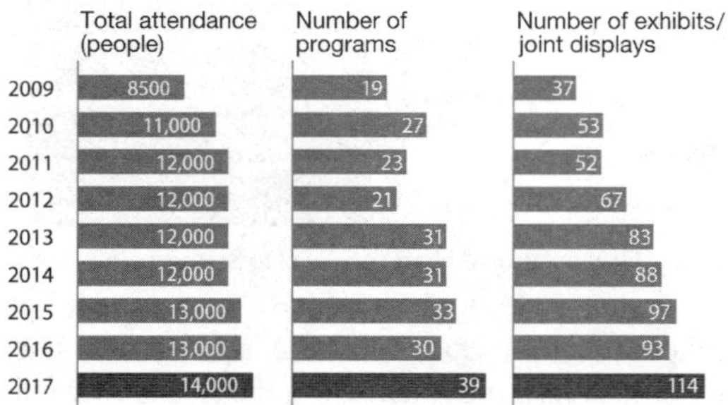

Fig. 3 Trends in numbers of participants and partners.

Table 1 Breakdown of exhibits and joint displays at the Hakodate International Science Festival 2017.

Category	Number of Exhibits Joint Displays
Schools	37
Political, economic, and cultural organizations	31
Academic and R&D organizations	11
Other education and learning support industries	9
Local public agencies	5
Professional service industries (not classified elsewhere)	4
Information service industries	2
Food and beverage retailers	2
Other retailers	2
Medical industries	2
Others (general construction industries, etc.)	9
Total	114

Table 2 The ways of contribution.

Ways of Contributing	Overview	Arranging the Venue	Expenses	Responding to Entry
PR partner	Post information of exhibitor's event as part of the science festival program to PR media	Exhibitor	Exhibitor	Exhibitor
Exhibitor's own project	Exhibitor presents original program at the festival; assumes CSR activities	SSH	Exhibitor	SSH/Venue facilities
Regular venue/ conventional style	Format is a science booth, science cafe, stage event, etc.	SSH	Exhibitor with support from SSH	SSH
Invited guest	SSH invites local and national performers	SSH	SSH	SSH
Others	Collaborate with SSH to implement an event	Negotiable	Negotiable	Negotiable

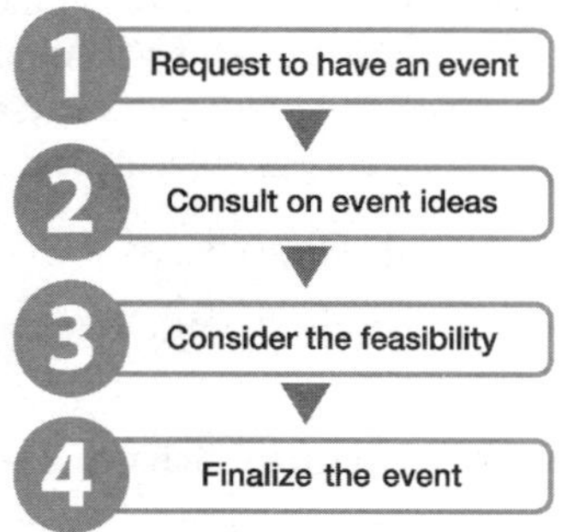

Fig. 4 Process for determining events for the science festival.

2) Training Course (Hakodate Science Terakoya)

Every year, the three-day intensive training course, Science Terakoya, is offered before and after the science festival for college students and citizens. The themes dealt with so far are as follows:

- Development of scientific events
- Information design
- Dissemination of information by media
- Science communication aimed at informing citizens about food safety in the region
- Linking science to everyday life based on an understanding of yourself
- Science demonstration by Dr. Nabe
- Workshop for communicators connecting sea, ship, stars, and science

3) Major Activities

The following are the major activities associated with expansion of the science festival:

(1) *One year prior to the science festival (until 2008; Distributed Activities Phase)*

Before the science festival was held, various science events were held in the city on weekends and holidays throughout the year. Organizers were individuals, groups of volunteer citizens, and organizations such as universities and research institutes. Their advertising methods were ad hoc and included distributing flyers and announcing events on their websites. As a result, information did not reach the targeted people properly.

(2) *Holding the science festival (from 2009; Networking Phase)*

SSH has been inviting individuals and organizations to conduct science events as exhibitors in the science festival. SSH advises them that festival participants enjoy an intensive week during the summer to reach citizens through science events. SSH provides the venue and publicity at no additional cost.

Emphasis is placed on the design of productions and exhibitor space. SSH engages the services of an art director and produces poster leaflets, program brochures, banners, etc., according to the theme of the festival. While such actions promote visibility to citizens, it also fosters unity among exhibitors.

As previously mentioned, SSH holds an intensive course—Hakodate Science Terakoya—every year to develop science communicators. The course is also open to citizens and college students.

(3) *Birth of citizens' groups (from 2012; Organizing Phase)*

Two civic groups have emerged during this period. One is a performing group and the other is the curating team.

a) Performing group

Upon attending Science Terakoya in 2011, a group of voluntary citizens known as Kagaku Tanoshimitai (Science Enjoyment Corps) was born (Fig. 5). The group is composed of people of varying ages and occupations, including business professionals, homemakers, retirees, municipal officials, teachers, and students. An evening meeting is held once a month as a space for learning, exchanging information, and sharing experiences. At the science festival, in addition to working as voluntary staff, the group also holds its own science shows and workshops. Although the group was formed for the members' mutual enjoyment of science, in recent years, it has started to receive occasional requests to put on science shows and workshops throughout the year.

Fig. 5 Kagaku Tanoshimitai.

Chikako is a female member of the group. She used to be a physics teacher. When she moved to Hakodate, she quit her job because of family circumstances. After learning of the existence of Kagaku Tanoshimitai from the newspaper, she began to take part in its activities. She is now in high demand as a science communicator in the Hakodate area (Fig. 6). She stages science shows and workshops for children and parents in venues that include local kindergartens and nursery schools, elementary schools, and shopping malls. Beginning this year, she has taken the rostrum once again as a part-time instructor at a high school.

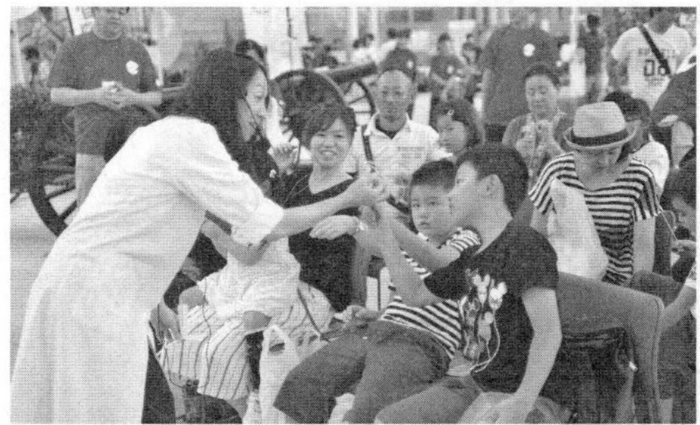

Fig. 6 Ms. Chikako's Show.

One of her charms is her approach. Using materials found in one's house, she tells children's mothers that 'you can also do this at home' and 'you can find these materials in any 100-yen shop' (the Japanese equivalent of a dollar store). As a result, ideas well up in their heads, making them want to experiment for themselves when they return home. The important thing is that instead of showing something off in a one-sided manner—like an 'amazing' magic trick— her stance is 'Let's do it together' and 'You can do it, too!' Her incredibly fun and energetic mode of activity has also been attracting new partners. Several other examples exist of regular citizens who became active as science communicators after being inspired at the science festival.

b) Curating team

Every year, the special exhibition in the festival

is conducted by SSH as the organize's message to the public and a challenge to make them aware of modern scientific issues. The exhibition was carried out with borrowings from Tokyo and other big cities for the first three years. Eventually, some people decided to carry out a special exhibition by themselves and organized a curating team. The members include writers, photographers, graphic designers, and researchers. In line with the festival theme for each year, the exhibition appeals visually to festival goers; topics have included the environment, food, and health—issues that have local as well as global ramifications (Fig. 7). To develop the contents of the exhibition, the curating team interviews local people, such as scientists, farmers, people working in the fishery industry, and government workers.

A workshop is held by the team to promote an understanding among citizens of the meaning of the exhibition on the evening preceding the opening. SSH creates a printed catalog of exhibitions, distributes them, and also publishes a digital version on its website. After the science festival, the exhibitions are available for use by schools, institutes, and city halls for free.

Fig. 7 The special exhibition.

(4) *Cycle and guidelines (from 2015; Integrating Phase)*

a) Learning cycle throughout the year

Since the science festival's eighth year, the annual cycle of activities has become more or less fixed (Fig. 8). First, in January, SSH puts out a general call for participants in what we call the 'kick-off.' Workshops are held to generate and discuss ideas about events for the festival. Participants' motivations are varied: some already have their own ideas and are seeking collaborators; some want to flesh out their concepts further; some have interests in themes like the environment, food, and health; and others are interested in education and community activities in general.

Fig. 8 Annual schedule of the science festival community.

Several groups are assembled to make the ideas generated by the kick-off reality. By June, all programs are fixed and announced to the press. In early August, the exhibitors come together for a social gathering during which we review various points to remember during the festival's run. Each group introduces its respective events and shares expert know-how; topics include methods for increasing attendance.

In the latter days of August, during the run of the science festival, we hold a social gathering for exhibitors and include guests who have traveled considerable distances. A month after the festival is over, we hold an exhibitors' meeting to share our experiences in a follow-up review of the year's science festival; additionally, we begin preparing for the following year.

What is noteworthy here is that participants in the social gathering held in early August end up participating in the science festival events introduced to them there. Participants at this social gathering are initially interested in a specific field of science, as they are the ones in charge of their own events. Through the social gathering, their interests broaden to include other fields and events, and by participating as guests, they operate as an interface, spreading their interests to the general public.

Getting ordinary citizens with little interest in science to take an interest in the subject based on attending a single event is difficult; even when they do take an interest, it is often only temporary. What the social activities described here have revealed is the effectiveness of expanding from people with a strong interest to people who are less interested. This result has been substantiated by the increase in exhibitors and joint displays, as noted in Fig. 3.

b) Building guidelines of the community

As the activities of the science festival—now in its seventh year—have continued to expand, so have the number and scope of activities we wanted to undertake. Though it is good for activities to expand, dilution of the characteristics of the science festival as a whole is a possible outcome. Therefore, the necessity of activity guidelines came up for discussion among the members of SSH. Looking back over our experiences to date, the activity guidelines listed below are in place. They were introduced to participants at the science festival kick-off event held in January 2015. These activity guidelines are still shared today whenever the opportunity arises (Fig. 9).

Fig. 9 Activity guidelines for Science Support Hakodate.

(5) *Collaborating with other communities (from 2017; Synchronizing Phase)*

In the ninth year, SSH began to receive offers to carry out events with groups not related to science, such as the Junior Chamber and the Hakodate Women's Conference. SSH has begun collaborating with these groups not only for events related to the science festival, but also for their activities, workshops, and learning groups throughout the year.

There is a symbolic story about connecting different communities. It began with a proposal from a member of Kagaku Tanoshimitai. He wanted to consider the hearing impaired in the science festival. As a result of discussions between Kagaku Tanoshimitai and SSH, it was decided to experimentally use UD Talk (http://udtalk.jp/en/), an application that visualizes communication support and conversations. Now, more than half of all events are using it with technical and financial support from developers in Tokyo. The proposal triggered a connection among tech people in the central city and the disabled community in the local city and SSH (Fig. 10).

Figure 11 shows a scene from the disaster prevention science show in the science festival, which originated with a proposal from the Hakodate Women's Conference.

Fig. 10 Using UD Talk and a sign language interpreter.

Fig. 11 Disaster prevention science show.

5. Discussion

Recently, in school-based education, 'active learning' has been attracting attention as a learning and pedagogical method. It represents a shift away from the conventional method of teaching with mass lectures. It is a project-based learning style in which students learn actively by discovering challenges independently and working toward solutions. Learning occurs not only in terms of knowledge and skills in related areas, but also with respect to things like teamwork and logical and analytical modes of thought. Even more important is how this experience is conceptualized as knowledge that can be applied in other contexts.

The Hakodate International Science Festival, with reference to the relationship between science, technology, and society, aims to turn people's attention from their immediate problems to the wider world so that they may more easily consider global problems as their own. Beyond the planning and organizing of events, a science festival that includes a cycle of learning, creating, and sharing is such a place of active learning.

Gaining science literacy also leads to the prevention of illness, the reduction of waste, and a lower likelihood of being deceived by unscrupulous business practices. Developing science literacy benefits individuals and reduces social costs in general. Moreover, besides learning about science and technologies related to healthcare, the environment, or food, scientific thinking—in terms of thinking analytically and logically and seeing things critically—is also important. Mastering this kind of knowledge and way of thinking could also affect the next generation, resulting in a positive chain of effects that could lead to the elimination of economic and educational disparities.

From the time it was first held in 2009, the Hakodate International Science Festival has had as its slogan, 'Thinking about the world from Hakodate, and thinking about Hakodate from the world.' By showing the connections between people's immediate problems and those of the world, we help people become conscious of global problems and their potential to solve them, thus putting their thoughts into action. We provide the chance to 'think globally, act locally'.

Regarding the six phases shown in section 3, 2), activities implemented by voluntary citizens' groups have begun. Furthermore, a science communicator emerged, an exhibition curating team was born, a cycle of learning was formed, activity guidelines were created, and contributors to the festival promoted the connection to the community (which is typically unfamiliar with science). These facts show the possibility of promoting science communication and improving science literacy in a way that is different from science museum activities and the local science festival.

It is perhaps necessary to point out here that these developments also seem to relate to the history of the city of Hakodate. Soon after Japan emerged from its period of national isolation, its first public park was built in Hakodate in 1879. Private citizens put up their own money and provided labor to bring the project to completion. Rather than waiting on the government to act, taking action on their own and working together to create a park that they would support and use was a more manageable and comfortable option.

Compared to science festivals in major cities that have large budgets and draw large crowds, a science festival carried out in a community as a grassroots activity is something vernacular—a commons—and may thus be said to be bringing the idea of the convivial society to fruition.

Though sporadic at first, as our networks expand, so too does the circle of resonance, with the whole eventually synchronizing and developing into a large-scale community as shown in Fig. 1. Designing not only science festivals, but any new learning environment in the public sphere, is extremely effective for laying the foundations of science communication activities and stimulating the development of citizens' science and social literacy.

A festival is a kind of cultural apparatus. As stated previously, the festival creates a place of learning and a place of connection in the context of a society dedicated to lifelong learning. The festival is not for appreciation, but for participation. The threshold for participation is low and open to everyone from children to adults and from laypersons to experts. A festival engenders social inclusion. The promotion of social inclusion leads to the transformation of social structures and the construction of a society that can withstand disaster; the aim is a society in which everyone can showcase their latent capabilities and make connections with each other.

Moreover, the method of the festival is one that can be implemented for a variety of subjects, and festivals organized around not only science but also music and books are already underway. More than anything, the festival is an enjoyable, exciting, and ceremonial place—indeed, a convivial sphere.

From local to global connections, festivals provide opportunities for us to be conscious of the world and take actions even as we think about traditions and regional specificity. The local science festival is a cultural apparatus that we change ourselves in terms of both style and substance to suit the respective history, culture, and circumstances of our individual communities.

6. Conclusion

More than 250 years ago, the philosopher and educational thinker Jean-Jacques Rousseau (1758) opined that holding open-air republican festivals in a rich community atmosphere, rather than establishing a luxurious theater, strengthens a community:

Plant a stake crowned with flowers in the middle of a square;
gather the people together there,
and you will have a festival.

Two-hundred and fifty years later, at the opening of the Hakodate International Science Festival, the following declaration in honor of Rousseau was made:

Bring science into a town;
gather the people together to communicate,
and we will have a future.

Local science festivals are designed and implemented in a context in which people with diverse backgrounds come together. As an adult mode of active learning, they create places to enjoy science, engage in dialogue, and learn, thus leading to future community development. Many cities in Japan and around the world face situations like that of Hakodate. I would like to share the lessons and design models that we learned through the implementation of the science festival as a cultural apparatus. I believe that local science festivals will contribute to a sustainable and resilient society in the future. The challenge to integrate festivals with learning and community development will surely continue.

Acknowledgments

Gratitude is expressed to all current and previous Science Support Hakodate staff and its supporters for their creativity and passion. This work was supported by JSPS KAKENHI Grant Number 18H01056.

References

Bollier, D. (2013). Ivan Illich and the contemporary commons movement. Available at https://www.resilience.org/stories/2013-08-05/ivan-illich-and-the-contemporary-commons-movement/ [Accessed September 11, 2018]

Illich, I. (1981). *Shadow work*. Salem, New Hampshire, and London: Marion Boyars.

Illich, I. (1973). *Tools for conviviality*. New York: Harper & Row.

Illich, I. (1971). *Deschooling society*. New York: Harper & Row.

Kurihara, A. (2006). To Iwanami contemporary library edition, in Shadou wahku [Shadow work]. Tokyo: Iwanamishoten. (in Japanese)

Lave, J. and Wenger, E. (1991). *Situated learning: Legitimate peripheral participation*. Cambridge, England: Cambridge University Press.

Mima, N. (2018). The challenge for higher education reform in Japan by seven samurai. International Conference on Blended Learning ICBL 2018: Blended Learning. Enhancing Learning Success. Cham: Springer, pp. 3-16.

Mima, N., Watanabe, M. (2008). Creating a platform for science literacy: Move out of the classroom, get into town. *Journal of Science Education in Japan*, 32(4), pp. 312-320. (in Japanese)

Mima, N., Yamauchi, Y. (2005). *Design for learning environments: Space, activity and community*. Tokyo: University of Tokyo Press. (in Japanese)

Rousseau, J. J. (1758/1960). *Politics and the arts: Letter to M. D'Alembert on the theatre*. Glencoe, Illinois: The Free Press. (Original) *Lettre à d'Alembert sur les spectacles*. Amsterdam.

Developing New Strategies for Science Museums in Increasing the Public Understanding of Science: Case Study from CSTM

Mo Xiaodan, Ma Yugang, Qi Xin

Science Research and Management Department, China Science and Technology Museum, Beijing, China

Abstract: Science museums have unique resources distinct from formal educational institutions, and play a very important role in bringing science and applied technology to children and adults of all backgrounds, to promote their understanding of science. China Science and Technology Museum (CSTM) has been constantly seeking strategies in the creation, conservation and communication of knowledge and identities. This study examines effective pathways and methods to stimulate the interest of visitors and meet public demand for scientific literacy. (1) Bolster audience participation through digital and intelligent service to make the exhibitions content richer and more interesting. (2) Rethink the role of CSTM in K-12 science education, and to improve the quality and influence of museum-school collaboration. (3) Use social media to expand the influence of the exhibition resources in a way that resonates with the young generation. As a leader in the 'Science museum system with Chinese characteristics', CSTM plays a big part in solving the problem of inadequate public scientific education services in rural areas by promoting the sharing of resources with multiple cooperation projects and platforms. Our findings suggest that science museums should make new and greater contribution to promote the development of a learning oriented society.

1. Introduction

The achievement and advances of science and technology (S&T) are deeply changed the life of general public. All technologies involve advantages and disadvantages that have had a broad and profound impact on the society. It will contribute to the sustainable society when the general public is cable of understanding and support the positive role of S&T. However, the rapid development and complexities of science and technology, such as nanotechnology, bio-medical engineering, new and renewable energy, and other discoveries could hinder a clear understanding of the public. To form an environment that the public understand of S&T. Education and communication strategies for the public should be developed in order to strengthen S&T education. This article is divided into three main sections. The first section introduces the role of science museums in increasing the public understanding of science. The second section reports the practice of China Science and Technology Museum. The final section provides reference and reflect on the path through the summary.

Corresponding author: Mo Xiaodan, China Science and Technology Museum, No.5 Beichen East Road, Chaoyang District, Beijng. moxiaodan@cstm.org.cn.

2. The Role of Science Museums in Increasing the Public Understanding of Science

1) The Science Museums

Science museums are defined as socially-involved institutions dedicated to science communication and education. According to Bernard Schiele's definition, science museum is a socially-involved institution dedicated to making the general public aware of the latest science discoveries and development of technology application. (Bucchi, 2014) Historically, the development of science museums has undergone four phases. The first phase of development (1683–1929) was dealt with the enrichment of the collections and displaying the history of technology. The second phase of development (1930–1959) kept pace with the times by showing contemporary science and knowledge distribution. The third phase of development (1960–1975) has a significant effect on the development of science museums led to a proliferation of knowledge. The latest phase of development (1976–present) is not without controversy, with some researchers viewing it as a period in which science and technology innovations fully interact with the society (Cheng, 2017) and respond to current-day challenges. To advance public understanding of science, learning in informal context has become a more acceptable part of science education. And one of the trend of museums and science centers is that these institutions have regularly had increased numbers of visitors during the last decades. Take Chinese Science and Technology Museum (CSTM) as an example, CSTM was opened to public in 1988, there were 21 million person–times total in the last two decades. And the new venue of CSTM was opened to public in 2009, there were 27 million person-times total in less than 10 years, which 2015, 2016, 2017, three years in a row, over three million person-times per year. A research done by China Research Institute for Science Popularization (CRISP), which issues such survey result of scientific literacy of Chinese citizens every other year, showed that the population who visited natural science museums is as follows: science center (22.7%), natural history museum (22.1%). (National Science Board science & engineering indicators, 2014)

Since 1990s, the growth of science museums is closely related to the developments of the information society. Also, the continuing world-wide trend is for a broadening of the subject range of science museums and an increasingly interdisciplinary approach to exhibition themes. The number of science museums have increased regularly during the last decade in China. There were 11 science museums in 2000, there were 192 in total in 2017 (Fig. 1). This phenomenon is closely related to the growing impact of science and technology in public's daily lives.

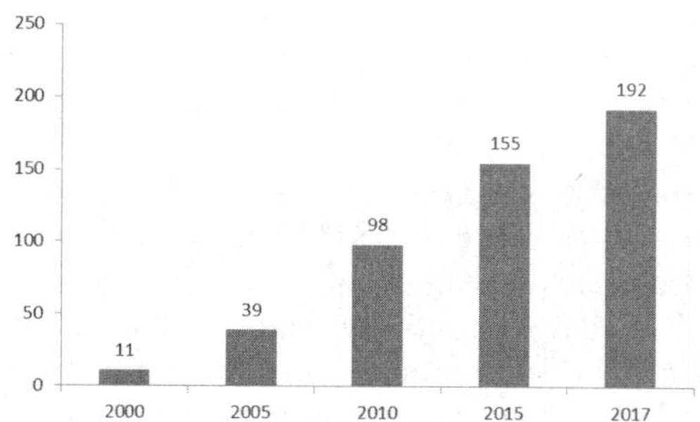

Fig. 1 Number of science center in China.

2) The Opportunities for Science Museums in the Development of Science Communication

Science museum is an important education institution to foster science communication and informal learning, as well as the platform for public engagement with science. As we know, the progress of science and technology is inevitable. New science and technology such as information sciences, biomedical engineering, Nano materials and technology, new and renewable energy, marine/geospatial technology and its applications have had a broad and bold impact on society. Public began to reflect, even to question the price we have paid for when we are enjoying the great conveniences brought by scientific and technological progress and economic development. It also brings ethical issues involved in the overpopulation, environmental pollution, food security, genome editing and etc., leads to what we call social issue with scientific context and scientific issues of social significance. To advance public understanding of science, new forms of education were actively sought. The International Committee for Museums and Collections of Science and Technology (CIMUSET) works to 'popularize and promote science and technology among children and young people all over the world'. Additionally, the increased focus on the visitor in the form of stronger ties the local community, prioritizing the viewpoints of the visitors over the 'language' of the object, and ultimately, attempting to create scientifically informed citizens, seems to be an indication of a broader tendency where institutions place their users and communities at the center of their functions. First of all, science museums need to integrate the state-of-the-art technological advances into their existing exhibitions and displays, as well as their educational programs, so as to avoid lagging behind. Secondly, science museums take the advantage of such high technologies as IT in their effort to seek innovation in terms of exhibition education, visitor services, and user experience, as well as to extend their services to visitors prior to and post of their visit, and to the public who cannot make it to the museum. Finally, science museums set up new platforms for scientific exchanges, bring into full play their unique features and advantages, and promote the public understanding of issues such as science and society and science and ethics.

In a conclusion, the global concerns about the sustainable development, social issue with scientific context and scientific issue of social significance are responding by natural science museums through the best practices to educate the public. Correspondingly, the public's new expectation for understanding and participating in science, these are the new needs, asking the natural science museums to respond. The following session is who and how to provide this kind of service deserved to discussion.

3. Science Communication in China Science and Technology Museums

1) Bolster Audience Participation Through Digital and Intelligent Service to Make the Exhibitions Content Richer and More Interesting

The popularization of intelligent terminals has brought an APP into all fields of people's life, brought convenience to the public, and also brought new approaches to museum education. A portable phone or tablet helps to the promotion of information related to museums, and can help the public better understand of exhibitions. CSTM carries out the projects of so-called 'smart venue' to improve the level of information construction. It develops public intelligent service platform which includes ticketing, exhibits guide, wireless network, passenger flow density and other information systems, integrates WiFi, Bluetooth and location-based technology, and provides personalized services to the public through combining mobile APP, applets and mobile terminals, and through information acquisition methods such as QR codes, iBeacon, etc.

In recent years, with the increasing audience, to provide convenience to the audience, CSTM carries out the overall reform of ticketing system, to add a variety of forms for purchasing tickets, implement that viewers can enjoy one-stop service of purchasing, changing and checking in after finishing real-name ticket purchasing, update the online ticket booking function, optimize online ticket purchasing process, and provide Alipay, WeChat, mobile payment and other third-party payment forms, which conforms to the development of modern science and technology and meet the demand of the audience for many kinds of payment forms. In order to further provide considerate services, CSTM also regularly maintains and updates public service facilities such as deposit boxes, phone-charging stations, baby carriages, automatic vending machines and hundreds of audience seats, etc., conducts regular audience reaction survey and timely provides feedback and deals with the problems reflected by the audience, and publicizes and forecasts the activities in the museum through various channels, such as the official website, the LED display, WeChat, Sina Weibo, etc. so that the audience can obtain the exhibition information in advance, which increases the comfort level of the experience.

To help the public understand natural selection and evolution, based on the '13 kinds of birds with different mouth (Darwin thinking)' exhibits of Hall B for exploration and discovery of CSTM, and combining with related exhibition resources, the popular science APP is designed and developed with 'one day of Darwin in the Galapagos Islands' as the theme, to realize online and offline participation and interaction of the public. After completing the task, the public can view the text or animation display of the relevant exhibits. The game rewards in the APP are designed based on the exhibits, which can stimulate the learners' strong interest in scientific inquiry in the combination of combination of emptiness and reality to achieve the effect of 1+1>2. Selection and design of all kinds of game elements fully reflect the concept and method of communications, adopting the design idea of situation introduction, role play, task-driven, motivation and problem-orientated and making use of the methods of pedagogy and communications to design tools for the communication and learning of scientific and cultural knowledge with both knowledge and being interesting.

CSTM proposes 'take the science and technology museum home' to share information of the exhibits online so that the visitors can search and learn from home through information technology. The website of China Digital Science and Technology Museum has more information about exhibits and exhibition items, which expands the service scope of CSTM and the audience's experience of the museum.

The education activities based on the exhibits and exhibition items focus on cultivating scientific thoughts and spirit of the audience. Through the exhibition content and the audience's participation and interaction with the combination of knowledge, entertainment, and science, they show the scientific principles and technical application to the audience and encourage the audience to explore and practice. In addition, while conducting the exhibition education, CSTM also organizes various scientific popularization practices and training experiment activities which enable the audience to deepen their understanding and perception of science through their own participation, and to improve their scientific literacy unconsciously. Two education concepts of CSTM: one is Maker Education. In order to enhance education effect and expand the audience, CSTM has launched a series of exhibition hall workshops such as 'Maker Dreamworks', 'Science SUBWAY', 'Play Science', etc. in public space. The second is to carry out the maker education project with STEM education concept to carry out education activities which are deeply loved by the majority of the young audience.

While popularizing scientific knowledge, it pays attention to promoting scientific spirit, advocating scientific methods, propagating scientific ideas, and spreading the process of science and stories of scientists. With the brand of 'base of combination of museum and school', CSTM leads the popular science services for teenagers, providing 150 primary and secondary schools with venue education activities, serving 30,000 teachers and students, and organizing special training of science and technology education for primary and secondary schools, serving 200 teachers. With the brand activity of 'custom travel', it leads personalized and differentiated popular science services for the public, carrying out the special customized 'scientific birthday party' activity for 12 sessions, and 150 customized group counseling sessions. It has developed 4 multi-activity resource bundles to serve different application environments including a physical museum, mobile museum, and popular science caravan. It has held the 'night of science and technology museum' that is the first large-scale evening activity of science education. It aims to build scientific education brand of 'Huaxia science and technology school', carrying out 40 themed education activities based on ancient science and technology exhibition and exhibits to tell the 'Chinese story' in a diversified way in 2017.

2) Rethink the Role of CSTM in K-12 Science Education, and to Improve the Quality and Influence of Museum-school Collaboration

Lifelong learning needs new practical forms, and the formal education can learn something from the informal, open learning environments like the science museums. The relation to science, technology and education are met through the cooperation between universities, science centers, schools, teacher education and school authorities. Science museum is located where science, technology and education meet. A science museum features all of these three fields. Out of school education often uses informal education sources for formal education. Science museum education is one form of out-of-school education. The education objectives under the background of the science and technology museum, and scientific and specific knowledge, skills, attitudes, and tendencies of the development of school are inevitably overlapped. Therefore, it can play a complementary role in promoting the development of the common cause between the informal learning environment and the school. How to closely connect with the formal K-12 curriculum; how to carry out off-campus teaching, service to school, resources loan, course development, teacher training and so on. In recent years, driven by the development of the times and under the guidance of national policies, the school education increasingly needs to cooperate closely with the venue education. By exploring and practicing the cooperation mode of the museum and school, abundant exhibition items and education activities resources in the venue are taken as the beneficial supplement to the existing education system, so as to further improve the education system. When learning in the venue, experiential teaching, inquiry teaching, and other teaching strategies should be adopted, and educational concepts such as STEAM, should be adopted to break the barrier formed by the imparting teaching method in the traditional school education, so as to build a broader platform for students to learn scientific knowledge, master scientific methods and pursue scientific spirit.

CSTM mainly implements its education functions through exhibition items and education activities. Compared with formal school education, education of the science and technology museum has the advantages of independence, interestingness, participation and interactivity, which can display the dull and abstract textbook knowledge in a vivid and intuitive form, encourage students to explore and practice by themselves and obtain direct experience, which can effectively make up for the deficiency of school education. For the cooperation between museum and school, for one thing, CSTM inherits and carries forward the excellent practices of science and technology museums at home and abroad, such as free visits of students, training of teachers in primary and secondary schools. For another thing, it launches a number of distinctive projects, such as students' practical experience materials, 100 courses of themed scientific practice, customized travel to science and technology museum, going over the test for senior school entrance, innovative methods training, etc.

'The first lesson' is an important ceremony to start the new term. In order to thoroughly implement the National Scheme for Science Literacy (2006–2010–2020), and actively respond to requirements of Ministry of Education for primary and secondary schools to do a good job in the first class, it implements the action of science literacy education for teenagers, vigorously carries out the science and technology education activities both inside and outside the school, and regards 'the first class' as an important part of ideal and belief education, popular science education and school education at the beginning of the term. The exhibition education center of CSTM takes advantage of its unique advantages and makes full use of the exhibition resources in the museum to elaborately design 'the first class' for primary and secondary schools throughout the city to provide students with a distinctive opening experience through vivid and abundant education forms.

The theme of 'the first class' of 2016 is 'space exploration, building a strong country'. The past year of 2016 is the 60th anniversary of the founding of China's space industry. At the beginning of the new term, it coincides with the moment when the transformation of the exhibition hall 'space exploration' of CSTM is completed. This activity leads students to experience the mysterious wonders of space. They can also play the role of astronauts, and in the laboratory, students can simulate the dialogue between heaven and earth, learn the knowledge of space plant cultivation, and watch scientific experiments of astronauts. They can experience frontier space technology and the latest achievements that Chinese people are proud of. In the space show, the science instructor will prepare making on your own hand's activity 'a small rocket to soar into the sky' and rocket-themed science shows for students.

Teacher training comes again shared discussion and collaboration help to make progress—The second phase of science and technology teacher training of 2017 was successfully held

Since 2016, the exhibition education center has launched a series of science and technology teacher training for primary and secondary school teachers in Beijing, which has been widely praised by teachers in the school. On March 22, 2017, the exhibition education center completed the first phase of 2017 science and technology teacher training in the form of open class, and communicated with the attending teachers about related issues of teacher training after the meeting. Based on teachers' feedback, in order to further promote carrying out activities of the combination of museum and school, make teachers more in-depth understanding of education resources of CSTM, promote students to visit the effect, the exhibition education center, on May 9, conducted the second phase of 2017 science and technology teacher training for junior middle school physical teachers in Beijing, and nearly 300 teachers from more than 100 schools in Beijing attended the training. It centered on the topic 'how to use the science and technology museum resources to carry out the junior middle school physics teaching' to conduct a special topic lecture.

3) Use Social Media to Expand the Influence of the Exhibition Resources in a Way that Resonates with the Young generation

The Internet has fully penetrated into education, service, management and others of the science and technology museum, making it a smart science and technology museum; Its core is to provide new ideas and methods for the science education of science and technology museum based on experiential practices through Internet concepts and application of technology to promote the design and implementation of exhibitions or education activities to be more open, cooperative, shared, accurate, and more attractive, and realize the innovation and change of science popularization mode of science and technology museum.

The traditional mode of science and technology museum is that staff of the science and technology museum develops exhibits and education activities, and the audience interacts with the exhibits and education activities in the venue. The science and technology museum in the network era has added network platform. The public can interact with the website through the network, or with the physical museum through the network, and there is a connection between the entity and virtuality. The public has the opportunity to participate in the construction of physical and network popular science, so that information flow and workflow have been expanded in a wide range, fully reflecting the Internet concept of open, collaboration, sharing and shared work.

As a state-level public welfare science popularization service platform, China Digital Science and Technology Museum (CDSTM) has been committed to creating high-quality and original science popularization content, and providing high-quality science popularization resources for the public, especially children. In recent days, the video playback amount of 'Science, open the door', which is an original audio program for children of CDSTM has exceeded 10 million on the Himalayan FM that is the largest audio we media platform in China. In just half a year, it has doubled from 4.98 million at the end of last year to 5 million. The program started in April 2016, and up to July 26, 2018, it has been broadcast 129 programmes, with a total duration of over 800 minutes. And the highest single broadcast volume reached 193,000. With a total of 35,000 subscribers, the program has a completion rate of 69.46%, ranking the seventh in the Himalayan children's science popularization channels.

CDSTM has conducted an in-depth cooperation with Baidu Baike, Baidu Zhidao Daily, Baidu Wenku, Baidu Chuanke, and other columns, and has achieved good results. Through strategic cooperation, on the basis of good cooperation in the past, CDSTM will enter Baidu Zhidao as a science popularization professional institution, continuing to cooperate in multiple columns and channels, providing users with solutions to their scientific problems from a professional perspective, popularizing scientific knowledge, and interpreting scientific hot spots.

CDSTM works hard to dispel poor creative micro-video. Centering on the focus of popular science work, focusing on topics with public concern and social concern such as PX, waste incineration and nuclear power, etc., it pioneers a combination of matting technology and interesting animation. It teaches through entertainment and guides the public to reject rumors and be objective and rational through creative forms of Internet thinking such as inviting WeChat call by popular science persons, delivering popular science newspaper, opening up popular science adventure game, situational presentation, etc. The highly innovative work has been recommended by China Popular Science, CDSTM and other WeChat official accounts.

It will create high-quality and featured network popular science resources, and adopt 'informatization + exhibits exhibition + education activities' to build all-media exhibits exhibition and activity database, and build a demonstration interactive learning experience center. It will promote the application of virtual reality technology, adopting 360-degree panoramic virtual roaming technology to allow the public to visit the popular science venues with a feeling of actually being there without leaving their homes through panoramic shooting and interaction design. It will develop and promote 300 mobile terminal virtual reality projects and popular science mobile games, providing the public with novel and convenient popular science resources.

4. Conclusion

Through science learning in the designed scene, various projects and media can satisfy different demands of the public on science, stimulate their interest in science, build their scientific knowledge and skills, help the public to cope with science more freely and confidently, and construct and understand scientific learning experience.

References

Cheng, D. (2018) Opening Up a Bright Future for the Development of Natural Science Museums under the Belt and Road Initiative. *Journal of Natural Science Museum Research*. 3(01):17-26.

Massimiano, B., Brian, T. (2014). *Routledge Handbook of Public Communication of Science and Technology, Second edition*. London: Routledge.

Research on Science and Culture Communication Strategy of Science and Technology Museum's Based on Visitor Needs

Liu Yuhua, Ma Yugang

Scientific Research Management Department, China Science and Technology Museum, Beijing, China

Abstract: As a public scientific and cultural service facility, the STM should not only disseminate scientific knowledge, ideas, methods and spirits, but also improve the public's ability to understand science, innovate and apply technology, and enhance the integration of science and culture. In view of the different scientific and cultural needs of different visitors, the adoption of different scientific and cultural communication strategies is a guarantee for the realization of STM's function of public cultural service. For example, for preschool children, the 'Science Park' is set up to realize scientific enlightenment education mainly through games and experiences; for students in school, the 'combination of STMs' and schools' education activities are set up in combination with the science curriculum standards; for young people with innovative needs, the 'scientific and technological innovation' exhibition area and related training courses are offered; for the parents who accompany the children to visit, parents-children interactive education activities are set up; for the elderly visitor, health and nutrition lectures are offered. In a word, the STMs in various places continue to explore and innovate in the dissemination of science and culture, and play a more and more important role in promoting the improvement of scientific and cultural literacy of the whole people.

Keywords: Science and Technology Museum; Visitor; Demand; Scientific Culture; Communication

On March 1, 2017, the Law of the People's Republic of China on the Guarantee of Public Cultural Services (hereinafter referred to as the Guarantee Law) formally implemented. The law clearly points out that public cultural facilities mainly include libraries, museums, cultural centers (stations), art galleries, science and technology museums, memorial halls and so on. The main purpose of the Safeguard Law is to take the people as the center and meet the basic cultural needs of citizens, which is also the goal of the science and technology museums. The promulgation and implementation of the Safeguard Law further confirms the legal status of science and technology museums as a public cultural facility, and provides a basic legal guarantee for the dissemination of science and culture of science and technology museum.

Scientific culture communication is to popularize scientific knowledge, advocate scientific method, disseminate scientific thinking, promote the scientific spirit, in order to improve the scientific literacy of the public.[1] As a public scientific and cultural service facility, modern science and technology museums should not only disseminate scientific knowledge, methods, thinking and spirit, but also constantly improve the public's ability to understand science, innovate and apply science, and enhance the integration of scientific culture and humanistic culture.[2]

To meet the scientific and cultural needs of the public, science and technology museums need to provide different scientific and cultural communication strategies according to different needs of different visitors. From the perspective of the visitor structure of science and technology museums, children, adolescents, adults and the elderly are all parts of the visitor of the museum. Different people have different physiological and cognitive development charac-

1 Huang Jibing. (2007). Strengthening the dissemination of science and culture to promote social civilization and harmony[J]. Journal of Xihua University: Philosophy and Social Sciences, 26(6): 70-71.

2 Xu Shanyan. (2015). Thoughts on science and culture communication in the Science and Technology Museum [J]. Science Education and Museum, 1:320.

teristics. It is necessary to study and analyze the characteristics and needs of different visitors so as to provide exhibitions or activities which they need.

1. Setting up Paradise for Children

Children's concepts are mostly formed in life and play, and new knowledge is acquired in play. They seek physical contact and participation in hands-on exhibits, and acquire new knowledge in contact and participation. Children tend to discover multiple senses, which are the natural way to understand their surroundings. Multisensory knowledge is also a means for children to grasp and test things. Children, especially those aged 3 to 6, can repeat their favorite exhibitions and experimental activities and enjoy them. They visit science and technology museums for both play and study.

Play is the main way of children's participation exhibition in Science Park. According to incomplete statistics, about half of science and technology museums in China have specialized children's exhibition halls or exhibition areas. According to the observation in the science and technology museums work and field research, no matter in the off-season or peak season, Science Park is one of the most popular exhibition areas with a large flow of people Science and Technology Museums.

The exhibition hall of Children's Science Paradise of China Science and Technology Museum, which was the earliest completed in China, opened in 2001. It is warmly welcomed by children and parents. It received an average of 380,000 visitors a year. In the eight years from 2001 to 2009, it received more than 3 million visitors and won the National Museum System Most Popular Award for Visitors.

The exhibition hall of Science Paradise in the new China Science and Technology Museum takes children's growth needs as the basis, exhibits scientific contents suitable for children's physical and mental characteristics, adopts a variety of teaching and exhibition methods based on games and exploratory interactive participation, encourages children to experience and think positively, pays attention to the interaction between children and parents. In the exhibit and play, children can experience the pleasure of inquiry, which are stimulated curiosity and cultivate a passion for science.[1]

When designing the exhibition area of Rainbow Children's Paradise, Shanghai Science and Technology Museum thinks that the most important activity in childhood is 'play'. Children learn and imitate through play, and then construct their own ideas and thinking. For children aged 3 to 10, play is an important way for them to learn. Therefore, the activities of the children's science exhibition area are planned and designed in accordance with the characteristics of children's behavior and the way of games.[2]

With the increase of children's vocabulary, there is a sensitive period of speech, reading and writing. However, language development is still influenced by context and depends on context.[3] It is one of the methods of exhibition design in Science Park to create a situation so that children can master more scientific knowledge in the situation experience. According to the characteristics of the children in this age group and relevant educational theories, the Children's World in Guangdong Science Center can be understood through interactive games, role playing, situational experience, scientific inquiry and other colorful forms from the perspective of children's experience, observation and cognitive development. Simple scientific knowledge can enrich children's childhood experience, keep them curious, stimulate imagination and creativity, and arouse interest in learning and exploration desire.[4]

2. Setting up Activities of 'Combination of Museums and Schools' for Students

Visitors of science and technology museums are mainly students. Young students from primary schools, junior middle schools and senior middle schools constitute the core of the current visitor of science and technology museums. Teenagers at the learning stage have already contacted some basic knowledge in school, but they have not

1 Wang Songduan, Qi Xin. (2009). Science world children's paradise—Design thought of theme exhibition hall of science paradise[J]. Science and Technology Museum, 5:40.

2 Xinge, Song Xian, Wu Weihao. (2011). Creating the most beautiful rainbow—Exploration and Practice of Children's Science Exhibition [J]. Popular Science Research, (2): 85.

3 Zhou Jingjing. (2017). An analysis of the expressions and educational motivations of children's exhibition interpretation[J]. Museum of Natural Science, 2(2):65.

4 Guangdong Science Center. (2010). Children's World Pavilion, [on-line]. http://www.gdsc.cn/cgnr/cszl/201001/t20100121_17674.html.

reached a comprehensive, professional and deeper level. The interactive exhibits of science and technology museums can practice what they learned in school, understand the application of knowledge in life, so that they can grasp and use scientific knowledge faster and better.

Zhengzhou Science and Technology Association, Zhengzhou Education Bureau and Zhengzhou Civilization Office jointly issued a paper to promote the work of Science and Technology Museum Activities into Campus. The three units jointly formed an office and established a long-term mechanism of 'combination of museum and school'. During the activities, Zhengzhou Science and Technology Museum also undertook the work of training courses for primary and secondary school teachers, science course design contest and so on, and promoted the combination of resources of the science and technology museum and school education.

Since 2015, Chongqing Science and Technology Museum has carried out a comprehensive project of combination of museum and school, which includes theme visits, exhibition hall theme activities, happy popular science drama, interesting scientific experiments, and hands-on training in the Science Dream Workshop. All activities are designed on a class basis. The school can arrange and calculate the teaching content by ordering dishes according to the students' time in science and technology museums. By September 2018, Chongqing Science and Technology Museum independently has developed 125 curriculum resources, signed 108 primary and secondary schools, a total of 1557 reception classes, 3268 class hours, benefiting more than 140,000 students. Chongqing Science and Technology Museum has also designed and manufactured 'Activity Manual' which combines museum and school services and has put into trial use more than 30,000 copies.[1]

China Science and Technology Museum established the project of 'school based on museum-school combination' in 2017. It provides the first batch of 200 cooperative schools with five services, including venue activities, innovative personnel training, school-based curriculum development, science and technology teacher training, and the activities of the CSTM into the campus. In the venue activities, in addition to the exhibition hall visits, but also for the students to come to the museum to provide a 'first lesson' 'high school entrance exam string talk' 'custom Science and Technology Museum tour' and other brand activities.

Combining with the science curriculum standards, we carry out various activities of combining museum with school, invite teachers and students to come in, design various visiting routes and experimental courses which are combined with the curriculum standards, and provide scientific practice activities for students; at the same time, the science experiment and curriculum of Science and Technology Museum 'go out' make the science course of science and technology museums enter the science classroom in school.

3. Setting up Parent-child Interaction Activities for Adults

Survey data show that in science and technology museums the most is the age of 26 to 45 years old visitors, accounting for 47.1%, mostly visit with children. So parents constitute the largest number of visitors in science and technology museums. In addition to student group visits organized by schools, children and adolescents and other minor visitors are mostly accompanied by parents to visit science and technology museums, and even two or three parents accompany with a child, especially in science parks where parents are more than children. So, is the parents looking at the museum during the visit, or do they take the whole process to guide their children to visit?

Brown observed the parent-child interaction at the Science Center and found that 'parental response' can be divided into eight categories: (1) custodial type; (2) maintain order; (3) aid type; (4) with head type; (5) deputy type; (6) partner type; (7) leadership type; (8) demonstration operation. From the first to the fourth parental behavioral roles are classified as negative, while the fifth to the eighth are positive. Because subconsciously they do not regard themselves as the target visitor of the science and technology museums, but only with the idea of accompanying the visit, the number of parents in science and technology museums is very large, but the degree of participation is relatively low, some just follow the children behind to remind the children of safety, and some even sit beside as spectators, in the process with the children. Play and study

1 Chongqing science and Technology. [on-line]. http://www.cqkjg.cn/school/

as the focus, ignoring their own knowledge of science and technology learning, visit the process of children's guidance is not enough attention. Therefore, there are more negative parents in science and technology museums.

Many science museums abroad attach great importance to parent-child group visitors, and many family projects emphasize parent-child interaction, such as 'grandchildren building a house' and other game forms of parent-child interaction activities have been widely used abroad. It is worth studying how to better play the role of parents and increase the participation of parents. In China, science and technology museums pay more and more attention to parent-child group visitors. When developing the exhibition activities, we should tap the potential of accompanying parents to participate in the interactive experience activities of science and technology museums.

During the National Day of 2015, China Science and Technology Museum held a Parent-child Science Carnival. During the seven-day holiday, 98 events were held, attracting 1548 children and parents from 619 parent-child families. Based on the rich educational resources of science popularization activity room and laboratory, the activity integrates three interactive science games and three science inquiry activities into a carnival theme activity. By setting up diversified parent-child interaction projects, parents and children are allowed to play, compete, produce and experiment in different forms. Explore science and experience pleasure in a relaxed and joyful holiday atmosphere.

Parent-child interaction can better mobilize parents to participate in the exhibition activities of science and technology museums, so that they can not only guide their children to visit science and technology museums, but also experience science in parent-child activities, and enhance parent-child relationship.

4. Setting up Health Education Programs for the Elderly

At present, China has already entered an aging society. According to the statistics in 2005, China's elderly population over the age of 60 has accounted for 11% of the total population, and is growing rapidly at the rate of 3% a year. The demand for cultural life of the elderly is increasing day by day. With the implementation of the policy of free opening of science and technology museums, more and more elderly visitors come to the museum, especially during non-holiday period, the proportion of elderly visitors in S science and technology museums is relatively high.

Except for accompanying grandchildren, most of the elderly visitors to Science and Technology Museum belong to visiting science and technology museums independently. For many reasons, such as retirement at home, the elderly have more free time to spend than adults. They want to visit science and technology museums to understand the state of social development. The development of science and technology has reached a certain degree. After all, many new technologies and theories are produced every day, hoping to learn from the exhibits of science and technology museums. We live in a different, new, and progressive environment than we have ever experienced. Of course, exhibitions closely related to life may attract the attention of older visitors, which makes it easy for them to relate to the reality of life, and hope that through visits to understand the state of their lives, to understand the common sense of life, to improve and improve their quality of life.

At present, science and technology museums do not pay enough attention to the elderly visitor. Some science and technology museums organize special visits to the elderly on the 9th of September of the lunar calendar. The elderly show great interest in the interactive and experiential exhibits of science and technology museums, and have a high degree of participation and experience. However, science and technology museums have not yet developed a large number of educational activities for the elderly visitor, some health lectures are still traditional lectures, lack of innovation, rigid form. It is suggested that the interactive science exhibition and experience activities should be developed to meet the health, nutrition and health needs of the elderly, so that the elderly visitor can get more scientific and cultural contents of their concern in science and technology museums.

According to the different scientific and cultural needs of different visitors, different strategies of scientific and cultural communication are the safeguard measures for science and technology museums to realize its public cultural service function. Science and technology museums have been exploring and innovating in the dissemination of science and culture. This paper studies the scientific and cultural needs of the visitors in science and technology museums from the perspective of age, and analyzes the innovative exploration made by science and technology museums according to the different needs of different visitors. It is believed that with the deepening of the research on the needs of the visitors, science and technology museums will play an increasingly important role in promoting the scientific and cultural literacy of the whole people to innovate teaching activities according to the different needs of different visitors and improve the effect of scientific and cultural communication.

Representing Science as Culture in Museums

Li Xiang

National Academy of Innovation Strategy, CAST, Beijing, China

Abstract: The function of science and technology museums to promote public understanding of science has been increasingly valued by the academic community. In recent years, the scientific centralization of science and technology museums has become distinct, including the increase in the number of science centers and the transformation of traditional science and technology museum into science center. By taking the UK, i.e. the museum power and the cradle of public understanding of science, as research case, this paper reviews the scientific centralization progress of science and technology museum and analyzes the essential characteristics of science center as well as science and technology museum. By summarizing the situation in the UK, the view of science contained in science and technology museums is put forward, and then the refection on construction and development of science and technology museums in China is proposed.

1. Introduction

From the exploration on relationship between the traditional functions and new functions of science and technology museums to the reasonable review on scientific centralization trend of contemporary science and technology museums, we shall trace back to the background of the birth of different functional traditions, and consider the connotation displayed by museums[1] and science centers. In recent decades, as the public science education function of science and technology museums has been increasingly concerned, museums have paid more attention to wide public influence. [1, 2] With the increasing importance of science education function, promoting the interactive display mode of science center has become one of the main development directions of science and technology museums. [3-6] However, although the increasing growth of science center has expanded the way of public participation in science, it has also impacted the concept of communication science of traditional museum. Domestic and foreign scholars have proposed various functional issues at different levels for science and technology museums.[7-13] Throughout the birth and development of science and technology museums, the collection of industrial heritage and the research on the history of science and technology once dominated. The educational function of British science and technology museums, as a derivative, only drew public attention after the rise of public understanding of science. [2, 14]

2. Scientific Centralization of Science and Technology Museums

The scientific centralization of science and technology museums has broken through the original museum tradition and strengthened the function of science education. Since the birth of museums, the exhibition and display of natural science has never stopped. After the successive emergence of natural history museums and science and technology museums, the emergence of science center has brought a new display mode, thus promoting changes in the museum industry.[11, 15, 16] The Palace of Discovery in Paris, France is currently recognized as the first science center in the world. Since then, the science centers have been built at a rapid pace in about 100 years. The construction of science and technology museums in China in recent decades may also be considered as a microcosm of the development of science centers around the world. Some scholars refer to this process as science center movement.[4, 6, 17] The Exploratorium in the United States and Ontario Science Centre in Canada further have highlighted the interactive display on the basis of the Palace of Discovery, while the City of Science and Industry of La Villette in Paris has created a more relaxed environment to allow visitors to experience science in leisurely culture atmosphere. Compared with natural history museums and science and

1 Unless otherwise stated, the museums ireferred to in this paper refer to science and technology museums.

technology museums, the most distinct feature of these emerging science centers is that they no longer regard physical collection and historical research as their main functions. From the significance of display, it means that the original elements of a story are no longer a necessity, that is, the story of science center can be completely created by exhibit design.

The science centers have strengthened science education while also impacted the traditional museum concept. Since the publication of 'Bodmer' report in the UK in 1985, the role of museums for public understanding of science has been increasingly concerned, and the display and teaching method of science center has become the learning object of many museums.[18-20] In 1986, UK established the first batch of science centers and had continued the construction for 30 years.[21] According to the report of 'UK Association for Science and Discovery Centres', there were more than 60 non-formal education institutions such as science and technology museums and science centers in the UK by 2014, of which about 15 were science centers, and most of them were completed after 2000.[22] 3,5 At the same time, the existing science and technology museums have gradually joined the interactive display and teaching method of the science center, and even some science and technology museums have set up separate exhibition halls as micro-science centers. However, compared with North America, the boundaries between the science center and the science and technology museum in UK is clearer, and the former is mostly targeted at children.

As a new type of non-formal education institution, the science center is not naturally regarded as a science and technology museum. Some museum visitors have no idea what the science center is.[1]

Now, science center has a more pronounced impact on the science communication of British museums. On one hand, the newly-built science center has attracted a large number of children to visit in a family form, and the update of display technology has also expanded the form of participation of visitors. On the other hand, museums are also influenced by the concept of science center. Breaking down information barrier and attracting public participation has become one of the most important tasks. At the same time, the emergence of activities such as science week and science festival have extended the display and education of museums and science centers, and the differences between them are further blurred in these activities.

3. Two Functional Traditions: Case Analysis on Museums and Science Centers in UK

In the late 1980s, the Committee on the Public Understanding of Science (COPUS) set up working groups in important fields of public understanding of science. In view of the increase of science centers in UK, COPUS decided to further research the role of museums in public understanding of science and established a museum working group in 1990. In 1992, John R. Durant summed up the research findings and raised the core issue in research on science and technology museums: what is the distinctive role of museums in public understanding of science? He also pointed out that the focus of this issue is to understand the difference between museums and science centers, and proposed the main differences between them: [23] 7-11

> *science centers are dedicated to making the public understand science through display and interaction, while museums are provided with a collection function;*
>
> *museums display a complete scientific story, while science centers display fragmented scientific knowledge and principles;*
>
> *science centers displays science to be discovered, while museums displays the steady progress in a steady walk in nature.*

Obviously, it has been noted here that they focus on different aspects of science. In contemporary, although both museums and science centers are committed to public participation in science, and the scientific centralization of science and technology museums is intensified, the differences in contents between them have not obtained enough attention. By review on differences between them, we may see different views of science behind the two display modes.

1) Science Center: Promoting Public Participation in Science

Science center emphasizes practical participation of visitors, and this concept has been carried forward with the changes of the times. If

1 The view comes from interviews with visitors at the London Science Museum and the Museum of Science and Industry in Manchester.

the initially born science center is to get rid of the shackles of the museum, which not only makes museum exhibits, usually displayed as cultural relics, become more intimate and participatory due to more exposure, but also makes the new museum no longer lacking the necessary materials caused by restriction on the number of collections, most of the existing science centers in UK are born without such shackles, so they are more free in exploring new ways of public participation, and their targeting at young people also makes the form of activity more important than exhibits. The most important task of a science center is to stimulate the public interest in science. The knowledge, principles and applications of science can be selected and arranged in order to accomplish this task, and finally contained in exhibits to be presented in front of visitors. Therefore, a science center cannot display isolated scientific knowledge, but strives to put science displayed in contexts that are closely related to public life, and to evoke the resonance of visitors through the connection between science and life.[1]

Most science centers in the United States are named as 'museums'. Many well-known science centers are finally aimed at comprehensive museums rather than relatively single-function science centers.[2] The situation in UK is quite different, and the 'museum' and 'center' have their respective clear references. The clear positioning makes the objective of the science center more singular, and the identity separated from 'museum' is no longer essential for all-round representation of science, and the tool attribute of science is also highlighted under such a target positioning. From scientific explanation of various phenomena in life, hot scientific issues widely concerned to certain industry with high scientific and technological content, the display of the science center must always reflect the social situation in which science and life are inseparable, so as to maintain attractiveness to the public.

The core objective of a science center is to guide visitors to cultivate a 'scientific way of thinking' through the experience of scientific connotations in life context, especially in the hope that they will still maintain such a way of thinking after leaving the museum. From the perspective of decision-making, the displayed content of a museum is usually decided by the administrator and the academic committee while that of a science center is directly in the charge of the curator, with influences by the opinions of the funders from time to time. [3]Therefore, the definition of 'scientific way of thinking' is largely influenced by administrators and funders. The science center puts scientific content into context in order to promote public understanding, but this artificial context created to achieve a certain understanding is far from the real context where science itself is located, and 'useless' science naturally abandoned during the selection of displayed contents. As Durant said, what can be seen in a science center is always the 'science to be discovered', while other scientific content unnecessary for the public to understand will not be put in a science center.

2) Science and Technology Museums: Carry History and Describe Science

The long tradition of museum in the UK has given the public a clear understanding of it, and history-oriented science display has become one of the deep-rooted expectations of visitors to science and technology museum.[4] The differences of display ways proposed by Durant are still the criteria for judging whether a museum is a science and technology museum or a science center.[5] However, the difference between the two is not only reflected in the display ways and the function of collection, but also the background of the birth. The duty of science center is to communicate science, while most of science and technology museums in the UK are derived from the remains of industrial heritage, and the

1 This view comes from interviews with Susan Meikleham, a display and teaching staff from Glasgow Science Center, and Raj Bista, a display and teaching staff from @Bristol.

2 This view comes from the interviews with Rob Semper, the deputy curator of Exploratorium, and Chris Cardiel, a researcher from Oregon Museum of Science and Industry.

3 In English, the word Curator does not have an accurate translation in Chinese. It is intended to refer to a group engaged in research or management and can determine the direction of the museum. They serve in a way similar to members of an academic committee. The current Chinese translation of Curator is obviously inconsistent with the original meaning. The view on museum comes from interviews with Timothy Boon, head of the collection department of London Science Museum, while the view on science center comes from interviews with Susan Meikleham and Raj Bista.

4 This view comes from interviews with visitors at London Science Museum and the Museum of Science and Industry in Manchester.

5 The interviewed working personnel of museum, working personnel of science center and researchers in the field of museum and science communication in UK agree this standard.

collection agencies are gradually opening up to the public for their own survival, eventually forming a public-oriented work style.[1] The emphasis is that these exhibits are not collected for the purpose of display, but more for judgment of historical value. Compared with art museum and natural history museum, science and technology museum is not very special in museums,[24] but has the functions of cultural relics collection and historical research previous to science communication.[25] Although traditional museums pay more and more attention to public science education, this derivative function has not subverted the museum tradition.

Collection and historical research are the unique functions of science and technology museums, but these traditional functions determine the inevitable difference in communication content between museum and science center to some extent. The public education of museum cannot be separated from the display of existing collections, which determines that even if the museum is committed to making science related to contemporary life through display, a series of statements on certain discipline or technology are necessary. As an essential element of exhibits, the collections limit the curatorial range and story context, while the historical research on collections further limits the narration style of museum based on historical facts. A collection is given the historical value of science by museum in collection process. The presentation of such value always needs to return to the historical situation in which it is located, which makes all the scientific knowledge displayed in museum become historical knowledge and lead visitors to view science historically.

Another main content characteristic of science and technology museums comes from the characteristics of collection work. Science center can select scientific content for exhibition or activity design to convey the knowledge and idea they wanted to express. Global science and technology events can be the exhibition objects of any science center. Science and technology museums have gradually introduced the science of other historical traditions into exhibition hall, but the characteristic of collection work relying on accumulation enables the local scientific tradition still the main content of museum. The Museum of Science and Industry in Manchester demonstrates the contributions of Manchester to industrial revolution. The London Science Museum is dedicated to demonstrating scientific and technological achievements of UK, therefore the country's scientific image has become an important goal in the process of exhibition. Similar to the history research of science and technology, any kind of diversified view of science always requires an in-depth understanding of certain traditional science prior to reviewing the history through change in perspective. Therefore, the science collection of science and technology museums based on local cultural traditions is the important basis for expanding global perspective.

The functional orientation of science and technology museums is clearer and the form is more mature. Although science center has a relatively clear definition, it is more like a product of conception. Even the mature science centers in the world are still making new attempts to explore the connotation of science center. Museum is an important reference frame when judging the exploration of science centers.[2] Such relation enables science center more like an extension of museums rather than a complete separation from museums.

3) Different Views of Science Between Science and Technology Museum and Science Center

The views of science embodied by science and technology museum and science center are different to some extent, and such difference also affects the public understanding of science.

(1) *View of science of science and technology museum*

The key difference between science and technology museum and science center is to represent two kinds of sciences, that is, to represent science from different evolvements. In terms of collection function, the age-old artifacts have the value of cultural relics, which is not different from the collection of other categories of museums, and is one of the symbols of museums. In terms of historical value, the collections, as unique historical data, still have high representation value, and can provide unique visual impact or providing

1 This view comes from the interviews with Timothy Boon, the head of the collection department at London Science Museum, and Kate Campbell-Payne, the head of the marketing department at the Museum of Science and Industry in Manchester.

2 This view comes from an interview with Sharon Macdonald, a professor of sociology at University of York.

model for replica.[26] 56 Therefore, the significance of collection for representation is irreplaceable: if the stories of science museums in connection to the ancient and modern allow people to see the way evolution of understanding nature and transforming nature, collections makes the story provided with irreplaceable historical basis rather than just being told.

Certainly, the story told by museum is more close to the reality not because of representing the science with historical evolvement. Since the wide variety of museum collections, some donated collections are eventually accepted by museum because of the collector's personal preferences, resulting in no uniform standard for historical value of collections. However, the diversity of history itself enables the exhibition of science and technology museums not limited to conclusive historical stories, causing possibility of statement in various ways instead so that the museum can make open description on science. If the donated collections reflect the collector's historical understanding on science, the museum gathering collections will become a historical data library integrating works of numerous historians. History is restored by combining the historical data, science is not only the power for promoting development of human society but also a monster which may bring disaster.

(2) *View of science of science center*

Since the rise of public understanding of science in 1980s in the UK, the importance of the public participation in science, right to know about science and participation right to related policy making have been emphasized continually.[18, 27, 28] In recent years, surveys on the public have further highlighted British emphasis on public participation in science, especially the public's interest and attitude on science have received continued attention.[29] The contemporary science that gradually develops into the cause of the whole nation certainly needs public support. The understanding and participation of the public also provide certain guidance for the progress of science; however, the science is limited to tool attribute by linking science with national development and economic construction. The UK's latest report on public attitude to science in 2014 shows:[30]

> *People do not consider science as a way of thinking in most case.*
>
> ...
>
> *The vast majority of the public believes that science is beneficial and can provide people with more convenient life.*

Obviously, the public understanding on science reflects the judgment of tool attribute, they believe that science is the means of providing convenient life rather than a way of thinking. The public participation in science and technology museums forming a contrast; as indicated in the report in the past year:

> *67% of the public participated in science-related culture or recreation activities, among which, 40% have visited nature reserve, 39% visited zoo or aquarium, 23% visited science museum, and only 13% visited science center.*

It can be seen that the public is neither disinterested in exploring nature, nor down-hearted on historical display of science, they just don't think that above are constituents of science. The image of science in their minds is clear and single, and it doesn't require much attention as a tool to improve life; people are more inclined to spend their leisure time on the exploration of nature and the attention to history rather than link these contents with science.

4. Summary on UK Case

1) Evolution of View of Science for Science and Technology Museums in UK

The summary of changes in content and form of science and technology museums in UK can be regarded as the evolution of view of science in the past few hundred years ultimately. At the beginning of birth, science and technology museum is similar to other types of museums, which is a comprehensive space for cultural relics collection, historical accumulation and wisdom gathering. The science in museum is reflected as the achievement gathering of great ideas. Therefore, the displayed content in early stage is not different from the stacking of monuments, and the scientific communication based on this highlights the important role of science culture relative to other cultures. With the continuous integration of science and technology, the tool attribute and economic value of science have become increasingly prominent, making knowledge dissemination an important factor for affecting innovation and controlling comprehensive national strength. Under such era background, science communication not only plays a role in spreading a culture called science, but also makes more people understand and participate in science, and achieve the utilitarian appeal of improving national strength through the maximum use of social resources. When science and its communication are increasingly being rethought by researchers, the rationality of communication itself and the selection of communication content are re-examined. Just as the London Science Museum, it begins to pay attention to non-western traditional science, such change reflects the regression to the cultural attribute of science. With such view of science, the history-oriented science display is no longer as old and boring as the scholars who emphasize public participation, but possesses a new cultural connotation through repeated examinations of history, so that people can view the science culture of UK regardless of the scientific tradition of UK.

What is noteworthy is that most of the early science centers were born in cities that already possessed science and technology museums. Science center emerged in an augmented manner, at least it did not result in subversive impact on the view of science of traditional museums at the beginning of birth. Figure 1 shows the classification of natural history museums based on represented co content and labeling is made according to the order in which each type of museum is born. Science center and science and technology museum are often referred to as science and technology museums in the industry, and the sum of garden museums including natural history museum, zoo and botanical garden, and aquarium is called natural science museum.[1] As a new form of museum, science center naturally occurred in the corresponding historical

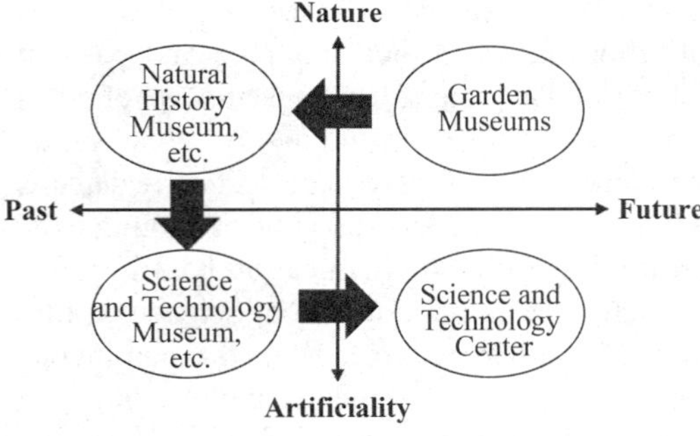

Fig. 1 Classification of content-based natural science museums.

1 This view comes from the interviews with Xu Shanyan and Li Xiangyi, the founder of China Science and Technology Museum, and Bernard Schiele, a professor at University of Quebec.

stage, adding the science content represented in museums. The relative perfection of other representation forms is an important prerequisite. The scientific centralization of science and technology museums can only form the filling of the fourth quadrant in Figure 1 under such prerequisite.

2) In the UK: Two Views of Science Coexist, and the Types of Science and Technology Museums Vary

Seen from the situation in UK, the emergence of science center has expanded the displayed content of museum, allowing the visitors to see a more diversified science. The science and technology museum built at the end of the 19th century shows the physical culture of science to the public, but it prefers to show the benefits and inspirations brought by the science rather than considering whether the public understands it. Visitors are deemed as accepting the words of scientific authority rather than interacting directly with nature itself. Such a tradition makes the science and technology museum in UK tend to describe the achievements obtained in the past while ignoring the contemporary science which has been changed. The lack of change in representation way of science not only leads to a one-sided statement of science, but even overstates the authority of the museum itself. Definitions and choices are always contained in science communication. Contents exposed to the public not only include ‘facts’, but also those considered scientific (by some people). [31] In this sense, the science center brings at least some other contents considered as scientific (by others).

The main problems of science and technology museums in UK are: first, the collections, as museum-specific physical materials, limit the representation way of science to a certain extent while providing materials, and the emergence of scientific center is obviously a supplementation to this limitation. For example, Frank Oppenheimer was influenced by the Children’s Gallery in London Science Museum, prompting him to lead the establishment of Exploratorium many years later; however, the Science Museum itself was limited by too many collections, and it was difficult to make major reconstruction in space. Second, the collections largely guarantee the authenticity of scientific characterization, but when the collection-based display becomes an authority, it will also cause the deviation from scientific communication content in the museum.

Although the scientific centralization trend of science and technology museums in UK is becoming more and more obvious in modern times, the boundary between museum and science center is still relatively clear. New science centers, represented by Glasgow Science Center and @Bristol, are mostly deemed as an activity center and convention center for teenagers rather than a museum. Science and technology museums, represented by London Science Museum, Museum of Science and Industry in Manchester, and Thinktank in Birmingham, have arranged exhibition halls in the existing space as the children's activity center or interactive experience area without changing the original major display areas. For the UK, which has large-scale collections, the scientific centralization of science and technology museums can be regarded as a balance to the authority of traditional museums. The presentation of two views of science enables the museum and science center to jointly provide a more complete scientific picture, thus forming a more diversified view of science.

3) Outside the UK: Relationship Between the View of Science and the Type of Science and Technology Museum

In China, the thinking to functional orientation of science and technology museums and the understanding of scientific culture need further investigation. By 2014, the number of science and technology museums in the whole country had reached 409, and it is still growing at a faster rate. [32] 31 The science and technology museum in China is equivalent to the science center in western countries, [33] 52 corresponding to the fourth quadrant in Figure 1, while the science and technology museum corresponding to the third quadrant has developed slowly. In terms of the growing demand for science popularization in China, although the science and technology museums, as important infrastructures, have achieved remarkable achievements for the promotion of public understanding of and participation in science, the reflection of view of science by museum has not received enough attention behind the functions of display and education. If the display and teaching method of science center provides a more efficient means for public participation in science, the science and technology museum still holds an important position to show a distinctive view of science in a certain geographical or cultural context. For a museum in the name of science, the view of science shall not stop at the simple copying of the cognitive style of scientific community, nor portray science as merely a tool for achieving purpose, but be based on the deep thinking of research approaches such as history of science, philosophy of science and sociology of science as well as the accurate mastering of local science and culture. However, unlike the attributes of tools, these dimensions of science have not been naturally considered by the constructors of science and technology museums in China, instead, they have been repeatedly ignored due to deliberate misinterpretation of western advanced cases.

The science and technology museums in UK show various milestones that can represent the scientific image of the UK, which is based on the recognition and understanding of the unique scientific culture of the UK, that is, the reflection on view of science born in the context of British culture. By analogy, the lack of foundation of industrial revolution is not a key factor hindering the construction of science and technology museums in the backward countries; the lack of excavation of treasures in their own cultural context is an insurmountable gap in the construction of science and technology museums; a view of science stemming from continuous research and exploration rather than from following the herd is more precious than the industrial heritage that China lacks. Only with such a view of science, can the science and technology museums and science centers continue to spread more comprehensive, vivid and real science and lead the public to understand, participate and reflect the science, rather than just 'allowing' the public understand science. Only with such a view of science, can the scientific centralization of museums endow the original intellectual space with new vitality.

5. Conclusion

According to the multi-angle cognition to scientific centralization of science and technology museums in UK and reflection on their influences, and in combination with the current research achievements of scientific communication theory and the analysis on current situation of the British empirical investigation, the researchers believe that with the deepening of scientific communication research, it is very necessary to research the science and technology museums, as well as specifically research the roles of museums and science centers in science communication, which is of great significance for us to understand the issue of science communication more comprehensively. However, the research on science and technology museums and science communication at present needs further study in the following aspects.

First, in the context of the scientific centralization of science and technology museums, the emphasis on attributes of scientific tools is far greater than that on the attributes of culture. As science and technology museums become more and more important in science communication, such a single communication orientation will hinder the public to understand the science comprehensively and deeply. Second, as far as the practices of science and technology museums, it can be seen that the construction and development of domestic science and technology museums (science centers) have received sufficient attention, while those of other types of museums such as natural history museums and science museums have received less attention. On the contrary, in the UK, the birthplace of public understanding of science, the communication at different levels like scientific culture, scientific thought and scientific spirit has received sufficient attention from science and technology museums; the parallel development of various museums also provides a basis for the communication of diversified scientific culture. Finally, the supreme goal for the science communication of the museum is always meeting the demands of the public. However, with the improving scientific quality of citizens, the increasingly diversification and rapid changes in public demands have brought more challenges to science and technology museums.

When we recognize that the science and technology museum plays an important role in shaping the public cognition to science, we will pay more attention to the scientific centralization of science and technology museums, make a research on corresponding museum history and scientific communication, and propose the advantages and communication effects of various science and technology museums on the basis of theoretical research, thus making it conform to the scientific communication concept advanced with the times and become a more ideal scientific communication mechanism.

References

[1] Sharon Macdonald. Behind the scenes at the Science Museum[M].Oxford: Berg, 2002.

[2] Peter J. T. Morris. Science for the nation: Perspectives on the history of the Science Museum[M]. London: Palgrave Macmillan, 2010.

[3] Fiona Cameron, Ann Deslandes. Museums and science centres as sites for deliberative democracy on climate change[J]. museum and society, 2011, 9(2): 136-153.

[4] Damian White, Josephine Anne Stein. Museums and science centres in the UK: Interactivity, infotainment and viability[J]. 2002.

[5] David A. Ucko. Science centers in a new world of learning[J]. Curator, 2013, 56(1).

[6] Ecsite-uk. The impact of science & discovery centres: A review of worldwide studies[R]. 2008.

[7] Larry Bell. Engaging the public in technology policy: A new role for science museums[J]. Science Communication, 2008, 29(3): 386-398.

[8] Alison Kadlec. Mind the gap: Science museums as sources of civic innovation[J]. Museums & Social Issues, 2009, 4(1): 37-53.

[9] Museums' role: Pollen and forensic science[J]. Science, 2013, 339(6124): 1149.

[10] Brice Laurent. Science museums as political places. Representing nanotechnology in European science museums[J]. Journal of Science Communication, 2012, 11(4):1-6.

[11] Zhu Youwen. The Evolution of the educational function of science and technology museums[J]. Science Popularization, 2014(4):38-44.

[12] Zhu Youwen. Review on the development of science and technology museums in China[D]. Beijing: Science and Technology of China Press, 2005 .

[13] Liu Li. The development stage and trend of international science and technology museums and science centers and their enlightenment to China[J]. Science Education and Museum, 2015 (06): 401-404 .

[14] David Follett. The rise of the Science Museum under Henry Lyons[M]. London: Science Museum, 1978.

[15] Wang Heng. A brief history of the development of science and technology museums [J]. Chinese Museum, 1990 (02): 48-54.

[16] Zhu Youwen. Science & Technology Museum and science centers in China [J]. Science Popularization, 2009 (02):68-71.

[17] John G. Beetlestone, Colin H. Johnson, Melanie Quin, et al. The Science Center Movement: contexts, practice, next challenges[J]. Public Understanding of Science, 1998, 7(1): 5-22.

[18] Li Zhengwei, Liu Bing. The investigation and analysis for three important reports about PUS in Britain[J]. Studies in Dialectics of Nature, 2003,19 (5):70-74 .

[19] The Royal Soceity. The public understanding of science[R]. London: 1985.

[20] House of Lords Committee on Science and Technology. Science and society: Third report of session 1999-2000 [M]. Beijing: Beijing Institute of Technology Press, 2004.

[21] Ulrkie Felt, et al. Optimizing the public understanding of science: Survey of science popularization in Europe[M]. Translated by: Committee on Compilation and Interpretation of this book. Shanghai: Shanghai Scientific Popularization Press, 2006.

[22] UK science and discovery centres: Effectively engaging under-represented groups[R]. Bristol: UK Association For Science and Discovery Centres, 2014.

[23] John Durant. Introduction[M]//John Durant. Museums and the public understanding of science. London: Science Museum in association with the Committee on the Public Understanding of Science, 1992.

[24] John Durant. Science museums, or just museums of science?[M]//Susan Pearce. Exploring science in museums. London: The Athlone Press, 1996:148-161.

[25] Edward P. Alexander. Museums in motion: An introduction to the history and functions of museums[M]. Plymouth: AltaMira Press, 2008.

[26] Dominique Ferriot. The role of the object in technical museums:the Conservatoire National des Arts et Metiers[M]//John Durant. London: Science Museum in association with the Committee on the Public Understanding of Science, 1992.

[27] Li Zhengwei, Liu Bing. Theoretical study on public understanding of science: the Deficit of John Durant[J]. The Influence of Science on Society, 2003 (03):12-15.

[28] Liu Bing, Li Zhengwei. Brian Wynne's public understanding of scientific theory: Reflexivity model [J]. Studies in Science of Science, 2003 (06):581-585 .

[29] Department for Business Innovation & Skills. 2010 to 2015 government policy: Public understanding of science and engineering[R]. London: 2012.

[30] Sarah Castell, Anne Charlton, Michael Clemence, et al. Public attitudes to science 2014[R]. London: 2014.

[31] Jane Gregory, Steve Miller. Science in public: Communication, culture, and credibility[M]. New York: Plenym Press, 1998.

[32] Ministry of Science and Technology of the People's Republic of China. China science popularization statistics 2015 Edition[M]. Beijing: Scientific and Technical Documentation Press, 2015.

[33] Ren Fujun, Li Zhaohui. Report on development of China science popularization infrastructure (2011) [G]. Beijing: Social Sciences Academic Press (China), 2011.

Science and Technology Museums Serving for Urban Development Keep Science around Us

Qi Xin, Ma Yugang

Science Research and Management Department, China Science and Technology Museum, Beijing, China

Abstract: Science and technology museums have important social education and public service functions for urban development. The history of science and technology museums worldwide is closely related to economic and social development, scientific and technological progress, and urban development. Science and technology museums are faced with three new challenges: sustainability, unbalanced development, and technological innovation and its interaction with society in promoting urban development. Modern science and technology museum system with Chinese characteristics greatly makes up for the unbalanced regional distribution of national science and technology museums and improves the utilization rate of science popularization resources. It extends public science popularization services to cover all parts of the country and people from all levels of society, and promotes the science popularization service of science and technology museums to be fair, beneficial and efficient to make 'science around us' from a concept gradually become a reality.

No matter in the world or in the country, science and technology museums in each city play an important role in promoting scientific spirit, popularizing scientific knowledge and spreading scientific culture in local cities. And they benefit local people, especially children, with science and technology to enrich their daily life and their future. In the process of interacting with the public, science and technology museums pay attention to stimulating their interest in science, enlightening their scientific thinking and cultivating their scientific spirit to imperceptibly help the public to foster correct world outlook, views on life and values; In the process of interacting with the society, the science and technology museums strive to promote technological innovation and its interaction with the society to enhance citizens' scientific literacy and social responsibility, and thus to promote the sound and sustainable development of cities.

1. The Development of Science and Technology Museums Worldwide

After more than 100 years of development, science and technology museums worldwide are always echoed by the development of social economy, politics, science and technology, and education. And they complement the development of cities so that the social education and public service functions are constantly expanded and strengthened.

In the early 20th century, as the First Industrial Revolution deepened, great scientific and technological inventions and creations emerged one after another, and countries held world fairs or set up museums to collect and display the fruits of the industrial revolution. Deutsches Museum (1906) represented the transformation of the science museums from a storeroom with researching objects through science to a powerful place to explain scientific facts to the public through physical education, and especially the way of participatory display and audience participation became the most vital part of the science and technology museums.

In the middle of the 20th century, with the Second Industrial Revolution developing rapidly, science and technology played a more important role in promoting the development of productive forces, and science and technology museums were popular for enhancing the public's understanding for science and technology. Palais de la Découverte (1936), Exploratorium in San Francisco, USA (1969), Ontario Science Centre (1969), etc., they carried out scientific education for the public by specifically using exhibits transformed from scientific instruments, emphasizing direct experience, experiencing science and exploring science. And they developed exhibition modes and educational concepts with 'inquiry learning' as the core. Later, science and technology museums around the world took interactive exhibits as the main form of science education, and dug deep into and further developed education content, education form, and public service. And they have developed more flexible education resources including small exhibition and exhibits, experiment courses, and science performances, and they have extended single interior displays to schools, communities, and streets in the cities through temporary exhibitions or tours.

After the 1980s, the new technology revolution greatly encouraged the change of the content of science and technology museums. Science and technology museums with physics and other subjects as the main content have gradually turned to the popular science places which reflect high and new technology and frontier science. The life science, environmental science, and information technology have become the important contents that science centers exhibit. The science and technology museum in this period was represented by City of Science and Industry in Villette, France (1986). It took modern science and technology as its main content to reflect the relationship of mutual penetration between science and technology as well as various disciplines, and to highlight the interrelationship between science and society.

After entering the 21st century, with the rapid development of the Internet, science and technology museums worldwide made full use of network, digital and other information technologies to expand their audience from the inside to the outside. In recent years, the functions of science and technology museums have been continuously extended, gradually combining venue education with school education, social education, and individual lifelong education to have a wider impact. And they have been playing an increasingly important role in the improvement of scientific literacy of the public and have gradually becoming communication centers of urban science and technology culture.

Thus, in the process of development, science and technology museums worldwide have gradually established and deepened its concept and functions to make it become nonprofit social education and public service agencies, open to the public, especially teenagers to carry out work and activities related to the popularization of science and technology, with experiencing science, stimulating interest, promoting exchanges and inspiring innovation for the purpose, the public's participation in the experiential exhibition exhibits and education activities for the main form, and autonomous learning based on scientific and technological practice for the main characteristic.

In a sense, the expansion and transformation of social education and public service functions of science and technology museums as well as the deepening and renewal of the public's understanding of science and technology museums are the inevitable outcomes of the city's own development and the progress of the times, and science and technology museums, in turn, continue to promote urban transformation and improve civic literacy. The interactive relationship between science and technology museums and the city has become one of the important features of modern society and the sign of the progress of the times.

2. New Challenges that Science and Technology Museums are Facing in Promoting Urban Development

The science and technology museums are not ivory towers, but urban public infrastructure providing scientific communication services and important platforms for the public to participate in science. On one hand, the development and application of science and technology, such as information science, bioengineering, new energy and new materials, marine technology and space science, etc., have brought great benefits and convenience to us; on the other hand, they also have brought a series of problems including population expansion, environmental pollution, food security and so on. The public has begun to reflect, and even question whether the price we pay for it is worth. Therefore, the called social problems with a scientific background and scientific problems with social significance arose, which triggered the thinking and changes in the field of science and technology museums. To sum up, sustainability, unbalanced development, and technological innovation and its interaction with society are the most important new challenges faced by science and technology museums and the key to promote urban development in the future.

1) Sustainability

According to 2015 Annual Report of Worldwatch Institute, human needs are growing, and natural resources have been unable to support immoderately growing demand. To solve the problems in sustainable development, in September 2015, the UN Sustainable Development Summit adopted the ambitious Transforming our World: The 2030 Agenda for Sustainable Development (2030 Agenda) that proposes 17 Sustainable Development Goals and 169 targets, and the Goals and targets will be achieved in the next 15 years. At the global and national policy level, the smooth transition from the Millennium Development Goals to the 2030 SDGS has been achieved.

Based on this, the science and technology museums spare no effort to foster the public's interest in science, provide a platform for the public to understand science, promote the public to study science outside school, promote the innovation and entrepreneurship of the whole society, and cope with various challenges. Through these actions, the science and technology museums have sent a clear and strong signal to the whole society on the importance of science for sustainable development to provide an interactive space for the benign development of cities, and offer balance of information and practical help for the government, society and citizens to jointly address the problems of sustainable development of cities.

2) Unbalanced development

Over the past few decades, the world economy has been growing rapidly, but the resulting wealth and prosperity have been distributed so unevenly that it has been seen as a source of worsening social problems, political turmoil, and unbalanced educational resources in many cities. There is also an unbalanced development in the field of science and technology museums.

In November 2017, the Second World Science Center Summit was held in Tokyo, Japan, and the participants had an extensive and in-depth discussion on the current global problems of human society with the theme 'connect the world, achieve a sustainable future', especially on how to effectively achieve the balanced distribution of science popularization resources and, fair and beneficial science popularization services to seek strategic countermeasures with long-term effects.

At present, there is an unavoidable fact that science popularization resources provided by science and technology museums in underdeveloped areas are limited and lack of access to shared resources. In contrast, large and medium-sized venues in economically developed regions and central cities have relatively high-quality resources. In this case, the latter has the responsibility and obligation to provide resources and services for the former through the integration, sharing and flow of resources across regions and venues. It involves the interactive relationship between urban development and rural revitalization, and the science and technology museums of China have made a lot of efforts here, which will be offered more details in part three.

3) Technological innovation and its interaction with society

New scientific discoveries and achievements constantly drive the transformation of the world, and constantly refresh the way people understand science and the world as well. For technological innovation and its interaction with the society, there are three aspects: the first is the basic education of science which breaks through the core of the science and technology museums, and timely reflects the progress of frontier science and technology, and the latest technology in the exhibition. The second is to adopt advanced technology as a tool and means to effectively improve the quality of exhibitions and education activities. The third is to promote social problems with scientific background and scientific problems with social significance into the science and technology museums, and build a communication platform through the interaction between scientists and the public.

Although scientific and technological progress is irreversible, it brings us many challenges while making great achievements. For this reason, the science and technology museums must: (i) integrate the latest scientific and technological achievements into the existing exhibition contents and education activities to avoid the lag; (ii) make full use of new and high technologies such as information technology to seek innovations in educational exhibition, service for audience and user experience, and expand its services to the audience before and after the visit, and to people who cannot visit the scene; (iii) establish a new platform for scientific exchange, give full play to its own characteristics and advantages, and strive to enhance the public's understanding and participation in scientific and social issues as well as scientific and ethical issues.

In a word, the science and technology museums must timely respond to social concerns about sustainable development through its exhibition content and education activities, elaborate social problems with scientific background and scientific problems with social significance, and satisfy the public's new demands and expectations for understanding and participating in science.

3. Solution: Science and Technology Museum System Keeps Science Around us

It is a challenge, but more an opportunity for science and technology museums on how to cope with new challenges and problems from sustainability, unbalanced development, and technological innovation and its interaction with society. As mentioned above, science and technology museums are the product of urban development, but one of the important signs of the progress of the times is the equalization of public rights in urban and rural areas. People living in less developed areas and rural areas have the same or even more urgent needs with urban citizens in terms of enjoying scientific popularization public services. In response to the above problems, science and technology museums in China have been exploring ways and practices to realize public science popularization resources and services to benefit to all, making science and technology museums gradually extend to cities and towns across the country.

In 2012, China Association for Science and Technology started from the situation of insufficient and unevenly distributed public science popularization facilities and resources to integrate physical science and technology museums which have been developed for nearly 30 years, popular science caravan introduced in 2000, China digital science and technology museum launched in 2005, and China mobile science and technology museums which started in 2011. And it launched the building of a modern science and technology museum system with Chinese characteristics (hereinafter referred to as the 'science and technology museum system'): physical science and technology museums should be built in places where conditions permit; In the places that do not have the conditions, the exhibition tour of the mobile science and technology museums should be carried out in the towns, and the popular science caravan activities should be carried out and rural middle school science and technology museum should be built in remote areas; the Internet-based digital science and technology museum website should be developed. Since the start of the construction, the science and technology museum system has aimed at extensive coverage and practical effect, the constructions of physical science and technology museums, mobile science and technology museums, popular science caravan and digital science and technology museums have been accelerated, the abilities of development and sharing of science popularization resources and services have increased gradually, and a wide range of services has been achieved, making 'science around us' from a concept become a reality.

1) With the rapid development of physical science and technology museums, capability of science communication has been significantly enhanced

(1) With the rapid growth of the scale of venues, social benefits are prominent

In the 1980s, China built and opened the first batch of science and technology museums represented by China Science and Technology Museum, which started the construction of science and technology museums in China. After 2000, science and technology museums came into blossom all over the country. Between 2000 and 2017, the number of science and technology museums around the country that have been built increased from 11 to 192, and China has become the country with the fastest growth in the number of science and technology museums in the world in the 21st century. In 2017, the national science and technology museums served a total of 56,982,000 visitors, 1.7 times more than in 2010. And the total number of visitors to China Science and Technology Museum and Shanghai Science and Technology Museum both exceeded 3 million in 2017.

Since May 2015, science and technology museums have carried out free opening of pilot, and the number of visitors to science and technology museums that have been included in the pilot program has grown rapidly. In 2015, 92 science and technology museums were included in the pilot program, serving 26.58 million visitors. and it increased to 123 in 2016, serving 37.22 million visitors. The free opening of science and technology museums has greatly improved the financial difficulties in the central and western regions, and small and medium-sized science and technology museums in China, and promoted public science popularization services to be fair and benefit to all in underdeveloped areas.

(2) *The functions of exhibition and education have been steadily enhanced, and the content and form have been continuously enriched*

In 2007, the total area of permanent exhibition of science and technology museums in China was 1.214 million square meters, an increase of 94.6% over 2010. Science and technology museums around the country pay attention to exhibition planning and design, and there are various forms for the exhibition, such as theme expansion, storyline, knowledge chain and subject classification, etc., and the exhibition content of frontier technologies such as new energy, aerospace, information technology, bioengineering and new technologies including VR and AR are constantly emerging so that interactivity, inspiring innovation, and special features of the exhibits are increasingly enhanced.

Science and technology museums around the country have also intensified the development and introduction of short-term exhibitions. In 2017, national science and technology museums held 835 short-term exhibitions, serving 24. 551 million visitors, a four-fold increase compared with 2010.

Education activities are in full swing in science and technology museums around the country. Science and technology museums in China conduct 71,850 education activities like popular science training (activities) and popular science report (lectures), serving 5.631 million visitors, an increase of 2.2 times over 2010. In addition, science and technology museums around the country pay attention to carrying out themed and serialized education activities in combination with exhibition and exhibits, and launching a large number of high-quality activities such as scientific performance, summer camp, etc. for the public. The number and type of education activities have significantly increased, and the level and quality have also been significantly improved.

2) The mobile science popularization facility has achieved remarkable results, and its service coverage has been gradually expanding

(1) *Mobile science and technology museums enrich the county popular science resources*

Relying on the provincial science and technology museums, the mobile science and technology museums carry out tour exhibition services in county-level administrative regions where the science and technology museum has not been built. By the end of 2017, a total of 364 sets of exhibition had been produced and distributed, and 2,339 tour exhibitions had been conducted, serving 87.51 million visitors, and successfully achieving the expected goal of ‘covering counties (cities) for four years’. In September 2017, the second round of national tour exhibition was officially launched. The mobile science and technology museum has been recognized by all sectors of society for its popular science communication mode with a small investment and large benefits, which has greatly enriched the popular science exhibition and education resources in the central and western regions and improved the utilization rate of popular science resources.

(2) *Popular science caravan promotes grassroots popular science work*

Relying on basic level Association for Science and Technology to carry out exhibition services, the popular science caravan is an important carrier for the last kilometer for ground application of addressing popular science work. By the end of 2017, 1,445 vehicles had been allocated to the whole country, with nearly 34.24 million kilometers of driving distance, and nearly 196,000 activities had been conducted, serving a total of 215 million people. With its flexible feature, it meets the needs of the grassroots public for popular science, affectionately known as ‘popular science Hussar’, which strongly promotes the grassroots science popularization work, especially rural science popularization work.

(3) *Rural middle school science and technology museum promotes the balance of rural popular science education resources*

The project of the science and technology museum in rural middle schools is supported by China Association for Science and Technology, and Foundation for the Development of Science and Technology Museums in China raises funds from the society, makes use of the existing places in rural middle schools, allocates popular science exhibits to build science and technology museums in schools, which achieved good results in improving the scientific quality of grassroots youth, promoting the equalization of popular science resources and promoting the industrialization of exhibits in science and technology museums. By the end of 2017, a total of 539 technology museums had been built, serving more than 2.06 million youth directly. Among them, 384 science and technology museums for middle schools were built in poor areas, accounting for 71.24%. Rural middle school science and technology museums promote the combination of science popularization work and the construction of science and technology museums, and national poverty alleviation, education improvement and prompting ambition.

3) Digital science and technology museums have been booming, and their resources and influence have been greatly enhanced

In 2005, China digital science and technology museum, as the national infrastructure platform of science and technology project of the Ministry of Science and Technology, was officially launched. In 2010, China Science and Technology Museum took full charge of its operation and management. In 2011, it became the only popular science platform jointly identified by the Ministry of Science and Technology and Ministry of Finance, and listed by the state among the 23 national infrastructure platforms of science and technology. By the end of 2017, the total official resources of China digital science and technology museums were 10.4TB. The website had an average daily PV of 3.13 million. ALEXA China ranked highest at 76, with an average stable around 100.

At the same time, China digital science and technology museum gives full play to its advantages in serving the science and technology museum system to actively develop and explore the construction of science popularization platforms and digital science popularization resources which are complementary to physical science and technology museum, mobile science and technology museum, popular science caravan and rural middle school science and technology museum. At the same time, virtual reality and other new and high technologies were introduced into the construction of science and technology museum system. In 2017, 25 science and technology museums were set up in the country, and 726 projects were configured.

It can be seen above that the construction and development of science and technology museum system greatly makes up for the unbalanced distribution among the national science and technology museum regions, improves the utilization rate of science popularization resources, enables the public science popularization service to cover all groups in all regions of the country, and promotes the science popularization service of science and technology museums to be fair, beneficial and efficient. Looking forward to the future, the science and technology museum system will continue to focus on solving the problem of unbalanced and inadequate development, promoting the transformation from the growth of quantity and scale to the development mode of quality and efficiency, realizing the innovation and upgrading of the science and technology museum system to further give play to the unique role of science and technology museums in promoting public science popularization services to be equitable and benefit to all.

Research on the Influence of Data Journalism on the Diffusion of Scientific Culture

Ren Ruijuan[1], Huang Chuxin[2]

[1] Journalism and Communication College of Hebei University, HBU, Baoding, China
[2] Journalism Studies Department of Chinese Academy of Social Sciences, CASS, Beijing, China

Abstract: Starting with the function and significance of scientific culture to the scientific community, this paper analyzes how the scientific cultural ideas and norms spread from the scientific community to the masses, combining the value of information and the law of communication, it finds that the data journalism coincides with the function of scientific culture and the historic mission through the analysis of the idea, structure, processes, attributes and final story telling of the data journalism. Data journalism has the function of accelerating the spreading speed of the scientific culture conception and institutional norms from the scientific community to the masses, and it is convenient to popularize the scientific culture conception and institutional norms of the common body to the public in the mass community. Therefore, this article attempts to apply the conception of scientific culture and institutional norms to the design ideas of data journalism through cases study, and penetrates them to the whole process of data collecting, data processing, data visualizing, data interacting and other production of data journalism. The article tries to present them with the words and charts, in order to widely disseminate and communicate with the public through data journalism. Finally, this article concludes that the data journalism can be the bridge tower of the dissemination of scientific culture, and the data journalism is an excellent information carrier and communication tool for the promotion of the scientific cultural literacy among the public.

1. Introduction: The Characteristics of Information and the Law of Information Dissemination

1) The Characteristics of Information

(1) *The objectivity and universality of information*

The existence of information is objective, everything in the objective world is constantly moving and changing, and shows different characteristics and differences. Information is everywhere and is always there.

(2) *Convertibility and transferability of information*

Any information must be represented by a material carrier. The invariance of information in transforming its carrier makes it possible for information to be transformed from one form to another. In order to realize its value, information must go through the movement process of starting from the source, relying on the information carrier to be spread out, until it reaches the sink.

(3) *Information sharing and exploitability*

Information sharing mainly refers to the same content can be mastered and used by two or more users at the same time. For the same information, different value and utility can be found because of the different experience, status and knowledge level of the recipient. Therefore, information also has certain exploitable nature.

2) The Essence of Information Dissemination

(1) *Information dissemination is the way to realize information value*

Information science claims that information, as a unity of meaning and symbols, spiritual content and material carrier, cannot be expressed without symbols, while symbols leaving meaning are just some inexplicable substances, neither of which can cause social interaction alone. The function and function of information can only be realized through this kind of social interaction, which is essentially a communication activity. That is, information must rely on communication activities to realize its value.

(2) *Information dissemination is the symbolic expression of information content*

Information dissemination, as an activity, mainly includes information content, information symbols, information sources, information lodging, channels, information dissemination methods and other elements. Information content reflects the nature of information. It is the core of information dissemination, the core of information value, and the first element of information dissemination. Information symbol is the carrier of information content and the substitute and conversion of information content. The advantages and disadvantages of the symbol system will affect the expression of information content.

2. Data Journalism Coincides with Scientific Culture

1) Scientific Culture

In the three levels of culture: the level of physical appliances, the level of institutional norms, and the level of ideological concepts, this paper should pay attention to the level of institutional norms and ideological concepts. Scientific culture embodies the spiritual temperament of science and scientific community, and is the cultural standard and symbol of science. Scientific culture is the scientific life form and attitude of the scientific community in scientific activities, or the aggregation of the attitude and paradigm of scientific activities that the scientists consciously and unconsciously follow. Scientific culture takes science as the carrier, contains the natural endowment and nature of science, and reflects the spiritual temperament of the scientific field and the scientific community which is cultivated together in this field. Therefore, scientific culture is the cultural standard and symbol of science. Compared with sub-cultures such as art and religion, the history of science and culture is much shorter. However, science and culture are deeply embedded in science and obscure in the world. It can imperceptibly infiltrate people's thoughts and psychology, and even shape public thinking mode. Therefore, the promotion of scientific culture can promote the rational thinking of the public and promote the progress and perfection of human society.

2) Data Journalism

Data journalism is an extension of 'computer aided journalism report' and 'accurate journalism'. In the current large data horizon, assisted journalism reporting is far more than a computer, sensors, UAVs and other assisted reporting tools. In the 'Internet+' information environment, online and offline integration, User Generated Content (UGC) has greatly increased, which has changed the mode of production and dissemination of traditional journalism. If accurate journalism is more from the professional perspective of the media, it is seeking to use data to enhance the accuracy and accuracy of reporting. Obviously, the data journalism is more from the perspective of users, to explore the meaning of data for users; and users from the passive acceptance of journalism to active access, participation in production, they occupy an important position in the production of data journalism. From the scope of influence, the concept of accurate journalism was proposed by Philip Meyer. Its dissemination and practice mainly in the United States, but also in other countries around the world, but the scope and scope of influence is limited. Data journalism is not a concept proposed by a single scholar. It springs up among journalists around the world and has a broader impact (Ren Ruijuan, Bai Gui, 2015).

The essence of journalism is information, and data journalism is a new way of making and presenting journalism. Under the 'Internet+' information environment, data journalism must also have 'Internet+' propagation characteristics under the information environment.

(1) The essential attribute of journalism is its information attribute. According to the objective attribute of information, journalism also has the objective attribute. Using the natural objectivity of data and emphasizing the information attribute of journalism, the paradox of journalism objectivity in journalism can be corrected and interpreted. Information sharing nondestructiveness and polymorphism lays the data foundation for the multi-channel and diversified dissemination of data journalism, and is the premise and basis for vividly interpreting data.

(2) The visual expression of data in data journalism, the viewpoints or stories presented are more in line with the psychological habits of human beings to read and receive information, which makes data journalism more easily accepted by the audience. The visual narrative mode of data journalism will make the presentation of journalism content more concrete and vivid. The combination of scene personalized sharing and other elements will make data journalism insert wings.

(3) The polymorphism and non-destructive

sharing of information provide theoretical basis for the interactive, multi-channel and multi-touch network chain transmission of data journalism.

3) 'Internet+' Propagation Path Characteristics of Data Journalism in the Environment

The one-to-many broadcast situation in the mass communication environment is the one-way non-feedback mode of mass communication. In the 'Internet+' information environment, the release of information presents the characteristics of multi-source: the decentralization of information sources, and the interactive way to make multi-touch chain communication widely sought after. The way users access information mainly includes thematic communities and friends communities. The core reason is the self-organization of information and individualization of information users' needs under the current information environment. The self-organizing overlap of personalized needs and information forms the network interactive transmission of information, that is, the network interactive multi-point chain transmission, which has the characteristics of real-time dynamic and multi-directional interaction. The dissemination of data journalism is summarized as follows:

(1) *Two-way low feedback transmission*

The disseminator disseminates information to the recipient and receives a small amount of feedback. That is to say, the producer of the 'transmitter' data journalism releases the data journalism to the 'recipient', and the audience reads the journalism and feeds back a small amount of information to the media. The most common is to publish data journalism directly on their own websites for the audience to read, and provide comments, comments, messages and other interactive feedback functions.

(2) *Multidirectional interactive propagation path*

Information communicator and receiver interact with each other. The earliest and most influential Pro Pulica journalism organization is a non-profit online journalism site and an outstanding pioneer in data journalism. Unlike other media, where data journalism is mostly represented by charts and data maps, Pro Pulica focuses on journalism applications. It is good at collecting data with users. In 2013, Pro Pulica launched a survey called The Price of an Internship, which collected a large amount of research-worthy user data. Such interactive dissemination, dissemination of information will allow the audience to quickly accept and timely feedback to the effect of the disseminator. Such circular dissemination not only achieves the pre-dissemination effect, but also allows the disseminator to enrich and improve their own dissemination content. This kind of interactive communication breaks the traditional linear and single static transmission characteristics of the report, so that the audience from passive acceptance to active participation, to create a custom space for the audience.

(3) *Customized propagation path*

In the 'Internet+' information environment, information is carried out between specific communicators and recipients. Customized communication is based on the needs of the audience to provide suitable content for the audience. To maximize the different objectives and resources of different customers, the Guardian's customization business also needs to share revenue. These orientations are generally targeted at specific groups of people, research scholars, media reporters and enterprises. The advantage of this kind of communication is that both sides can meet their respective needs to the greatest extent, and the communication effect is ideal.

3. Data Journalism Coincides with Scientific Culture

(1) *Authenticity and objectivity*

Data journalism includes the process of topic selection, data acquisition, analysis and presentation. Once completed, data journalism contains not only the original material of journalism facts, that is, the factual information (data) of journalism's free and non-disseminated state, but also the factual information of journalism which is reflected through the cognition of the disseminator. After that, it enters the state of acceptance of communication—the visualization of data after the process of analysis, mining and front-end design of journalism fact data, the communication state of network interaction or personalized push, and even the interactive design. And the audience's personal data or opinions directly as journalism material—data, real-time participation in production and presentation, that is, the state of non-dissemination and dissemination of two in one. This two in one deduction explains the essence of information attribute as data journalism. In this way, the theory of journalism reaction (subjectivity) has been greatly weakened, and the attribute of information (objectivity) has been greatly enhanced, thus the objectivity and persuasiveness of data journalism have been greatly increased. The object and content of scientific culture are real rather than illusory. Scientific culture is rooted in the real object of study and the careful logic of research in the field of science.

(2) *Skepticism and criticism*

Science and culture are particularly strong rational and positive culture; doubt and criticism are the life of science and culture and the internal driving force of the development of science and culture; reliability (i.e. credibility) reputation is the primary personal asset in science and culture. Data journalism requires not only official data but also UGC data to provide storytelling material (data), and requires the declaration and annotation of true and reliable data sources and the presentation of data cleaning process, data cleaning tools and methods require transparency, not hands-on; data analysis requires scientific and reasonable. We should not make assumptions and speculations, and objectively explain the relationship between great journalism events and individuals. So the two coincide.

(3) *Diversified interpretation*

Front-end design of data journalism, such as visualization, personalization, scene, is an important element in the presentation and dissemination of data journalism. The data objectively presents the unprocessed material of journalism content in the natural state, unlike traditional journalism which is directly summarized or described by reporters' reasoning and deduction, reporters are the descriptors or narrators of traditional journalism's unprocessed state. Sowing state. In data journalism production, the reporter's role is transformed into the analyst and digger of journalism facts (data), and becomes the narrator or narrator of journalism stories. This journalism story is obtained on the basis of data, or the story is narrated with data analysis as the carrier or presentation tool. Data journalism transforms journalists' identities into interpreter analysts who influence journalism factual information. The change of the role of journalist undoubtedly changes the subjectivity of journalism communication, and this change is based on data as the premise or carrier, objectivity greatly enhanced, traditional reporters from story tellers to data interpreters, leading the public to understand the objective facts (interpretation of data) faithful guide!

To sum up, science and culture are universal, public and shared; science and culture are autonomous, initiative and non-historic; science and culture are the arena of diverse opinions and their interpretation, the contexts of argument and refutation institutionalization; reliability (or credibility) is the primary personal asset in science and culture. Peer review is the key system of scientific culture, which has some ethical implications, especially honesty first. All these coincide with the data journalism, or the data journalism coincides with the concept of science and culture, which is convenient to condense and explain the logical symbols and ideas of the process of scientific and cultural verification, and meets the various needs of the presentation and dissemination of science and culture. The essential attribute of data journalism corresponds to the symbolic expression of scientific and cultural transmission. Therefore, the dissemination of data journalism must conform to such a law of dissemination, as is the dissemination of the data journalism entity itself, which contains the content of scientific and cultural dissemination.

4. The Spread and Diffusion of Scientific Culture from Scientific Community to Public Group

Before you begin to format your paper, first write and save the content as a separate text file. Keep your text and graphic files separate until after the text has been formatted and styled. Do not use hard tabs, and limit use of hard returns to only one return at the end of a paragraph. Do not add any kind of pagination anywhere in the paper. Do not number text heads-the template will do that for you.

Finally, complete content and organizational editing before formatting. Please take note of the following items when proofreading spelling and grammar.

1) The Essence of Dissemination of Scientific Culture

The core idea of the diffusion and dissemination of scientific culture is to transplant the scientific and cultural concepts and institutional norms to the public, that is, to let the public understand scientists and scientific activities. It emphasizes that the public can understand, communicate and question science as a human cultural activity. Therefore, the diffusion and transplantation of science and culture need to rely on the interpersonal communication between specific groups of scientists and the public and other types of communication practice.

2) The Spread of Scientific Culture and its Legal Protection

The diffusion and dissemination of scientific culture emphasizes the diffusion degree and dissemination effect of the concept and institutional norms of the scientific community in the public. That is, the purpose of scientific and cultural communication is to promote public understanding of the cause of science, break the barrier between the cause of science and the public, and promote understanding and communication between the public and scientists in terms of scientific ideas and ideas, norms and methods. This is the core content stipulated in the science popularization law of our country. In fact, these not only conform to the principles of science popularization activities, 'carry forward the spirit of science, disseminate scientific ideas, introduce scientific methods and popularize scientific knowledge', but also conform to the principles of proximity and importance in the five nature of journalism value. They are also good applications of the proximity and innovation principles in communication.

3) 'Internet+' the Spread of Science and Culture of Traditional Media in the Media Environment

Dissemination of science and culture has rich practical basis and content. The progress of modern media technology, especially the role and status of the public in science popularization activities under the Internet+ 'information environment' is increasingly prominent. From the perspective of communication effect, or from the perspective of interaction between the public and scientists in the process of communication, the media are the main body and hub in the dissemination of scientific culture.

In the 'Internet+' information environment, almost all the major media in the world today have created their own network platforms and their APP or other media applications. The information mirroring of their journalism reports or media reports occurs almost simultaneously in the 'Internet+' information environment. M. McLuhan, a famous Canadian communication scholar, put forward the idea that 'media is information' as early as the 1960s. As a link between science and the public, media representative organizations (media) play an extraordinarily important role. In Europe and the United States, the birthplace of science communication and the main emerging countries, science communication is mainly done by the media and scientists. For example, the New York Times has a lot of coverage of scientific activities, reporters often interview scientists with questions and their own understanding, reveal scientific events, the stories behind scientific achievements and their social impact through personal interviews, as evidenced by Hainan's New York Times 50 Scientists. However, there are also scientific and cultural communication programs in the major domestic media into the pursuit of mysterious errors, such as CCTV's Water Monster Survey and Phoenix Satellite TV's Chinese Fengshui Culture and other programs.

5. Data Journalism Carries the Advantages of Scientific and Cultural Dissemination and Diffusion Mission

The current 'Internet+' information environment, data journalism coincides with the concept of scientific culture and the way of communication. Its visualization, interaction, multi-directional interactive communication with the communication and communication between the communicator and the audience advocated by science and culture, as well as the interaction between the audience and the audience, are becoming more and more scientific and cultural. The important work of communication is not just a part of the past as a detail or tool. Another important change is that the application of data journalism to new technologies, compared with traditional journalism, changes in communication channels, interactive technology development and other changes in communication patterns caused by changes in the dissemination of science and culture and promote more influence. Thirdly, based on the discovery of the implicit relationship in the data and the intersection of the past and the future, nature and society, science and technology, data journalism is an ideal tool for communicating scientific culture with the public and helping the public understand science. The following case illustrates four distinct advantages of data journalism in disseminating science and culture.

1) The Visualization Presentation of Data Journalism Reduces the Public's Difficulty in Understanding Science

For a long time, science and technology journalism is facing problems that are difficult to understand and cannot be understood by the public. Data visualization makes up for the shortcomings of difficult reading caused by obscure content of popular science reports. Visualization is one of the distinct characteristics and advantages of data journalism.

Visualization can integrate different stages of data and present it in a graph, which is conducive to mining the details of journalism reports. Using data visualization reporting method can conveniently display the relationship between journalism facts, use graphics to interpret journalism facts in depth and shallowly, and enhance the readability and appeal of journalism. The visualization of data journalism expands the way of journalism narration, making boring numbers interesting in the spread of science and culture, and using data journalism to disseminate hot issues in the field of science and culture to the public has a significant advantage. As the Guardian reported on the riots in London in August 2011, the timeline shows how a protest spread from northern London to all of England into violence. Reporters use Google maps and data analysis to fuse data, plotting where the riots occurred to show the details of the incident, and then exploring why a parade evolved from a controversial shooting spree to a nationwide riot. The riot map in the data journalism constitutes data journalism. Important clues.

Using data visualization to report journalism can make it easy for the audience to understand the process of event change and save a lot of space. There are many vivid examples, such as 'a picture that teaches you to read...' Series. Another example is Tencent's special report on influenza H7N9. For example, Encyclopedia Illustration, Influenza Notes describes in detail the symptoms and characteristics of influenza in graphical form. In order to show the distribution area of influenza more clearly, Tencent has produced a distribution map of H7N9 avian influenza in China. In the map, the distribution area of influenza is marked, and the number of symbols is added after the text description. The icon enables people to have a more intuitive understanding of the virus's rampage.

2) Data Journalism Platform can Combine Data Analysis and Mining to Better Explain Scientific Knowledge and Scientific Laws

Net Ease Number Read and 'The Shadow of Coal Dependence: More than Mine Accidents and Mist' introduced in detail the serious consequences of coal dependence on health related knowledge. What the public really needs is the understanding and practice of knowledge. In the case, through the mining and processing of a large number of data through the website, the sublimation from information to knowledge can be completed. Only in this way can the new discoveries and new technological achievements in the field of science and technology be analyzed thoroughly and the public's scientific and technological literacy be improved. Fully reveal the haze, coal dust and smoke will bring lung cancer, heart disease, stroke and other health threats. The article also quotes Greenpeace, an environmental group, as issuing a survey on the health effects of coal-fired power plants on residents, pointing out the number of deaths caused by PM2.5 in 2011. It is reported that the more the death toll is, the larger the circle area will be.

The advantage of data journalism is to show the change of the amount of journalism events. The application platform or data analysis makes the horizontal or vertical comparison of similar events, so that the report is in-depth and superficial. The platform can integrate different stages of data and present them in a graph. The data journalism network platform is helpful for mining the details of scientific communication reports. For example: Xinhua data journalism Cold Knowledge: Watch Law at the beginning of the monkey cartoon to tell the public what is the watch law, instead of the previous definition of the first reporting model, so that the public easier to understand and accept. Then, with the specific example of parent-child education, the Enlightenment of watch law is drawn, and the transformation from abstract to concrete is completed.

3) Data Journalism can Effectively Combine Hot Topics with Predictive Reports

At present, more and more media use data journalism to discuss hot topics, and forecast the future situation of events. The dissemination of science and culture requires that the report itself undertake the task of disseminate scientific and technological achievements and predict the future state. Therefore, in the field of science and culture dissemination, data journalism has obvious advantages. Xinhua data journalism geeks say the blockbuster article The Future, These Dark Technologies Are Coming Near You combines the world released at the Third World Internet Congress

6. Conclusion

Data Journalism has many advantages in the dissemination of science and culture, but there are few data news advertisements, long production cycle, technical requirements and other constraints. The website should increase the investment in the data news science and technology sector, change the click-through-only business philosophy, and look at data news in a long-term perspective. It is advantageous to apply the visual narrative expression of data news to the dissemination of science and culture; to expand the rational use of data, analysis and mining of data in the news data platform, and to reveal the connotation and laws of science to the public; and to apply the predictive function of data in the dissemination of science and culture, it is more convenient for the public to make predictive reports. Effective use of interactivity enables data journalism to successfully stimulate public participation in the spread of science and culture. In a word, data news has become the bridgehead of scientific and cultural communication and its role in the dissemination of scientific and cultural.

References

Bai Gui, Ren Ruijuan. Comparison and reexamination of traditional news and data news[J]. Yunnan Social Sciences. [J]. 2016.1 (1): 186-188.

Fang Jie. An introduction to data news: Operational ideas and case analysis[M]. Beijing: Renmin University of China Press, 2015:2.

Ren Ruijuan, Bai Gui. Theory and practice of data journalism: Model, discovery and thinking. Journalism University[J]. 2015 (3): 65-72.

Rogers Simon. Trends in data news: Unleashing the power of visual reporting[M].

Sheng Yitao. Application of data visualization in science and technology dissemination: A case study of BBC and New York Times Ebola virus report[J]. Science and Technology Dissemination, 2014 (20): 36-37.

Yu Guoming, Li Biao, Yang Ya, et al. The big data age of news communication[M]. Beijing: Renmin University Press, 2014:12.

A Restricted View over a 10-year Research Trend of International Science Communication

——Analysis based on science communication during 2008 and 2017

Yue Mengmeng

Department of Science and Technology Communication and Policy,
University of Science and Technology of China, Hefei, China

Abstract: We analyze 354 papers published from 2008 to 2017 in the international authoritative journal Science Communication in this study. The temporal and spatial distribution, author and coauthors, affiliation, citations and keywords are analyzed in details. The application of diagnostic tool VOSviewer reveals the research hotspots and evolution trend in the field of science communication. That is, in the past ten years, international science communication research has experienced from the framework of news reports, traditional media, scientific literacy, biotechnology and other traditional issues, to new topics such as science communication from the perspective of new media, visual communication, the forefront science and the public, risk in new technology, etc., which are becoming new hotspots in the international research field of science communication in recent years.

Keywords: Science Communication; VOSviewer; Bibliometrics; Hotspot; Research Trend

1. Introduction

As a branch of communication, science communication has the characteristics of multidisciplinary intersection, and it has been received more and more attention in the increasingly development of science and technology. Since the 1980s, research on science communication has gradually started in China, and for a long time, it has attracted more attention from scholars of science and technology history and philosophy of science and technology. Although scholars in the field of traditional journalism and communication, as well as natural sciences, STS (science and technology and society), and sociology of science have turned their attention to science communication. But on the whole, domestic cohesion to scientific communication research is low. The research results are scattered in the journalism communication science, the history of science and technology, science and technology philosophy, scientific sociology, public relations and other Chinese journals, and the voices in the field of international science and technology communication are extremely weak. Therefore, paying attention to the research status in international high-level journals has a positive reference for the scientific communication research of domestic scholars.

International journals with high influence in the field of scientific communication include Science Communication (SC), Public Understanding of Science (PUS), Journal of Science Communication (JCOM), etc. Science Communication is a leading journal of scientific communication research with a high international perspective, which is the dissemination of science and technology in the public and the diffusion of scientific knowledge. SC and PUS are the only two SSCI journals in the field of science and technology communication. In the past, some scholars in China have focused their attention on PUS, but few people pay attention to SC, which is also a world-class journal. The predecessor of SC was created in 1979 by Knowledge: Creation, Diffusion, Utilization, and in 1994 its name was changed to Science Communication. So far, there have been no papers published in the journal as the first author of the Chinese mainland. The two academic journals, Science Popularization and Science and Technology Communication, which are closely related to scientific communication sponsored by Chinese institutions, have not yet been included in CSSCI source journals. Among them, the purpose of Science and Technology Communication is to 'improve the ability of the country to spread science and technology and serve the social and economic development'. It is issued 24 times a year, focusing on the content of technology application. The bi-monthly Science

Popularization published by the China Science Research Institute covers the fields of science and technology communication, science education, science exhibition, science and culture. The purpose of this journal is to promote the study of theory of science popularization and promote the development of science popularization. And it has played an important role in leading China's science popularization study.

In the 1930s, science communication attracted attention in the international arena. Bernard first mentioned the term science communication in Social Function of Science. In recent years, the role of science and technology in the development of human society has become increasingly prominent. The focus on scientific communication issues has generally increased. Science communication is the link between science and social relations. TW Burns defines science communication as 'using appropriate methods, media, activities, and dialogue to trigger one or more reactions of individuals to science, including consciousness, pleasure, and interest., opinion formation and understanding' [1].Nowadays, more scholars agree that science communication is 'the process of knowledge sharing between different individuals through the spread of scientific and technological knowledge information across time and space' [2]. Scientific communication covers a wide range of subjects and has interdisciplinary research attributes, including communication, psychology, sociology, and philosophy. In the study of the concept of scientific communication, Bernard mentioned 'the whole issue of scientific communication, not only including the exchange of questions between scientists, but also including the issue of communication with the public' [3]. Wu Guosheng said that science communication is a new form emerged from the public understanding of science, and it is a continuation and expansion of the previous stage [4].The analysis of these concepts reflects the continuous evolution of the connotation and extension of scientific communication in the development process, and the research content of the corresponding scientific communication field may change accordingly. Chen Fajun [5], Zhang Ting [6], Zhou Yanqi [7], Qi Jianxun [8], Zhu Qiaoyan [9] and other scholars have carried out bibliometric analysis of PUS or SC from different aspects, of which PUS literature analysis is relatively new and the literature analysis of SC was mainly before 2013. In addition, comparing domestic and foreign research literature, it is not difficult to find that domestic research generally lacks international perspective, and empirical research is relatively weak. The problems of scientific communication are mostly limited to the Chinese issues and the Chinese perspective, and have not yet entered the international mainstream discourse system. Therefore, the innovation of this article is to keep track of the latest data of SC. Through VOSviewer's topic clustering and time series analysis of keywords in the last ten years, we try to reveal the latest changes and trends of research topics in SC, and provide international reference for domestic science and technology communication research. While actively paying attention to the international research trends of science communication, it is of positive significance for domestic scholars to understand the international science communication discourse system and as soon as possible and realize the leap-forward improvement of research.

2. Research Methods

In order to describe the development trend of scientific communication, the research uses SC bibliography as the object, and uses the bibliometric method to conduct statistical analysis on the journal articles published in the field of scientific communication. According to the Journal Citation Reports (JCR), the impact factor of SC has been on the rise since 2008, and the impact factor in 2017 is 2.032. It ranks 19/84 in SSCI's communications category and has been in Q1 for many years.

Therefore, the analysis of SC is highly representative in the field of scientific communication.

In terms of data acquisition, this paper uses the core database of Web of Science (hereinafter referred to as WOS) under the Thomson Reuters Group to obtain data. The time span is set from 2008 to 2017, and the publication name is searched by Science Communication. As a result, 354 articles were retrieved. Based on this, we use the bibliometric method to quantitatively analyze 354 articles from the aspects of spatial distribution, publishing institutions, author background, highly cited papers, keywords, and research hotspots.

At the same time, the VOSviewer software is used to process the data to describe the research hotspots and future evolution directions of the international field of scientific communication.

VOSviewer is a free bibliometric analysis software developed by Nees Janvan Eck and Ludo Waltman of Leiden University in the Netherlands in 2009. It is used to achieve co-occurrence analysis in various fields [10], and has been used by many scholars in literature analysis research. The software has four display modes: label view, density view, cluster view and scatter view. Compared with other similar software, the visualization effect is more prominent, and the analysis function is more diverse.

3. Basic Literature Measurement Analysis

Science Communication published 354 articles in 2008–2017, an increase of 127 from 1998 to 2007. Among them, the number of articles issued in 2008–2011 was basically kept at about 30 articles per year. After 2012, it was changed from quarterly to bi-monthly, and the Commentary section was added. The number of articles was increased to about 40 per year. At the same time, the analysis shows that SC's papers show an increasing trend of collaborative research. Among them, 254 original research articles (article) cohesion rate was as high as 72.05%.

1) Literature Spatial Distribution Analysis

Through the statistics of the sources of the SC ten-year literature (all classified according to the country of the first author), 354 articles were found from 44 countries, followed by the United States, the United Kingdom, the Netherlands, and Germany. However, the global distribution of articles on scientific communication research is extremely uneven, and there is a phenomenon described by Pareto's law that 20% of countries/regions publish 80% of papers[11]. Figure 1 clearly shows that North American countries and European countries have a higher say in the study of scientific communication. The United States has a maximum of 180 articles (50.85%), followed by 33 in the UK (9.32%).), 19 in the Netherlands (5.37%) and Germany (4.8%).

In the past ten years, China has only published six articles (1.69%), including one from the mainland, Hong Kong and Macao, three from Taiwan, and the first authors of six articles are from Cornell University and The Chinese University of Hong Kong, School of Science, Technology and Humanities, University of Macau, I-Shou University, Chengchi University, Taiwan Normal University Science Education Center. Only one paper in mainland China was

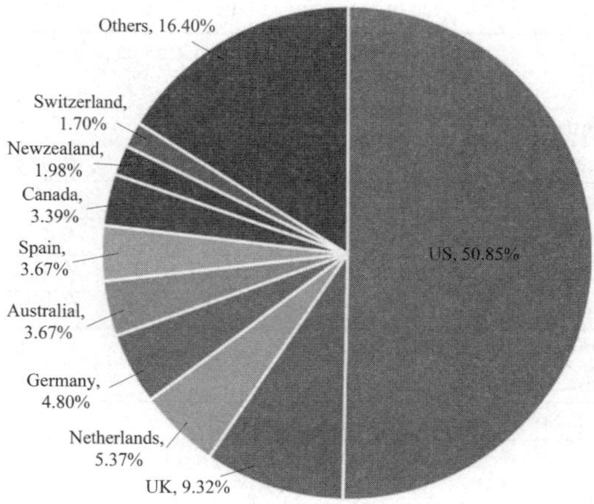

Fig. 1 Country/region distribution percentage of SC literature.

published entitled Encountered but Not Engaged: Examining the Use of Social Media for Science Communication by Chinese Scientists. The first author is Dr. Jia Hepeng, a PhD candidate at Cornell University. The author was included in Sun Yat-sen University in the second unit. China's mainland region is in a weak position compared with developed countries in the field of international scientific communication, and has not yet issued a clear voice in the field of international science and technology communication.

By counting the organizations of the authors of 354 articles, these articles were found in 323 institutions, including 123 in the United States (38.08%), 31 in the United Kingdom (9.60%), 16 in Spain (4.95%), and 15 in Germany (4.64%) (Table 1).

The number of papers published by the top ten institutions of international scientific communication research accounted for 26.6% of the total number of papers issued in 10 years, and 9 of them were American institutions. It shows that the United States has accumulated strong scientific research strength after years of research in science communication. Of course, one important factor is that the magazine is sponsored by the United States. At the same time, 268 institutions in 323 institutions are universities and colleges, and the rest are science centers, museums, foundations, scientific research institutions, etc. It can be seen that scholars from universities and colleges have played a pivotal role in scientific communication research. At the same time, the United States, the United Kingdom, the Netherlands and other countries with a large

Table 1 The top ten organizations in the number of SC posts.

Institution Name	Number of Posts (pieces)	Organization Type	Country
University of Wisconsin	16	University	United States
Cornell University	12	University	United States
George Mason University	12	University	United States
University of South Carolina	11	University	United States
University of Texas at Austin	10	University	United States
Colorado State University	8	University	United States
Michigan State University	7	University	United States
University of Nevada, Reno	7	University	United States
University of Twente	6	University	Netherlands
Iowa State University	5	University	United States

number of papers, as well as the University of Wisconsin, Cornell University, George Mason University, etc., may still make a huge contribution to the field of science communication for a long time to come.

2) Analysis of Highly Cited Papers and High-yield Authors

Being quoted by other articles can be used as one of the indicators to measure the quality of an article. Highly cited articles can highlight the development level and research direction of science communication, and are important evidence for exploring the hot topics and research frontiers of international scientific communication research. The papers of the top ten SCs cited in the past ten years are as follows (Table 2).

Table 2 Distribution of the top ten papers cited by SC in 2008–2017.

Serial number	Title	Author	Time	Cited Times
1	'Fear Won't Do It' Promoting Positive Engagement with Climate Change Through Visual and Iconic Representations	O'Neill, Saffron,etc.	2009	249
2	The Rise of the Knowledge Broker	Meyer, Morgan	2010	147
3	Reorienting Climate Change Communication for Effective Mitigation Forcing People to be Green or Fostering Grass-Roots Engagement?	Ockwell, David,etc.	2009	98
4	Constructing Communication: Talking to Scientists about Talking to the Public	Davies, Sarah R.	2008	95
5	A Two-Step Flow of Influence? Opinion-Leader Campaigns on Climate Change	Nisbet, Matthew C,etc.	2009	85
6	After the Flood: Anger, Attribution, and the Seeking of Information	Griffin, Robert J,etc.	2008	78
7	Science-Media Interface: It's Time to Reconsider	Peters, Hans Peter,etc.	2008	60
8	Meaningful Citizen Engagement in Science and Technology: What Would It Really Take?	Powell, Maria C,etc.	2008	59
9	From Public Understanding to Public Engagement An Empirical Assessment of Changes in Science Coverage	Schaefer, Mike S.	2009	46
10	The Matilda Effect in Science Communication: An Experiment on Gender Bias in Publication Quality Perceptions and Collaboration Interest	Knobloch-Westerwick, Silvia,etc.	2013	44

Published in 2009, 'Fear Won't Do It' Promoting Positive Engagement With Climate Change Through Visual and Iconic Representations, as the most cited article, was cited 249 times by the end of 2017. The article through empirical research proves that although fear appeal has the potential to attract public attention to climate change issues, it does not motivate individuals to truly participate. Conversely, non-threatening visual and symbolic expressions that are closely related to people's daily emotions and concerns often increase public awareness of climate change issues. With the emergence of climate change issues, the Copenhagen World Climate Conference was held in 2009, and many countries discussed global emission reduction issues. The number of publications on climate change in SC was as high as seven in 2009 and the most frequently cited article in the decade is also the issue of climate change. This reflects that important social issues have led to academic research in related fields, and research that reflects major social issues is likely to become a hot spot of concern.

633 authors (including the second and third authors) published in 354 articles from 2008 to 2017, and 11 of them published four papers (Table 3). The authors published 3 and 2 articles respectively, with 14 and 58 respectively, and 550 authors who have published only one paper.

Table 3 Authors of more than 4 articles issued by SC in 2008–2017.

Author	Number of Posts	Author Unit
Priest, Susanna Hornig	16	School of Communication, University of Washington, USA; Department of Communication, University of Nevada; Department of Communication, George Mason University, etc.
Valenti, JoAnn Myer	12	Brigham Young University of communication
Besley, John C	7	School of Journalism and Mass Communication, University of South Carolina
Brewer, Paul R	5	University of Delaware School of Political Communication
Brossard, Dominique	5	Department of Life Science Communication, University of Wisconsin
Dunwoody, Sharon	5	Department of Life Science Communication, University of Wisconsin-Madison
Hart, P. Sol	5	University of Michigan, Ann Arbor School of Communication
Feldman, Lauren	4	School of Communication and Information, Rutgers University, USA
Ho, Shirley S.	4	School of Gold and Communication and Information, Nanyang Technological University, Singapore
Mulder, Henk A. J.	4	Institute of Science and Society, University of Groningen, The Netherlands
Scheufele, Dietram A	4	Department of Science and Communication, University of Wisconsin-Madison

Professor Susanna Priest, who has published 16 articles, is the current editor-in-chief of SC. She has been transferred to the Department of Journalism and Communication at Washington University, University of Nevada, and George Mason University. Her main direction is communication, and her research focuses on the role of science in American society and culture, such as mass media coverage of science, science policy, and public opinion. Professor JoAnn M. Valenti from the School of Communication at Brigham Young University in the United States has published as many as 12 papers. Her main research direction is journalism and communication. She is not only a member of the editorial board of the SC journal but also a member and council representative of the American Association for the Advancement of Science. Dr. John C. Besley has seven papers and as an associate professor at the School of Journalism and Mass Communication at the University of South Carolina. His research interests include science communication and public participation in science. According to statistics, the authors with the highest rankings are all from the United States, which is consistent with the situation that the number of US publications is the highest in the decade. It indicates that the United States has occupied an important position in the field of international science communication. It is worth noting that the analysis of the authors in Table 3 shows that most of the international high-yield scholars in the field of science communication are the main force of the journalism and communication departments, which is in stark contrast with the distribution of scholars of science communication in China. According to the statistical analysis of scholars in the past, most of the authors of the most prolific and cited scientific communication papers in China are from the history of science and technology and philosophy of science and technology. This phenomenon remains the same today. And its research is based on concepts and texts, and less empirical research methods are used.[12] This shows that there is a big difference between the scientific communication research in China and the international research in the same field in terms of talent reserve and research status.

The above analysis of the SC article from 2008 to 2017 found that the United States and the United Kingdom and European countries have absolute advantages, which leads to the extremely uneven distribution of international scientific communication paper sources. The background of the authors is diverse and the institution is university-led. Journalism and communication related Majors are the main force in scientific communication research.

4 .Science Communication Research Hotspot Map Analysis

1) Keyword Analysis

The key word is the product of the author's concise article theme, which is a high-level summary of the core of the full text. Through keyword search for articles published in SC for ten years, vocabulary with weaker relevance to scientific communication is removed, and synonyms are combined (such as climate change and climate changes, mass media and mass-media, risk and risks, etc.) , the main keywords are shown in Fig. 2 word cloud.We counted the keywords with frequency more than 30 times. The highest frequency is communication 354 times, which can be an important label in the field of science communication. The other keywords are science 102 times, public engagement 69 times, climate change 61 times, Knowledge 51 times, mass media 48 times, media 42 times, information 42 times, news 38 times, science communication 36 times, risk 34 times, perception 31 times, attitudes 30 times. The higher frequency keywords indicate that the research direction has received more attention. The new and old keywords alternately highlight the changes in the research interests of scholars in the field of international science communication.

Fig. 2 SC 2008–2017 keyword word cloud.

2) Research hot knowledge map

Key words co-occurrence can reflect the close relationship between words. Based on VOSviewer software, the distribution of all keywords in the SC article in 2008–2017 is counted, and the hotspot map of international scientific communication field is drawn to reveal the hot research topic and evolution trend.

First, collect SC keyword data from 2008 to 2017 from WOS. The format is a txt format recognizable by VOSviewer, and the analysis range and threshold are set and imported into the software. The weight of the element determines the size of the label and the font. The larger the weight, the larger the label and font, and the more important it is in the field. At the same time, the element spacing can also reflect the relationship; the closer the relationship between elements, the shorter the distance between them, and vice versa. It can be seen from Fig. 3 that the degree of polymerization of the elements is more dispersed. This shows that scientific communication research has a wide range of fields and a large interdisciplinary relationship. Different colors represent different themes, and circles with the same color belong to the same cluster. The collection of different elements in Fig. 3 forms a total of seven clusters, which are classified into five themes by analyzing the keywords. We will mark them as A, B, C, D, E. These five themes are the media impact on public understanding science (theme A), the analysis of news content (theme B), the study of public cognition and attitude (theme C), the study of climate change and risk (theme D), and the study of visual communication(topic E). Finally, a research hotspot knowledge map in the field of science communication will be generated, as shown in Fig. 3.

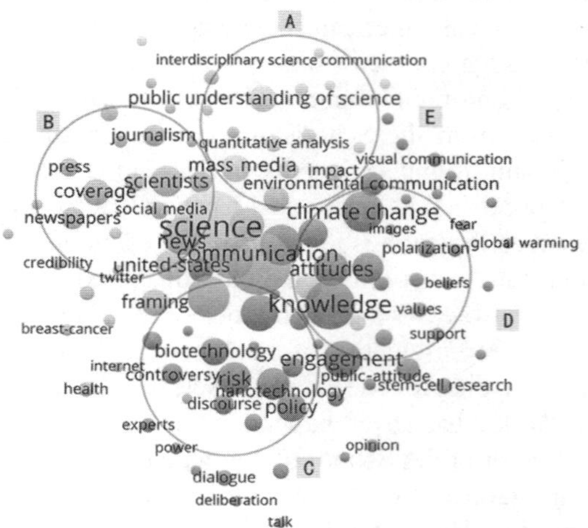

Fig. 3 SC science communication research hotspot clustering knowledge map.

High-frequency keywords can reflect the hot spots in the field of science communication. But some low-frequency keywords may contain innovations in research. According to the density view generated by VOSviewer (Fig. 4), the closer the color of the element is to red, the closer the study time is, and the closer the color is to blue, the farther the study time is. But as time goes on, the related research interests have also changed.

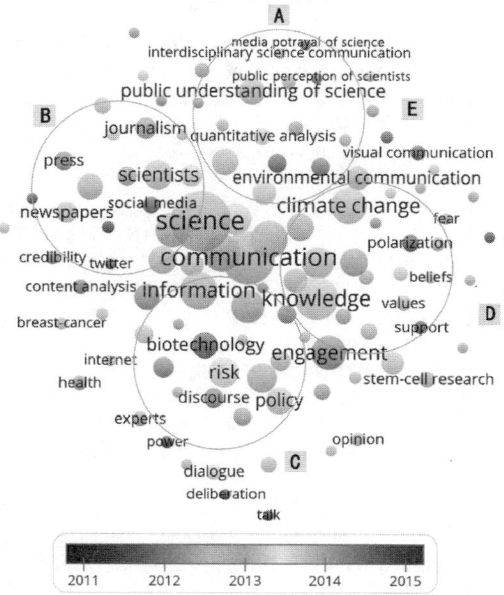

Fig. 4 SC science communication research timing density map.

Theme A focuses on the relationship between the media and the public understanding of science. So the frequency of keywords such as mass media, television, and mass communication theory is generally higher. The frequency of science is up to 102 times, and other keywords are distributed around it, mainly including mass media (48 times), public understanding science (16 times), television (12 times), and influence (11 times). From Figure 4, it can be found that the media's depiction of science and the scientists in the eyes of the public are emerging topics, and quantitative research has gradually attracted the attention of the academic community. It broke the previous situation of scientific communication research based on critical reflection.

Theme B is the content analysis of news reports. The most frequently used keywords are communication (354 times). Other key keywords include news (38 times), framework (29 times), news reports (22 times), and journalism (18 times). In the ten-year sample, framework theory has been used more frequently. The news media used the cost-benefit framework, risk framework, political framework, and contradictory framework when reporting on specific events. Related articles include Finding a News Media Framework on Science and Technology Progress, Risk, and Regulatory Stories (cited 33 times), The Science and News Framework Beliefs for Audience Evaluation of Embryo and Adult Stem Cell Research (cited 16 times). It can be seen from the density view in Fig. 4 that social new media is a hot topic of recent research. Nowadays, the scientific communication research about we media and social media is prevalent at home and abroad, and how to continue to play the role of science communication in the new media environment has received much attention.

Theme C is a public cognition attitude study. The frequency of key word public participation is 69 times. Other keywords are risk (34 times), nanotechnology (26 times), biotechnology (22 times), controversy (14 times), and the public opinion (11 times) and so on. Science communication has undergone three stages of development. The status of the public has been greatly enhanced, and public attitudes and prejudices have received attention. In the past, scientific communication was dominated by scientists, but now it has become to a multi-directional interaction between communicators, scientific information producers and audiences. At the same time, with the issue of scientific communication following the trend of scientific development, research focuses on cutting-edge technologies such as nanotechnology, transgenic technology, stem cell research, etc. How the public understands that emerging science and technology profoundly affects the relationship between science and society has inspired research enthusiasm for attracting public participation in the topic of scientific communication.

Theme D is climate change and risk research. The frequency of keyword knowledge is 51. The remaining keywords are climate change (61 times), attitude (30 times), crisis spread (21 times), and global warming (4 times). With the 2009 Copenhagen Climate Conference and the 2014 UNEP UN Climate Conference in South Africa, research on climate change has intensified, and the number of related papers has increased in 2009 and 2014 respectively.

Theme E is a visual communication study. It has only six elements and is the least of all clusters. The frequency is in turn for scientific communication (36 times), visual communication (6 times), and images (5 times).The most frequently cited article in the SC in ten years describes the role of visual expression of images and symbols in increasing public participation in climate change issues. According to the search for WOS, there are 22 papers on images in the ten-year sample, and 15 papers on visual communication. And in the density view, it can be seen that the color of the two elements of image and visual communication tend to be red. It indicate that it is a hot topic.

The analysis of the hotspots of international scientific communication research shows that scientific communication research has gradually changed from the traditional framework of news reporting, the public debate on biotechnology, the impact of television on public cognition science, gender and scientific literacy to the scientific communication research under the new media perspective, scientific communication research that appeals to vision, promotion of public participation in cutting-edge scientific and technological issues, and research on risk communication etc.

5. Conclusion

Since its start publication in 1979, SC has become the top journal in the field of international scientific communication, which embodies the most influential research results in the field of science and technology communication. Science communication is concerned with the mutual influence of science and society. Although China has many outstanding scientific and technological achievements in recent years, there is still a big gap in China's scientific communication research from the international level. Through the combing of the development trend of the SC in the past ten years, it also reflects that if we want to break the barrier between science and the public, and make the field of scientific communication research into the international mainstream, we still have a long way to go.

References

[1] Burns T W, O'Connor D J, Stocklmayer S M. Science communication: a contemporary definition[J]. Public understanding of science, 2003, 12(2): 183-202.

[2] 刘兵, 侯强. 国内科学传播研究: 理论与问题[J]. 自然辩证法研究, 2004(05): 80-85.

[3] Bernal J D.《科学的社会功能》[M]. 陈体芳, 译. 桂林：广西师范大学出版社, 2003: 341-355.

[4] 吴国盛. 从科学普及到科学传播[N]. 科技日报, 2000-09-22.

[5] 陈发俊, 史玉民, 徐飞.英国《公众理解科学》(Public Understanding of Science)文献计量研究[J].中国科技期刊研究, 2007, 18(03): 398-401.

[6] 张婷. 科学传播研究热点的演进——基于科学知识图谱的可视化研究[J]. 科普研究, 2010, 5(01): 17-27+47.

[7] 周雁翎. 科学传播研究的国际学术理路——对 Science Communication 期刊论文的计量研究[J]. 自然辩证法研究, 2013, 29(06): 50-54.

[8] 褚建勋, 倪国香. 科学传播领域关于气候变化议题的研究现状分析——以 SSCI 期刊 Science Communication 为分析样本(2002—2013)[J]. 科普研究, 2014, 9(04): 54-58.

[9] 朱巧燕. 国际科学传播研究: 立场、范式与学术路径[J]. 新闻与传播研究, 2015, 22(06): 78-92+128.

[10] Van N J, Waltman L. Software survey: VOSviewer, a computer program for bibliometric mapping. Scientometrics[J]. 2010, 84(2): 523-538.

[11] 陈积银, 刘颖琪. 国外新媒体研究 16 年发展脉络分析——基于 SSCI 期刊《New media & Society》1999 年至 2014 年的实证研究[J]. 新闻大学, 2015(06): 120-128.

[12] Xu L, Biaowen H, Guosheng W. Mapping science communication scholarship in China: Content analysis on breadth, depth and agenda of published research[J]. Public Understanding of Science, 2015, 24(8): 897-912.

Mapping the Contemporary Science Communication Landscape in Canada

Michelle Riedlinger[1], Alexandre Schiele[2] and Germana Barata[3,4]

[1] Communications Department, University of the Fraser Valley, Abbotsford, Canada
[2] UQAM East Asia Observatory, Université du Québec au Montréal, Montreal, Canada
[3] Laboratory of Advanced Studies in Journalism, State University of Campinas, Campinas, Brazil
[4] School of Publishing, Simon Fraser University, Vancouver, Canada

Abstract: The new media landscape presents challenges and opportunities for science writers and communicators that have not yet been fully realised. This paper presents the findings of collaborative work conducted using emerging new media research tools, including Altmetrics and traditional survey tools to identify the growing social media communicators engaging Canadian publics with science. Using an online survey tool, we compared survey responses from these social media science communicators to members of Canada's professional member associations—the Canadian Science Writers Association (CSWA) and the Association des communicateurs scientifiques du Québec. We found that Canadian social media science communicators were younger, paid less (or not at all) for their science communication activities, and had been communicating science for fewer years than other kinds of science communicators. They were more likely to have a science background (rather than communication, journalism or education background) and were less likely to be members of professional associations. These communicators tended to communicate with each other through their own informal networks. These findings provide professional science communication organisations in Canada with empirical grounding to help them develop training, support and outreach activities to improve the quality and ethical standards of public engagement with science in Canada.

1. Introduction

The Science Writers and Communicators of Canada (SWCC) has recognised over the last ten years that economic and technological changes have fractured the mainstream media landscape, creating changes to their membership base. SWCC first formed in 1971 as the the Canadian Science Writers Association (CSWA) to support professionalism in Canadian science journalist and broadcasting. However, the organisation recognises that a growing number of Canadian communicators, working as freelancers or in unpaid capacities, are engaging publics with science. The Association des communicateurs scientifiques du Québec (ACS) recognises that French-speaking science communicators face similar challenges, and a new challenge of establishing online presences in the overwhelmingly Anglophone social media environment of Canada. Both organisations wish to know more about the challenges and opportunities for science communication in the current media landscape, including demonstrating where sustainable career paths in new and emerging science communication professions may exist, and pointing to new ethical challenges for science communication in the changing social media environments.

2. The Science Communication Landscape in Canada

1) Canada's Developing Science Communication Landscape

Canada's modern science communication landscape developed out of the professionalization of science writing and broadcasting that emerged after World War II, and Canadian government investment in science communication efforts from the 1980s until early in the 21st century. Government efforts during this time recognized that economic and social progress could not be separated from science and technological progress (Schiele et al., 2012). During the latter part of the 20th century, science communication efforts

Corresponding author: Michelle Riedlinger, University of the Fraser Valley, 33844 King Road, Abbotsford, BC, Canada, V2S7M8. michelle.riedlinger@ufv.ca in Canada focussed on informing the public, promoting scientific careers to support economic development, and increasing science literacy in the canadian population. In the early 21st century, Canadian science communication efforts focussed on public engagement, knowledge co-creation (Einsiedel, 2008), and a 'science in culture'—or thinking about how society talks about science.

The science communication landscape in Canada is rapidly changing. Federal and Provincial government support in Canada for science and science communication has reduced in recent times (Boon, 2017). And Canadian journalists are seeing a change in how publics prefer to receive news. In 2016, around 40% of the Canadian population spent more than three hours online daily and nearly 60% of Canadians use social media and the Internet to search for news and current events (CIRA, 2016). Canadians prefer to be informed through the media but on their own timelines and with little or no cost to themselves (The Shattered Mirror, 2017). Scientific researchers in Canada are also increasingly pressured to demonstrate research impact as part of the current 'audit culture' of universities and research institutes. Many are taking on roles as scientist-communicators in social media settings. Added to the social media dissemination of science are the blurring boundaries between journalism and communication systems (Rollwagen et al., 2017), with many former science journalists working in freelance operations. In 2015, Rollwagen and her colleagues surveyed Canadian journalists to look at the implications of the changing media landscape. They found that some of the greatest impacts on the work of journalists were procedural—including information access, journalistic ethics, media laws and regulations, available news gathering resources, and time constraints of the job. Many of these procedural influences are quite feasibly different for science communicators working outside of mainstream journalistic systems. Bucchi (2013) argues that these conditions, which include a declining role for filters and processes to guarantee the quality of information, can be summed up as a 'crisis of mediators'. Social media communicators may have different interpretations of information quality, accountability and professionalism. If trust in science relies on trust in science communication (Weingart et al., 2017), then addressing what quality science communication looks like in Canada's fractured media landscape is an important focus for research efforts. Bucchi (2017) argues that declining quality of science communication via the Internet and links to declining trust in science have yet to be established—but he does call on researchers and institutions to take greater responsibility for producing quality science communication and critiquing science where it is needed. If discussions about science in society are to be democratic (Bucchi, 2017), then greater recognition and support may be needed for the growing number of people working outside of intuitional structures who communicate about science in the contemporary media landscape.

2) Roles for Professional Member Associations

Professional member associations in Canada have recognised that they play important roles in supporting critical innovation in science writing and communication. The Canadian Science Writers and Communicators of Canada (SWCC) is paying close attention to these issues. After much debate occurring over many years, the organisation recently changed its name from the Canadian Association of Science Writers (CASW) to recognise their changing membership base, and the important roles for science communicators in Canada. Both the SWCC and the Association des communicateurs scientifiques du Québec (ACS) are keenly interested in who is communicating science in Canada and how they are doing it so they can better develop new policies and practices.

This collaborative project has two parts:

(1) identifying and mapping social media science communicators in Canada;

(2) surveying Canadian social media communicators and members of SWCC and ACS and comparing their responses to those of social media communicators identifed through social media mapping methods.

The focus of this paper is the second part—the comparison of professional association members and this emerging, possibly independent group of social media communicators.

3. Methods

1) Identifying and Mapping Social Media Science Communicators in Canada

We started by identifying Canadian science communicators on Twitter and Instagram, two of the most popular social media platforms in

Canada. For Twitter, we obtained data from Altmetric.com, which produces alternative metrics to track how papers are shared on social media and other online platforms as Wikipedia, news outlets, blogs. We identifed Twitter science communicators with the tag 'science communicators' using geolocation to narrow the population down to Canadian communicators. We gathered 855,016 Tweets between 2015 and 2016 and then tracked kewords and #hashtags related to 'science communicator', both in French and English, on the biographies of Twitter handles. After cleaning the data we identified a total of 197 Twitter unique IDs.

We used Netlytic software, developed by Ryerson University, Canada to identify Instagram communicators. General geolocation on Instagram is reported to be around 6%, so we extracted different sets of data using four hashtags: scicomm, commsci, vulgarisation and sciart. We also tracked keywords related to Canada in the biographies of Instagram users as a way to identify their geolocation (i.e. provinces, capitals, as well as #cdn, Canadian). After cleaning the data, we identified 59 Instagram science communicators posting from Canada.

Findings from this identification and mapping work are presented in another paper in these proceedings (Barata et al., 2018). However, we note here that the geographical patterns of science communicators affiliated with the two professional associations and those identified through social media methods were similar, with strong concentrations in the provinces of Ontario, Quebec and, to a lesser extent British Columbia.

2) Surveying Science Communicators in Canada

We invited science communicators identifed through Twitter and Instragram, and members of SWCC and ACS to complete an online survey. This survey was designed to compare the demographics of these populations, their activities related to science writing and communication, their attitudes towards science writing and communication, and their social media practices. We sent out an initial invitation and three reminders. We received responses rates of over 25% for each population group (143/524 or 27% of SWCC members; 87/309 or 28% of ACS members; and 74/256 or 29% social media communicators identifed through Altmetric and Netlytic).

We conducted quantitative data analysis on the responses we received using IMB SPSS 24 statistics software. Qualitative data from the open-ended questions was coded using NVivo 11 qualitative analysis software. Data is reported below as descriptive statistics and associations through cross tabulations (chi square).

4. Findings

1) Demographic Comparisons

The demographics of the SWCC and the ACS groups of survey respondents were found to be relatively similar. The ACS group had the highest proportion of respondents who were paid employees, and the social media respondents we identified through social media mapping had the lowest proportion. Over 40% of social media group respondents identified through Altmetric and other online mapping tools were unpaid for the science writing and communication work they did.

Compared to professional association members, the social media communicators we identified through new media mapping were more likely to be female ($\chi^2(1) = 4.97, p = 0.026$) and younger (less than 30 years of age) ($\chi^2(2) = 35.64, p = 0.000$). SWCC respondents were more likely than respondents from the social media group to indicate that they earn more than CAD$50,000 per year for science writing or communication work ($\chi^2(3) = 25.69, p = 0.000$) and were more likely to have ten or more years experience in the field when compared with respondents from the social media groups ($\chi^2(3) = 40.79, p = 0.000$). There were no significant differences between SWCC members and the respondents from the social media group identified through social media mapping in terms of respondents' main source of income. However, SWCC members were more likely than social media group respondents to state that science writing or communication was their primary occupation ($\chi^2(1) = 5.73, p = 0.017$).

Respondents from the social media group were less likely to have a professional background in areas other than science (i.e. journalism, communication, education, public relations or marketing) ($\chi^2(1) = 18.05, p = 0.000$). However, there was no significant relationship between these groups in terms of the level of training they had received (certificate, degree, graduate study etc.). The number of social media respondents who belonged to a professional science writing or communication association was much smaller than the number of respondents from the SWCC group. We could not demonstrate a statistical difference because we only asked respondents to name the professional associations they belong to, so numbers were based on counting those respondents who nominated a professional association. Only 14 respondents from the social media group indicated that they belonged to a professional science writing/ communication association. However, social media group respondents were more likely to be involved in an informal network of science writers or communicators than SWCC members ($\chi^2(1) = 10.18$, $p = 0.001$). These networks included informal social media groups, local events or groups that got together face-to-face, and connections with trusted colleagues and fellow alumni from university and college courses.

2) Communication Purposes and Challenges

The main purpose SWCC members gave for science writing and communication was increasing public awareness, while ACS members stated that their main purpose was helping the public form opinions. Social media group respondents nominated quite mixed purposes but when the three groups were compared, the patterns were not significantly different at $p < 0.05$. Any differences we identified in the patterns of findings associated with key challenges to science writing and communication were not statistically significant. However, all science writers and communicators identified major challenges associated with funding and time.

3) Science Communication and Quality

In an open-ended question we asked respondents to describe 'What makes "good" science writing or communication?' For all three groups, 'accuracy' was nominated as one of the top five categories. ACS members were the only ones to use the term 'vulgarised', which refers to the use of non-specialised language. We also asked respondents to cite up to three groups or individuals who, in their opinion, were engaging in good science writing and communication practices. In comparison to the social media group of communicators, SWCC and ACS respondents focussed more on mainstream journalists, traditional publications and broadcast mediums. Respondents from all groups included communicators and organisations that aimed their activities at schools and youth, including museums and science centres, and outreach activities from universities and research institutes. Respondents from the social media group identified through social media mapping particularly emphasised the work of 'scientist' communicators who aimed their communication at adults through social media and broadcast media. SWCC and social media respondents both also noted the good practices of recently-emerging English-language independent online publications aimed at all ages, such as the blogging platform, *Science Borealis* and *Hakai* magazine. They also recognised science communication aimed at adults through YouTube and podcasts.

4) Social Media Communication Practices

Not surprisingly, respondents identified through the social media mapping work were more prolific users of social media for science writing and communication compared to professional association members. We could not identify statistically significant patterns in the use of particular social media platforms because numbers of respondents were small. However, preferences for all groups were divided between Twitter and Facebook; social media group respondents appeared to actively engage in the range of social media platforms we asked them about: Twitter, Facebook, YouTube, Instagram, Google+, Tumblr, Pinterest and Snapchat. SWCC member group respondents who used social media were more likely to receive funding for social media activities than respondents in the social media group identified through social media mapping ($\chi^2(2) = 9.00$, $p = 0.011$) and ACS member respondents. Around 60% of ACS/SWCC members stated that their social media activities were unfunded or self-funded. Over 75% of respondents from the group identified through social media mapping stated that their social media activities were self-funded or unfunded.

Finally we coded the responses to an open-ended question asking respondents in the three groups to talk about why they used social media for science communication. Of the top five responses, the following three were common across the three groups:

(1) The broad reach of social media.

(2) Targeting or engaging with particular interested audiences or communities.

(3) Being able to see social media users interact with each other.

5. Discussion

In this study, we set out to compare survey responses from professional member association respondents to responses from science communicators who we identified through social media mapping. Our findings confirm pre-existing understandings that many science communicators in Canada are not captured by the professional member organisations. We found that the respondents from the social media science communicator group were demographically different to professional association members in many ways. Scientist-communicators and others who did not work as professional science writers or communicators were participating in science communication activities in unpaid or self-funded capacities. They also used a wider range of social media platforms than professional association members. Social media communicators currently have a greater reliance on informal networks to connect with others doing similar work, indicating concurrent streams of activity with little overlap between professional and informal networks.

We also identified some commonalities across the groups. We found no significant differences in the patterns of challenges faced by social media communicators and those who belonged to professional member associations—time and funding were the biggest challenges for all groups. Accuracy in science writing and communication was a key value for respondents for all groups, as were traditional journalistic values (i.e. relevance, accessibility, storytelling, independent research, and credible/trusted sources of information), and language choices (style) associated with user engagement.

SWCC have indicated that they will use this information to encourage a diversity of science writing and communication practices (without conflating science journalism and communication), tap into informal networks, provide better networking opportunities for science writers and communicators in Canada, and support independent science communication enterprises and the work of freelance science writers and social media communicators. They have indicated that they will work more closely with university-employed scientists, science communication consultants, science writers working for independent media outlets, and social media communicators to better understand their professional practices and career trajectories. They will use this information to help develop training courses for members and themes for future SWCC conferences.

Of particular interest is investigating how writers and communicators are balancing concerns of generating income and managing the increasing demands on their time, with producing accurate and engaging science-related content. SWCC has developed a new Ethics Committee to examine professional norms and conduct (e.g. hype, accuracy and conflicts of interest) in the contemporary media landscape, which will be informed by findings from this project.

Acknowledgments

This study was supported by a Social Science and Humanities Research Council (SSHRC) Partnership grant (892-2017-2019) held by researchers, Juan Alperin at Simon Fraser University (SFU), Michelle Riedlinger at the University of the Fraser Valley (UFV) and the Science Writers and Communicators of Canada (SWCC). The funders had no role in study design, data collection and analysis, decision to publish, or preparation of the manuscript. We are grateful for the continued support of Juan Pablo Alperin from the School of Publishing at SFU, and the assistance of Shelley McIvor, Janice Benthin and Tim Lougheed from SWCC and Stéphanie Thibault from Association des communicateurs scientifiques du Québec (ACS). We would also like to thank Chantal Chapman from UFV for her assistance with data analysis. Finally, we would like to thank everyone who participated in the survey. Without their input we would have nothing to report here. Their contribution to this project is significant and gratefully appreciated.

References

Barata, G., Riedlinger, M., Schiele, A., et al. (2018) Using social media metrics to identify science communicators in Canada. *Science & You*, Beijing, China. 15-17 September.

Boon, S. (2017). Science funding in Canada should include science communication. Watershed Moments: *Thoughts from the Hydrosphere* [blog]. Retrieved from https://snowhydro1.wordpress.com/2017/09/13/science-funding-in-canada-should-include-science-communication/.

Bucchi, M. (2013). Style in science communication. *Public Understanding of Science*, *22*(8), 904-915. doi: 10.1177/0963662513498202.

Bucchi, M. (2017). Credibility, expertise and the challenges of science communication 2.0. *Public Understanding of Science*, *26*(8), 890-893. Doi: 10.1177/0963662517733368.

Canadian Internet Registration Authority (2016). Internet use in Canada. *Cora Internet Factbook 2016*. Retrieved from.https://cira.ca/factbook/domain-industry-data-and-canadian-Internet-trends/internet-use-canada.

Einsiedel, E. (2008). Public participation and dialogue. In M. Bucchi, B. Trench (Eds.), *Handbook of public communication of science and technology* (pp. 173-184). London/New York: Routledge.

Rollwagen et al. (2017). Professional role orientations and perceived influences among Canadian journalists: Preliminary findings from the Canadian Worlds of Journalism Study. 2015 Proceedings of the Journalism Interest Group of the Canadian Communications Association. Retrieved from http://cca.kingsjournalism.com/?p=400.

Schiele, B., Landry, A. (2012). The development of science communication studies in Canada. In *Science communication in the world* (pp. 33-63). Springer, Dordrecht.

Weingart, P., Taubert, N. (Eds.). (2017). The future of scholarly publishing: Open access and the economics of digitisation. Cape Town, SA:African Minds.

Exploring the Function of Agenda-Setting in Science Communication in the Era of the New Media

——Taking smog reports of the Web portals as examples

Li Na, Qian Shuangshuang

University of Science and Technology of China, Hefei, China

Abstract: Reviewing the development of science communication, we know that the media plays an important role. However, in the era of the new media, various science communication elements have undergone tremendous changes, and science communicators need to take a new perspective on the science communication. This article takes the smog reports of major portals as examples. We collected the smog news reports of Phoenix Network, Tencent, Sina and NetEase Network from the year of 2015 to 2018, and then analyzed the framework of the smog news, mainly from the aspects of the number, types, keynotes and topics of reports, to find out whether and how the function of agenda-setting exists. Then, we explore methods of scientifically setting up the smog news agenda. In the end, we express views and expectations on the smog report in the era of the new media, in order to contribute to the development of science communication.

1. The Concept of Science Communication and Agenda-settin

The development of science communication and media is closely related. In the era of mainstream media communication characterized by newspapers, radio and television. It is difficult for audiences to interact with the media on information due to the limitations of technology. However, in the era of new media, various elements of communication have undergone tremendous changes, and science communicators need to take a new perspective on the science communication. The theory of agenda-setting is an important field during the studying of communication effects, and even a non-negligible researching area of scientific communication.

McCombs and Shaw who are American communication scholars published an article entitled 'The Agenda-Setting Function of Mass Communication' in the book of the Public Opinion in 1972, in which firstly proposed the concept of 'agenda-setting' ,that gradually involved into the so-called theory of 'Agenda Setting. The great communication scholars viewed that the mass media had the function of setting the agenda for the public, and also information communication activities such as news reports gave different degrees of distinctiveness to various issues, affecting people's judgments about the major events and their importance in the world around them (GUO Qingguang, 2011).' The agenda-setting theory consists of two levels of meanings: the level object and attribute agenda-setting. The level of object refers to information communicating activities of media by reporting or not, affecting people's judgments about the events and their importance of the surrounding events. This is the first level of concept of the agenda-setting. McCombs believed that the theory of agenda-setting had turned to the second level, that was, the attribute agenda-setting, that not only 'introducing' various topics to us, but also guiding us to care about attributes of these objects through describing attributes of them.

In recent years, the communication of the smog news has attracted a lot of attention in the field of science communication. In this article, we focus on the function of agenda-setting in science communication in the era of the new media, and we take smog news of the portal websites as samples, and then collect and analyze datas, in order to know scientifically spreading scientific information in the era of the new media.

Corresponding author: Qian Shuangshuang, Jinzhai Road, Hefei, China. 1157875223@qq.com

2. The Selection of Samples and Analysis of the Framework

1) Selection of Research Methods and Objects

In terms of research methods in this paper, we mainly adopt the methods of content analysis and literature research. We select the smog news in the scientific communication with great concern in China as the research objects. According to the list of news media websites published by the webmaster's home, we find that Phoenix, Tencent, Netease, and Sina are ranked 8th, 9th, 10th, and 15th respectively. Taking into account factors such as flow and credibility of the websites, we think that the four major websites can be selected as samples.

2) Selection of Time Period

Since 2013, severe smog has appeared in some areas, and smog has become an object of spit in people's mouths, which not only seriously affects people's normal daily life, but the sulphur dioxide and nitrogen oxides contained in the smog are harmful to people's health. At present, smog is already a 'tumor' in the health field. In 2015 when Chai Jin's Under the Dome was launched, the public sentiment of the smog incident was upgraded again. For a time, relevant government departments and petroleum fuel companies were pushed to the forefront. The documentary sparked thoughts about the haze. Therefore, we select the smog news reports from 2015 to 2018 in this article.

3) Definition of Smog Weather Reports

Searching for the keyword 'haze' on the four major websites, we define the haze weather reports with the following criteria. Firstly, we define news reports with the word 'haze' appeared in the title or text. Secondly, although the title and body of some articles have the word 'smog', the content of the news is completely linked to the advertisement which is not defined. Thirdly, reports related to haze weather are selected.

4) Choice of Reporting Framework

McCombs argued that clarifying a more general theoretical structure was necessary to describe the frameworks and attributes that were important to the communication process (Stanley Baran et al., 2004). The concept of the framework is derived from Bateson, who believed that the framework was a set of messages or action with meaning, with the development of the concept of the framework analysis, it has gradually extended into a two-dimensional measurement framework theory, the purpose of which was to analyze the way of media constructing and highlighting the significance of the news from different angles during the entire period of news development.

When the concept of agenda-setting was firstly proposed, it was mainly applied in the political field. Attributes were only related to politics, such as the image of political figures and the interests of voters. Gradually, the concept of agenda-setting is aimed at the topic itself, and the attribute agenda-setting breaks through this limitation. In other words, it focused on the image of the political candidate on the media. At present, that is to say the same topic and event have many attribute characteristics that are different from other things. When some features are emphasized, then the more prominent features that are highlighted, the greater the probability of being remembered and discussed. Relevant research results show that when the media reports an object, or when people talk about or think about an object, some attributes of the object are highlighted, while other attributes are passed or ignored, and the highlighted attributes are highlighted. The agenda will affect our understanding of the object (SUN Hui, 2017).

With the support of the above theory, we analyze the number of reports, the genre of reports, the tone of the report, and the topic of the report in the form of framework analysis. Then, we observe distribution of the smog news reports for 12 months, if there is any quantitative regularity, to prove whether the theory of agenda-setting exists or not. In the meanwhile, the genre, tone, and subject matter of the reports are classified into the scope of attribute agenda setting. By sorting out the frequency of occurrence of keywords, we can draw relevant conclusions.

3. Conclusion Analysis

1) Number of Reports

In this article, samples we selected are from the four websites of Phoenix, Netease, Tencent and Sina from the year of 2015 to first half year of 2018. The total number of reports are 652, respectively are 166, 194, 168, 124. According to the chart above, reports numbers of the four major websites on smog have gradually increased, and reached the highest peak in 2017, although the current data for the year of 2018 is only the first half, the trend in the line has shown a gradual upward trend for the smog news reports. Among them, the report of Phoenix Net has reached 78 peaks from the total of 22 articles in 2015 to 2017, and there has been 51 reports in the first half of 2018.

Judging from the monthly distribution of the smog reports, we find reporting of the four major websites show seasonal changes. Reporting numbers are concentrated in January, February, March, November and December, which means the concentration period of smog reports is basically consistent with seasons of autumn, winter and early spring, when haze days are frequent. This shows that the mass media will set the agenda according to the current life and the hot issues of concern.

However, since 2017, the smog report is not only concentrated in the smoggy autumn and winter seasons, but figures shown that in the period from June to September there are also many related reports such as aspects of daily governance or environmental protection. Before the change, the report only focused on the autumn and winter seasons, but in the other seasons, haze days were ignored. It shows that in the past two years, the four major websites have increasingly reflected the daily trend of environmental science-related issues in the agenda setting .

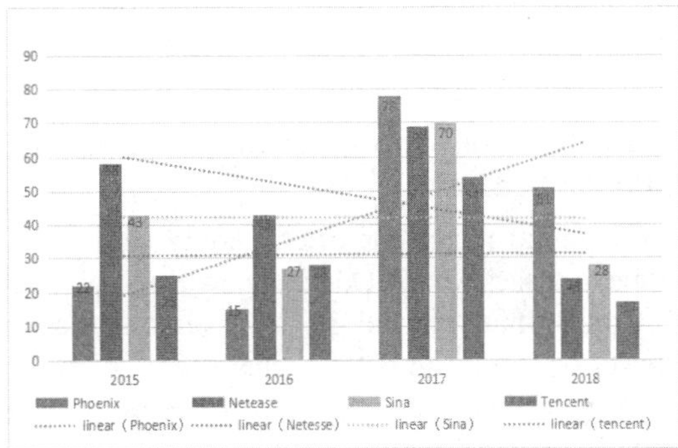

Fig. 1 The total number of smog reports of the four major websites from 2015 to 2018.

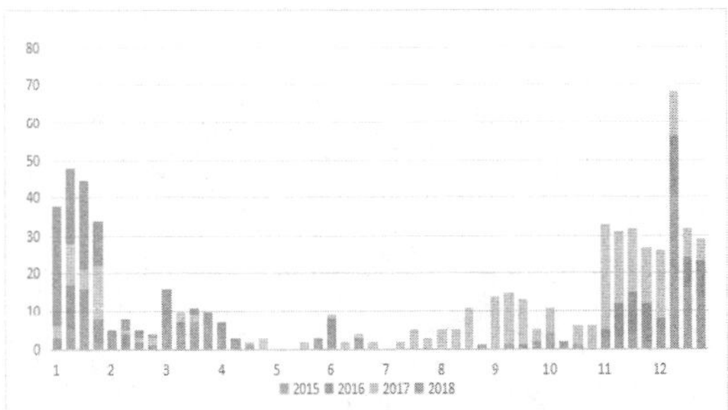

Fig. 2 The monthly distribution of the smog reports in the first half of 2015–2018.

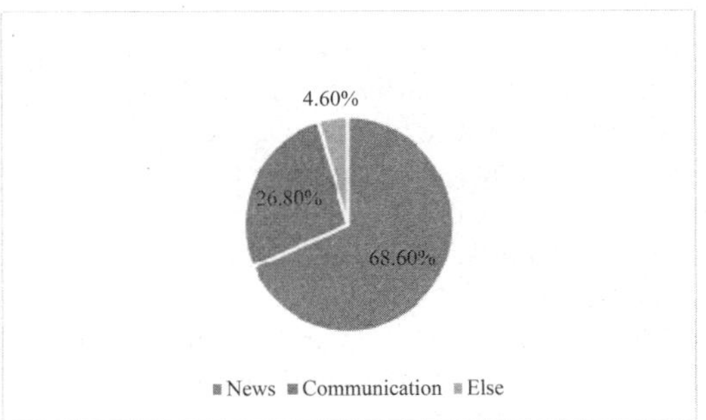

Fig. 3 Smog report genre chart of the four major websites in the first half of 2015–2018.

2) Report Genre

According to the data, the total number of reports on the smog coverage of the four websites is 652, of which 447 are news, accounting for 68.6% of the total, and communication topics are 175, accounting for 26.8%. The remaining 4.6% includes news reviews, in-depth reports, images, and more.

3) The Tone of Report

We analyze keynotes of the four major websites from 2015 to 2017, among them, 226 positive reports, accounting for 42.5% of the total, and 160 neutral reports, accounting for 30.1% of the total. 146 negative reports, accounting for 27.4% of the total. In the overall report, the tone of the reports of the four major websites on smog tends to be positive and objective neutral. From the perspective of annual trend changes, the trend of the keynotes tends to be positive, especially the number of positive reports in 2017 reached a small peak within three years, reaching a 51% ratio in the total report base of the year. Besides, considering the changing trend in reporting theme and keywords in 2017, we find media reports began to focus on the governance of smog and the protection of the environment. Such an agenda-setting is not only a change in the issue itself, but also a switch to attribute agenda-setting.

4) Reporting Theme: Analysis of Keywords

According to the 652 news reports on the four major websites during the period from 2015 to early 2018, we analyzed the theme of the

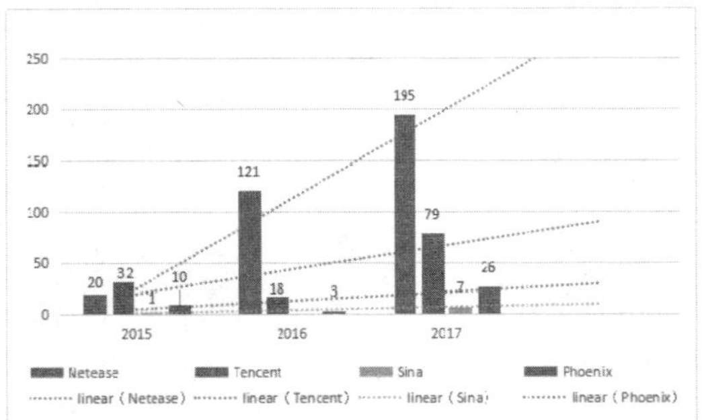

Fig. 4 The frequency of occurrence of the word 'governance' in the four major website reports.

report and combined the keywords. The topics covered are mainly divided into the following five aspects: smog dynamic tracking reports, reports on smog governance, knowledge reports on the causes of popular smog, green life advocacy reports, and residents' life information reports.

From analysis of keyword frequency, it can be seen that during the period of 2015–2017, the main keywords of major reports were 'haze', 'pollution' and 'air quality', etc. The keyword 'haze' appeared 3527 times. Pollution has also reached 2,307 times. Since 2017, keywords such as 'governance', 'purification', 'green' and 'governance' have repeatedly appeared in the coverage topics and reports of these medium, and the frequency of the words are on the rise. From this trend change, we see that the framework of covering on smog has changed. Obvlously, the changes in the focus of media coverage are also brought about by changes in the environment. The current framework for smog has changed with the calling for the policy of re-adjusting the country, the further development of governance, and the effectiveness of the treatment. These are concerns of the public and problems that need to be solved urgently. The medium's issues, while affecting public issues, are also updated as public issues change.

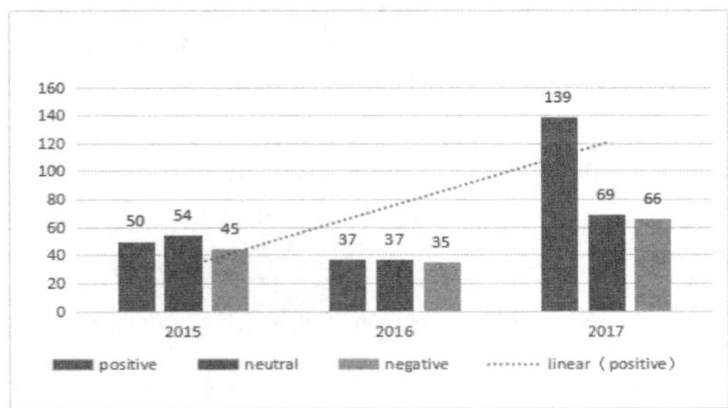

Fig. 5 Smog report keynote annual trend chart of the four websites.

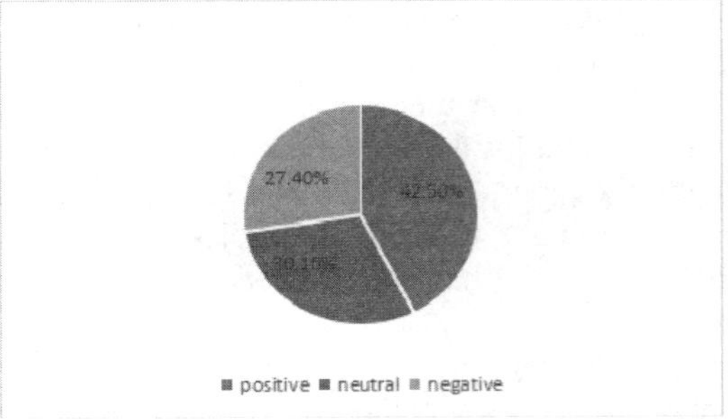

Fig. 6 Proportion of report tone on smog report in the four websites.

4. Discussion on Setting Agenda

1) Long-term Battle: Setting Daily Issues

From above analysis, we can see that the current smog reports in China present a phased and seasonal feature. But as an environmental issue, the report of smog should be a persistent reporting campaign. The media should regard smog report as a permanent issue in the daily reporting framework, rather than just making a large scale of report in the smoggy season. In the spread of environmental issues, media must always give the public a wake-up call, scientifically spread smog news and other environmental protection issues in daily life, and scientifically disseminate causes, hazards, project progress, governance methods, and so on. At the same time, it must be subtly impersonated into the public life, which can make the public realize that this is a 'long-term battle' with smog, and the battle must start from personal life and environmental protection.

2) Highlight the Topic Attribute: Being the Public Leader

The public agenda are often heavily influenced by media issues, which Donald Shaw and Maxwell McCombs had proved in 1968. In the daily reports on environmental issues such as smog, when media sets the issue, it is necessary to set tone of the report, determine subject of the report, and don't be malicious, sensational, and subjective. Besides, media should pay attention to the guidance of public opinion, and create a relevant public opinion field through daily news business in order to attract different circles to participate (XU Jiabiao et al., 2014).

3) Science Communication: Scientific Issues

As the recipient of public power, media needs

to bear a high degree level of social responsibility. The moment when masses give their rights to the media, it means that media should be ready all the time. The public has the right to know, and the media is a tool to better meet the audience's right to know. In the setting of issues, especially on the agenda of public health, media reports, media should not be limited to report on the actual situation of smog, but also be scientific about the causes and prevention of smog. Reporting at the same time, it is also possible to report on the measures of local governance to reduce the panic of the public and prevent the possibility of rumors(ZHOU Jie, 2014).

5. Notes

Although we list smog news reports of the four major portals, there is a lack of smog news reports from the we-media. In addition, data analysis does not provide a comprehensive summary of the types of reports. Therefore, this gap will be filled in the next smog news research.

References

Guo Qingguang [M] (2011). *Journalism and Communication.* BeiJing: China RenMin University Press.

Stanley Baran, Dennis Davis [M] (2004). *Mass Communication Theory: Foundations, Ferment and Future.* BeiJing: TsingHua University Press.

Sun Hui [J] (2017). Exploring the Theory of Attribute Agenda Setting from Framework Theory. *Art Science and Technology*.

Xu Jiabiao, Han Zhaowei [J] (2014). Legalization of Real Narration: Research on the Framework Strategy of Haze Report—Taking 'Hua Shang Bao' and 'Xi'an Evening News' as Examples. *Contemporary Communications*.

Zhou Jie [J] (2014). Practice and Reflection on the Report of Haze in the Field of Health Communication. *Chinese Journalist*.

Using Social Media Metrics to Identify Science Communicators in Canada

Germana Barata[1,4], Michelle Riedlinger[2], Alexandre Schiele[3], Juan Pablo Alperin[4]

[1] Laboratory of Advanced Studies in Journalism, State University of Campinas, UNICAMP, Campinas, Brazil
[2] Communication Dept., University of the Fraser Valley, UFV, City, Canada
[3] UQAM East Asia Observatory, Université du Québec au Montréal, UQAM, Montreal, Canada
[4] School of Publishing, Simon Fraser University, SFU, Vancouver, Canada

Abstract: As part of a project mapping the new landscape of science communication in Canada, we recognize that social media has allowed non-traditional science communicators to participate as content producers. Yet their practices are not visible to the mainstream science communication community because these social media communicators build networks independently of professional member associations. This project identifies and maps science communicators on Instagram, Twitter, and blogs in Canada using data gathered for creating social media scholarly metrics (Altmetrics) and from social media platforms directly (with the help of the Netlytic tool). Using a combination of automated methods (querying the database of Altmetric LLC and application programming interface of Instagram) and manual searches, we identified hashtags, keywords and users, and then filtered them to those who were geolocated in Canada. We identified 56 science communicators on Instagram, 196 on Twitter and 60 bloggers. Our findings show a rich picture of activity on social media within Canada, mainly in Ontario, Quebec and British Columbia. Identifying these accounts is a first step towards documenting the innovative practices of these communicators so that associations such as the Science Writers and Communicators of Canada (SWCC) and the Association des communicateurs scientifique du Québec (ACS) can provide better professional support and improve policies related to science communication practices.

1. Introduction

Social media has changed the way we communicate, including the ways in which we communicate science (Brossard et al., 2013). This is unsurprising given the broad reach of the largest social media platforms. Worldwide, Facebook has reached more than two billion users, YouTube 1.9 billion, WhatsApp more than 1.5 billion, Instagram one billion, and Twitter 355 million (Statista, 2018a). Of these, Instagram is growing the fastest. Twitter, although having the smallest number of active users, has a lot of visibility. Anyone interested in sharing scientific information with the general public can use social media to freely speak, as well as co-create content with their audiences (Rollwagen et al., 2017). As such, social media has the potential to foster public dialogue. It is rich for direct communication between scientists, experts, universities, NGOs and society, without the need for mediation by journalists.

The broadband connection to the Internet in Canada reaches 86% of homes. The majority of Canadians spend between three to four hours daily on the Internet and around 64% of them enjoy engaging on social media (CIRA, 2018). Facebook is the second main source of news for Canadians, just after television (Abacus Data, 2017), which demonstrates the growing relevance of this social media to Canadians' daily lives. Among Internet users in Canada, 80% use Facebook, 39% Instagram and 35% Twitter (Statista, 2018b).

Social media has drawn the attention of information scientists who have analysed ways of measuring the circulation of research on social media—so called altmetrics (Priem et al., 2010). While altmetrics can be collected and captured by anyone, a British company, Altmetric, is one of the most widely used sources of altmetrics data (Robinson-García et al., 2014). Altmetric tracks how scientific articles and

Corresponding author: Germana Barata, rua Seis de Agosto, 50. Cidade Universitária, Campinas_SP, Brazil. germana@unicamp.br documents are shared online by looking for links and mentions on various platforms, including Twitter. Researchers have proposed using these engagements as alternative indicators to measure the impact of science outside the walls of academia (Bornmann, 2014). Although altmetrics have limitations related to the 'heterogeneity of social media acts, users and motivations' (Haustein, 2016, p.4), and the social activity tracked cannot be compared to social impact, it provides rich data to track and analyze publics interested in consuming scientific information. Therefore, we suggest that social media traces of research can be used to study science communication, science communicators and writers, and to produce indicators that might motivate scientists and institutions to track the attention of their public outreach activities.

To begin work on developing social media indicators of public outreach, we first needed to identify the extent to which science communication activities are taking place on social media. As such, in this paper we identify and map science communicators on social media in Canada using altmetrics and social media platforms as a source of data. Our goals are to understand the new landscape of science communication in Canada, provide visibility to social media science communicators, and stimulate networking among them. Moreover, we believe such work may assist Canada's professional science communication organizations, the Science Writers and Communicators of Canada (SWCC) and the Association du communicateurs scientifique du Québec (ACS), in providing better training and support to their members, and to promote and improve policies and practices related to science communication.

2. Methodology

We identified and mapped science communicators on Twitter and Instagram by looking at user self-descriptions (biographies) for mentions of science communication. Twitter, has the highest coverage for scientific papers on Altmetric (Robinson-García et al., 2014) and so we used it to conduct our analysis in two steps. We: 1) identified groups of users whose accounts were geolocated in Canada using a list of relevant keywords and hashtags in their biographies; and 2) looked for keywords in blogs to identify science bloggers.

1) Twitter

We searched a local copy of Altmetric data for all research articles shared on Twitter between January 2015 and June 2016. We then used Altmetric's 'science communicators' category to filter only those accounts that included any of the following keywords: news service, magazine, editor-in-chief, editors-in-chief, correspondent, Jornalista, reporter, journalist, journo, scicomms, science communicator, science communication, editor, blogueiro, blogger, librarian, Bibliotecari, journal, society, proceedings. Although limited (e.g., does not include any French keywords), it offers a reasonable starting point. From this set of data, we searched for French keywords relevant to science communication including: commsci, comunication scientifique, vulgarisation, culture scientifique, journalism scientifique etc. We subsequently filtered these users by those accounts that were geolocated in Canada.

This method identified 855,016 self-identified science communicators in the Altmetric database who shared at least one research article on Twitter from 2015 and the first half of 2016. However, this method had limitations for identifying those from within a given country: only around 56% of all tweets are geolocated (Haustein, 2018). Therefore, we may have missed science communicators who: a) used those keywords in their biographies (self-identified as science communicators) but did not provide geolocation; b) did not use or used different keywords/hashtags to describe their work with science communication. One other important limitation is that Altmetric does not use French keywords under their 'science communicators' category. Therefore, we could only track those who used English keywords/hashtags on their Twitter biographies.

To increase the number of French-Canadian science communicators contained in our Canadian dataset, we pulled data from the broader category 'Member of Public' from Altmetric.com. This included all Twitter users who did not belong to the other three categories: 'science communicators', 'practitioners' and 'researchers'. Since this was a larger sample, we were able to track all the accounts geolocated in Canada and, once again search for the French keywords/hashtags. The results of this work are still under analysis and will not be presented here.

2) Instagram

To identify Canadian Instragram science communicators, we used *Netlytic* software, developed by Ryerson University's Social Media Lab. This social network analyser pulls data from different social media, including Instagram. Although the software tool has limitations, we have found it useful for capturing Instagram user accounts by hashtags or geolocation. Geolocation on Instagram is made by picture, not linked to the ID handle. Therefore, we decided to extract different sets of data using four hashtags shared in different posts: #scicomm; #sciart; #commsci and #vulgarisation. While #scicomm appeared on 1,376 unique accounts, sciart was present on 479 unique accounts. Both are used by French-Canadian or English-Canadian science communicators and were the most used hashtags. We also identified 180 unique accounts using the French hashtags #vulgarisation and #commsci. As a way to identify Instagram handles from Canada we then searched the accounts we had identified for keywords related to Canadian geolocation (i.e. using the names of provinces and capitals, #cdn, Canadian). After cleaning the data, we identified 56 Instagram science communicators posting from Canada.

3) Blogs

From the samples of science communicators identified on Twitter and Instagram (above), we searched for science bloggers or potential bloggers, by looking for URLs and the keyword 'blog' and identified.

We have also pulled Altmetric data of papers with Canadian authors published on Web of Science (WoS) that have been shared on Twitter between 2014 and 2017. There were a total of 17,514 posts sharing paper link, among which 1,898 blogs worldwide (raw data). We have selected the blogs identified with location in Canada ('.ca') or including keywords related to Canada in their description.

We tracked a total of 60 blogs and identified 52 science blogs after excluding the non-active or non-related to science.

3. Results and Discussion

Through this approach, we identified and mapped 197 science communicators on Twitter, 56 on Instagram and 52 science blogs (Figure 1). These science communicators are based in nine of Canada's ten provinces and in two of the three territories. There is a significant concentration in Ontario, Quebec and British Columbia, the three most populous provinces and also the ones with a longer record of investment in science communication policy, training and public outreach activities (Schiele, et al., 2011).

Although Twitter is not the most frequently-used social media platform worldwide, nor in Canada, the opportunity for science communicators and publics to engage with researchers on this platform may be on the rise as it becomes more widely used by scholars (Noorden, 2014). Given that science communicators often communicate across multiple platforms, the effectiveness of our approach to locate science communicators on Twitter may prove valuable for identifying them on other platforms.

Instagram is more widely used than Twitter, yet we only identified 56 science communicators on this platform using *Netlytic* software. More than two thirds of these accounts were located in Ontario and Quebec (28 and 11 respectively). Notably none were located in the northern territories.

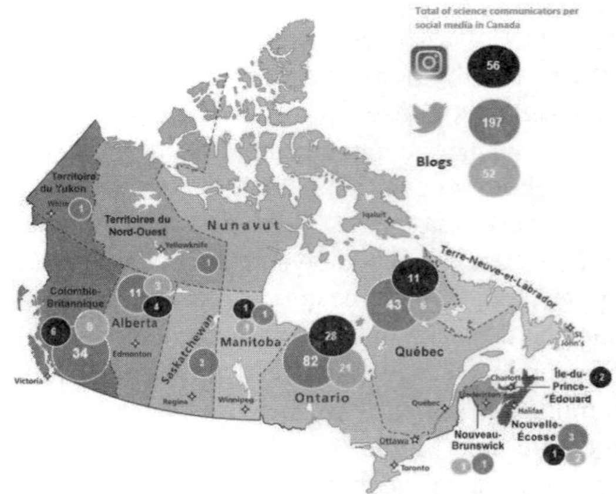

Fig. 1 Map of science communicators on Twitter (blue circles), Instagram (black circles) and science Blogs (pink circles) in Canada among provinces and territories. Some science communicators were geolocated in Canada but not at a province or territory, therefore they are not on the map. The total of science communicators per social media is above on the right.

We also identified a similar number of blogs, which were located in fewer provinces and we identified none of the territories. However, we did not include in this procedure the Science Borealis blog hub which contains more than 130 science blogs spread across the country, even though some of its blogs were tracked by our methodology, which means that only the ones sharing a paper link were identified by Altmetric.

The accounts and users included researchers and research groups, members of formal organizations and informal communities, and individuals operating for both commercial and non-commercial purposes. We include four examples of science communicators from each platform:

1) Twitter

- @AnatomySupply: Maria Romanova calls herself a medical illustrator and animator and shares #scicomm in her biography. She has more than 1,130 followers and is based in Toronto, ON. She joined Twitter in 2014.
- @DrummerBoy2112: Brian Wagner is a Chemistry Professor who works with fluorescent Chemistry in Charlottetown, PE. He joined Twitter in 2011 and has more than 14,000 followers and includes 'scicomm' in his biography.
- @JeremyBouchez: Jeremy Bouchez is a 'communicateur scientifique' in Montreal, QC. He has 1,241 followers. He started tweeting in 2011 and mainly communicates in French, but also in English.
- @WhySharksMatter: David Shiffman is a post-doctoral fellow at Simon Fraser University (SFU) in Vancouver, BC. He is a marine biologist expert in sharks with more than 38,000 Twitter followers. He describes himself a science writer. He started his Twitter account in 2009.

2) Instagram

- @oliviarozema: Olivia Rozema is a full-time science artist. She has 454 followers and is from Regina, SK. She has used #sciart in her bio.
- @petridishpicasso: This is a student organization from the University of Calgary using bioart for public outreach and communication in Calgary, AB. They have 2,101 followers and have used #sciart and #scicomm in their bios.
- @pineappleswhalesci: Daisy and Chloé are two science artists from Winnipeg, MB, who draw infographics about ecology and evolution. They have used #sciart and #scicomm and have 175 followers.
- @science.sam: Samantha Yammine is a neuroscientist PhD candidate at University of Toronto, Toronto, ON. She considers herself a 'science storyteller'. Her communication is mainly about neuroscience and STEM. She has more than 31,000 followers.

3) Blogs

- http://www.alternativesjournal.ca: Alternative journalism focused on environmental issues. Located in Kitchener, ON. This blog was located on Twitter.
- http://www.drsharma.ca/: Health expert who writes about obesity in Edmonton, AB. This blog was located on Twitter.
- https://impactethics.ca/: Run by a team of researchers from Faculty of Medicine at Dalhousie University in Halifax, NE. It deals with bioethical issues. This blog was located on Twitter.
- https://www.sciencepresse.qc.ca/blogues: Hub of science blogs located in Montreal, QC gathers French-Canadian science bloggers from different fields of knowledge. This blog was located on Twitter.

4. Conclusion

Data gathered from mentions of research on social media from Altmetric LLC and social media data gathered with Netlytic have proven to be useful for tracking Canadian science communicators, even in cases where users did not include geolocation information in their accounts. Our approach involved much manual work and suffers from other limitations already noted. However, we believe our social media mapping work demonstrates that the new Canadian science communication landscape is rich and can serve to promote connections between new and emerging science communicators in Canada. That is, the map has the potential to empower the science communication community, whom we hope see value in contributing to and updating the data map we produce. This map could help inform policies to promote science communication activities on social media and help professionalize and enlarge the community within Canada. It may also promote collaboration among social media science communicators and raise public interest in public outreach activities on social media.

Acknowledgments

This study was supported by the Social Sciences and Humanities Research Council of Canada through Grant (892-2017-2019) to

Juan Pablo Alperin and Michelle Riedlinger. The funders had no role in study design, data collection and analysis, decision to publish, or preparation of the manuscript. We would like to thank the Science Writers and Communicators of Canada (SWCC) for their partnership in this project. In particular, we are grateful for the continued support and assistance of Shelley McIvor, Janice Benthin and Tim Lougheed from SWCC, and Stéphanie Thibault from l'Association des communicateurs scientifiques du Québec (ACS).

References

Abacus Data. (2017). Matters of opinion 2017: 8 things we learned about politics, the news, and the Internet survey. Summary available at: http://abacusdata.ca/matters-of-opinion-2017-8-things-we-learned-about-politics-the-news-and-the-internet/ (accessed 31 October 2018).

Bornmann, L. (2014). Do altmetrics point to the broader impact of research? An overview of benefits and disadvantages of altmetrics. *Journal of Informetrics*, *8*(4):895-903. doi: 10.1016/j.joi.2014.09.005.

Brossard, D., Scheufele, D. A. (2013). Science, New Media, and the Public. *Science*, 339(6115), 40-41. doi: 10.1126/science.1232329.

CIRA - Canadian Internet Registration Authority. (2018). Canada Internet Factbook 2018: Canada's source for current Internet data. Available at: https://cira.ca/factbook/canada%E2%80%99s-internet-factbook-2018 (accessed 1 November 2018).

Haustein, S. (2018). Twitter in scholarly communication. *Altmetric Blog*. 12 June 2018. available at: https://www.altmetric.com/blog/twitter-in-scholarly-communication/ (Accessed 31 October 2018).

Haustein, S. (2016). Grand challenges in altmetrics: heterogeneity, data quality and dependencies. Scientometrics 108 (1):413-423. Doi: 10.1007/s11192-016-1910-9.

Noorden, Richard Von. (2014). Online collaboration: Scientists and the social network. *Nature*, *512*(7513): 126-9. doi:10.1038/512126a.

Priem J., Taraborelli D., Groth P., et al. (2010), Altmetrics: A manifesto, 26 October 2010. Available at: http://altmetrics.org/manifesto (accessed 31 October 2018).

Robinson-García, N., Torres-Salinas, D., Zahedi, Z., et al. (2014). New data, new possibilities: exploring the insides of altmetric.com. El Profesional de la Información, 23(4). arXiv:1408.0135 [cs.DL]. Retrieved from: http://arxiv.org/abs/1408.0135.

Rollwagen et al. (2017). Professional role orientations and perceived influences among Canadian journalists: Preliminary findings from the Canadian Worlds of Journalism Study. 2015 Proceedings of the Journalism Interest Group of the Canadian Communications Association. Retrieved from http://cca.kingsjournalism.com/?p=400.

Schiele, B., Landry, A., Schiele, A. (2011). Science communication in Canada: An inventory for the major PCST initiatives carried out in Canada. Research report to the Korean Foundation for the Advancement of Science and Creativity. Montreal: CIRST-UQAM, University of Quebec at Montreal.

Statista (2018a). Most popular social networks worldwide as of October 2018, ranked by number of active users (in millions). Available at: https://www.statista.com/statistics/272014/global-social-networks-ranked-by-number-of-users/ (accessed 1 November 2018).

Statista (2018b). Percentage of internet users accessing selected social media platforms in Canada as of May 2018. Available at: https://www.statista.com/statistics/468482/selected-social-media-user-share-canada/ (accessed 1 November 2018).

Comparison on Construction of Communication Context in Traditional Artistry and Culture Between Space of Hypermedia and Space of Traditional Communication

——Case study in Xuan Paper in Museum of Xuan Paper of China

Cheng Xi[1,2], Tang Shukun[2]

[1] School of Public Administration, University of Science and Technology of China, Hefei, China
[2] The Research Centre of Science Communication, University of Science and Technology of China, Hefei, China

Abstract: one of main means in science communication, museum burdened the mission to preserve traditional culture with respect to traditional technology and artistry. Except offline visit, many museums add online service to visit like offering immersive technology to visitors from anywhere to experience on-site visiting, which formed hypermedia space as it beyond limit of distance in traditional channel. Space of hypermedia provide brand new context of communication opposite to offline visit. Not only as an important part consisted of dialogue, context also played an important role in formation of content and means. Different context of communication tended to form different content and means of communication as communication is a kind of dialogue between communicator and receiver.

Xuan Paper, one of Four Treasures of Study, represented ancient and traditional artistry and culture of China. As the first systematic and integrated museum on Xuan paper in China, Museum of Xuan Paper of China, opened in 2015, taken both missions on introduction of classic skill and craft on paper making, and communication and preservation of culture in Xuan Paper. The museum offered VR visiting on official account in Wechat recently instead of visiting online based on site choosing from a map in previous official webpage of Cultural Park of Xuan Paper. It provided communication channels both in online and offline. As to explore that how to build traditional artistry and culture in different context especially in space of hypermedia, the paper is planning to deconstruct and compare different context from online and offline from three viewpoints—field of discourse, keynote of discourse, and mean of discourse—through specific experience examples in two visiting channels, including analysis of common and distinctive features of two contexts from three viewpoints respectively. At last, path of culture building in Xuan Paper from two contexts will be concluded.

Keywords: Traditional Technology and Artistry; Communication Context; Hypermedia; Museum

If the communication is regarded as the communication between the communicators and the recipient, then the context becomes the composing content of the communication environment, while some scholars believe that the environment of the media premises constitutes the context of the media transmitting information, and different communication contexts form different communication characteristics. The intervention of new media has changed the mode of communication and the communication context, so the communication context is of great significance to the analysis of modern media communication, such as the influence of contextual theory on the transmission of information in the context. The communication context caused by the new media, the French thought home Is Depo compared it as a 'landscape society', that in this society, the virtual and real boundaries have been completely blurred.

The new media virtual context also emerged. Unlike previous concerns about the consensus and communication between the communicators and the recipients, people began to focus on the impact of the media tools themselves on the context of communication, and to excavate their creative communication contexts in terms of their characteristics and practical experiences. Different communication tools also should be shipped different content. In recent years, a variety of resources characterized by new media technology have been enriched, especially in science and technology and art, and technology as a new development potential is constantly courting new media technologies. As the essence of technology and the beauty of art, the artistry of the intangible cultural Heritage has been paid more and more attention by the State and society.

Xuan paper, as the country's first non-material cultural heritage and traditional handicraft representatives, its inheritance and protection of urgency. As a representative of the rice paper artistry and cultural synthesis, China Xuan Paper Corporation is not only the largest rice paper production enterprises in China at present, but also in the efforts to build a production base of rice paper, paper production technology-based line, Xuan paper cultural experience as the theme of the Rice paper culture industry tourism. At present, it has been built with the paper culture garden as the main part of the tourism Experience project with different papermaking bases for the points, and the soon-to-be-built Rice Xuan Impartation Base (Culture Park two) will be integrated into rice paper ancient books printing, museums, four treasures and other upgrading projects, more omni-directional for visitors to show the culture and charm. In order to allow tourists to better experience the rice paper art and feel the rice paper culture, different types of new media are integrated into the rice paper spread, between the traditional and modern bridge to promote the digital process of traditional technology, while the maturity of new media technology is also a long-term cultural heritage preservation, widely disseminated another form and needs. According to the current use of new media tools in the dissemination of Xuan paper technology, the new media methods adopted and to be developed are:

Adopted New Media	New Media Under Develop
Digital pictures, digital video Xuan Paper Culture Park website Site Perception Media Xuan Paper map Virtual reality and game Simulation	Digital magazine, Digital newspaper WeChat, Weibo, public account Interactive museum Holographic projection …

Different new media features differ in the above table, and the virtual context of new media technology has its own characteristics. As a new media form commonly used in museums and exhibition halls in recent years, locative media has been widely used and developed by Xuan Paper.

With the continuous popularization of the application, the orientation media develops into the virtual context characteristic of the personalized new media technology which belongs to the paper propagation under the joint action of the characteristics of the media itself and the different platform features of Xuan paper propagation.

1. Locative Media Skills

locative media or location-based media, also known as Place media or Location Media, is based on Wikipedia's interpretation of a media tool that relies on locative positioning. The multimedia and other content associated with it is sent to the mobile media device based on the location of the user, so it is also seen as a digital media that is applied to the real world and causes real social interaction. The art of 'site Perception Media Art' (locative media art), which is presented as carrier media, Christian Paul the concept in context network: data, identity and collective production, and then develops the network Locative media art project, which is a new form of public art, which can be cross-regional.

It can also be a specific place to enhance or enhance the physical space with contextual information that can be stored and/or retrieved. According to the extension of the definition of the site Perception Media Art Project (locative media art project), the display of the technique as a carrier can be regarded as the 'locative media skill' project. Xuan Paper Culture Park is currently developing locative media for Xuan paper production technology display, through the equipment induction visitors to the rice paper different craft workshop, show the relevant content. On the network platform, Xuan Paper Map Project is under development. Xuan paper map, that is, on the map to mark the production of rice paper each production point, such as the production point of the Liao and Liao grass, green tan tree production points. The paper map shows the content of the rice paper making space by sensing the place where the visitor clicked.

Compared to the cultural park to visit the orientation of the media in the automatic orientation, the network platform in the paper map positioning through the active choice of tourists, but both are based on the 'positioning' on the basis of the dissemination of production technology, so or can be seen as the two forms of media technology transmission.

2. The Virtual Context of Locative Media Communication

In recent years, the virtual context of new media technology has been paid more and more attention, and the definition of 'a human-computer context based on digital communication technology and virtual reality' is applicable to the virtual context created by locative media in communication technique.

However, compared with other new media forms, its context presents its own characteristics, so on the basis of the combination of generality and individuality, this paper focuses on the 'virtual context of locative media communication' in the technology and cultural communication of the new medium. Context from the perspective of both global and local perspective is related to the position, from the global context is about location, enrich specific local characteristics, and from a local context is about activities and agents, and the location of the ability to connect. The virtual context of locative media communication is the embodiment of contextual change from position. The virtual context of locative media technology refers to the virtual context which is constructed when the technique of bearing media is displayed. Because of the dependence of the locative media on the actual position, the content of the display is switched by the change of the media position, so it is more dependent on the reality context, the Environment of Speech Act is closely related to its content dissemination, even the scene deduction of its content, because of the virtual context under other new media technology. The virtual context of locative media communication can be regarded as the product of the combination of realistic situational context and new media virtual context.

The virtual context of the site intuition Media communication is no longer prominent, as if the communication gradually away from the context of the reality of the new media context to return to the reality, and verify that all the new media context has a specific realistic context of the reference, but also shows the reality of situational context for speech behavior understanding of the indispensable. Locative media is now popular in all kinds of exhibitions, its form is still constantly being innovated, mainly because of its in the dissemination of the real environment and virtual environment combination of the practice by tourists favor.

Locative Media communication Virtual context not only provides the accessibility of contextual information in the real environment to tourists, but also makes its understanding and feeling more realistic and accurate, and takes the virtual context as the refuge to exert the imagination of tourists, so that it can participate in the communication Process and Speech act, and construct its own understanding context.

3. The Characters of the Chinese Media Transmitting Virtual Context in Xuan Paper

Malinowski the context into contextual context and cultural context, and the virtual context of locative media communication is different from other new media technology virtual context in the context of communication, so the analysis and comparison of its characteristics will be based on the context of the situation. There are many factors that influence context, in context, according to Gregory and Carol, the scope of discourse, the tone of discourse and the way of discourse are the three main factors that affect contextual context, and the language variants produced by these three factors are the domain in systemic functional linguistics.

The spread of Xuan paper is currently carried out in the form of field-visit locative Media and network platform, both of which have similarities and differences in the construction of situational context in the communication, and the following will be carried out from the analysis of three factors that affect situational context.

1) The common and personality of communication situational context in different conversation topics

As one of the content of discourse scope, conversation topic influences the construction of the virtual language of locative media communication. At present, the paper is mainly divided into two parts: Art and culture communication, and these two parts belong to the technical topic and non-technical topic under the topic range according to their content characteristics. Technical topics include research reports, medical guidelines, and more. Technical communication content belongs to the technical topic, the technology topic often uses the specialized vocabulary, the concrete expression is 'the Liao skin, the Liao grass, the skin, the tread material' and so on. The use of proprietary vocabulary requires information senders and users to create new conversations further to the explanatory text. There is less use of professional vocabulary in the dissemination of rice paper culture and history more in line with non-technical topics using everyday language and essays and other features. In the content dissemination, regardless of the network platform or the field visit site Perceptual media its content roughly similar, the language is more formal, therefore the topic context does not have the difference. Only slightly different in individual aspects, such as the expansion of the network platform to read more rich, and the field visits in the technical content more realistic. The normative presentation of each process in the field tour is also part of the content of the communication, which is lacking on the network platform.

At the same time, in order to reflect the interaction of Speech Act, visitors can choose to experience the production process of rice paper, which is also the feedback to the effect of locative media communication, and network users can experience it through virtual reality technology.

2) The common and personality of the context of communication situation under different conversation venues

In order to let tourists better understand the art and culture of Xuan paper, Xuan paper spread respectively with 'rice paper production process and technical visit', 'Green tan skin tree planting environment experience', 'Liao Fur production process inspection' and 'rice paper products show.' One of the 'rice paper production process and technical visit' and 'Liao Skin production process inspection' mainly for the dynamic flow process, the field visits in accordance with the process of different production workshops, the content of the site is different, and the site is also one of the content of the discourse, so the dissemination of the specific context of each has different. In contrast, the 'green Tan skin tree planting environment experience' and 'rice paper finished display' More inclined to static growth environment and product physical experience, do not have dynamic scene switching, so the overall context in its communication is not as complex as the other two contexts. The context of the four items in the field visit is obvious, and the difference in the network platform is mainly manifested in the number of Web pages, which also weakens the process and fluidity in the 'paper making process and technique' and 'the process of the production of Liao Pi', and the user can only interact with the interface according to different process. But the number of pages more than 'green tan skin tree planting environment experience' and 'rice paper products show'.

The communicative process is the process of the continuous construction of the context, and the constant switching of the site scene makes the virtual context in the Locative media communication project constantly being constructed.

3) The common and personality of the context of communication situation under different discourse modes

The accuracy of information transmission is higher because it relies on location positioning to convey information. The media or channel used in language activities is called discourse mode, which is regarded as one of the three factors affecting situational context, so the virtual context in the Locative Media communication project is different. The art of Xuan paper and the culture communication mainly take the field visit and the network platform two forms mainly, the way is the locative media and the network Xuan Rice map. Due to the continuous updating of technology, the current stage of the site to use the location of the media limited to 'Xuan Paper Culture Park', paper production process involved in other production process cannot be displayed in the cultural park, such as Liao grass production in Jing County Zheng Cun, green tan tree planting base in Jing County Tsai Village, but these sites can be accessed through the Web platform of Xuan paper.

Therefore, the contextual range shown by the two locative media is not the same. The different exhibition rooms of Xuan paper Culture Park are distributed according to the different production processes of handmade papers, so visitors entering different exhibition areas are confronted with different context of fixed communication situation. For example, rice paper Culture Park's fishing workshop by 6 workers standing in the 3 fishing paper station respectively, at this time, the orientation of the media perception of visitors into the workshop, will automatically play about the fishing paper on the skills and culture, while the workers are not on-the-spot fishing paper, while the workshop wall will also be equipped with pictures and physical exhibition to spread the Viewing the real-time operation not only improves the knowledge of the visitor's understanding of the orientation media, but also stimulates the visitor's questions arising from the constant observation. From the rice paper Culture Park set up to return to the original scene of the tourists in the heart of the paper to understand the understanding of the template answer, reduce its based on the text image of the space for its imagination, but the context of the set-up of the spread of the context of the environment of the visitors in the creation of contexts is weakened, which is different from other new The network map, although the visitor chooses the location to provide the information voluntarily, but at present the technical restriction also needs the visitor to assist oneself constructs the context when understands through the picture, the video, the animation and so on multimedia means dissemination content. There may be some discrepancy between the context of the communication and the official context, but the choice of location information has been selected to some extent to the effective information, which is one of the advantages of map design.

The context of the communication situation of the network platform is not completely unrestricted, the pictures, videos and animations provided are the result of subjective selection and processing, and the communication situational context which the network platform constructs itself has certain subjective restriction compared with the different viewing angles which can be chosen on the spot.

4) The common and personality of communication situational context under different personal tone

In the virtual context of new media technology, new media works are often the product of complex collaboration between visual and performing artists, programmers, scientists, designers and others, so specific information senders cannot be identified. New media works for the mass communication, the sender cannot be sure to accept the specific characteristics of recipients of information, so in the new media technology virtual context of communication between the two sides have ambiguity, and then affect the relationship between the two sides of the judgment. The individual relationship between the two sides of communication is the personal keynote that influences the tone of discourse. One of the three main contents of the foreign language knowledge in the context--mutual understanding between the two sides, also known as mutual knowledge, refers to the knowledge shared by both parties, and the other party is required to judge whether the other side has such knowledge, because in the new media technology virtual context, communication can't be specific to the individual.

Therefore, the information sender can only through the pre-judgment of the tourist background knowledge of the general planning discourse content, background knowledge of the individual differences can't be taken into account, mutual knowledge can't be specific, to a certain extent, influence the communication of discourse effect. In order to make up for the influence of both sides on the discourse effect, site visits set up staff or in the location of the media set 'problem query' and other functions to help visitors answer questions, it becomes the new media communication agent, on behalf of the 'New media', the real information sender and their own to answer questions. According to the context of the content information and communication in the new media, the staff members answer according to the different representative official position or individual position and the official position. For example, in the drying paper workshop, visitors asked 'Master, I just saw you have been using the edge of the nail to pull the paper, what the reason is for this', the staff will say: 'This is loose paper, the main purpose is loose sticky paper, convenient for me to Peel paper.' 'At this time the staff is the visitors really want to ask the object, you can stand in personal position for the visitors to answer questions.' If a visitor asks for information about the content in the orientation media, the staff member should discard their personal position to answer it.

For example, the tourist asked 'Hello, please explain the machine on this paragraph I do not understand, can you excuse me', because the content or is not being questioned by the staff to write, so staff need to figure out the true message sender's meaning, standing in the official angle to answer. The agency has become a new feature derived from the new media, which includes not only acting as agent on behalf of others, but also the freedom to create, change and influence institutions and events. Network platform, the user through search engine search related knowledge, online inquiry and comments and other ways to feedback. The development of the search engine can satisfy the user's questions about many knowledge, and also become one of the most commonly used ways to find doubts. Content knowledge of rice paper network platform users may not be able to fully understand, through the knowledge point link or search engine to create new knowledge dialogue, not only make up the background knowledge, but also make mutual knowledge more consistent. The search engine shows the function of its agent through the free creation of events and agent identities.

Online inquiry and message as a classic interactive form, the staff of the new media to answer the user's feedback information, the nature of the site is similar to the answer. Feedback helps the next speech act to be completed successfully. In the field visit, the staff to answer questions and visitors to open a one-to-one face-to-head dialogue, communication between the individual relationship became clear. Although the staff is not a true sender of information, but can be through dialogue to understand what tourists think, and then feedback to the real sender of information, the characteristics of tourists with the information collected gradually clear, conducive to the further improvement of mutual knowledge.

In the network platform transmission, the information collection can be realized, but the distance limit, one-to-one face-to-face dialogue still cannot realize.

5) The common and personality of situational context under different function tones

When people communicate with others, they should not only consider their relationship with each other, but also try to realize their communicative intention, that is, the keynote of function. As a kind of discourse tone, functional keynote is another important factor in contextual context. Although the communication between the two sides can't establish an individual, but the new media technology in the propagation of Xuan paper the nature of communication in the virtual context is predictable, that is, the message sender belongs to the Xuan paper, and the recipient is happy to understand the group of Xuan Paper, so the two purposes of dialogue can be judged as 'display Xuan paper skills, spread rice paper culture'. At the same time, through the formal level of language can pre-sentence the functional tone of discourse. The functional function of discourse is divided into various kinds, and Gregory has divided it into several types, such as greeting, explanatory, persuasive and teaching, descriptive. In the technical and cultural communication of Xuan paper, the content of each part is set up for different purposes, so its functional keynote can be divided into different categories. For example, in the production skills and process content demonstration, Xuan paper spread to promote tourists understand the purpose of rice paper production, focusing on scientific and comprehensive interpretation of various techniques, functional keynote can be judged as illustrative. In order to make visitors more comprehensive feeling of rice paper culture, Xuan Paper group to show the history of Xuan paper, paintings and celebrity stories as content for tourists to tell the story behind the paper, the function of the keynote attribution for descriptive. In the finished display of rice paper, excellent quality has become the driving force to attract customers and become one of the ways of advertising communication, so it can be regarded as the embodiment of persuasive function tone.

Site visits and web platforms are built for the same purpose, so that the tone of the function is not changed by presentation.

4. Summary with the Continuous Innovation in the Form of Locative Media, the New Features of Situational Context in the Communication will be Constantly Emerging, and Better Assist Visitors to Understand the Project

The application of the Chinese media of Xuan paper to the media has become a window to open the project, which has gradually matured in the fields of science popularization, art and so on, and attracts the eyes of the outside field, so the virtual context under the transmission of the locative media has the space of further exploration. The birth of new media has led to the need to re-return to the media tools themselves. Locative Media is only a representative of many new media communication tools in the spread of Xuan paper, and new media are constantly creating a communication context for people to experience, and the settings of these contexts will vary depending on the target population, for example, children prefer to experience the fun of knowledge in the context of interactive games. Young people are more inclined to follow the information dissemination context provided by micro-WeChat. The communication context created by new media is integrated with its communication system to build a communication platform for visitors, such as public accounts, where visitors can subscribe to update information and interact with the communicators.

This convenient communication platform not only has the authority, but also constructs the virtual space for the dissemination, at present the dissemination gradually tends to build ‘the communication platform’ the form, but the Xuan Paper group also actively constructs own communication platform.

References

马格·乐芙乔依, 克里斯蒂安·保罗, 维多利亚·维斯娜. 语境提供者[M]. 北京：金城出版社, 2012.

张君昌. 超媒体:下一代传媒竞逐的制高点——关于“超媒体”概念的学理阐释[J]. 南方电视学刊, 2018(1): 43-46.

吴飞. 超媒体时代的国际传播战略思考[J]. 新闻与写作, 2016(1): 31-34.

A Research on China's Science Communication Ecosystem under the Background of Media Convergence

Sun Xiaocui, Jin Xiaoli

School of Information Management, Wuhan University, Wuhan, China

Abstract: Media plays a connecting role in the three-dimensional structure of 'communicator – media – receiver' in the whole science communication ecosystem, and media integration changes science communication ecosystem to a certain extent. After analyzing the relationship between science communication and media, this paper further systematically studies the iterative process of 'point – line – surface' in science communication brought about by media reform. Finally, this paper focuses on sorting out the innovative operation mode of science communication ecosystem brought about by media convergence — media think tank mode for communicator, whole network multi-screen mode for media and community wisdom mode for receiver.

1. Introduction

The Internet and its innovations in the information age have affected all aspects of social and economic life in unprecedented breadth and depth. The most significant impact on traditional social and economic life lies in the fact that on the one hand, the Internet has become the engine of the development of new modes, and on the other hand, it has played an innovative and creative role in the traditional mode. This kind of innovation is mainly featured by 'media convergence', broadening the development space of science communication, promoting elevation and rationalization of the relationship dynamics between science communication and media, and then accelerating the optimization of the innovative operation mode in science communication ecosystem.

2. The Relationship Between Science Communication and Media

The media has the function of connecting the three-dimensional structure of the 'communicator – media – receiver' of the entire science communication ecosystem, and the media bears the function of carrier and intermediary in science communication. The core of science communication is to popularize science knowledge and information to the public, and to realize the transfer and acceptance of information from 'communicator' to 'receiver' through multiple organization methods and channels of communication. There are two main types of media: one is the science knowledge and information conveyed in the form of pictures, texts, audio and video. It needs to be attached to paper, silk, stone and electronic paper. This carrier is the media (the organization method). The second one is the carrier carrying science knowledge and information, which needs to reach the audience through book, periodical, radio and television, and Internet terminal equipment. These devices are the media (the means of transmission).

The evolution involving the technology and development of the media is also part of the science communication. All the research of the technology and development of the media is also the science knowledge to be transmitted and popularized by the science communication. 'The media is the message' – McLuhan's famous opinion points out that the influence of application of media technology is more profound and thorough to the audience than the science knowledge and information carried by the media at some point. Specific to the relationship between science communication and media, changes in social and economic life brought by media technology and its development is non-negligible content of science communication.

3. Evolution of the Science Communication Ecosystem Brought About by Media Change

In the ecology theory, the ecosystem refers to a dynamic equilibrium system formed by

Corresponding author: Sun Xiaocui. sunxiaocui0315@ 126.com

the circulation and flow of matter and energy between the biocenosis and its surrounding living environment within certain time and space. Specific to the ecosystem of science communication, more attention is paid to the three-dimensional structure of 'communicator – media – receiver', which involves dynamic equilibrium system of circulation and flow of knowledge and information based on the production, transmission and consumption of science knowledge and information between people and media, society and other elements.

The evolution of the science communication ecosystem is accompanied by constant changes in media technology and its applications. Traditional science communication presents the characteristics of 'one-way communication', mainly in the stage of individual media and mass media, and is the 'point – line' level of communication ecosystem. Later, with the further promotion of Internet information technology, the science communication in the digital media stage is characterized by 'two-way communication'. The interaction between the communicator and receiver of science knowledge and information is strong. The initiative and discourse power of the media gradually increase, and more involve the dynamic balance of the 'face' on the subject and object in the science communication ecosystem.

1) Individual Media Stage: Point

The concept of individual media is mainly featured by the function of the producer in science knowledge information — the individual. The individual media stage mainly existed in the primitive society. The emergence of early social labor and division of labor not only generated basic science knowledge and information such as individual labor experience, but also created language — the most primitive and basic science communication medium. Therefore, the two most critical elements of the science communication ecosystem in the individual media stage — science knowledge and information and media (language) were created. This laid a solid foundation for the science communication of 'peer-to-peer' through individual 'mouth-to-mouth'.

In this historical stage, the individual represented the media necessary for 'mouth-to-mouth' — the user of the language. Therefore, this stage is called the individual media stage. In the ecosystem of science communication at this stage, as the most important single factor, the individual presented the circulation and flow of science knowledge and information at micro level between 'point' and 'point'. Therefore, in the individual media stage, large-scale science communication was usually difficult to achieve. Although in the science communication ecosystem, an individual (point) spread to multiple individuals (multiple points), which was only a simple sum of numbers, and not substantial changes in essence.

2) Mass Media Stage: Line

The mass media is relative to such niche media as individuals. The main difference between mass media and niche media is that the former can achieve large-scale science communication. Many scholars believe that the mass media in the traditional sense mainly includes media types such as book, newspaper, periodical, radio, television and internet. In this paper, the network media is classified as a digital media type, which is mainly considered to be of great significance in promoting the realization of the 'two-way communication' of science communication. In the mass media stage, science communication finally realized the 'point-line' type of ecosystem function — the 'point' represented by the media could form a strip of communication 'line' (path) through the open dissemination of science knowledge and information.

In the mass media stage, according to different characteristics of science communication, its ecosystem can be divided into subject and object elements such as primary producer, secondary producer, intermediary, first-level receiver and secondary receiver of science knowledge and information. Applying these different links to the practice of science communication in the mass media stage is the realization of the communication path of 'author – editor – mass media – consumer – recycler and user'. Similar to the individual media stage, the science communication system in the mass media phase also focused on 'one-way communication'. Although it achieved the improvement of the efficiency of science communication in the individual media stage to a certain extent, it was not the best state of the science communication ecosystem because of the lack of interaction between the 'communicator' and the 'receiver'.

3) Digital Media Stage: Face

The continuous development and application of mobile information technology is driven by high-tech such as big data, cloud computing and artificial intelligence, making digital media play an increasingly important role in the science communication ecosystem. The integration of mobile social network concepts has promoted the innovation of science communication forms (Chen et al., 2018). Digital media has achieved a major breakthrough. Through the Internet and excess database, the audience can query, receive and publish information at anytime and anywhere. The 'hyperlink' technology provides great convenience for information mining (Shi et al., 2015). Through these digital media technologies across time and space, the science communication ecosystem at the digital media stage frees the receivers of science knowledge and information from the point and line constraints of traditional individual and mass media.

In the digital media phase, the science communication ecosystem has achieved a 'face-to-face' evolution, the most critical feature of which is media convergence. In the digital media stage, a wider range of circulation and flow of knowledge and information between subject and object in the science communication ecosystem can be realized. The integration of the media will inevitably bring about the integration between the 'communicator' and 'receiver', forming an ecology network of science communication and achieving integration on the 'face'. The network externality in the digital economy will be perfectly realized in the science communication ecosystem of the digital media stage, and the scale effect will affect the benefits of the science communication ecosystem — the widest range of communication of science knowledge and information.

4. Innovative Operation Mode of Science Communication Ecosystem under the Background of Media Convergence

As mentioned above, the change of the media has a very importantly decisive influence on the way and effect of science communication and the public's understanding and application of science knowledge and information. The integration of new media represented by network media, digital media and terminal devices such as mobile phone, and traditional individual media and mass media is the driving force for the innovative change of the science communication ecosystem operation mode. Combining with the three-dimensional structure of the 'communicator – media – receiver' of the science communication ecosystem, this paper summarizes three aspects of the innovative operation mode of the science communication ecosystem under the background of media convergence, namely, the media think tank mode for the communicator, the whole network multi-screen mode for the media and community wisdom mode for the receiver.

1) Communicator: Media Think Tank Mode

The innovative operation of media think tank mode for communicator in science communication ecosystem is mainly influenced by the two elements — 'subject generalization' and 'function realization'. The former is mainly reflected in the influence of multiple business forms of communicator's identity brought by the media reform, while the latter is the differentiation of function in different identities of communicators in the science communication ecosystem.

(1) *The media think tank mode can improve the communicator's benefits of 'subject generalization'*

As the initial link in the science communication ecosystem, communicator mainly plays a role in producing and providing science knowledge and information. In the Internet economy era, with the emergence of industry convergence and media convergence forms, communicator in science communication ecosystem is not only one of the traditional science research workers, social organizations such as China Association for Science and Technology and government agencies. More and more ordinary people and enterprises obtain 'communicator' status through UGC and content startup. Under the background of media convergence, communicator in the science communication ecosystem shows the trend of 'subject generalization'.

How to gather these generalized communicators to constantly fulfill the needs of the receivers through producing and providing science knowledge and information. The 'clustering' effect of the media think tank mode is needed. The modern think tank was originated from Europe, and rapidly developed in the United States. Its main feature is to further study national policies by analyzing science knowledge and information, which is independent and non-profit. Specifically, the media think tank mode is a further upgrade of the 'communicator' function in the science communication ecosystem under the background of media integration. It is also an important path for the country to upgrade from a policy think tank to a new type of think tank with Chinese characteristics.

(2) *The media think tank mode is influenced by the communicator's 'function realization'*

For communicators, the most important thing is to give full play to their functions in the science communication ecosystem. In the context of media convergence, communicators, who are represented by researchers in the form of PGC and ordinary people in the form of UGC, can generate and provide more extensive science knowledge and information resources. By strengthening the pooling and mining of science knowledge and information data related to economic, cultural and social fields through connecting resource advantages between professional communicators and mass communicators, the collision of different science views can be realized, and the most authentic and comprehensive think tank reports and policy recommendations can be provided for the government and its functional departments.

The communicators in the science communication ecosystem are mainly divided into three different media think tank modes, which are industrial, professional and platform mode, according to their different fields and links in the ecological chain. The industrial media think tank mode requires communicators to pay attention to more macro knowledge and information in the field of science communication, such as institutional reform, policy formulation and industrial structure adjustment. The professional media think tank mode mainly focuses on the meso-level science knowledge and information such as policy evaluation and interpretation, and performed by science research institutions and colleges and universities. The platform media think tank mode is more suitable for non-professional research institutions such as the general public and enterprises, involving the research perspective at micro level, such as the application of advanced technology.

2) Media: Whole Network Multi-screen Mode

Media, as an intermediary link in the science communication ecosystem, is the most critical intermediate force between 'communicator' and 'receiver'. In the history of science communication, the media, as the middle course of a river, plays a connecting role (Liu et al., 2017). The background of the media convergence has the most profound influence on innovative operation mode of the media: on the one hand, in the form of science communication, the fusion trend of individual media (human body and its extension on tools), and mass media (book, newspaper, periodical, radio and television, etc.) and digital media (individualized Internet and mobile terminal equipment) is formed. On the other hand, on the content of science communication, the circulation and flow pattern of science knowledge and information content in the ecosystem of science communication has been changed.

(1) *The whole network circulation of science communication*

The whole network circulation of science communication mainly reflects the integration of science knowledge and information communication in the form of media under the background of media convergence. This kind of integrated communication not only shows that the ways of science communication are diverse which involve traditional network, PC and mobile media. It also shows the seamless integration between different media. Through information technologies such as big data and cloud computing, explicit interactions between subjects and objects in the science communication ecosystem can be realized, and mutually reinforcing relationships between them can be evolved. Especially, with the circulation and flow of science knowledge and information between different media in the whole network, the subject and object in the science communication ecosystem are continuously empowered to promote the survival and development of its ecosystem.

A larger media system can be formed in the whole science communication ecosystem through the whole network circulation of science communication. All kinds of media under different network backgrounds can be classified into a group in the science communication ecosystem, such as smart phones, tablets and smart watches in the mobile Internet. The multiple networks involved in the whole network circulation are multiple communities in the science communication ecosystem. The communicators, media and receivers among different communities form a strong covering capacity of relationships through the circulation and flow of science knowledge and information. All parties in the science communication ecosystem, with the help of the whole network circulation, form explicit or implicit interactive behaviors, and further improve the strength of 'advertising' and 'popularity' of science communication.

(2) *Science communication content presented by multi-screen*

The development's breadth and depth of the media for the content of science communication provides the basis for its multi-screen presentation. Among them, the breadth of development is mainly reflected in the expansion of the scope of presented content, including not only the current science content that can be truly experienced by the senses, but also the historical and future science content. The depth of development is mainly reflected in the diversification of presentation form of a certain theme content, as shown in figure, text, audio and video. Under the background of media convergence, media technology in the science communication ecosystem has been continuously strengthened in the development of content, which is reflected in the flat, three-dimensional and other types of applications on the screen. For example, the application of AR and VR media technology in the development of science communication content can be realized through professional eyepieces and handles in the three-dimensional 'screen' of virtual space.

The multi-screen presentation of science communication content needs to be differentiated according to the content theme. For some content topics with strong demand for context and interactivity, such as medicine and dressing, three-dimensional screens that can achieve 'immersive' reception and learning are needed. Science communication content, such as language, which focuses on a single sensory effect, needs to be developed with the help of media technologies such as 'speech recognition' and presented on relatively flat screens. To sum up, in the selection of media technology and screen, science communication contents in different disciplines and fields should focus on the comprehensive weighing of the subject and object involved in the science communication ecosystem.

3) Receiver: Community Wisdom Mode

Under the background of media convergence, especially in the development and application of self-media in digital media, there is no doubt about the transformation brought by science communication ecosystem. In the era of self-media, networked users (receivers) — networkers, whose identity is both field and tool, present a core change which reveals the opening of a new paradigm of science communication (Chen, 2015). Since the end of the 20th century, the development of Internet in China has brought the information field of community mode for the receivers in the science communication ecosystem. This mode has developed from the network chat room and BBS community in the early PC era to the clients such as WeChat and microblog in today's mobile Internet era, forming a community wisdom mode of different opinions and public discussion in the science communication ecosystem.

(1) *Influencing factors of community wisdom mode for the receiver*

The continuous update of media technology is a key factor influencing the innovative operation of science communication ecosystem for the receiver. In the early days of the development and application of the Internet, Internet chatrooms and BBS communities in the PC era won the favor of the receivers with their technological advantages across the space and time. Later, with the technological development of mobile Internet and smart terminal equipment, clients relying on mobile terminal devices such as smart phones began to be widely used. The convenient release and real-time sharing of science knowledge and information enables the receivers in the science communication ecosystem to migrate between different media in the form of community — a direct manifestation of network externality in the Internet economy.

The need to share and communicate among intelligent people in science communication system is also the key to influencing the paradigm shift from the receiver to the community wisdom mode. In the science communication ecosystem, receivers can realize the real-time and 'fresh' resource uploading, searching and analyzing science knowledge and information by relying on the mass intelligence interaction in the community. In the context of media convergence, the receiver can obtain the 'popular science' information of science communication to the largest extent. The community and its public wisdom have become the center of the public opinion field of science communication, where the identity boundary of the receiver and the communicator is continuously blurred, and their identities overlap in different fields.

(2) *Evolution of the community wisdom mode of the receiver in the science communication ecosystem*

The community wisdom mode with social relations as the core provides a good mechanism for the personalized information satisfaction of receivers in the science communication ecosystem. In the process of media convergence, the rise of social media such as self-media weakens the central position of traditional media, namely, the one-way communication mode of 'individual' and 'mass' media with 'point' and 'line' is more and more unsuitable to the science communication needs of the receivers. The decentralization evolution in the science communication ecosystem is accompanied by the rise of the community wisdom mode of the receiver, which is also the process of the survival and development of the media self-organization in the ecosystem.

In combination with the interrelationship between species in the process of community succession in ecology and the characteristics of the phase of science communication, the evolution process of community wisdom mode in the ecosystem of science communication can be divided into the stages of creation, expansion, maintenance and evolution. Currently, the pattern is in the extension phase. The rapidly updated science knowledge and information in the era of knowledge economy requires far more breadth and depth of science communication than any previous era, and receivers need more convenient and real-time media technology support. With the help of the community wisdom mode, the appropriate ecological niche can be selected in the context of media convergence, and the advantages brought by the 'symbiosis' in the science communication ecosystem can be fully obtained to further promote its continuous expansion.

5. Summary

The continuous development of information technologies, such as the Internet and smart mobile terminals, has brought a significant impact and innovation – driven to the 'communicator – media – receiver' involved in the science communication ecosystem. As a connecting medium in this three-dimensional structure, itself is the content of science communication. Under the background of media convergence, the 'face' of science communication in the digital media stage is the evolution of the 'point' in traditional individual media stage and 'line' in the mass media stage.

Under the background of media convergence, communicators and receivers in various science communication ecosystems face the challenge of innovative operation and exploration. The traditional science communication media model is weakening the power of content production, and the control of communication channels is being decentralized. Integration has become the theme of the times. The media think tank mode for the communicator, the whole Internet multi-screen mode for the media and the community wisdom mode for the receiver are all direct manifestations of 'decentralization' in the science communication ecosystem. Under the background of media convergence, the reconstruction of science communication ecosystem has already come.

References

Chen C Y. (2015). 'Online Crowd' Communication: A New Paradigm of Science Communication in the Era of 'Self-Media'. *Northern Thoughts*, (6), 64-67.

Chen Y, Xin G. (2018). A Brief Analysis of the Science Communication Strategy of Planetarium in the Context of Media Convergence. *Science Communication*, (9), 187-190.

Liu Q, Shi W. (2017). Research on the Dilemma and Countermeasures of Scientific Communication under the New Media Environment. *China Collective Economy*, (7), 103-104.

Shi W, Yang Y. (2015). Media Technology Innovation and Science Communication Model Change. *Science and Technology Communication*, (1), 62-63.

Application and Exploration of New Media Technology in Agricultural Science and Technology Communication

Wang Yiqing

Science and Technology of Communication and Policy, University of Science and Technology of China, Hefei, China

Abstract: Agricultural informatization is an important strategic plan for agricultural development in China. In recent years, the application of new media technology in agricultural science and technology dissemination has played an important role in agricultural science and technology promotion.However, the traditional agricultural technology promotion system is still difficult to be replaced,there are many problems and challenges.This paper studies the current situation of the application of new media technology in agricultural science and technology dissemination in China,analyzing the shortcomings of the application of new media technology, and giving some suggestions and countermeasures based on the situation of our country comparing with the advanced experiences and models domestic and overseas.

Keywords: Agricultural Informationization; New Media Technology; Agricultural Science and Technology Communication

1. Introduction

As the biggest developing country with the largest agricultural output in the world, the process of agricultural modernization and informatization is not only related to the promotion of 'agriculture, rural areas and farmers', but also determines the firmness of the foundation of the modernization of the national economy. In 2018, the No. 1 Document of the Central Government showed us the task of implementing the rural revitalization strategy, which is 'by 2035, the rural revitalization shall has made decisive progress, and the agricultural and rural modernization will has basically been realized.'[1] To achieve this goal, the achievements of agricultural science and technology research must be timely conveyed to agriculture to turn them into actual productivity. With the development and popularization of new media based on digital information now, we can use new media technology to spread agricultural information, to popularize agricultural technology, and to establish an efficient and convenient agricultural science and technology promotion and dissemination system. We believe that this is the only way to modernize development.

According to the definition of both domestic and overseas researchers' studies, Peng Lan defined new media as[2]: An interactive and convergent media form and platform based on digital technology, network technology and other modern information technology or communication technologies, and new media technology is a kind of media technology means of providing information services. In this paper, new media applied to agricultural technology communication mainly include digital magazines, newspapers, radio, digital TV, broadband TV, smart phones, mobile networks, and various websites. The understanding of agricultural science and technology communication is literally the process of transmitting and disseminating information, such as agricultural knowledge and technology. Agricultural science and technology communication can be considered as a cyclical communication activity that agricultural technology communicators use various communication means to convey agricultural science and technology to the specific fields of agriculture, rural areas and farmers. It is aimed to promote the application of science and technology in agriculture and rural areas, and to form a feedback mechanism.[3]

The traditional agricultural science and technology communication model has long been unable to meet the needs of agricultural development. Since traditional agricultural science and technology communication is mostly a top-down and word-of-mouth model, this makes the dissemination of agricultural information and the promotion of agricultural science and technology very lagging behind the constraints of time and space, and more importantly, with the rapid development of agricultural technology, the knowledge storage of traditional agricultural science and technology communicators cannot meet the needs of producers, such as how to use new agricultural equivalents after the introduction of large-scale modern equipment from domestic and foreign, the marketing management of special agricultural products and the data collection and processing technology of dynamic information of the market. But because the new media has the features such as timeliness of dissemination, accessibility of information, mass information storage, and vividness of communication, the gap between time and space can now be broken, then greatly enhance the efficiency of communication, at the same time meet the needs of producers for various agricultural knowledge and information. And its unique participatory and interactivity characteristics enable agricultural technology communication to achieve full exchange of information and high interaction between groups in an open and diverse environment.

In this context, this paper studies the characteristic of the communicator and audience of agricultural science and technology communication from the perspective of new media technology, and extracts relevant cases from investigations and interview notes in relevant project work that I participated, comparing traditional media and new media, and the application status of new media technology in agricultural science and technology communication from home and abroad, inferred the problems and challenges in the application of new media technology in agricultural science and technology communication, reach a conclusion about countermeasures and suggestions.

2. New Media Technology in the Spread of Agricultural Science and Technology

At this stage, China's agricultural science and technology communication is making active attempts to apply new media technologies. The integration of traditional media and new media is gradually changing the way of dissemination of agricultural science and technology, and there are successful cases such as the Farmer Mailbox in Zhejiang Province. However, the development of agricultural technology has increased the demand for science and technology communication, but the traditional agricultural technology communication model still occupies a dominant position. At the same time, the full-time high-quality science and technology personnel reserve is insufficient, so it is urgent to promote the application process of new media technology. In addition, the construction of agricultural informatization in developed countries has certain reference significance for the dissemination of agricultural science and technology in China.

1) Agricultural Science and Technology Communicators, Audiences and Media

(1) *Communicators in the spread of agricultural science and technology*

The communicators of agricultural science and technology communication refers to those who have the ability to understand and practice the object in the field of agricultural science and technology communication, or the issuer of information and action of agricultural science and technology communication. The communicators of science and technology communication in rural China under the new situation is diverse, according to the nature of the subject, it is divided into official subjects and non-official subjects. The official subjects include the China Association for Science and Technology, Ministry of Agriculture and Rural Affairs of the People's Republic of China, the National Development and Reform Commission, China Women's Federation, China Disabled Persons' Federation, China Communist Youth League, etc. In addition, major research institutes, universities, agricultural associations, and the media. Non-official entities include Agricultural enterprises, 'Bellwether', civil society groups and individuals.

As of 2015, there are 222,000 professional science communicators in China, and the average number of full-time science and technology personnel per 10,000 people is only 1.6.[4] In this context, the full-time staff of agricultural science and technology communication is even more lacking. Agricultural researchers and technicians are mostly far from farmland. On the other hand, the scientific knowledge and technology of grassroots science workers have not kept pace with the development of agricultural science and technology, and it is necessary to have an efficient and convenient agricultural science and technology extension system to support the development of first-line agriculture. Therefore, the new media technology is the main body of science and technology communication. The application in it is even more important.

In recent years, major agricultural science and technology communication subjects have been trying to use new media technologies to promote agricultural science and technology, but they have not formed a mature and standardized system. During the 'Twelfth Five-Year Plan' period, China's agricultural science and technology achievements conversion rate is about 50%, while developed countries have more than 80%[5]. Whether it is official or unofficial agricultural extension workers, the demand for new media technology is urgent, but and the ability to use new media technology carries out the dissemination of agricultural knowledge and technology, and transform agricultural science and technology research results into actual productivity is not strong. At the same time, there is a shortage of specialized and versatile talents that integrate scientific research capabilities, popular science capabilities and new media technologies. The extent of integration with new media is insufficient, and there is no profit promotion model integrating production, education and research, or a perfect training system for talent team construction.

(2) *Audience in agricultural science and technology communication*

The audience of agricultural science and technology communication refers to individuals, families, groups or organizations that accept agricultural knowledge and technology that mainly engaged in agricultural production and services, and they are the most important service target of agricultural science and technology communication.

The audience of agricultural science and technology communication in China is hierarchical and different. The scientific quality of science and technology of traditional agricultural producers is generally not enough, and the ability to accept new media technologies is relatively weak, the participation in popular science activities is not enthusiastic, anyway the new professional farmers are modern with scientific and cultural qualities, modern agricultural production skills, and certain management capabilities, with agricultural production, management or service as the main occupation. Agricultural workers are living in rural areas or towns with agricultural income as the main source of life, specifically, the new business entities is a new agricultural management systems gradually formed on the basis of household contract management, with large professional households, family farms, farmer cooperatives, and leading enterprises in agricultural industrialization as the backbone, other forms of organization as complement. They are becoming the main builders of modern agriculture and have certain agricultural knowledge and technical capabilities and new media literacy.

At present, the traditional agricultural science and technology dissemination objects are developing toward new professional farmers and new business entities. There are about 14 million emerging professional farmers in China nowadays. There are various forms of agricultural technology dissemination for new professional farmers. The development direction of agricultural science and technology communication is to using the media technology to pilot online training courses, docking agricultural information platform and agricultural information platform, agricultural information service platform and resource database, mobile internet service and full-course tracking guidance, to achieve online and offline integration.[6]

(3) *Media in the spread of agricultural science and technology*

According to market research firm Zenith's data survey, China's mobile phone users will exceed 1.3 billion in 2018, ranking first in the world. [7] Agricultural technology communication is undergoing a transition from traditional media to new media. New and old media are very different from the perspective of communicators, media and audiences., traditional media is still playing a considerable role in the spread of agricultural science and technology. The biggest feature of new media is that it breaks the obstacles of time and space, enriches the form of communication, and leads to the spread. great difference in effect.

Table 1 Comparison of traditional media and new media.

Prepagation Mode	Manifestations	Advantage	Disadvantage
Traditional media	Newspapers, radio, books, CDs, etc.	Complete infrastructure construction, perfect legal system and rich experience	Not targeted, slow to update, costly human resources, material and financial resources, fixed forms

(Continued)

Prepagation Mode	Manifestations	Advantage	Disadvantage
New medium	Mobile phones, internet, digital TV, internet TV, etc.	Update transmission is efficient, convenient, interactive and participator, rich in content, and big data to accurately target users	High media literacy requirements for communicator s and audiences, inadequate regulatory systems, and more spam

2) Application Research of New Media Technology at Home and Abroad in Agricultural Science and Technology Communication

(1) *Application Status of Domestic New Media Technology in Agricultural Science and Technology Communication*

At present, the rural technology communication media formed by relying on new media technologies mainly include agricultural websites, smart mobile terminals, Internet TV, mobile TV and so on. The official agricultural science and technology communication subjects have new media projects or platforms in China now. For example, the 'Country e-station' project of the China Association for Science and Technology is a rural science and technology O2O comprehensive service system built in rural areas, relying on agricultural resources stores, farm stores, agricultural stations, etc. through PC, APP, WeChat, hotline, Weibo, precision. popular science push system, Zhongke Yun Media, etc. to build a practical technology learning platform, remote interactive training platform, instant information query platform, rural e-commerce startup platform, expert online service platform, etc.[8] In addition, each entity also has an official website, Weibo platform, etc. to publish important policies, technologies, news, market and other information, and provide agricultural database platform and user online consultation and expert connection services, set WeChat public number regularly push related articles and e-commerce trading platforms to provide online trading venues.

Unofficial agricultural science and technology communication subjects also have a rich form of media, using new media technology to release information to the public and establish contacts with them. The more successful agricultural big data platform named The Agricultural Circle operated by Yangling Agricultural Circle Information Technology Company, is based on the Northwest A&F University, providing information services to farmers. Its Internet application platform gains more than 3 million information readings per week.[9] But at this stage, relative to the official subjects, there are still many problems in the dissemination of authority, science, professionalism and vitality of the unofficial subject.

(2) *Application Status of Foreign New Media Technology in Agricultural Science and Technology Communication*

The development of world agricultural informatization began as early as the 1950s. Among them, the developed countries and regions with the United States, the European Union, Germany, and Japan and other developed country have built a new media agricultural science and technology communication system based on the agricultural information network technology, agricultural database technology, and multimedia technology.

The agricultural technology communication system in the United States is highly marketized. Based on its strong agricultural information network and data collection system, it collects first-hand information and information from all farmers who receive government subsidies, builds a huge agricultural information database, and then process it and release it through various new media channels, and receive feedback in time; EU implements Agricultural Informatization and Communication Technology Project[10] to train and educate farmers on agricultural informatization and new media technology, and use new media and network technology to build efficient and convenient e-commerce platform; Germany's agricultural technology promotion system pays great attention to agricultural information infrastructure, uses advanced remote sensing and monitoring technology and remote diagnosis system to evaluate crop growth in real time, and the government attaches great importance to the education of Internet technology and new media technology promotion; Japan's industrialization development and education system is a solid foundation for its agricultural technology communication. Farmers have higher computer technology and new media literacy, stronger ability and motivation to receive new technologies, and most of the technology research and development funds come from government support. The rural communication network is well-established, and the use of new media for agricultural information and technical exchanges and cooperation is very ample; the development of agricultural science and technology in France is largely attributed to the government's large amount of financial support, in order to strengthen the application of the network in agricultural production, the government releases free new media equipment to farmers and the development of a large number of application software which have greatly promoted the new media technology application of agricultural production.[11]

(3) *Comparison of the application of new media technologies at home and abroad in agricultural science and technology communication*

As the foundation of the national economy, agriculture has a certain commonality in the application of new media technologies in various countries. From the perspective of agricultural science and technology communication subjects, the application and promotion functions of new media technologies are mainly undertaken by the government. At the same time, non-official entities carry out part of the training work on the basis of product promotion. From the perspective of policy support, both domestic and foreign governments are giving strong support to policies to ensure the promotion and application of new media technologies through policy tilt and investment. From the perspective of new media technology applications, all countries attach great importance to agricultural information infrastructure construction and new media technology education. Hope to rely on the development of media platforms or software and technical training for farmers to strengthen the application of new media technologies in agricultural production.

At the same time, due to different national conditions, the application of new media technologies at home and abroad also has great differences. From the perspective of capital investment, most of China's fiscal investment is in the form of direct economic subsidies, paid to farmers or the process of new media technology promotion, while the financial support of developed countries is mostly through the improvement of agricultural production level and agricultural informationization to enhance the income of farmers, and thus further promote the development of agriculture[12]. From the perspective of the relationship between the two sides, the new media technology information of China's agricultural science and technology dissemination is generally a single-line transmission from the sender to the recipient, and spread by the professional agricultural science and technology of the government departments with technical knowledge, industry trends and other information, while the US model is generally a closed loop of two-way information transmission. Farmers provide real-time agricultural information data to the government, and then the government integrates these data into a database to provide Farmers.

3) Analysis of the Characteristics of New Media Technology in Agricultural Science and Technology Communication

(1) *Individuation of the media*

The new media relies on big data technology to collect and analyze user behavior data, multi-dimensional insight into user habits and preferences, and through the message push and other means to send the latest relevant information to the user terminal instantly and accurately, this personalized information release will largely solve the problem of not targeted in the dissemination of agricultural science and technology. Agricultural producers will obtain personalized information in different professional fields and production links in the first time, greatly improving the efficiency of production and sales.

(2) *Formal diversity*

The new media has variety forms of communication, including video, pictures, sounds, animations, animations, etc. which not only improves the user's practical interest, but also enhances the user experience. The vivid expression of the image enhances the information transmission efficiency even among farmers with lower level scientific literacy can also accept it well.

(3) *Participative and interactive*

The emergence of new media has reduced the proportion of interpersonal communication in mass communication to a certain extent, but it has made people's online communication and interaction more frequent and active. Farmers and farmers, experts, products, services supply and demand sides, and technology extension personnel, etc. Can achieve full contact on the Internet which not only enhances the rate of information dissemination, but also brings unprecedented communication and interaction experience to farmers and improves entertainment.

(4) *Information release dynamic instant*

Agricultural production has strong time-saving requirements in weather forecasting, pest control and market information. The immediate release and acquisition of information is of great significance to farmers. The application of new media technology in agricultural science and technology communication can effectively solve this problem. Time is the benefit.

(5) *Case: 'Farmer Mailbox'*

Zhejiang 'Farmer Mailbox' is a new agricultural media network service platform jointly constructed and managed by Zhejiang Provincial Department of Agriculture and Provincial Mobile Company. It has two forms of website and mobile APP, providing users with government and business information, knowledge base and experts. the library agricultural consultation, industrial technology team docking, agricultural science activity information, agricultural credit guarantee, interactive communication and other services. At the same time implement real-name registration, provide users with a certain electronic questionnaire collection information to the municipal department, and timely collect user feedback.

The platform has information subscription and regular push function. Farmers can subscribe and select related services according to their own needs, and release information such as weather, market, agricultural product finance and technology to users in the form of videos, pictures and texts in real time, providing consultation and interactive communication. The e-commerce platform encourages users to share market information in real time, clusters experience wisdom and expert resources to solve technical problems and other issues, and enhance farmers' participation and interactive experience.[13] At present, the farmer mailbox system has 2.8 million registered users, Daily visits is up to 2 millions, the annual number of text messages sent is about 500 million, and the collection and storage of various types of knowledge information data is about 13T.[14] It is a relatively successful agricultural new media technology application practice in the eastern coastal areas of China.

3. Problems and Challenges in the Application of New Media in Agriculture

(1) *Lack of effective cooperation mechanism between research, promotion and new media talents*

Most of the agricultural technology researchers, new media technology workers and agricultural technology extension workers focus on their respective fields, lacking a multi-functional talent, and there is no effective connection and cooperation system among the three. Researchers have no promotion experience of agriculture, the technology promoters have insufficient new media technology capabilities, and the combination of scientific research promotion and new media technology is not close.The technology conversion rate is low, which directly affects the contribution of scientific and technological achievements to agricultural production. At the same time, scientific and technical workers are unable to maintain dynamic contact with grassroots agricultural technicians, which leads to the lack of pertinence and practicality of agricultural technology education and training supply. It is easy to cause the disconnection between agricultural science and technology communication and actual production.

(2) *Low new media quality of agricultural science and technology communication audience*

Farmers ability to learn and accept new media technologies is relatively low, and their scientific quality are poor. Most farmers are more accustomed to using new media tools as a means of daily leisure and entertainment rather than a platform for information acquisition and reception, even for emerging professional and technical farmers. With a certain new media literacy, the ability to learn new media technology is not enough, and the ability to acquire information needs to be improved. On the other hand, the regulation of new media information is not yet mature, and agricultural science and technology information is highly professional, so it is difficult to identify for farmers.

(3) *The infrastructure of agricultural informationization is weak, and the data of agricultural science information platform needs to be improved and enriched*

The new media infrastructure is still lagging behind in the entire rural areas, and rural informationization and modernization are still at the development stage. Most of the agricultural science popularization information platforms at the grassroots level are immature, the development level is uneven, the information collection capacity is limited, the information integration is not enough, the resource database data is insufficient, and the data quality is difficult to guarantee. Many website public numbers are basically idle. The information is updated slowly and cannot be connected with actual needs. The more mature agricultural websites are mostly large-scale websites built by the state. The service targets are too broad and not professionally targeted. The homogenization of service content is serious and the actual production needs are not understood. The feedback link is emphasized in the construction of new media channels, but the actual feedback efficiency is insufficient, and the interaction between new media users and experts is not strong.

(4) *The agricultural new media platform is less market-oriented*

At present, most of the new media technology application platforms for agricultural science and technology communication in China are public welfare, relying on the government's financial investment to build and maintain, and the entry threshold for private capital is high, the non-official entities are not strong because of financial constraints and uneven resource allocation, lead to the less market-oriented of profitability, loss of human resources and idleness.

4. Conclusions and Recommendations

1) Establishing an Fusion System of Agricultural Science and Technology Communication of Production, Education and Research

In the training of new media talents, colleges should not only pay attention to the output of theoretical knowledge, but also pay attention to the education and training of new media technologies, train a group of applied talents with popular science and new media technology, and establish cooperation and exchanges mechanism of agricultural science and technology communication, to realize the dynamic connection between agricultural technology and new media technology researchers and science and technology extension personnel. Based on the understanding of the actual problems of farmers, the research work will be carried out in a targeted manner, and the interaction and feedback of the industry-university-research system will be strengthened efficiently.

2) Improve Farmers' New Media Literacy, Improve the Information Supervision and Evaluation Mechanism

Focus on strengthening the training and education of farmers' new media literacy and technology, so that farmers can independently use new media technology to obtain information related to production and achieve dynamic contact with researchers and agricultural experts; On the other hand, the government should increase the strength of the information supervision of the new media platform, establish an effective information release supervision and evaluation mechanism and build false information punishment provisions to ensure the scientific and professional nature of agricultural technology information.

3) Promoting the Infrastructure Construction of Agricultural Information, Enriching and Improving the Content Quality of New Media Platforms

The infrastructure of agricultural informatization is the basis for the dissemination of new media in rural areas. It should invest a large amount of funds, formulate landings and specific policy guidance and support. At the same time popularize and educate farmers on new media technologies, and provide policies for farmers to purchase new media equipment, establish a sound rural information network; enrich the expression of science and technology communication, and add new forms of communication such as short video, GIF animation, Flash and HTML5, pay attention to the information construction of new media platforms, set up full-time posts for information collection, establish contact mechanism with farmers and farmland through '3s' and remote sensing technology, directly collect agricultural production information and materials from farmers and allowance certain subsidies, and then the data processing will be integrated and processed by the dedicated information processing personnel and released to the new media information platform for users to interact.

4) Enhancing the Marketization Level of the New Agricultural Media Platform

The market-oriented development of the new media platform for agricultural science and technology communication should learn the experience model of other industries, relying on investment, flow, advertising and e-commerce services to achieve commercial operation of the new media platform through various online and offline profit models; The state should open up agricultural-related databases and rationally allocate resources. The government will take the lead in focusing on policies, funds, and human resources to establish and grow a number of agricultural science and technology communication enterprises and organizations that form industries, formulate detailed market plans, and form mature market operations, establish an agricultural information supervision and evaluation system and a talent team establishment and incentive model, thereby radiating and promoting the commercial development of the grassroots agricultural new media information platform.

References

[1] Ministry of Agriculture and Rural Sciences of the People' Republic of China. Document No. 1 of the Central Committee of 2018 [Z]. http://www.moa.gov.cn/ztzl/yhwj2018/zyyhwj/, 2018-02-05.

[2] Peng Lan. Three clues to the definition of new media concepts[J]. Journalism and Communication Research, 2016(3): 125.

[3] Zhou Meng. Research on audience feedback mechanism in agricultural science and technology communication—Taking Zhoulaizui Town of Jianli County as an example[D]. Huazhong Agricultural University, 2013.

[4] Wang Kangyou. Science popularization: National science popularization development report (2006~2016)[R]. Beijing: Social Sciences Academic Press, 2017:32-33.

[5] Sand Law, Hu Lianglong, Zheng Wei, et al., The Transformation of agricultural machinery science and technology achievements in China from the perspective of rural revitalization strategy[J]. Chinese Journal of Agricultural Mechanization, 2018, 39(7): 90-93.

[6] Ministry of Agriculture: By 2020, China's new professional farmers will reach 20 million people [N]. People's Daily, 2018-01-28, http://news.sina.com.cn/c/2018-01-28/Doc-ifyqyqni4054779.shtml.

[7] Foresight. In 2018, smartphone penetration will reach 66%. Chinese users will break 1.3 billion in the United States and subsequently [Z]. 2017-10-19, Http://news.cnfol. com/chanyejingji/20171019/25494643.shtml.

[8] Phoenix Finance, 'Popular China Village e Station' will be built 2,600 [Z]. 2016-05-26, http://finance.ifeng.com/a/20160526/14424614_0.shtml.

[9] Large Agricultural Circle - Network gathers the power of agricultural people. The official website of the large agricultural circle enterprise [Z]. http://www. danong quan.cn/index.html. 2018-02-06.

[10] Li Denghua, Liang Danhui. New experiences in foreign agricultural informationization and its implications for China[J]. Agricultural Science and Technology Outlook, 2015(5): 59.

[11] Li Shiyue. Application of new media in agricultural science and technology promotion [D]. Hunan Agricultural University, 2016: 36.

[12] Huo Yunting. How to realize China's agricultural informationization from foreign experience [D]. Jilin University, 2012: 8.

[13] 360 Encyclopedia. Zhejiang Farmer Mailbox [Z]. https://baike.so.com/doc/3690764-3878767.html, 2018-03-18.

[14] Caizhao. Zhejiang Agricultural Information Center on Zhejiang farmers' mailbox system function expansion and basic operation support project—basic operation support project single source procurement publicity [Z]. Http://www.bidcenter.com.cn/newscontent-52547033-1.html, 2018-06-21.

PhD Students Involved in Telling Science: The Possibilities of Comic Medium in Science Communication

Bordenave Laurence

Stimuli Association, Paris, France

Abstract: The approach to science communication through storytelling should be closely examined, as its assets of attractiveness and efficiency are not without constraints. The French Stimuli association supports devices and research on the impact of comic language on science communication and education, giving an active role to the PhD students. Two devices are presented and discussed: i) the 'comics & science workshops', in which a PhD student provides a scientific presentation to a group of teenagers invited to create a one-page comic strip from the scientific theme provided, ii) the 'Science through comics' trainings throughout young researchers are informed how to transpose his/her research into the semiotic codes of graphic narration. In both cases, the PhD students are guided to adapt their discourse to the public, to be watchful on the visuals, languages, and metaphors they deliver and/or they validate. If they assume the role of co-mediator in the workshop, they become co-author during the training, fruitful experiences in term of appropriation of scientific narratives. The specificities, the benefits and the limits of both situations are presented according to their impact on young researchers' implication in sharing science to different audiences.

1. Introduction

At a time when scientists are becoming increasingly involved in sharing knowledge with society, the possibilities of engagement of PhD students in outreach projects are variable and eclectic. Their engagement may depend on different parameters among which the government's policies (Trench et al., 2012), the institutions and laboratories (Courty and Dematteis, 2015), the disciplines (Jensen et al., 2007), the personal involvement (Daelli et al., 2014). Whether it is voluntary involvement in dissemination of scientific information, or enhancing their research through institutional projects, young researchers are willing to engage with the public with an attitude of openness to dialogue and participation (Cerrato et al., 2018). This new approach to science communication, far from the old-fashioned top-down model, is not without constraints, on the contrary.

One of the challenges is that this current dialectical tendency between researchers and the public expectations supposes effective scientist's skills in science communication. In this perspective, a wide range of media trainings has been offered to PhD students (Frey-Klett et al., 2015; Trench et al., 2012). Actually, dialoguing and participating with the general audience involves science communication competences like knowing the public prerequisites, translating the specific jargon into useful and careful metaphors (Taylor et al., 2018), or mastering the storytelling techniques.

In this last domain of communication, the narratives and storytelling tools constitute a mode of sharing the science that underlie an additional challenge for the professionals of the science communication, and therefore for the PhD students. Dahlstrom (2014) explains in a review of literature on narrative communication that 'science communicators have to decide when and how narratives can effectively and appropriately help them communicate to non-experts about science'. According to the study, narratives may facilitate audiences to comprehend scientific content and help the public find them more engaging. Even so, narratives are intrinsically persuasive, and may represent a non-ethical influence on the non-experts depending on the theme, the medium and the target public.

For these reasons, the French Stimuli association that produces and supports devices in science communication and education with the help of comic storytelling invite the researchers involved to be guided and/or trained to the specificities of the comic medium.

Corresponding author: Bordenave Laurence, Stimuli Association, 75020 Paris, France. bordenave@stimuli-asso.com

Two postures are proposed to young researchers with regard to the storytelling of their research. One of them places the young researchers as trainee-mediators during a workshop where teenagers are led to transcribe PhD research or scientific content associated into comic strips. In this first situation, PhD students are 'escorted' all along the process by a science communicator and a comic artist. The other, corresponding to the 'science through comics' Stimuli's training, gets them into position of future co-writer of comic strip projects in which they will have to find their role within a team, choose the target public, define the subject and characterize the treatment of their project in this context. In this second situation, PhD students have to collaborate and overtake the traditional costume of the scientific expert: they become co-authors.

The purpose of this communication is to present and compare the conditions of engagement and the parts of the PhD students involved in science communication storytelling projects consecrated to comic medium. The communication starts with the devices presentations including portrait, role-play and feedback of the young researchers involved in each of configurations. Then the impact of the storytelling and design comic strip practices is discussed. Finally an open question is proposed on the balance between the PhD students' involvement as expert and commitment requiring the acquisition of professionalizing skills in the light of the experiences studied within the Stimuli association.

2. Comics and Science Workshops

1) Preamble

The French Stimuli association is a group of scientists, comic artists, teachers and science communicators involved in science and comics projects since 2007. The Stimuli group supports devices as workshops and classrooms for teenagers, webcomic series for science teachers, trainings in comic medium for practitioners. Stimuli is also partner of the science education research unit Laboratoire de didactique André Revuz (LDAR of Paris Diderot University, France) to lead research projects on the role and the impact of comic medium in science education (De Hosson et al., 2018).

2) Principle, Context and Objective

The comics & science workshop (C&S workshop) is a short, science-based comic strips dispositive in which a group of volunteers aged 10 to 16 years create a one-page comic strip about science in their leisure time or in classroom (Bordenave, 2012). In this creative workshop, the science is delivered to the teenagers by a scientist who presents and discusses on a scientific theme associated to his/her research. To create their comic strip, the teenagers are guided by a comic artist for the basics of the comic design (character, expression of emotions and feelings, composition of a frame, sequential movement, etc.) and by a science communicator for the basics of the storytelling (protagonist, exposure, climax, punch line, etc.) adapted to the scientific content (care of science visual representations, language and metaphors).

Between January 2011 and June 2018, 31 workshops have been conducted, including 16 co-facilitated by a PhD student or post-doctoral researcher (whom one of them interacted in 3 workshops). Other situations involved researchers, engineers, teachers or science communicators in the role-play of the scientist. The sessions involving PhD students took place in science and culture centre's (10), in classrooms (4) or in public libraries (2), during a 12-hour workshop led for a total of 116 teenagers aged 9 to 19 years old.

3) Portrait of the PhD Student in C&S Workshops

The profiles of the 14 C&S workshops PhD students are various, consisting of 9 females and 5 males. Most of them were engaged in the device during their first (6/16 workshops) or second (4/16 workshops) year of thesis. Three PhD students enrolled during their last year and 3 were post-doctoral researchers at the time of their participation. They spent 6.6 hours total (on average) in the workshop with a minimal duration of 2-hour and a maximal duration of 12-hour.

Table 1 shows that most of the PhD students are affiliated to natural science graduate schools, with a large distribution among the scientific domains of research. None of the selected workshops were devoted to human and social science in that selection. The obtained distribution of the major areas of research is the result of policies of institutions on which the development of a workshop by Stimuli association depends. Most of the workshops were requested and/or sponsored by French scientific organisms (Institut national de recherche médicale-Inserm, Commissariat à l'énergie atomique-CEA) or by scientific and cultural centres in parallel with a scientific cultural event. This organization had an impact on the choice of disciplines, and thus on the doctoral student research domain.

Table 1 Graduate school affiliation of PhD students involved in 16 C&S workshops between 2011 and 2018.

Domain of Research	Graduate School (number of PhD students)
Sciences of universe and matter	Science, technology and society (1)
	Sciences of universe (1)
	Astronomy/Astrophysics (1)
	Physics (1)
	Physics and chemistry of materials (1)
	Chemistry (1)
Life sciences	Life and health science (2)
	Neuroscience (1)
	Experimental and clinical pharmacology (1)
	Medicine, toxicology, environment (1)
	Cognitive science (1)
Mathematics	Computer science (2)

4) The Role of the PhD Student: A Co-mediator

The functions offered to PhD students involved in a C&S workshop are to present through a dialogue with the teenagers the elements of knowledge around a scientific subject based on a slide show. They are also invited to present their research project in the context of the scientific themes selected. The preparation of the oral presentation, the design of the slide show and the face-to-face with the young public are elaborated in collaboration with the scientific communicator ahead. In this context, PhD students were guided all along the activity by a professional of science communication that allows him/her to get into position of co-mediator.

Among the media supports, the slide show is an essential element as it constitutes for the teenagers the main visual support on the scientific content and, as a consequence, is considered as a reference source by the participants. A recent study led on 7 C&S workshops between 2014 and 2017 with the analytical support of LDAR unit shown that teenagers are largely influenced by the images showed by the PhD student (scientific images, drawings, diagrams) and some elements of the slides could be directly reinvested in the comic strips (de Hosson et al., 2018). Then the influence of their presentation is directly identifiable making the C&S workshop as possible educative tool for the PhD students.

In addition, the workshop is often the first opportunity for teenagers to meet a young researcher. This meeting visibly influences adolescents for whom the world of research is very abstruse and the stereotypes of the untidy hair researcher may still alive (Karaçam, 2016). This influence is clearly observed through the created comic strips in which teenagers frequently represent young researchers in their working environment. It is true that classical scientist stereotypes remain prevalent in some of the participants' stories, probably in connection with the nature of the comic strip medium in which the figure of the scientist is one of its cornerstones (Allamel-Raffin et al., 2007), and turns out a useful character to dramatize a story. Nevertheless, it is highly probable that this encounter with young adults, women or men, with whom teenagers can directly exchange, offers a new picture of the scientists to the young participants. By breaking with the usual stereotypes of the scientist as in the project of the '11 theses in comics' of the Lorraine University (Beck et al., 2016), we could argue that this science communication device in which PhD students are science representatives may help young public to apprehend the place and the real occupation of a scientific researcher.

5) Feedback from PhD Students after C&S Workshops

Five of 14 young researchers joined the Stimuli association after their first experience in a C&S workshop, either to participate to a new session, or to take part in other science communication activities of the group. Among them Olivier, PhD student in microbiology, expressed the importance of the face-to-face experience within the framework of informal discussion with non-experts: '*It brought me some distance by recontextualizing my thesis but also something more important: it made me breathe. The thesis is a rather solitary experience, and being able to discuss it with an interested public does a lot of good. I'd love to do it again.*' Other accounts (evaluation questionnaires) should be useful to evaluate the impact of this experience on the science communication curriculum of the PhD students involved.

3. 'Sciences Through Comics' Training

In a context of deployment of storytelling tools in science communication, comic medium has become a new tool for science communication institutions. Recognized as an attractive medium for the general public, and carrying of media qualities through its iconic-textual essence, the comic strip makes it possible to transmit scientific information efficiently (Tatalovic, 2009). However, few studies exist on this real informative scope, even less on educational impact, and on the future of scientific knowledge sifted through the constraints of graphic novel. In France, the majority of large-scale editorial projects, i.e. collections of so-called 'scientific' comics, although still endorsed or co-written by scientists, are carried by editorial structures whose intentions and challenges are not always those of science communication. Therefore, training young researchers in the understanding of the semiotics and use of the comic medium seems to be a crucial point to guarantee fruitful inter-professional collaborations between researchers and creators, and thus enrich the diversity and the quality of future comics and science communication projects.

1) Context and Objective

Proposed by Ombelliscience association, a scientific culture structure established in the north region of France, the PhD student training 'sciences through comics' is part of a broader training offer on the storytelling of research ('sciences through the tale'). It is within this framework that the Stimuli association developed, for the first time in France, a training programme on comic medium specifically intended for PhD students.

In partnership with two graduate schools (Human and Social Sciences on the one hand, Science, Technology and Health on the other) of the French University Picardie Jules Verne, the Stimuli association has designed and led a doctoral training on the graphic narratives of research. Since 2017, two sessions have been scheduled to train a total of 17 PhD students over a 3-days session. This training is financially supported by Hauts-de-France region and by an investment programme of the French State.

The session is designed to make young researchers aware of the semiotic codes of comics and to initiate them to collaboration with artists on future projects to promote their research through comics. The tutors are one specialist of science and graphic novels links, also scriptwriter, and one professional drawer.

The training program consists of theoretical and practical elements organized in 3 stages (Table 2): a state of the art on comics, manga and Franco-Belgian scientific comics with the tools for decoding its particularities (place of the scientific contents and of the scientists, visual representations, etc.), a focus on script writing methods and on the collaboration with a comic artist, and finally a practical design session to create a short comic strip on the doctoral student's research topic. This last step is the innovative educational point of this program, funded on the bias of the PhD students to put into practice the technical basics of comics through the realization of a short comic strip.

Table 2 Summary of the PhD training program 'science through comics'.

State of the art on scientific comics	Typology of scientific graphic novels Roles within authors involved Scientific themes and disciplines in science comics Articulation and formalisation of the contents
Script writing methods	Comic script writing tools Collaborate with a professional comic artist
Practice of comic medium	Write and produce a comic strip on research project (theme, script, storyboard) Reading, critiquing and improving other comics research storytelling projects

2) Portrait of PhD Students Trained in Telling the Science with Comic Medium

A total of 17 PhD students (11 females and 6 males) signed up to the 2 sessions of training, 12 registered in their first year of thesis, 4 in their second year and one registered in master degree. The distribution of their domain of research was eclectic with a predominance of life science and social science, that resulting of the scientific area research of the graduate schools associated to the project.

Among the participants, one of them was engaged in a PhD student conference programme

Table 3 Domain and discipline of research of PhD students trained in comic medium (2 sessions).

Domain of Research	Discipline of Research (number of PhD students)
Sciences of universe and matter	Physics (1)
Life sciences	Biotechnology (1) Neuroscience (3) Ecology (1) Biology (1) Health (4)
Human and social sciences	Psychology (2) Anthropology (1) History of communication (1) History of Art (1)
Mathematics	Computer science (1)

for general audience during which the comic strips of the PhD students were exhibited.

3) PhD Students' Account

At the end of the two training sessions, the participants answered an evaluation questionnaire designed by the Ombelliscience association, which makes it possible to identify their motivations for the comic strip medium as well as to evaluate the impact of raising awareness to the importance of promote their research through comics or other media. Nine of the seventeen participants answered to the questions.

There was no typical PhD student profile engaged in the comic medium training. The expectations and needs of PhD students are diverse. They may expect general knowledge like '*Getting a better understanding of the process of setting in comic book a scientific content*' or specific skills like: '*Learning how to make a comic book script, and find ways to disseminate my research*'.

Although that most of the participants had not particular skills for drawing or telling a story, they have all played the game of the creative practice on which the training is based. Some of them expressed the satisfaction of designing a one-page comic strip on their own: '*I discovered how good comics could be in promoting our research to the general public. This training gave me a new understanding of science communication, and I am really happy that I was able to follow it.*'

The questionnaire pointed out the possibilities of using concretely the comic medium in future science communication projects asking if the 'science through comics' training make them want to use or integrate comics in their presentations to a scientific or non-expert audience? Some answers drawn the attention of the real interest of the media and, at the same time, some restriction on the realization in concrete terms. Véronique, involved in anthropology domain and familiar to the use of drawing answered: '*Yes, it is currently envisaged that comic strips will be integrated into my thesis (as a presentation of the subject, or to illustrate writings). I also think of using it in other settings (for example, to exchange with the people I interview in the form of a comic strip.)*' On the reverse, Audrey, PhD student in environmental and biological research, with not effective skills in drawing techniques declared: '*This training actually made me want to use comics in my presentations. I would even like, if possible, to integrate a one or two-page comic strip in my thesis. Currently I have no drawing or scriptwriting skills. It is obvious that if I really decide to carry out this idea to integrate comics in my thesis, I will need collaboration with specialists. I'm not thinking about it right now though.*'

The training, although specific, proved useful to the majority of the participants in term of science communication skills. In the second training session, 6 (out of 7) PhD students answered that they could foresee reinvesting the content of the session into scientific or popularized presentations on their research topic. In addition, the possibility of insertion of short comics in their thesis manuscript, even if not generalized, indicates another application of the training that will have to be taken into account in the future.

4. PhD Students Engagements According to the Device

Through the two types of science communication devices presented in this communication, young researchers get experiences in practices of comics strip storytelling to share their research. The particularities of doctoral students' involvement in each of the science communication mechanisms show common points and differences that may be useful to highlight for future implications.

First of all, if the PhD students in both cases are always accompanied by science communication and comic medium professionals, their use of storytelling is different in the workshop and in the training. Placed as scientist and 'expert' in the workshop, they are not directly involved in the choice of the framing and/or the image setting, acts produced by the young participants. In this situation, the PhD student is, above all, co-mediator of the scientific knowledge (Fig. 1), and, in regards of the comic medium, raised awareness to the comics constraints, through exchanges with adolescents, and as critical reader of the one-page comic strip obtained.

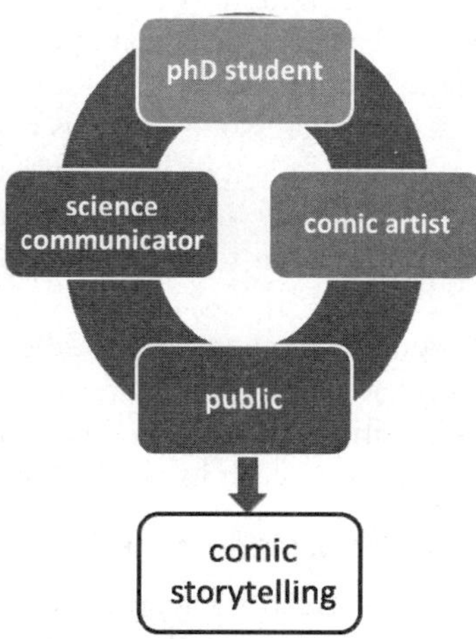

Fig. 1 Place of a PhD student in a comics and science workshop regarding the public (teenagers) and the medium (comic storytelling).

In terms of production, the one-page comic strips created by adolescents have a more or less informative scope depending on the workshops, disciplines, themes and participants (De Hosson et al., 2018). Their comic strips may deliver a more or less scientifically rigorous message, but, far from being a sign of incomprehension, the freedom taken in view of scientific content by young authors is a sign of a necessary reinterpretation to the specific constraints of comic medium (distance and time scales, choice of personification of an inert entity or concept, etc.). During the workshop, PhD students are thus indirectly confronted with the choices of representation of their own discourse and slide show made by the young authors. Then the 'one-page comic strip' constitutes an ideal storytelling object where one can experiment comic framing without the usual editorial constraints. For this reason, the C&S workshop may be considered as a congruent experience for the first engaged comic storytelling about science of young researchers.

Within the 'science through comics' training, PhD students are directly placed at the heart of the construction of the graphic novel (Fig. 2). In fact, after being introduced to the basics of the comic storytelling, they become author of a story they wish to tell about their own research. This guided practice of storytelling requires them to make choices that only the position of a complete author, i.e. scriptwriter and drawer, can provoke. Thus, although they might not be sole author in the future, but probably co-writer, this test induces an appropriation of the comics and science communication constraints: extracting a topic from a set of data, sequencing actions, defining a protagonist, caring of the visual representations, etc. Similar to an investigative approach adapted for the creation of media support, this training is a practical awareness that can be adapted to other types of storytelling medium.

Fig. 2 Place of a PhD student in a 'science through comics' training regarding the public (general audience) and the medium (comic storytelling).

5. Limits of the Comics and Science Communication Projects

These projects of graphic narration of science in collaboration with doctoral students can present some limitations. The first concerns the necessary curiosity of the researchers in comic medium that reduces the potential candidate number for that particular engagement. The second restriction could be in term of scientific disciplines or themes that could approach the limits of potential comic storytelling, because of an abstract or abstruse theme of research.

The Stimuli association experience indicates that there is not particular counter-argument to transpose into comics any scientific domain but rather more adaptable themes whatever the domain is. For instance, two C&S workshops on mathematics have been led with success probably owing to the theme of cryptography that suits efficiently to the structure of a story-telling (a coded message has to be transmitted between two entities that arouse a mystery). In consequence, it is recommended to engage young researchers considering this point. A third limit could be more specific to the training and the availability of young researchers in term of time and schedule. How to transfer their new science communication skills learned in training into concrete projects, often time-consuming and partners-dependent? This point appeared in the PhD evaluation of our training partner and should be taken into account for future science communication courses or outreach programmes of universities and institutions.

6. Conclusion and Perspectives

In short, the appropriation of graphic narrative by young researchers to disseminate science to general audience is possible and even profitable under certain conditions. The various experiences in C&S workshops or training showed that the one-page comic strip is a congruent model of experimentation for participants who are non-professionals of comics but are familiar with graphic novels as readers, and then capable of appropriate the language of comic medium easily.

Moreover, the experience of the Stimuli association, in workshops or training, let think that close supervision of PhD students is an essential condition for the successful implementation of the science communication projects. This professional support is fruitful in that it allows young researchers to get to grips with communication practices without assuming all the responsibilities (quality of the message, quality of the storytelling and the image settings). In return, this policy is based on a trust given to professional creators whose artistic vision and skills constitute the essential conditions for narrative projects that promote research at a high level of quality.

Paradoxically, published scientific comic books are frequently the fruit of collaboration of comic artists with experimented researchers that have not been trained on that specific media. What are the current perspectives for young trained researchers who would like to participate in editorial support? What are institutional avenues for research dissemination through comics to the general public outside the traditional editorial channel? Through these open questions, collaborative approaches are determining. Effective participation of researchers with science communication or storytelling professionals may address the concerns regarding the risk of distance between scientists and non-experts when only intermediaries choose and decide the message to be conveyed to the public.

Giving young researchers some of the keys to the media vehicle could reduce this distance and allow the scientific knowledge and the scientific method they carry to reach its destination.

Acknowledgments

The author thanks the Ombelliscience team for its permission to use the results of the evaluation questionnaires of the 'Science through comics' training course.

References

Allamel-Raffin, C. et Gangloff, J.-L. (2007). Le savant dans la bande dessinée: un personnage contraint. *Communication et langages*, n° 154, 123-133.

Beck, N., Bertrand, P.-E. and Heckler, A. (2016). Une BD pour raconter sa thèse. International Conference Telling Science, Drawing Science, Cité internationale de la bande dessinée et de l'image. 24-25 November 2016, Angoulême, France.

Bordenave, L. (2012). Comics-Sciences workshops by Stimuli collective. International Conference on Science Communication, 4-7 September 2012, Nancy, France.

Bordenave, L. (2016). Les arcanes du récit de science en bande dessinée. International Conference Telling Science, Drawing Science, Cité internationale de la bande dessinée et de l'image. 24-25 November 2016, Angoulême, France.

Cerrato, S., Daelli, V., Pertot, H. and Puccioni O. (2018). The public-engaged scientists: Motivations, enablers and barriers'. *Research for All*, 2 (2), 313-322.

Courty J.M. and Dematteis C. Helping the researchers to communicate. International Conference on Science Communication, Science and You. 1-6 June 2015, Nancy, France.

Daelli, V. and Rodari, P. (2014). Scientists and the communication of science: Institutional activities, personal involvement and training needs. 13th International Public Communication of Science and Technology

Conference. 5-8 May 2014, Salvador, Brazil.

Dahlstrom, M. F. (2014). Using narratives and storytelling to communicate science with nonexpert audiences. *PNAS*. 111(4), 13614-13620.

De Hosson, C., Bordenave, L, Daures, P.L., Décamp, N., Hache, C., Horoks, J., Guediri, N. (2018). Communicate science through Science'n Comics Workshops: the Sarabandes research project. Journal of Science Communication. 17(02).

Frey-Klett, P., Brun-Jacob, A., Beck, N., Leclère, P., Oualian, C. and Cussenot, M. (2015). Science and mediation, an original doctoral training program proposed to all PhD students of Lorraine University. International Conference on Science Communication, Science and You. 1-6 June 2015, Nancy, France.

Jensen, P. and Croissant, Y. (2007). CNRS researchers' popularization activities: a progress report. *Journal of Scientific Communication*, 6 (3).

Karaçam, S. (2016). Scientist-image stereotypes: the relationships among their indicators. *Educational sciences: Theory and Practice*, 16(3), 1027-1049.

Tatalovic, M. (2009). Science comics as tools for science education and communication: a brief, exploratory study. *Journal of Science Communication,* 8 (4).

Taylor, C. and Dewsbury, B. M. (2018). On the problem and promise of metaphor use in science and science communication. *J. Microbiol. Biol. Educ.*, 19 (1).

Trench, B. and Miller, S. (2012). Policies and practices in supporting scientists' public communication through training, *Science and Public Policy,* 39(6), 722-731.

Research on the Mode of Science and Culture Communication under the Context of New Media

Zhu Yun

School of Cultural Creativity, Communication University of Zhejiang, Hangzhou, China

Abstract: With the rapid development of information and communication technologies, new types of media have emerged. The development of new media technologies has provided the necessary conditions for the reconstruction of traditional mode of scientific and cultural communication. As a kind of culture that supports and standardizes the scientific institutional environment and social atmosphere, scientific culture brings challenges to the authority of information, the privacy of users, the public opinion response ability of the government, and the news production of traditional media. The substantiation of communication forms leads to the transformation of traditional communication forms to digital convergence and the emergence of intelligent communication forms such as virtual reality. These changes have led to four predicaments in the current scientific and cultural communication. Firstly, the distraction of discourse power leads to the authenticity of the content of communication. Secondly, the limitation of the integration of traditional physical communication and new media communication. Thirdly, the technical complexity of the existence of new media communication science culture; Fourthly, the public's suspicion of the acceptance of the science and culture of the new media.

1. Introduction

With the rapid development of information and communication technology, new media are constantly emerging. The development of new media technology provides necessary conditions for the reconstruction of traditional scientific and cultural communication mode. According to the author's literature search, the meaning of 'scientific culture' is different. The mainstream views are: one is to understand it as the scientific spirit and scientific philosophy of the scientific community; Second, a unique cultural form created in the field of science and technology from the perspective of subculture; The third, which is also the opinion of this paper, is that scientific culture is the institutional environment and social atmosphere to support and regulate science.

The term 'new media' was originally coined by Goldmark, director of the CBS Institute of Broadcasting and Television Technology. 'New media' refers to a new form of media which uses digital technology and network technology to disseminate information through network and satellite channels and uses computer mobile phones and other digital receiving devices as terminals after traditional media such as television, film, radio, newspapers and periodicals. As a product of the progress of communication technology, network media has become a new milestone in the development of media history with its great advantages such as digitization technology of information load capacity and hyperspace communication[1]. Since the 21st century, new media represented by the Internet has been applied to media communication at the speed of blowout, playing more and more roles in people's lives. The wide application of new media has greatly changed the way people use media. The main way for people to obtain scientific information is no longer the traditional way of buying science books or listening to radio, but the way of acquiring scientific and cultural information by means of Internet or streaming media with quick and intuitive information pictures. The emergence of the Internet has provided a new mode of communication for the dissemination of scientific culture, such as the collaborative innovation of scientific culture that is crowd sourced on the Internet and the interactive demonstration of scientific culture by new media. As a result, the communication path of scientific culture is no longer linear, but complex two-way and even multi-purpose. This also makes the channels for receiving scientific and cultural knowledge and skills more diversified and convenient, and the public can participate in the construction process of scientific culture.

2. Scientific Culture Communication Mode Based on New Media Context

1) The Decentralization of Transmission Subject

With the development of new media technology, the distinction between the traditional information communicator and the audience is gradually blurred. The emerging 'we media' creates the space for self-communication for the public, making the general public not only the audience of information, but also the information communicator. The low threshold of new media such as WeChat and Weibo enables everyone to have microphones and the power of discourse is continuously dispersed. The impassable relationship between 'communicator' and 'audience' in traditional mass communication is broken. At the same time, the rapidity and simplicity of new media bring challenges to the news compilation of media. The forwarding mechanism of micro blog enables the news to spread as quickly as ten to one hundred, and even the influence of tens of millions of bloggers exceeds the coverage of mass media. However, this is not the case for everyone. Ordinary people (except the big V with strong influence) are limited by the censorship system, number of attention and trust, and their communication content cannot match the traditional mass media in some aspects, such as influence and spread breadth.

The influence of new media on traditional media is obvious. It dissolves the advantages of traditional radio and television media in the past, grabs most of the market share of traditional radio and television media, and diverts the previously loyal radio and television audience. Moreover, with the popularity of smart mobile devices and the development of social media, everyone is a journalist, and mobile phone is the media. News reporting and information release are no longer the patent of media people, which bring challenges to many people or institutions. The specific performance is as follows:

(1) *Challenges the authority of information*

The threshold of new media information release is relatively low, and the post-post review mechanism of 'release first, review later' brings soil for the spread of many false information and bad information. In order to improve their attention and make use of people's relevance to hot events, many people sometimes spread unverified and improper information, which once caused concern information panic, and also increased people's voices of doubt about relevant personnel and institutions, which had a negative impact on them.

(2) *Challenges users' privacy*

Network security has always been a concern and never been solved. The privacy of any behavior on the Internet is lost, so critical information is often compromised and security is questionable. In addition, the fast spreading power of new media and the human flesh searching power of netizens can make an ordinary person become an Internet celebrity overnight. All the information is exposed to the public. Positive news may help the development of Internet celebrities, but negative news will bring great trouble.

(3) *Challenges to the government's ability to respond to public opinion*

One of the biggest characteristics and advantages of network media is strong interactivity, which can communicate with citizens in a timely manner and reply to everyone's concerns quickly. Interactive we-media such as micro-blog further magnifies this advantage. For each major event, new media is often the first to speak out. At the same time, everyone has the right to speak, which makes the information under the same topic so numerous that it is difficult to distinguish the true from the false. Moreover, the performance and practice of government websites and various new media application carriers in the specific use will affect people's evaluation of government credibility and restrict the guidance and mobilization ability of government information release. However, while citizens are keen to participate in and discuss politics, some government departments fail to make full use of this platform. With the continuous expansion of new media users, the change of the public opinion field of new media not only makes it more difficult to guide public opinions, but also brings challenges to the credibility of the party and the government.

(4) *Challenges to traditional media journalism*

Traditional media are good at making news. Although they are time-sensitive, they are far from new media. Nowadays, the rapidity of new media brings challenges to the timeliness of news of communication media, so the link of news compilation changes. Not only that, new media has also become the source of news collection of traditional media, and events with high popularity in new media will become news content in traditional media. In order to prevent the loss of the first right of news discourse, traditional media also set up their own publicity platform in new media. At the same time, the interactive mode of new media also provides a feedback platform for audiences to comment on traditional media. Compared with the previous 'injected' communication, the mass media cannot ignore the audience's opinions, and much news is more cautious in collecting, editing and broadcasting than before.

2) The Three-dimensional Form of Communication

(1) *The transformation of traditional communication mode to digital confluence*

Nowadays, mass communication can no longer achieve monopoly as before, and traditional media can no longer have monopoly power in system and technology. Digital communication is diverse in technology and form, which completely changes the way of human communication in mass communication and rewrites the living state of various media. Therefore, it must be in keeping with the trend of The Times to rebuild the media and traditional media industry and transform the media management mode. This also promotes the development of traditional planar media to digital convergence. Traditional media organization and information content will generate media platform convergence through modern media technology and guide the rapid development of new media era.

Compared with traditional media, new media has its unique advantages and characteristics. New media has a powerful dissolving power and huge amount of information. New media can dissolve the boundary between traditional media (such as TV, radio, newspaper, etc.), the boundary between people, society and countries, and the gap between information sender and information receiver. In addition, new media are presented in front of people by digitalization, with information that traditional media cannot match, and can correct wrong and inaccurate information at any time. Through different channels, various information sources can be timely communicated to a wide audience.

(2) *Intelligent propagation form appears*

The intelligent application of new media will turn our city into an intelligent city and our culture into an intelligent culture. For example, through touch screen technology, virtual reality technology and synthesis technology, virtual space and real space can be connected, bringing unprecedented cultural experience to people. By using the unique technology expression language of new media, the exhibition industry, tourism and cultural services, cultural protection and cultural facilities services and other industries are upgraded and upgraded. Through the adaptation to the physical space of new media — that is, based on the investigation of people's living habits, working methods and communication methods, reasonably settle different new media, give full play to the flexible features of new media, and serve for cultural communication; As well as the adaptation of virtual Spaces, such as the promotion and practice of mobile maps, mobile banking functions, and other value-added services constantly developed by terminals, people have the possibility of consumption and life based on smooth mobile communication platforms. The application of these strategies will certainly promote the dissemination of scientific culture. More importantly, the integration of new media technologies can also reduce the startup and operation costs of cultural enterprises and effectively save social resources, thus bringing about the healthy development of the cultural industry.

In the new media era, science and culture, with the help of new technologies and new forms of communication, not only achieved good effect of cultural communication, but also gained good economic benefits. Such as introducing multimedia fantasy drama space-time Shanghai circus city tour use high-tech means to build full of wonder and magic of magical effects, with original large water curtain holographic projection high-tech means such as multimedia applications across to the audience the unique experience of space and time, let the audience get the aesthetic experience full of surprise and joy, and achieved good social and economic benefits.

3. The Dilemma of the Mode of Scientific Culture Communication in the Context of New Media

1) *The decentralization of the power of discourse leads to the consideration of authenticity of communication content*

As early as 2000, UK pointed out in its official report science and society that contemporary science faces a strong crisis of trust. Although people like popular science, paradoxically, hobbies always accompany growing skepticism[2]. Distrust scientists' statements on all types of science-related policy issues. Currently, the reality is that of a more serious with the rapid development of social media, the original notion of 'scientists' of authority is in the network society and dispersed, there are many plausible scientific information in the network, especially some brand 'opinion leader' of the house of pseudoscience is wanton comments and some of the science and technology personnel's irresponsible remarks, often give the public some illusion, through the Internet to spread scientific knowledge, true and false difficult points, its authenticity and authority severely questioned. In addition, driven by certain interests, some social organizations and new media deliberately cooperate to create rumors to confuse the public and seriously undermine scientists' discourse authority. This is one of the dilemmas of how to promote the dissemination of scientific culture in the context of new media, especially we media.

2) *The limits of integration of traditional physical communication and new media communication*

Although the development of new media provides more choices for the communication of science and culture, it does not mean that the communication of science and culture in the context of new media can abandon traditional traditions. In fact, the traditional mode of communication with science and technology museums as the physical state should not be underestimated. It should actively demand how to cooperate with new media to realize the 'integrated communication' of scientific and cultural products. Then, the question of where new media communication should be located, which is more suitable for new media communication, and how new media communication should be spread needs serious thinking and exploration. In addition, many scholars in China actively encourage the use of new media to replace the dull and obscure symbolic expressions in the forms of art, music and literary works to disseminate scientific content, which will help the public better understand science and contribute to the dissemination of scientific culture. But how to realize the traditional material form of popular science knowledge works to the popular entertainment form also has the dilemma of technical limit.

3) *New media disseminates the technical complexity of scientific culture*

Due to the specialty of scientific culture itself, the profound and obscure scientific culture is hard to be clearly perceived and understood by the public, which requires us to think about how to make scientific culture more acceptable to the public. With the help of virtual animation, multi-dimensional 3d and sound and image technology, new media can theoretically realize the interestingness and vitality of scientific and cultural communication, but it needs more technical intervention. Moreover, not all scientific knowledge can be 'simplified' with the help of new media technology.That is to say, the complexity of technology is an inevitable problem for thinking about how to better disseminate scientific culture in the context of new media.

4) *Public acceptance of new media in disseminating scientific culture*

Due to the autonomy of new media communication, there is insufficient review and supervision of information authenticity and science, and it is difficult for the public to distinguish the authenticity from the massive information. It is an indisputable fact that the credibility of new media is questioned, which seriously affects the effect of new media as an emerging carrier to disseminate scientific culture. In addition, because the public itself has the scientific knowledge level of high and low points, also can't rule out part of the social group is the difficulty of the new media using the technology, which can lead to such group lose enthusiasm in new media, science and culture, in-depth understanding of science culture and participation in scientific activities will inevitably affected by the the communication effect. The conflict of technology application on environment, energy, ethics, morality and scientists' behavior and interests, through the rapid spread of developed electronic media, the contradiction between public culture and scientific culture will also lead to the impact of public opinion on the development of science and technology.

4. Conclusion

Scientific culture construction is an important part of characteristic culture construction. With the rapid development of information and communication technologies, new types of media have emerged. The development of new media technologies has provided the necessary conditions for the reconstruction of traditional mode of scientific and cultural communication. As a kind of culture that supports and standardizes the scientific institutional environment and social atmosphere, scientific culture brings challenges to the authority of information, the privacy of users, the public opinion response ability of the government, and the news production of traditional media. The substantiation of communication forms leads to the transformation of traditional communication forms to digital convergence and the emergence of intelligent communication forms such as virtual reality. These changes have led to four predicaments in the current scientific and cultural communication. Firstly, the distraction of discourse power leads to the authenticity of the content of communication. Secondly, the limitation of the integration of traditional physical communication and new media communication. Thirdly, the technical complexity of the existence of new media communication science culture. Fourthly, the public's suspicion of the acceptance of the science and culture of the new media.

References

[1] Wang Ming. Innovation in the supply and cultivation mechanism of online new media talents [J]. China collective economy, 2009, (4) : 136. (in Chinese)

[2] The house of lords special committee on science and technology(a), Zhang Botian, Zhang Donglin(eds). Science and society [M]. Beijing: Beijing Institute of Technology Press, 2004: 21.

Trend and Difficulties of Science Communication under New Media Era

——Take 'science media center' as an example

Shi Congyi

Peking University, Yenching Academy, Beijing, China

Abstract: As a cross-disciplinary subject, science communication is closely related to both popularization of science and communication are closely related, and the development of science communication is also closely linked to them. With the development of science and the progress of society, the attention of public to science popularization is rising, and the way of science communication has gradually changed from traditional science popularization to a brand new mode. This new mode is centered on new media and traditional science popularization plays an auxiliary role. Therefore, based on the new media mode, science communication will have a profound impact on public participation inevitably.

Though it started late compared to Western countries, science communication based on the new media voice has developed rapidly in China. However, on the one hand, the new media and the scientific communication under the audience have increased the public engagement; on the other hand, the pseudo-science and false reports based on the new media communication have also had a huge negative impact on the public. Therefore, this article will take the 'Science Media Center' of the China Science Research Institute as an example, and combine the current status and predicament of science communication to make a corresponding analysis. This is important for us to understand science communication.

With the advancement of society and the progress of science and technology, both the approaches and methods of communication are constantly changing to diversification, and an increasing number of disciplines are also coming out of such inexorable trend as well. Based on the condition, for one thing, traditional modes for communication has transformed dramatically from those including newspaper, radio, broadcasting, television, and so forth, of which are one-way communications for people, to the brand new mode, a two-way or even multi-directional communication, which the information is communicated via Internet instantly.

Keywords: Cross-disciplinary Subject; Science Communication; Science Media Center; Internet

1. Introduction

In this era dominated by cutting-edge technology and science with brand new conceptions, media and communication are changing dramatically based on this inevitable trend. For science communication—a part and newly-emerging branch during the progress of media and communication, it also shows different characters. As an outcome for this era when new media is everywhere, science media center has appeared as medium to link to scientist, journalist and the public. However, the information age today is destiny to face the challenge from the merging of traditional media and new media, while how to solve this dilemma is also an important to not only the scholars, but also the public.

2. What Is Science Communication?

As the emerging academic sphere deriving from both philosophy of science and history of science, science communication itself is inextricably connected with traditional popularization of science and communication individually. (Wu Guosheng, 2016) Nowadays, there is still no significant and exact definition in academia for science communication, but the division between science communication and traditional science is clear, but the difference between 'science communication' and 'public understanding science' is not clearly defined. Because the scientific communication we are talking about is based on the public understanding of scientific conception, and it is the brand of the

E-mail: rogerscy@163.com

public understanding of scientific characteristics. It is extension and continuation of the public understanding of the scientific movement. Scientific communication is a new form of scientific popularization, and its theory itself is still developing.(Yang Qingyuan, 2011)

In western countries, the conception of science communication has been separated from other human activities since 17 century. Generally, there are three periods for 'science communication' there in western world: namely the stage of popularization of science, public understanding science, and science communication respectively.

1) Traditional Popularization of Science

Since its date of birth, science has to face the question about how to gain recognition from the public. Hence, Science communication is a powerful medium for exerting enormous leverage on science. Occurred in the 18th century, the Enlightenment in French was the first climax of science communication, and this enlightenment formed a mainstream ideology based on Newtonian mechanics as a typical example. Prior to the 20th century, the premise of science popularization included personal judgment and rooted in common sense. After the 20th century, science has ignored these two premises. Under the domination of physics, science and the public are getting farther and farther.

2) The Three Periods for Communicating Science

There are three stages that science communication has ever experienced, including scientific literacy or traditional science popularization, public understanding of science, and science communication. (Yang Qingyuan, 2011)

(1) *The period for science literacy*

Contemporarily, there are another three designations for science communication in China: science, science and technology communication, and science communication, representing the three groups and three modes of science communication. The science popularization led by the China Association for Science and Technology is mainstream and orthodox. It has the three characteristics of nationalism, utilitarianism and scientism, and has been marginalized in the past 20 years. The researchers of science and technology communication are mainly communication scientists, mainly focusing on the means of communication and the efficiency of communication, and there is no conflict with the concept of science. The advocates of science communication are mainly science historians and scientific philosophers, challenging the three major ideologies of traditional science. Marked by the Science Communication Center of Peking University, it has formed a critical school of Chinese science communication. The three modes are still in the process of intense interaction and integration.

According to their historical long-distance ordering: 'Popular Science' (since the 1950s), 'Science and Technology Communication' (since the 1990s) and 'Science Communication' or 'Public Understanding Science' (since the 2000s) (Wu Guosheng, 2016). In the context of contemporary China, they all have specific meanings and are not used at will. They represent the three groups and three modes of science communication in China. Although these three groups cross each other, they cannot be completely distinguished. In spite of the three models are in mutual influence and mutual learning, they have not yet been finalized, but these three names can still be used as three categories to describe Chinese science in general. It is an overview of the cause of communication.

(2) *Public understanding of science*

After the Second World War, the traditional concept of science popularization exposed more and more drawbacks. The development of science also makes it necessary to carry out the necessary reforms. The public understanding of science is a scientific mass communication activity in the United States and the United States since 1980 with the scientific community. In the first place, more and more scholars have refined scientific divisions. Communication and communication between various disciplines has become a must. Scientists must learn from each other and understand each other. Traditional science audiences are transformed from a single ignorant to a broader public, even including technical experts. What needs to be popularized from the communication of content is not necessarily professional and specific scientific knowledge, but a better understanding of science. Secondly, the negative effects of science are more obvious and prominent. For example, the intensification of environmental pollution, the energy crisis, the application of nuclear weapons, the public has begun to doubt about science and technology, people begin to pay attention to and have the right to evaluate the positive and negative effects of science. It is a large-scale project for money of taxpayers and relying on state funding. Therefore, people have the right to decide whether their money is used to make spacecraft or to transform ecology to reduce pollution. The public should play an important role in the direction of science development and how to achieve sustainable development.

(3) *Science communication period*

There are three differences between science communication and popular science: First, science is a process in which scientists transmit scientific knowledge to the public in one direction, and science communication is a two-way interaction between the scientific community and the public. Second, the goal of science is to serve the scientific community, but the goal of scientific communication is the construction of civic culture. Third, the subject of science is science plus literature, because for literature, like sugar coating, it can help the audience to better accept science; the academic foundation of science communication is the history of science and philosophy of science, and the goal is to integrate science and humanities. It is the basic position of the article to oppose the scientism presupposition in popular science and promote a new scientific concept. At the same time, science communication is a broader concept than the public understanding of science and traditional science, and the former includes the latter. (Wu Guosheng, 2016)

3) The Transformation for Science Communication

For the gradual decline of the influence for traditional media, and the increase for the influence of new media as well, both of them are inevitable trends of media transformation and media convergence. For communication itself, communication paradigm also exhibits diversity and variability with the changes of the media.

As the main medium of scientific information communication, WeChat public account plays an important role in the modern social life. Therefore, on the basis of analyzing the communication utility of the WeChat public number regarding scientific information and knowledge, we will further analyze the relevant implicit indicators, and finally summarize the possible influencing factors of the scientific WeChat public number in the process of science communication.

With the traditional popularization for science somehow transferring to the comparatively new conception science communication, science institutions have been established in a handle of countries, such as American Association for the Advancement of Science (stands for AAAS) of United States, and Australian National Centre for the Public Awareness of Science (stands for CPAS), and so on, which are regarded as the professional and authorities. The mission of these institutions is to introduce the latest and cutting-edge scientific payoffs, promote dialogue between the public and scientists, improve communication skills of scientists, and encourage society to develop modern science. For the public to understand science, these centers and institutions have spared no effort on and have been engaged in scientific research, education and practical work in the international science communication.

4) The Influence for Characteristics of Traditional Media to Science Communication

Traditional media, which is also known as old media or legacy media, is the system of mass communication. Prior to the Information Age and popularization of digital media, such as the Internet, other medium, those information carriers have temporarily referred to it as traditional media. Particularly, the main categories of traditional media include books, print media such as newspapers and magazines, radio and television, or venues of which rely on certain physical spaces, and so forth. In spite of the use of traditional media for science communication and popularization has its natural advantages, with the advent of the information society, the amount of information has soared, and the public's diversified demand for scientific knowledge and technology has become inadequate. There are mainly three characteristics for traditional media.

(1) *Relative authority during communication*

Generally, traditional science communication is led by the government as well as the scientific community in some significant measures, and both of them utilize the official and professional medium respectively to control and screen not only the scientific information, but also knowledge in strict methods. Therefore, the processes of information production and standards for management system are relatively high, and it has relatively high authority.

(2) *Feature of centrality*

Traditional media is at the center of mass communication. It communicates both scientific and technological information and knowledge to the public in a condescending manner, and presents a one-dimensional communication. At the same time, it accepts the public's view and the public rarely has the opportunity to interact with it. The public's recognition of the official professional media is also high, further strengthening the centrality of its status.

(3) *Information in hysteresis*

Obviously, compared to new media, traditional media is inferior in the timeliness of science communication, and scientific and technological information is lagging behind. The communication of new scientific knowledge requires a series of procedures, from the collection and editing of scientific information to the publication (broadcasting) to the public. Information have to go through a long time difference, which is not proportional to the speed of information growth, nor can it meet the public's knowledge needs in time. In addition, the traditional means of communication also have certain requirements for the cultural level (especially the print media), and the audience needs to have the corresponding literacy and reading skills. At the same time, traditional science communication will also consume a great deal of manpower, material resources and financial resources, such as the construction of science and technology venues.

5) The Influence for Characteristics of New Media to Science Communication

Unlike the era of traditional media, time of new media has redefined the communication in a brand new way, including science communication as well. For this era of new media, there are four features.

(1) *Feature for convergence*

New media can use a variety of symbolic forms such as graphic, audio and video to spread scientific and technical knowledge, organically integrate it into a communication unit, so that the communication of scientific and technological knowledge has a better comprehensive, intuitive, and visual, more in line with the law of public thinking, thereby improving the effect of science communication.

(2) *Immediacy for information and news*

With the advantage of digital communication technology, the new media can realize the rapid, high-speed, high-efficiency, instant, remote and asynchronous transmission of scientific and technological knowledge, which greatly expands the spatial scope of science communication and breaks the limitation of time and space.

(3) *Interactivity during communication*

The most significant feature of new media is interaction. During the process of communication, it is not only passively accepting scientific knowledge, but also requires interaction between the people and the people, and people and science as well. In addition, the application of new media has enabled the communicator and the recipient to play a dual role for each other. The audience is not only passive receiver for information and messages, but also actively participated in the process of media and communication. The public and scientists exchange on an equal footing, realizing the popularization of science communication and popularization.

(4) *Multi-centered for communicating*

Diversified sources of content based on new media communication, weakening the status of 'gatekeepers', based on topic classification or communication with a 'Multi-centered' characteristic is achieved by different bodies. The traditional science communication is mainly based on a certain point. The towering of scientific information and knowledge makes the scientific knowledge increasingly hard to understand and the public cannot even communicate with it. In addition, multi-centered science communication can not only achieve a peer-to-peer mode, but also a point-to-point and multi-point cross-interaction communication.

6) Media Convergence and the Characteristic for the Era

Generally speaking, as an inevitable trend during the long course of media progress, media convergence refers to the integration as well as merging for both traditional media and new media. In the Information Age, media convergence also can be regarded as not only the products, but also the systems or institutions. In China, some scholars have defined the concept of media convergence, and namely, it should include both narrow sense and broad sense individually. The narrow concept referred to the convergence of different media is to form the different media status or forms as a integration, producing a qualitative change from quantitative change, such as e-magazines, blog news, and so forth, while the broad sense for 'media convergence' contains the combination, convergence and even integration of all medium and related elements, including not only the merging of media forms, but also medium themselves. The integration is regarding function, means of communication, ownership, organizational structure, and so forth. The generalized 'media aggregation' is a process of gradual development from low level to high level, and the narrowly defined 'media convergence' is the highest stage of development' (Ding Bing, 2011).

Driven by big data, traditional media has the access to connect with new media counterpart during this procedure. According to statistics, the presidents, prime ministers and related institutions of 125 countries around the world have registered accounts on Twitter, and more than 75% of the governments in the Americas have their Twitter accounts, which are about 60% and 56% in Africa and Asia respectively (Zhang Suqiu and Gu Jiang, 2015). Rested on multiple technological means of new media, traditional media will carry out integrated development in terms of relevant communication concepts, including communication content, communication channels, marketing channels and public opinion orientation. Thus, from the different stages for the merging and integration of both traditional and new media, there are three types of convergence: namely convergence of content, convergence of channel and convergence of market. (Zhang Suqiu and Gu Jiang, 2015)

In the first place, as for convergence of content, both new media and traditional media rely on this part to make audience and users paying more attentions on the added value, while the advantage for traditional media demonstrates in content production, especially in. As for new media, networking is the means of communication for it, and its advantages, compared to the traditional media, are more convenient and more open. The content of the old and new media naturally complements each other on the network platform.

For digitalization of content, new media is based on the encoding of high-quality content provided by traditional media through computer technology and programming, and the initial text, audio, and video into a storage form that can be transmitted by using new media. On the one hand, the quality content of traditional media can be stored and collected, which is convenient for file management and historical research; on the other hand, the digital content can be truly integrated with new media. It deals with the content of traditional media with the carrier and presentation of new media.

Moreover, networked content is a pattern of manisfitation of content on the medium of communication in this new era. Networking means connecting media content through the computer Internet. Generally, netizens, for current usage, who most commonly refers to Chinese internet users in English language media, describing a person actively involved in online communities or the Internet in general, are partly the audience of media content. (DeLoach and Amelia, 1996) Networking is the means of communication of new media, and its advantages are more convenient and more open. The content of the old and new media naturally complements each other on the network platform. This mode of content integration is mainly the main type of dependence on the initial integration of new and old media. That is, simply using modern programming technology to digitize and mediate traditional media content and new media content, and to achieve docking and integration on a public network platform, can be simply understood as spreading old media content with new media technologies.

7) The Changing for Communication Paradigm

As is well-known, communication activity is one of the characteristics of human beings. It has appeared with the emergence of human beings, always accompanied by human life, and exhibited different development characteristics at different stages of social development. Paradigm, which belongs to a theoretical system, refers to the collection of ideas, values, and technologies shared by groups in the scientific field. It is a standard of behavior and value standards that are shared by research groups in a certain scientific field. The communication paradigm is a pre-set media model that clearly defines the components of the communication system, including the communication process, the spread of symbolic grammar, rhythm and style (David L. Altheide, 2003). Throughout the history of human communication, the development of media has brought about a corresponding change in the paradigm of communication, and the different modes of communication in the period of human development have also presented different characteristics.

To some extent, the development for media has given impetus on the advance of communication, which is not only for the progress of ways to communication, but also for the changing of communication paradigm. For the latter, during the communication process, media is the basis of the communication paradigm. From the perspective of media changes, the history of communication is mainly divided into four major stages: namely verbal communication, handwriting communication, as well as print communication and electronic communication successively. These four stages for communication complement each other and overlap. As for emergence of language, it has opened up a new era of oral communication, and the invention of words has opened the curtain of handwriting. All these communication patterns, more or less, as a process of media convergence in different ways and methods, have shaped and demonstrated the changing of communication paradigm.

American scholar (Goldmark, 1967) first proposed the concept of new media, which is a new concept compared to traditional media. It is a new media form after newspapers, radio, television, and so forth, by using modern communication technology. Through various intelligent mobile terminals, the media form is providing information and services to the audience. In the era of new media, the traditional news paradigm is facing a severe test of deconstruction and restructuring under the impact of new information technology. The new media has abandoned the original peer-to-peer communication model and opened up a new face-to-face communication model. Anyone can post information through the terminal. The new media communication mode blurs the boundary between interpersonal communication and mass communication, weakening the difference between public communication and private communication.

8) The Influence for Science Communication: From Both Traditional Media and New Media

Strictly speaking, the communication for scientific information is a practice of 'educational' function. Although the scientific knowledge provided by the mass media is not as complete as the system of school education, for the general public who have left school, the mass media is indeed a convenient and unrestricted educational channel, and the general public obtains scientific information as well as the main source of knowledge. For the science communication and scientific knowledge, the function of the mass media is not only to communicate scientific information, but to a certain extent it is also reshaping the message of science. (Dimopoulos and Koulaidis, 2002) In another words, the mass media not only provides us with that of science-related news, in fact, but is also guiding the public how to understand it and pay more attention to it.

9) The Conceptions and the Development of Media and Communication

To some extent, as the prerequisite for science communication and the science media center as well, the related conceptions for both media and communication, which are also changing dramatically of the methods and transforming gradually in the pattern of manifestation, do not only influence the structures for both, but trace the source of the branch. Derived theories during the media progress and convergence is shown below.

The development for media has not only boosted the mode for communication paradigm, but also motivated some related theories. Some of these theories have laid the foundation for science communication as prerequisite. Especially in the new era dominated by new media, if the development and advancement for media has boosted the transformation of science in a substantial level, these theories support science communication in theoretical level, and ethnical level as well. As a consequence, science media center also becomes an inevitable outcome for scientists and science journalists to shoulder the responsibility from the public.

(1) *Theory of two-step flow of communication*

Communication science, namely the Communication studies, has originated in the United States since the 1930s. Most of the early communication studies only involved the relevant content of traditional journalism. (Wei Feng, 2013) Based on a study on social influence in 1940s, the theory of two-step flow of communication states that media effects are indirectly established through the personal influence of opinion leaders. For the majority of people, they would receive much of their information and are influenced by the media secondhand as well, through the personal influence of opinion leaders.[1]

(2) *Professionalism and the impetus for media and communication*

As a concept of journalism, the professionalism of news is independent of any authority, with a certain idealistic color and strong anti-authority spirit. It requires journalists to report facts in an objective, true and accurate manner, to find out the truth of the matter, and to present the original ecology of the facts to the readers. Its most prominent feature is that professionalism could objectively report news facts based on the standpoint of non-partisan and non-group. The goal of journalism is to serve the entire population, not a particular interest group (Guo Zhenzhi, 2000).

(3) *Theory of social responsibility*

The period between the end of the 19th century and the beginning of the 20th century, with the intensifying for monopolistic competition, commercialization, centralization, and simplification of the communication industry became severe increasingly. Meanwhile, the communication industry also abused the freedom of press and publication, infringed on the rights of citizens, and caused social dissatisfaction.[2]

As for the reasons, it is mainly because the three facts. In the first place, both the increasingly monopolistic communication industry and the commercialized communication industry are pursuing economic interests as their primary goal. Thus, it is impossible to provide a 'free market of opinions', but a market-oriented counterpart; secondly, the truth cannot be self-corrected by its own power, because people are not born with the ability to distinguish between right and wrong. Only through a sound social system, the good education would be able to guide and encourage people to be good, and social code could cultivate people's ability to distinguish between right and wrong.

It is recognized that communication industry without any limitation is nothing but harm to the society, and communication participants should be in charge of the social responsibility for educating the public to understand science. For the theory of social responsibility, it is always emphasizes on the concept of which the certain democracy will be accompanied by the certain responsibilities. Therefore, there is no irresponsible democracy and no unrestricted freedom. Journalism enjoys the right to freedom of the press and should also shoulder the responsibility for society. Based on this paradigm of media and communication, not only the science journalists, but also the scientists have to be in charge of the authenticity and authority for science communication to the public.

3. Science Media Center

The Science media center is an emerging organization in the 21st century which aims to strengthen the interaction between scientists, the media and the public. An important intermediary for promoting the interaction between scientists, the media and the public, the role of the Science media center in the spread of science communication and scientific knowledge has become increasingly prominent. Since the establishment of the Science media center in the UK in 2002, the United States, Canada, Australia and Japan have established their own science media centers and carried out diversified services and work in the unique environment of their respective countries.

Essentially, as a communication activity, Science communication is a form of communication in the first place, which requires sources, channels, and the audiences to accomplish. In the process of science communication, scientists should be the source of knowledge, the producers for lexical scientific achievements, and the communicators of scientific knowledge; as for the role media, it should bear the function of the channel, so that the scientists are able to transfer scientific knowledge, methods, attitudes and spirits to a wide audience through it, thus promoting the literacy of the majority public as the well-being. However, if there is only the source but without channel, or there is a channel without a source of information, science communication is difficult to achieve the effect what it is supposed to. Therefore, the cooperation

1 https://en.wikipedia.org/wiki/Two-step_flow_of_communication#cite_note-1

2 http://www.sohu.com/a/144712605_653748

between scientists and the media are capable to achieve the purpose of science communication and integrate science into not only the specific spheres, but also the daily life of the public.

The voices from Scientists are sometimes difficult to communicate to the general public, partly because of the lack of professional science journalists for the latter, or the deficiency for basic literacy to jargon and terms. Scientists worry that they cannot convey the meaning of themselves truthfully during this course; sometimes the media journalists even cannot find a qualified scientist to explain science stories by plain language, and therefore they occasionally suffer distortions for understanding and translations.

Above all, as a bridge between scientists and media journalists, providing the necessary support and support for both sides is also an important function of the science media center. Therefore, the science media centers of all countries have carried out training for scientists and media journalists, aiming to let the former and the latter understand each other in a better way, and not only does it enhance the ability of scientists to deal with the media, but also enhance the scientific literacy of media journalist, for better covering scientific reports during this process.

1) Science Media Center in Western Countries

In order to coordinate the stress tension and misunderstanding between scientists and science journalist, namely the relationship between scientific community sphere and media hub, as well as the public of understanding science, the first Science media center around the world appeared in the UK more than a decade ago. Its main purpose is to provide scientific and accurate information to the media to ensure the validity and correctness of science communication. So far there are more than 20 science media centers in operation and in existence in the world. In addition to the British Science media center, there are the Australian Science media center, the Japan Science media center, the American Science media center, the New Zealand Science media center, and the Canadian Science media center one after another. The Science media center has also become a bridge, or called bond to scientists and the media, making the public for science not understanding in a term level, but a in popular and easy-to-understand language. For example, in 2017, the Australian Science media center recommended more than 800 scientists to conduct science communication activities to relevant scientific journals at home and abroad.(Wang Dapeng, 2012)[1]

Based on their respective national conditions and the current condition of science communication, the existing science media centers can be roughly divided into three types. First, the traditional type headed by the British Science Media Centre, including the Australian Science Media Centre and the New Zealand Science Media Centre; for historical reasons, the Science media centers of these three countries have basically focused their work on the scientist side, through the establishment of an expert pool. Provide relevant news reports and materials for the media to help scientists increase their exposure in the news media. Second, the activity type represented by the Danish Science media center; the Danish Science media center was established with the support of the government. Therefore, it promotes the development of the Science media center mainly through various activities in which scientists interact with the media. Meanwhile, it also undertakes some functions of the government in science communication. Third, the training type represented by the Japan Science media center; the Science media center in Japan was established by relying on the university, and the relevant organizations of Japanese science journalists also played a certain role, and thus the training for media reporters was greatly improved. The scientific literacy and media literacy of journalists have enabled the harmonious interaction between scientists and media journalists and reporters.

2) Science Media Center in China

With the inevitable trend of media convergence and emerging for science media centers in the western countries, since 2007, the China Association for Science and Technology hosted a series of campaigns related to science communication, recommended excellent research papers for the media, and communicate scientific results in the form of scientific reports and popular science. At the same time, from the year of 2011 to 2014, the China Association for Science and Technology also hosted a series of face to face activities between journalists and

1 http://news.sciencenet.cn/htmlnews/2012/6/265698.shtm

media professionals, in order to build a platform for scientists and media reporters to interact with each other on hot scientific topics, and strengthen the public engagement as well. In particular, the 'Scientists and Media Face to Face' series of activities played a very good role. (Wang Dapeng and Zhongqi, 2017) At present, many local and related departments are still carrying out similar expert and media meetings.

3) Realistic Difficulties and Dilemmas for Science Media Center

Although the Science media center has to some extent bridged the tension between science and the media, the development of science media centers in various countries is uneven and faces various challenges. Although the Science media center is working hard to build a global network and support other countries to establish science media centers, there are currently not many science media centers in the network. In addition, although many countries have established or intend to establish science media centers, the exchange and cooperation between national media centers has not yet formed an effective mechanism.

(1) *A limited number of audiences*

In the era that science popularization and communication are mainly dominated by authorities or governments, namely the time of communicating news by mass media, most of the public has deferred to the one way communication for the authorities. However, with the advancement of media and communication, the public is increasingly divided into different groups, especially in new medium. As a cross border and mixture of both traditional and new media, science media center has not lost the dominant condition from the authorities, but also become less competitive to those mew medium counterparts, and therefore, it shows the public drain temporarily, in both entity and networking.

(2) *Discordance between scientists, journalists and the public*

The discordance regarding three elements in science communication constrains the impact and effect of science media center. Differentiate from those professional institutions concentrated on current breakthroughs and achievements in science, and many of science medium, which cover the science-related news for attracting attentions of the public, science media center is combined the two parts as a neutralizer, because all the time the relationship between science and media is the basis of both science communication research and core practical issue for it. For scientists, they prefer interpreting science from perspective professionally, but compared to the former, science journalists counterparts are always reporting science news in a more popular way to the public, accompanied with some false reports as well. Hence, scientists and science journalists are not compatible with a harmonious science rapport. On one hand, the former stands on the professional level to criticize the former lack of science evidence and bring the false reports and unscientific contents to the public, leading them not to concentrate on the process of scientific research, but to concern the results of science report excessively. On the other hand, science journalists always complain about the phenomenon that scientists are unwilling to be interviewed by them, and they scientists also do not understand the covering process of media either.

(3) *Subject to the literacy of the public*

Although the emergence of new media has conformed to the development trend of the times to a certain extent, due to the specialization of the development of modern science and technology, the diversion of scientific and technological information, and the increasingly differentiated functions of science communication, there are certain limitations in the scope of the new media science communication, especially in the communication and cognitive groups. On the one hand, it is inevitable that science communication will involve professional terms. Practitioners of Science communication tend to use public formulas, models, experimental data, and difficult academic terms to explain new scientific discoveries or prove scientific problems. For those non-professional audiences or those with low scientific literacy, such scientific knowledge is 'unattainable', which reduces the focus on scientific information. On the other hand, although the use and operation of new media is simple and convenient, not everyone is able to master the use of new media. In addition to the age-adjusted population, which is not good at learning and accepting new technologies due to the characteristics of the development stage, there are also many people in the youth who are not good at using new media.

(4) *Lack of well-trained professionals in science communication sphere*

To some extent, science literacy is not only the essential foundation for the public, but the compulsory course for both science journalists and scientists. For science journalist, they are lack of paying more attention to media coverage regardless of the truth of science and making false reports regarding science sometimes because of lacking science literacy, regardless of the result of communicating in this wrongdoing. For scientists, for one thing, they are always attaching importance to science itself and ignore the communication. For another thing, some scientists themselves are also lack of science training regarding science communication, not only in China, but also in the rest of world. In addition, valuable information almost cannot communicate to the public with the inadequate professional training for both.

(5) *Obstructed by superstition and pseudoscience*

The open nature of new media has lowered down the threshold for all types of subjects to enter the sphere of science communication field. This feature provides the public with a platform for participation and exposes many problems, strengthening the public engagement. Due to the weakening of the 'check-in' in the process of science communication in the new media environment, various exaggerated and false popular science products and services are ubiquitous and confusing, making it difficult for the public to distinguish between authenticity and authenticity. For example, a large number of unverified 'pseudo science' have been spread, many excellent original works have been plagiarized and reproduced without restriction, and intellectual property rights are not protected; superstitious activities like 'divination' and 'fortune telling' are also based on the communication characteristics of new media. And technical means, spread in disguise with the guise of science. It can be seen that the science communication in the new media background still lacks a standardized and effective market access mechanism, a sound supervision and management system of communication behavior, or a sound government support facilities, which is also a shackle to hinder a healthy development of new media regarding science communication.

(6) *The gatekeeper inefficiency and redundancy of information*

In the era of new media, as far as the subject of science communication is concerned, not only for the scientific elites and government agencies, but also for the 'primitive level' class as an important participant in the spread of science. The public's participation in science communication is increasing, and the cost of information communication is almost zero. They can all speak in the 'field'. Each scientific problem is completed by 'the rabble' through its own cognitive surplus, and its content accuracy and scientificity cannot be guaranteed. Although the above mentioned collaborative filtering, there are different degrees of time difference between the public speaking and the collaborative filtering. The speed and breadth of the new media are unmatched by collaborative filtering. While the new media promotes the spread and popularization of science, it also brings regulatory problems. It is impossible to filter and test the content of science communication by the new media itself. It is still necessary to set up 'gatekeepers' in terms of technology and systems. The participating entities conduct 'monitoring' and check the content.

Redundancy for information is another problem brought by new media. At all times, on new media platforms, such as online media, knowledge and information are proliferating, as is the content related to science. The public has faced with such a condition with considerable and complicated information ever before, and obviously the public is not fully aware of the information and cannot obtain scientific knowledge effectively. If you are searching for an explanation of a problem in some online question and answer, there may be a variety of even opposite statements. In traditional media, the public can only consume the content produced by 'gatekeepers' and information producers. In the new media, every public can change the original content according to their own experience and needs, so that the content is in quantity ever before in this era. With the increasing number of both the engaged public and news covered by them, science news are exposed to the audience increasingly and communicate in a far-reaching way. In this situation, the spread and popularization of science may lead to the proliferation of scientific information, from science to scientific generalization, and the public cannot effectively learn scientific knowledge, or cannot absorb important scientific knowledge.

4. Conclusion

The media is an important channel for the public to obtain scientific and technological information. Scientists are the ring at first of science communication. Therefore, how to maintain the relationship between the 'three main bodies', namely scientists, media journalists, and the public, is an important issue in scientific communication research and practice. Especially in an era when new media is becoming more diverse, it is even more important for scientists to send scientific and accurate information in a timely manner so that the media can better report science. The science media center, as a bridge between scientists and media reporters, should play a more important role. In order to strengthen our understanding of science communication in the context of new media, the focus of science communication research should be expanded to include online and digital media, while recognizing the traditional continuing agenda to set the nature of news sources. For related authorities, they should not only should strengthen the construction of new media systems in science communication, the supervision of the science communication of new media, and the protection and incentive mechanism of professionals, but also cultivate the consciousness of science communicators in a popular way. Based on the measures beyond, science media center will play a more important role during the new era of science communication.

Acknowledgments

I would like to express my gratitude to all those who helped me during the writing of this thesis.

My deepest gratitude goes first and foremost to Professor Zhang Xiaoying, my supervisor, for her constant encouragement and guidance. She has walked me through all the stages of the writing of this thesis. Without her consistent and illuminating instruction, this thesis could not have reached its present form.

Second, I would like to express my heartfelt gratitude to those scholars who have led me into the world of translation and have instructed and helped me a lot in the past two years.

Last my thanks would go to my beloved family for their loving considerations and great confidence in me all through these years. I also owe my sincere gratitude to my friends and my fellow classmates who gave me their help and time in listening to me and helping me work out my problems during the difficult course of the thesis.

References

David, L. Altheide, Shao Zhize (2003). *Ecology of Communication*. Hua Xia Publishing House.

DeLoach, Amelia (September 1996). What Does it Mean to be a Netizen? Retrieved 6 June 2015.

Dimopoulos, K., & Koulaidis, V. (2002). *The Socio-epistemic Constitution of Science and Technology in the Greek Press: An Analysis of Its Presentation. Public Understanding of Science*, 11, pp 225-241.

Ding, Bing. (2011). Media fusion: Concept, cause, advantages & disadvantages. Social Sciences in Nanjing.

Guo, Zhenzhi. (2000). Public opinion supervision, objectivity and journalism professionalism. *TV Research* (3), 70-72.

Wang Dapeng, Zhong Qi. (2017). Issues related to science communication from the development of scientific media center. *Democracy and Science*. (4), 41-43.

Wu Guosheng. (2016). Science communication in contemporary China. *Journal of Dialectics of Nature*, 38(2), 1-6.

Yang Qingyuan. (2011). From traditional science to science communication. *Chinese and Foreign Entrepreneurs* (14), 233-234.

Wei Feng. (2013). Discussion on the science communication in the new media era. *Modern Communication (Journal of Communication University of China)*, 35 (9), 159-160.

Zhang Suqiu, Gu Jiang. (2015). The characteristics, power and path of the fusion of traditional media and new media in the era of big data. *Modern Economic Research*, No.467 (11), 50-54.

Smart Society and Smart Education for 2035

Zhang Yanxin, Chen Rui, Zhang Li

National Academy of Innovation Strategy, CAST, Beijing, China

Abstract: Along With the development of artificial intelligence technology, the form of society has changed from an information society to a smart society. The platform of smart society covers four aspects: intelligent life, intelligent production, intelligent culture and intelligent governance. Among them, intelligent education, as an important part of intelligent life, has undergone significant changes. Artificial intelligence changes all aspects of education, including the concept of education, the ecological environment, the education method, the teacher-student relationship, and the family relationship. These profound changes can subvert the notion of traditional education. While all countries in the world have made forward-looking plans around the new generation of artificial intelligence technology, China has also upgraded these concepts to the national strategic level. In July 2017, the State Council of China promulgated the 'New Generation Artificial Intelligence Development Plan', which made specific deployment for the development of intelligent education. The plane mainly contained, the teaching method reform, the construction of intelligent learning and interactive learning, the new education system and the use of intelligent technology, which can accelerate the promotion of personnel training mode, and so on. In this context, on the basis of the strategic background of the intelligent society, the definition and characteristics of the intelligent society was teased. The form of future education from the perspective of intelligent society, as well as the change and influence of future education were further explored in this paper. Finally, an outlook of the future education was also provided.

1. Introduction

Xi Jinping pointed out in the report of the 19th CPC National Congress that it is necessary to 'improve the socialization, rule of law, intelligence and specialization of social governance.' 'Good at using Internet technology and information technology to carry out work.' Promote deep integration of the Internet, big data, artificial intelligence and the real economy. The report of the 19th CPC National Congress has explicitly mentioned the smart society and is juxtaposed with strategies such as Digital China. When artificial intelligence meets education, it has caused profound changes in education. The State Council issued the 'New Generation Artificial Intelligence Development Plan', which regards intelligent education as the key task of 'building a safe and convenient intelligent society', emphasizing the use of intelligent technology to accelerate the promotion of personnel training mode and teaching method reform, and to build intelligent learning and interactive learning. New education system, carry out intelligent campus construction, promote artificial intelligence in teaching, management, resource construction and other process applications; develop integrated teaching grounds, online learning education platform based on big data intelligence; develop intelligent education assistants, build intelligent and fast comprehensive education analysis system; establish a learner-centered education environment, provide accurate education services, and realize daily education and lifelong education customization.

In order to implement the goal of 'ensure quality education in inclusiveness and promote lifelong learning for all', the 38th UNESCO General Assembly adopted the Education 2030 Action Framework in November 2015. At the same time, UNESCO has also released an important education report, 'Rethinking Education: The Transformation of the Idea of "Global Common Interest" ', which proposes to redefine knowledge, learning and education. Education should be based on humanism and should be emphasized. It is necessary to focus on the transformation from personal interests to attention to common interests. [1]

2. Definition of a Smart Society

A smart society is a more advanced social form after the agricultural society, the industrial society, and the information society. The new round of digital, network and intelligent revolutionary technology and industrial revolution is the fundamental driving force for the development of smart society. The key component of intelligence becoming the core competitiveness of enterprises, industries, regions and even the country is the important feature of the arrival of smart society. [2]

Since the first industrial revolution in the 18th century, every fundamental and versatile technological advancement has greatly increased the productive capacity of human society, which has led to subversive changes in industrial structure and production organization, and the creation of its own economy. The basic and cultural characteristics, promoting the innovation and improvement of lifestyle and level, bring about the breakthrough and progress of the whole society, and promote the development of human society from agricultural civilization to industrial civilization, and then from industrial civilization into information civilization and future wisdom and civilization. Different social forms have different developmental characteristics.

1) The Intelligent Evolution of Iconic Technology

From the agricultural water conservancy technology that focuses on agricultural production in the agricultural society, to the large-scale manufacturing of the industrial society, to the information society, it is enough to support the bursting of information technology in the new era. Expand to the ubiquitous connection between intelligent people and humans, between people and objects, between objects and objects.

2) Critical Infrastructure Enters the Virtual Space from the Physical World

From the most basic windmills and mills in the agricultural society to the roads, railways, and hydropower stations necessary for the large-scale production of industrial society, to the information society to handle information interconnection and base stations and data centers, and then to build a human society, the physical world, and the information space. The ubiquitous information physics system of the meta-structure completely realizes the fusion interaction between virtual and reality. [3]

3) The Main Production Features Have a New Trend of Full Customization, Flexibility, Real-time Synergy and Multi-subject

From the one-house production of agricultural society to the automation and centralized mass production of industrial society using mechanical equipment, to the information society to build information networks and platforms to achieve global collaborative research and development, collaborative manufacturing of small batches, multiple batches production, and then develop into a fully customized, flexible production of smart social supply and demand in real-time synergy, and break the boundaries of the organization, extending the production subject from the factory and the enterprise to everyone.

4) Man and Nature Move Toward Symbiosis

From the natural power of the agricultural society and the small-scale, small-scale transformation of nature, to the industrial society, the large-scale development of nature by means of machines, even to the extremes of plundering nature and destroying nature, and then reflecting on the information society, including information and networks, biological, energy and other emerging scientific and technological means to protect nature, control nature, and then develop into a smart society, human clothing, food, housing, travel and other aspects of all-round and natural further close and coordination, build a green, environmentally friendly, energy-saving people and nature An ecological civilization system with harmonious coexistence and a virtuous circle.

5) The Social Governance Approach is Moving toward a Multi-dimensional Synergy of Transparency

From the agricultural society, it is controlled by a small part of the elite system with limited productivity and limited resources, and the consciousness of group governance is driven by the industrial development. However, it still corresponds to the 'pyramid' industrial production organization structure, and the feedback mechanism of policy mechanism is not flexible enough.[4] The one-way governance, and then the various governance methods and objectives of the information society are completely open on the network platform, transparently supervised by all parties, and then developed into a smart society. Furthermore, it has developed into multi-dimensional collaborative governance in which smart communities and the public are engaged in smart technologies and platforms to collect clusters and seek common management solutions.

6) The Cultural System Is Increasingly Encouraging the Search for the Fun and Meaning of Living Individuals

From the closed cultural system that favors moral evaluation under the strict suppression of agricultural social hierarchy, to the open cultural system biased by technology under the collision of multiple ideas, concepts and methodologies of industrial society, and the positive feedback effect of information society network Underneath, there are virtual symbols, idols, achievements, and the network culture system with traffic as the core. Then the individual consciousness of the smart society becomes stronger and stronger, and the cultural forms and achievements can be multi-presented by various intelligent technologies, and more attention is paid to the pursuit.

3. Characteristics of a Smart Society

It summarizes the eight development characteristics of the smart society. First, the iconic technology is evolving intelligently. Second, the important infrastructure enters the virtual space from the physical world.[5] Third, the main production features appear full customization, flexible, real-time collaboration, and multi-subjective. Fourth is the development of the three elements of productivity from natural endowment to bit endowment. Fifth is that social relations are increasingly breaking the space and time constraints. Sixth is the symbiosis between man and nature. Seventh is the way of social governance towards transparency Multi-dimensional synergy and progress. Eighth is that the cultural system is increasingly encouraging the search for the fun and meaning of life individuals.

4. The Impact of Smart Society on Future Education

The intelligent society represented by the Internet, big data and artificial intelligence provides a transformative idea for future education.

An intelligent society represented by the Internet, big data, and artificial intelligence can not only provide efficient tools for the implementation of future educational strategic goals. Its popularity and penetration will also change the ecological environment in which some of our major strategies are implemented, thus providing transformative ideas and challenges to the implementation of these strategies.

1) Intelligent Society Can Accelerate the Fair and Balanced Development of Education

Quality digital education resources are shared and balanced. Achieve cross-regional collaboration and mutual assistance to promote educational equity. Promote the professional development of teachers.

2) Intelligent Society Is Conducive to Cultivating All Kinds of Innovative Talents Needed for Economic and Social Development

Training innovative talents provides strong support. Improve the mode and mechanism of innovative talent training.

3) Intelligent Society Can Promote Educational Innovation and Help Solve the Key Problems of Educational Development

The new generation of information technology represented by the Internet, big data, and artificial intelligence will transform the basic process of the education business, reconstruct the entire education system, and promote educational theory and institutional innovation.

5. Intelligent Social Change, Future Education

Future education patterns under the intelligent society process.

Core: Focusing on promoting regional education equity, quality school development, and individualized student growth.

Elements: Information technology as an infrastructure and innovation.

Content: Systematic changes in the education mainstream business such as the education environment system, the education content system, and the education public service system.

Objective: To build a new, educational, flexible, open, life-long, and educational ecosystem that addresses the critical issues of education in the information technology revolution.[6]

1) Opportunities and Challenges for Future Education in Smart Society

Promote education equity and improve the

quality of education. Challenge traditional education and weaken the function of educating people.

2) Intelligent Social Transformation, Various Elements of Future Education

Including learning environment, course, teaching, learn, evaluation, management, school organization teacher development.

3) Intelligent Social Transformation, Future Education Field

Including higher education, basic education, preschool education, teacher education, distance and education.

4) Intelligent Society to Build a New Type of Future Education Ecology

The intelligent society has knocked on the closed door of education and accelerated the self-evolution of education. Everyone is a producer of education, and everyone is a consumer of education. This open, innovative, integrated, and diverse new educational ecology will inevitably be more adapted to social development.

5) Intelligent Social Change, Individual Development in Future Education

Including lifestyle change, change in learning style, negative impact study and cognitive change.

6. Status of Future Education Development in a Foreign Intelligent Society

1) American Smart Education Development Strategy

In December 2015, the United States enacted the 2016 National Education Technology Plan (NETP2016) with the theme 'Learning for the Future: Reimagining the Role of Technology in Education'. NETP2016 is a programmatic document for the development of educational technology in the United States. Its main content is learning, teaching, leadership, assessment, and infrastructure. The plan hopes to implement universal design principles in all educational institutions; enhance technology-enabled assessments, enable assessments to be embedded in digital learning activities and resources, provide real-time feedback on learner learning processes; and build sound infrastructure to meet current and future goals Unicom goal.

2) Development of EU Smart Education

For the first time in the EU's Seventh Framework Programme (FP7), the application of information technology in the field of education has been elevated to the height of research topics. In order to continue the FP7 program, the EU launched Horizon 2020 (FP8) in early 2014, proposing that the three major spiritual axes cover 12 key areas, including future networks, automation, robotics, etc. The international wisdom education development strategy promotes information technology in learning and cultural resources. In the role of action, to meet the needs of personalized learning, and ultimately improve the level of intelligence in EU education.

3) Korea's Smart Education Development Strategy

In October 2011, the Ministry of Education, Science and Technology (MEST) released the 'Promoting Smart Education Strategy Implementation Plan', which prioritizes the development of smart education as a strategic focus of national informationization, transforming smart education, transforming classrooms, and improving technical support learning. The effect is to cultivate innovative international talents that will adapt to the future information society.

4) Japan's Smart Education Development Strategy

The Japanese government has proposed e-Japan, u-Japan and i-Japan three national informatization development strategies, from infrastructure to network interconnection to information application, covering a wide range and covering a wide range, fully reflecting the Japanese government's educational information. The emphasis on the top-level design of the policy.

5) Singapore Smart Education Development Strategy

The Singapore government announced the iN2015 plan (Intelligent Nation 2015) in June 2006, led by the Infocomm Management Development Authority (IDA) of Singapore. It plans to invest tens of billions of dollars in an enterprise to build an active and advancing trend in the decade from 2006 to 2015. The information and communication ecosystem has transformed Singapore into a smart country and a global city supported by the information and communication industry, where information technology is everywhere, enhancing the country's economic competitiveness and innovation capabilities.

7. Prospects for future education in a smart society

1) Information Technology is More Widely Used in Education

The influence of the fourth industrial revolution has become increasingly prominent. Information technology represented by the Internet, cloud computing, big data, Internet of Things, artificial intelligence, etc. has become more and more widely used in the field of education. The teaching business has become intelligent, automated and digital. .

2) Innovation is the Theme of Future Education

The scope of innovative concepts is far from being limited to new teaching tools, revised goals, original teaching methods, and redefinition of traditional roles at all levels of education.

3) Teacher's Traditional Role will be Resolved

Education will become more personalized and the traditional role of teachers will change. The teacher-student relationship has undergone a fundamental transformation, and the role that teachers fully dominate is guiding the development of the classroom to teachers and students.

4) The School will Become a Classroom-based Global Communication Network

The future school is a learning environment with social interaction. The school has become a physical place where students arrange their own learning. At the same time, online learning and group learning are not mutually exclusive.

5) Professional Qualifications and Technical Skills will Challenge Traditional Degrees

Future education should be liberated from a method of training people in traditional disciplines, and more emphasis on things that are operational and applicable... The degree may be a necessary condition for future survival, but it is no doubt not a Sufficient conditions.

6) Mixed Mode Based on Public and Private Funding will Be the Best Choice for Education

In future education, state finance will no longer be the main source of education funding. In contrast, schools will be funded mainly by private companies and families. Employability and market power will have a direct impact on the school curriculum.

References

Preparing For The Future of Artificial Intelligence [EB/OL]. [2016-12-20]. http://www.whitehouse.gov/sites/default/files/whitehouse_files/microsites/ostp/NSTC/preparing_for_the_future_of_ai.pdf.

Grover, S., Pea, R. Computational thinking in K-12: a review of the state of the field[J]. Educational researcher, 2013, 42(1): 38-43.

Farag, H. E., El-Saadany, E. F., Seethapathy, R. A two ways communication-based distributed control for voltage regulation in smart distribution feeders[J]. IEEE transactions on smart grid, 2012, 3(1): 271-281.

NMC Horizon Project Wiki contributors. Affective Computing [EB/OL]. [2016-11-24]. http://china.nmc.org/index.php?title=AffectiveComputing& oldid=3573.

Sleeman, D., Brown, J. S. Intelligent Tutoring Systems: An Overview[M]. New York: Academic Press, 1982:1-11.

Stone, P., Brooks, R., Brynjolfsson, E., et al. Artificial Intelligence and Life in 2030[EB/OL]. [2016-11-25]. https://ai100.stanford.edu/2016-report.

An Empirical Research to Explore the Factors Affecting Adolescents' Scientific Creativity

Li Mei, Shi Lei

National Academy of Innovation Strategy, Beijing, China

Abstract: Creativity is a significant capability for scientific research. For adolescents, their potential scientific creativity is developing, while it already shows differences between individuals. This study explored the differences and the influential factors. Data was collected from 809 middle school students in Beijing. Based on whether they received a reward in adolescents science and technology innovation contest or not, data was collected from two groups: high creativity group (HCG) and the common group (CG). Results show that HCG shows more interest in science, which implies the internal motivation promotes more exploratory behaviors in science. Additionally, HCG favour reading books, viewing websites, reading papers on science, and participating in science club. In contrast, CG are inclined to like watching TV programs and listening to broadcast related to science. HCG students do better in observation, the ability of raising research questions in natural science, and academic achievements. Boys propose more research questions than girls in natural science, but they don't show significant high ability in observation and raising social scientific questions. Families of HCG make more communication in scientific topics, and could tolerate more mistakes of children. Individual and family environment factors both make contributions to the development of scientific creativity of adolescents.

1. Introduction

Creativity is an essential part in the progress of science and technology innovation. Creative people and their products made great contribution to human civilization. They often propose new questions, find new method and instruments to solve problems, expand more thinking space for people to explore, and make inventions.

Although there has been no consensus with regard to the definition of creativity these years, many researchers approximately agree that 'new and useful' is the most important part in this connotation.

Given the large amount research directions in this domain, it is necessary to organize them in some frameworks. The most widely known theoretic framework is 4P model, which is firstly published in 'An Analysis of Creativity' by Mel Rhodes in 1961(Rhodes,1961). Continuous modifications were made to this framework during these decades. On the whole, four dimensions were identified, which were Person, Process, Product and Press/Place. Creative person refers to personality, such as curiosity, flexibility, openness, self-confidence and risk tolerance. Creative process often looks to how the insights and new ideas are found, which mainly refers to creative thinking. Creative products means performances, or some tangible form converted from idea. Lastly, creative press is originally presented to mean the environment factors. The term press is often called place in recent years, which also refer to the potential environment influence effecting on individuals. This study would design the measure instruments basis on the 4P framework.

Adolescents are being placed in the period of mental and physical growth and development. Creativity is also developed and changed during these years. China has realized the significance in cultivation of adolescents' creativity, and made effort to provide resources and platforms for them. Several significant contests were held every year, such as China Adolescents Science and Technology Innovation Contest and the Awarding Program for Future Scientists. It can be demonstrated in literature that there is a tendency in the course of change, both in east and west cultures. It showed an roughly upward trend, while the rising is fluctuated. For instance, originality appears a downtrend around the age of 11 (Claxton et al., 2005). Imagination was found to be gradually rising during junior middle school ages, and sunk to a low ebb after a peak at the age of 15 or 16 (Sun Peng et al., 2014).

For adolescents, creativity is developing in a process interacting with environment, which mainly includes family fostering, school education and social culture atmosphere. Given the earliest and most important part for individuals, researchers showed lasting interest in family environment. On the basis of a qualitative study, we explored the attitude towards the scientific exploring behaviors, the parenting style, and the communication related to science, to find the relation between family environment factors and scientific creativity for teenagers.

In this study, we examined the creative personality and the ability of raising creative questions of adolescents in China, and factors related to the family environment, for the purpose to describe creativity and explore the specific environmental factors affecting it.

2. Method

1) Participants and Procedure

To examine the differences between creative individuals and the others, this study was conducted within two groups, high creativity group (HCG) and the common group (CG).

The questionnaire were designed on-line and hyperlinks were sent to the teachers in middle schools, who organized students to answer it later. Data were collected by the website platform, and could be exported to the researcher's computers.

The criterion to screen creative adolescents is that whether or not the student was awarded a prize in Beijing Adolescents Science and Technology Innovation Contest and the Awarding Program for Future Scientists in China. Qualified students from 22 middle schools in Beijing were invited to participate the program. Finally 122 valid data were collected as HCG.

Data of CG were collected from 13 middle schools in Beijing, including 7 key schools and 6 common schools. Seven urban schools and six common schools were included. Finally 687 valid data were collected as CG.

Totally 809 data were collected and analyzed in SPSS 23.0. Descriptive statistics were performed to show the demographic characteristics of the sample, and *t*-test were conducted for each creativity score for testing the differences of individual and family environmental factors.

Males in HCG accounted to 59.84%, while females accounted to 40.16%. Males in CG accounted for 51.53% and females 48.47%. Therefore gender factor is proportioned in two groups. The average age of HCG is 16.29 ± 1.21, and the average age in CG is 15.42±1.55.

Table 1 Numbers of male and female in HCG and CG.

	HCG	CG	Total
Male	73	354	427
Female	49	333	382
Total	122	687	809

2) Instrument

Creative personality Chinese Adolescents' Creative Personality Inventory (CACPI). It includes 9 dimensions and 72 items assessed on a 5-point Likert scale, to describe creative personality characteristics for adolescents. Items were drawn from previous established questionnaires. Self-confidence, curiosity, and risk-taking items were drawn from the Creativity Assessment Packet; norm-doubt and independence items from the 16PF Questionnaire; internal motivation items from the Need for Cognition scale; openness items from the NEO PI-R Personality Inventory; persistence items from the Creative Tendency scale; and self-acceptance items from the California Psychological Inventory.

Parenting style Parental Bonding Instrument (PBI). Parker et al. (1979) developed the original PBI which was revised by other researchers. This questionnaire consists of two parts, one for mother and the other for father, with 23 similar items for each. It was designed to assess parenting styles in terms of care, overprotection and autonomy. Cronbach's alpha coefficient was calculated for each dimension and the whole scale, with a result as follows. Maternal behaviors: care (α=0.872), overprotection (α=0.870), and autonomous behaviors (α=0.857). Paternal behaviors: care (α=0.876), overprotection (α=0.897), and autonomous behaviors (α=0.881).

Items of the ability of raising creative questions were designed referring to Scientific Creativity Test (Hu, 2002). The design of some variables describing the family creative atmosphere were based on the results of the qualitative study previously.

3. Results

1) Creativity in HCG and CG

Comparing scores of creative personality

between HCG and CG reveals several significant differences (see Table 2). HCG showed significantly higher scores than CG in total score (p<0.001), curiosity (p<0.001), internal motivation (p<0.001), openness (p<0.001), self-acceptance (p<0.05), risk-taking (p<0.01), and persistence (p<0.05).

Table 2 The differences of creative personality between HCG and CG.

	HCG (n=122)		CG (n=687)		t
	M	SD	M	SD	
self-confidence	18.98	5.124	21.78	5.774	−5.020***
curiosity	33.43	4.008	31.15	5.545	5.409***
internal motivation	50.81	7.502	45.61	9.565	6.752***
norm-doubt	14.47	3.189	15.34	3.684	−2.464
openness	40.81	4.198	39.15	4.868	3.935***
self-acceptance	30.64	4.047	29.73	3.941	2.341*
independence	17.23	2.923	17.21	3.172	0.069
risk-taking	31.89	3.912	30.70	5.105	2.945**
persistence	15.17	4.067	14.21	4.295	2.306*
Total	253.43	15.913	244.87	20.121	5.239***

* p<0.05; ** p<0.01; *** p<0.001

The score of self-confidence showed opposite result that HCG scored significantly lower than CG.

HCG students scored significantly higher than CG students in the ability of raising creative questions in natural science (p<0.001), but the same result was not found in social science.

Boys propose more research questions than girls in natural science(p<0.001), but they don't show significant high ability in observation and raising social scientific questions.

Table 3 The differences of the ability of raising creative questions between HCG and CG.

	HCG (n=122)		CG (n=687)		t
	M	SD	M	SD	
raising creative questions in natural science	2.55	0.814	2.14	0.858	4.843***
raising creative questions in social science	2.16	0.817	2.06	0.916	1.128

*** p<0.001

2) Individual and Family Environmental Differences Between HCG and CG

HCG students showed significant more interest than CG students (p<0.001, see Table 4). The former also significantly performed better than the latter in school academic achievement (p<0.001). In a class including about 30~40 students, HCG ranked averagely the top 16.85%, while CG ranked averagely 23.92%.

Table 4 Scientific interest between HCG and CG.

	HCG (n=122)		CG (n=687)		t
	M	SD	M	SD	
feel happy in learning science	4.27	1.076	2.95	1.637	11.419***
be willing to solve scientific problems	4.43	0.908	3.85	1.160	6.201***
hope to explore unknown world	4.57	0.679	4.17	0.993	5.645***

* p<0.05; ** p<0.01; *** p<0.001

HCG students favoured reading books, viewing websites, reading papers on science, and participating in science club. In contrast, CG were inclined to like watching TV programs and listening to broadcast related to science.

As for family environment, there are a lot of interesting findings. The education level of HCG parents was found to be significantly higher than CG.

HCG parents were more willing to invest in their interests (p<0.001), while there is no significant difference in annual income between HCG and CG (p>0.05).

The parenting style also showed significant differences. In mother questionnaires, HCG scored higher than CG in care (p<0.001) and autonomous behaviors (p<0.01), and lower in overprotection (p<0.001). The same results were found in paternal behaviors in autonomous behaviors (p<0.01) and overprotection (p<0.001). However, no significant difference was found in father's care dimension.

Table 5 Parenting style differences between HCG and CG.

		HCG (n=122)		CG (n=687)		t
		M	SD	M	SD	
Mother	care	27.15	5.523	24.38	6.617	4.935***
	overprotection	4.32	3.331	6.08	4.828	−4.982***
	autonomy	14.47	2.988	13.64	3.825	2.685**
Father	care	24.48	7.387	23.17	6.802	1.946
	overprotection	3.15	3.608	5.05	4.904	−5.064***
	autonomy	14.57	3.201	13.79	4.022	2.402**

* p<0.05; ** p<0.01; *** p<0.001

The communication topics were found to be more likely related to science in HCG students' families (p<0.001). Moreover, when teenagers made mistakes in exploring the world, parents of HCG were found to be more tolerable (p<0.001).

4. Discussion

This study examined the differences of creativity and family environment through comparing data between HCG and CG. Results show that HCG students did better than CG students in creative personality characteristics and the ability to raise creative questions in natural science. And, HCG students showed more interest in science, which implies the internal motivation promotes more exploratory behaviors in science.

The score of self-confidence showed opposite result to the other dimensions of creative personality may indicated that highly creative teenagers adopted a more cautious attitude in the process exploring the world and making conclusions.

Parental care refers to parents' affection and emotional warmth, empathy, and closeness, parental overprotection refers to parents' over intrusion and control, while autonomous behaviors mean allowing teenagers to make their own decisions (Ngai, 2015). Previous researches involving PBI have demonstrated the association between parenting behaviors and adolescents' development. It has been found that parenting behaviors had positive associations with adolescents' greater life satisfaction, adaptive functioning (Park and Peterson, 2006), academic success, and overcoming adversity (Scales et al., 2000). This study added to these findings that creativity was also related to parental behaviors. HCG teenagers live in a more democratic family atmosphere. Their discourse power and feelings are better respected.

And their parents often communicates scientific topics at home, which potentially influences children to sense the value and joy about exploring in scientific domains. Mistakes seems to be not that terrible to HCG families, in which mistakes were more prone to be tolerable and forgiven.

It is not surprisingly to find that HCG parents had a higher education level. Maybe the attitude they treated their children relates to education background to some extent. But it is interesting that the results indicate socioeconomic status of a family is not a key factor related to creativity. However, HCG parents were found to be more willing to invest in children's interests. Thus, it can be inferred that it is not the real income of a family but the valid investment which used for teenagers' interest decided the discrimination of their development of creativity.

In a word, individual and family environment factors both make contributions to the development of scientific creativity of adolescents.

References

Claxton, A. F., Pannells, T. C., Rhoads, P. A. (2005). Developmental trends in the creativity of school-age children. *Creativity Research Journal*, 17(4), 327-335.

Ngai, S. S. Y. (2015). Parental bonding and character strengths among Chinese adolescents in Hong Kong. *International Journal of Adolescence and Youth*, 20, 317-333.

Park, N., Peterson, C. (2006). Moral competence and character strengths among adolescents: The development and validation of the values in action inventory of strengths for youth. *Journal of Adolescence*, 29, 891-909.

Parker, G., Tupling, H., Brown, L.B. (1979). A parental bonding instrument. *Br. J. Med. Psychol.* 52, 1-10.

Rhodes, M. (1961). An analysis of creativity. *Phi Delta Kappan* 42: 305-310.

Scales, P. C., Benson, P. L., Leffert, N., et al. (2000). Contribution of developmental assets to the prediction of thriving among adolescents. *Applied Developmental Science,* 4, 27-46.

Weiping Hu., Philip A. (2002). A scientific creativity test for secondary school students, *International Journal of Science Education,* 24(4): 389-403.

孙鹏, 邹泓, 杜瑶琳 (2014). 青少年创造性思维的特点及其对日常创造性行为的影响: 人格的中介作用. 心理发展与教育, 30(4), 355-362.

Volunteer Participation to STEM Education: Guidelines for a Co-creation Platform

Ahmet Süerdem[1], Soydan Soylu[2]

[1] Istanbul Bilgi University, Turkey
[2] Middlesex University, London, UK

Abstract: One of the major goals of basic education is increasing the scientific literacy of the citizens besides preparing students for science-based vocations. Making conscious decisions about science and technology issues is essential for democratic participation. Scientific literacy in Turkey, as measured by PISA scores, is among the lowest within the OECD countries which is due to a great extend curricula designed for passive learning. Active learning encourages students to engage, construct their own knowledge, and put learning into practice. This can be achieved by creating an ecology that encourages active participation of various stakeholders. Recently, organisations have started to interact extensively with their external environment to explore and exploit knowledge that is needed to generate innovative ideas. This perspective has great potential for harnessing the power of co-creative teaching. Open innovation has great potential for providing an active learning ecology as it provides tools and methods for managing knowledge flows within a co-creative system. The aim of this study is to share our recent field experiences about the existing practice in informal learning at TEGV (Turkish Education Voluntaries Foundation) and to suggest some future directions for creating an open social innovation learning environment for STEM teaching.

Keywords: Knowledge Co-creation; Knowledge Management; Scientific Literacy; Action Research; Socio-technical Systems; Informal Science Education

1. Introduction

We are witnessing a surge in open online learning resources such as Coursera with a great potential in engaging non-expert individuals to learning about scientific subjects. However, these sites are usually limited to the already competent parts of the society and hence carry the risks of opening a gap between these groups and those who are not endowed with sufficient skills to understand and voice their concerns about scientific controversies (Irwin, 2002). This gap is usually filled by informal virtual organizations such as newsgroups, Wikipedia, chat rooms, social media communities and online fora as a source of everyday science knowledge acquisition. These sites are widely used by lay people because of the easier readability of their language compared to expert language (Brossard, 2013).

These sites provide an important peer to peer alternative to the authority of professional experts. They facilitate non-expert individuals to co-create knowledge by sharing information and experiences, asking questions, seeking emotional support, self-helping or expressing their opinions in public. However, besides the opportunities for educating the citizens, these sites also create risks and challenges for users since the knowledge produced by often anonymous contributors are not governed. Because of their viral capacities, these sites facilitate the epidemic dissemination of pseudo-scientific claims, post-truth, manufactured facts and conspiracy theories.

Science shapes the world which we live in and it is important that people from all backgrounds develop the skills necessary to understand the consequences of scientific developments on their life. This understanding is a prerequisite for the democratic expression of ethical, cultural and societal concerns about the future effects and risks involved in scientific research and innovation. Making people more informed about science and forming them as capable individuals to participate in decisions concerning scientific issues run alongside for the functioning of a democratic society. People from socio-economically disadvantaged, minority ethnic groups, less developed regions, elder ages and

Ahmet Süerdem, Ahmet.suerdem@bilgi.edu.tr;
Soydan Soylu, S.soylu@mdx.ac.uk.

women are generally underrepresented in participation to citizenship activities and particularly in engagement to science because of their low scientific literacy levels. This makes them particularly prone to these threats since their access to other means of knowledge acquisition is somehow limited (Jolley et al., 2014). Despite its importance for conscious consumption and citizenship decision making, how the virtual co-creation volunteer organizations should be governed and what impacts this quality have on scientific citizenship formation on is a less studied area of research. Fluidity of the boundaries in these organizations makes the creation, sharing and management of knowledge a challenging task.

In this paper, following a review of the literature on scientific literacy, science popularisation in contemporary media, we discuss the governance of the complexities arising from the tension between know-what and know-how through developing agreed guidelines for the design of an online platform engaging the education volunteers to share their innovative ideas and collaborate with education experts.

We will first review scientific literacy and civic participation in today's societies, with an emphasis on the present efforts for linking in and out of class science. We will then review science popularisation in contemporary media. We will than discuss the new media such as internet, cloud computing, smartphones and their role in societies' engagement with science. Following that, we will discuss our action research findings regarding the issues TEGV volunteers encounter during knowledge sharing. These findings would guide us to understand potential communication problems when these stakeholders interact within a common platform.

2. Scientific Literacy and Civic Participation

Scientific literacy increasingly becomes an essential element of civic participation as it provides the knowledge and skills to navigate effectively in a society where science and technology (S&T) penetrate our lives (Jenkins, 1997). The term refers to the capability both for understanding and evaluating the scientific content in publicly available information sources and relating the scientific developments to one's everyday life issues (NAS, 2016). In contemporary societies, consumer and citizenship decisions can be highly deficient when scientific literacy levels are low (Bingle et al., 1994). Dialogical and participatory efforts for citizen and stakeholder involvement in science and technology will not accomplish their ends if the citizens' interest and information levels are not sufficient to judge the consequences of scientific developments (Jasanoff, 2003). Endowment of citizens with scientific literacy would provide a necessary setup for improving the innovation environment, the quality of policy decisions, consumer awareness and interest to science careers.

Despite its significance for informed citizenship global efforts to improve scientific literacy mainly through the formal education system has not achieved what it was supposed to do for certain sectors of society (Miller, 2004). Much of these deficiencies are due to the increasing gap between the textbook and practical science knowledge (Bauer et al., 1994). Formal education falls short to meet the cognitive burden instigated by relating increasingly complex scientific issues to real-world applications. STEM (Science Technology, Engineering, Math) educational activities aim to close this gap by integrating separate disciplines into a cohesive learning paradigm based on real-world applications (Honey, 2014). These activities usually aim to address the decreasing interest in science subjects leading to a shortage of employees with the necessary qualifications which are highly in need by the industry. However, scientific literacy cannot be limited only to enhance career qualifications but suggests a capacity for relating the scientific developments to citizenship issues. In this respect, EU adopts a more social approach to STEM, setting the targets for science education not only for making science careers more attractive for younger generations but also for broader societal goals such as increasing the means for personal development, boost social cohesion and safeguarding social achievements (European Centre for the Development of Vocational Training, 2017).

Present Efforts for Linking in and out of Class Science

Providing a direct link of personal experience between STEM in the classroom and the science in the outside world would be essential to develop more effective processes of public engagement and learning that could result in significant societal outcomes (Dillon, 2016;

Dillon et al., 2016). From this prospect, STEM *Enrichment and Enhancement (E&E)* activities aim to bring outside science into the classroom by enabling students to engage in real-world practices such as designing cars or computer modelling natural disasters or bringing a real scientist, engineer or mathematician into the classroom (http://www.score-education.org/policy-themes/curriculum/enhancement-and-enrichment). These activities do not only aim to bring science into the classroom but also encourage outside the classroom activities such as field trips, service learning and community engagement in science, living labs, science centres, science museums, experiential learning and place-based learning.

3. Science Popularisation and Media: Science Learning as a Lifelong Activity

Although E&E and online inquiry-based activities may help individuals with those skills associated with independent information seeking, synthesis and evaluation, their scope is limited to students, teachers and academics hence not covering a large portion of the society. Quest for scientific literacy should be a lifelong pursuit in a world of quick technological change. In that respect, science popularisation can be an important source of knowledge and information for a great majority of population since it can help to re-contextualise scientific discourse and transform expert knowledge into lay knowledge (Lewenstein, 1992). Mass media can play an essential role in science popularisation. Reformulating scientific discourse produced by and for experts to everyday language can facilitate the lay audience to integrate expert knowledge with their life worlds. On the other hand, science popularisation should not be considered as a one-sided communication process from experts to public. Culturally dominant view perceives science popularisation as a 'low status educational task of appropriate simplification' or as a distortion of scientific matters of fact by outsiders (Hilgartner, 1990). However, science in popular culture is also the manifestation of 'vernacular science knowledge' (Wagner, 2007): a cultural coding that symbolically reframes scientific concepts in a way to mirror hopes and fears residing in the public imaginary. These symbolic elements render hard scientific facts digestible for the lay people by translating them into metaphoric and iconic representations and help them to make sense of the scientific discourse. By creating a new language, they anchor scientific ideas in acceptable and legitimate belief systems.

In this respect, mass media is an important element of science popularisation as it plays a significant role during the formation of people's beliefs about specific events and cultivation of their concepts and worldviews through socialisation and enculturation (Gerbner, 1969). Mass media are a major source of information about science and the impact of science on society for most adults (Rennie et al, 2003). They also have an important role for informing citizens about their rights and responsibilities and empower them to make their voices heard. Democratic media is essential for encouraging all social actors to engage in decision making processes about issues concerning their lives. Science-related news stories have great potential as a resource for promoting scientific literacy and equipping citizens to critically engage with science since they form a bridge between expert knowledge and the 'real world'. Science stories as presented by the media can encourage learning by entertainment as they can fire the interest, motivation and the imagination of the audience regardless of their expertise in the subject. Individuals may change their dietary habits; consumption practices and communities may form policy advocacy groups on environmental issues based on the information acquired from the media.

However, despite its potential to bridge expert and lay, textbook and everyday knowledge science popularisation by means of traditional mass media have some shortcomings. Those who have the power to control the media outlets may filter, shape and manipulate news and information before it is disseminated to the public. Media can be used to disseminate questionable or potentially fraudulent or dangerous claims and practices with the pretext of educating the public. For a democratic functioning of the system each member of the society needs to have a certain level of understanding and active participation in the creation of knowledge and exchange of information. Regarding socio-scientific issues, when the scientific literacy is low for some disadvantaged sectors of the society, media stop being a source of information for most of the masses but splinter the society into an elite actively participating in the content creation and a public only passively consuming it (Fraser, 2009). Without proper skills, competencies, and knowledge to access, analyse and assess the media content, it would be impossible to put the information in a meaningful context.

New Media and Engagement to Science

Increasing use and democratization of new media such as internet, cloud computing, smartphones and social media brought forward some potential to manage this problem by empowering lay people to participate both to the consumption and production of the media content. In an ideal situation, new media would allow people to participate in the media production and generate and share the content that they believe to contribute to the solution of controversies. New forms of information channels and participation such as organising online campaigns carry the potentials to help the actors with fewer resources, such as small NGOs or individual citizens, to make informed decisions and mobilise around policy advocacy groups. In this respect, informal new media resources carry the potential of playing an important role for science popularisation and engagement to science by providing sites of knowledge co-creation. For example, individuals with common interests and concerns increasingly assemble in the virtual 'peer-to-peer communities' such as newsgroups, mailing lists, chat rooms, social media communities and online forums to share information and experiences, ask questions, for emotional support, self-help or to get their opinion public instead of relying on the authority of professional experts (Eysenbach, 2003).

However, theories of post-truth challenge the role of new media as a means for reviving the engagement of citizens to civic activities. As more people turn to new media as a primary information source, their knowledge transportation capacity does not only promote civic participation through fluid information flows but also create a propensity to epidemic dissemination of conspiracy theories. Half-baked facts resting on anecdotal evidence can overshadow scientific evidence because of the viral power of the new media. For example, while scientists were unable to find any causal link between the symptoms of autism and the vaccine preservative thimerosal, a considerable amount of people believe that vaccines are responsible for autism due to the viral effect of new media (Shermer, 2009). Considering the enormous amount of information available in the online environment, it would be a huge task for a lay citizen to critically assess the relevancy and quality of information retrieved from these sources. Gerhards and Schäfer (2010) have found only minimal evidence in favour of the new media for creating a democratically governed environment compared to the traditional media. According to their study, both media are equally ineffective in terms of encouraging the engagement of larger masses to the socio-scientific issues. Gate-keeping function of elites is replaced by the technical characteristics of websites which exclude less prominent voices by the search engines' algorithms.

4. Creating Guidelines for the Democratic Governance of Knowledge Co-creation Platforms

Governance of the knowledge co-created in virtual sites and citizen science platforms is a complex problem with many controversies. These activities on the one hand emphasize the teaching and learning aspects of science education and on the other hand have a formative impact on citizenship through helping immersive, sustainable, and empowering continual civic engagement. They facilitate the formation of citizenship know-how through voluntary participation to experiential procedures of performance rather than just focusing on teaching of formal know-what.

In this paper we will address the governance of the complexities arising from the tension between know-what and know-how through developing agreed guidelines for the design of an online platform engaging the education volunteers to share their innovative ideas and collaborate with education experts.

5. Method

1) A Meta-design Perspective

The features of present knowledge co-creation systems are designed by experts according to the needs of the institutional clients. Such systems develop features according to the aggregate needs of the average user without much considering the social and cultural diversities. Different factors motivate different groups to engage in citizen science and understanding why some groups continue their engagement and others do not is important for designing a sustainable and culturally acceptable system.

Adopting a participatory action research perspective actively engaging with the stakeholders throughout all the stages of the project lifecycle may help us to understand the expectation of different groups from the system. Encouraging different stakeholders' contribution is essential for a self-generating and self-maintaining system design. Contrary to the top-down methods of system engineering, this system needs to assemble the social and technical aspects of the system from a meta-design perspective. Information systems (IS) are socio-technical systems including both technical and social elements. While technical elements can be engineered for delivering predictable and consistent interactions between users and system functionalities, social elements are evolving complex systems containing many parameters endogenous in their interactions. This complexity can only be managed when the users continuously participate to the design of the system objectives, techniques, and processes. To this end, meta-design perspective does not offer fixed solutions but design guidelines and a framework. It involves all stakeholders to contribute both to the development of technical features and to the evolution of the social parameters such as organisational change, knowledge construction, and continuous learning (Fischer et al., 2011).

In this respect, action research links basic and applied research to contribute both to the practical concerns of people in an immediate problematic situation and to the goals of research process by involving the stakeholders within a mutually acceptable ethical framework (Rapoport, 1970). Open social innovation literature indicates the importance of conducting action research in facilitating knowledge co-creation and in enhancing its outcomes (cf. Brandsen et al., 2006). Action research engage the agents of social change to the knowledge production process with the researcher, rather than being passive subjects of the research (Brandsen et al., 2006). The authors point to the potentials of bridging the critical theory with action research for the transformative social research with an impact. This approach emerged as an interdisciplinary and problem- solving based way of doing research organized around concrete project groups such as work places. Action research supports collective action and change (social innovation) while at the same time producing new knowledge. Action research methodology helped us to understand the awareness, attitudes, values and practices of the various stakeholders concerning the knowledge co-creation platform. For delineating the guidelines for designing the platform from an action research perspective, we accomplished the following activities:

- Interviews with different managers and education experts for revealing what roles they are expecting to be accomplished by the volunteers and interviews with the volunteers to understand their capacity, motivation and awareness for participating in a knowledge co-creation platform. These can help us to identify the problems generated because of the lack of interaction between different stakeholders.
- Focus groups to reveal the specific capacity building actions needed to develop a common language between experts and top management and volunteers. These can help us to understand potential communication problems when these stakeholders interact within a common platform.

2) Research Design and Setting

We have conducted our research in TEGV. TEGV an education-oriented NGO in Turkey which was founded in 1995 by a board of trustees who believe that the main reason for some of the major issues that Turkey is facing is lack of education, and that it wouldn't be possible to reach the level of contemporary civilization without improving this. TEGV is focused on providing 'out-of-school education' support to elementary level students. Over the years, TEGV became the foremost non-governmental organization in its field of activity. In 2009, TEGV was announced by the National Assembly as one of the foundations allowed to organize charity collection without prior permission.

3) Data Collection

The interviews were conducted at three different locations in Istanbul and in Samsun with 15 volunteers from various TEGV units from İstanbul, Ankara, Sivas, Samsun, Rize, Çaycuma, and Erzincan. The interviews were all audio recorded and then transcribed verbatim. Data analysis was conducted by the help of NVivo software.

4) Data Analysis

We followed 'thematic networks' framework of Attride-Stirling (2001) to analyse the qualitative data in a systematic manner. NVivo 9, a computer-assisted qualitative data analysis package, was used for the analysis. Preliminary codes were applied to the textual data to dissect them into meaningful and manageable segments (Attride-Stirling, 2001) to facilitate comprehension of the emerging findings. These codes were collated into 'basic themes' and then were revised to be non-repetitive. Following this step, basic themes were collated under 'organizing themes' that reflected a broader level of meaning. In the final step, organizing themes were assembled under 'global themes' on the basis of similarities. Global themes are the core metaphors that encapsulate the main points in the text (Attride-Stirling, 2001).

6. Findings

Thematic analysis of the interview and focus group results revealed two global themes regarding the issues regarding the governance of knowledge co-creation in TEGV, namely 'existing feedback mechanisms' and 'program delivery'.

1) Existing Feedback Mechanisms

Our analysis identified two main streams of feedback running through formal and informal channels utilized by TEGV volunteers. These are formal feedback mechanisms such as surveys, meetings, observations and Skype meetings; and informal feedback mechanisms such as Facebook, Whatsapp, conversations, email/phone, and breakfast/lunch meetings.

Most of our interviewees mentioned that they regularly interact through at least a couple of mechanisms from each stream. Most referenced tools for providing feedback are Surveys (23) and Meetings (16) within the formal stream and Whatsapp (16), Facebook (11), and Conversations (14) within the informal stream. The data regarding existing feedback mechanisms is analyzed by exploring the issues and challenges identified by the interviewees.

Issues and challenges in existing feedback mechanisms

Formal Feedback Mechanisms: One of the main challenges in feeding back through surveys is identified as the long lag time between the start of the education program delivery and the survey. Education programs are delivered by the volunteers in a 10-week period and the survey is then conducted to get their feedback on the program delivery. This creates a challenge for remembering any key issues to report back to the organization about problems they encountered in the past 10 weeks.

There are also meetings organized every few years where volunteers from various TEGV units get together to provide feedback to TEGV management about whatever they deem necessary. These meetings are considered very valuable by the informants, yet not sufficient as they are very few and far between. Additionally, there are smaller meetings organized under a Volunteer Governance Initiative introduced last year. Individual TEGV units organize these meetings once or twice a year across specific education programs, such as Mathematics or Science. Although, again these meetings are considered to be very valuable as they can share issues/problems arising around specific programs in specific TEGV units, there can be attendance problems due to lack of time or because the units are very small in size and there is not enough people to attend. Furthermore, only a few participants mentioned other formal feedback mechanisms such as Skype meetings and Observation sessions. Skype meetings are considered very valuable as they provide instant and 'face-to-face' feedback and opportunities to have discussions with the program managers on a national level. In addition to very few informants being aware of these formal feedback mechanisms that are available, these also share the major issue mentioned by almost every informant about the formal feedback mechanisms overall. Although these mechanisms are considered very valuable, the volunteers are not exactly sure why as these are almost exclusively one-way knowledge transfers from the volunteers to TEGV. TEGV collects data/information from volunteers very efficiently, and volunteers do not consider these a burden. However, there is no top-down feedback loop from the organisation in response to the bottom-up data collection. Rather than an across stakeholder knowledge transfer opportunity, these feedback mechanisms thus remain only as linear and intermittent data collection instances by the organisation.

Informal Feedback Mechanisms: Whatsapp and Facebook are mentioned as the most widely used platforms for informal feedback between volunteers. Almost all informants claimed to be actively sharing knowledge on one or both these platforms. These platforms are generally used to setup small groups of volunteers from specific TEGV units and/or specific education programs. Another widely cited informal feedback mechanism is informal chats and conversations among volunteers and also between volunteers and the unit managers. The unit managers here replace TEGV as the authority and through these informal feedback mechanisms volunteers feel that the issues/problems they raise with managers are being delivered to the right people. Whether formal or informal, feedback mechanisms that are employed at TEGV are not considered a burden at all and informants are willing to do more. The key driver for being involved in knowledge transfer emanates from volunteerism. Volunteers are doing what they are doing because they believe in TEGV's cause and feel a certain belonging to the organisation. Yet, one thing that is disillusioning them is the lack of reciprocity from the organisation. They are willing to share their knowledge and ideas, but only if they feel that they are being listened to and taken seriously. This is a crucial point, as TEGV needs to keep these feedback mechanisms active in order to understand the issues and challenges that volunteers are facing in their local context. There are various issues specific to local realities impacting the delivery of the education programs, which TEGV develops centrally and expects to be delivered as efficiently and effectively possible everywhere in Turkey. However, Turkey is very wide and varied in terms of local realities, necessities, and resources and therefore local delivery of programs do not always go as planned and hoped for by TEGV. Next, we will talk about these local issues and challenges as identified by the interviewees.

2) Program Delivery

Due to various local issues, volunteers at TEGV encounter problems whilst delivering the programs to children. Most cited problems are identified in the data set as child disinterest (16), content suitability (14), and lack of resources (10) followed by others. One of the major issues identified by almost all informants is child's interest, or lack thereof. It is sometimes very difficult for volunteers to put the children into learning/education mode after they had school all day, when they come to TEGV's cite. Not only that, some children require special attention due to their specific qualities such as their level of keenness to learn, or interest in the subject, or even their demeanour and whether they are hyperactive or sluggish.

Another widely mentioned problem stems from the content not being suitable either for the children or the local context. The third most cited problem encountered by the volunteers is lack of resources. These problems reflect the wide and varied local realities and their sometimes stark differences from the content prepared far and away at TEGV central headquarters by specialist consultants. Although there are pilot programs running before the widespread delivery nationwide, these seem to be not enough to finetune for every local issue. Therefore, volunteers have to find/generate solutions to these problems there and then. These solutions are innovative and creative instances of tacit knowledge generation addressing the local problems, which the centrally created and top-down delivered explicit knowledge written down in program handbooks cannot handle or even foresee.

Ways in which Volunteers deal with these issues/ challenges locally

In terms of resource problems, there is not much volunteers can do other than finding outside sources for funding or materials. Some of the volunteers seem to be very resourceful in doing just that, yet some constantly feed this back to TEGV without much success. For all other problems that were cited by the informants, we have identified two courses of creative reactions, one through performative innovations and another through content creation.

Yet most of the informants were also hesitant in playing with content provided by TEGEV. First, they believe that experts have created the content and they have trust in their knowledge and expertise. Second, they feel there exist a hierarchical structure where they as volunteers are at the bottom of this structure, obliged to obey whatever is pushed down on them from the organisation.

Although, this is a common barrier for generating new content by the volunteers, they still do so because they feel they have to as the situation requires this from time to time. So, somewhat reluctantly they still generate their own content and find and apply innovative solutions whilst delivering their programs. But, what do they think about the value of this tacit knowledge and sharing it with others.

7. Concluding Remarks

In this paper, following the review of scientific literacy and civic participation, science popularisation in contemporary media, the role of new media such as internet, cloud computing, smartphones in societies' engagement with science; we argued that a meta-design methodology encouraging all the knowledge that a community might bring in to solve the challenges in promoting scientific literacy.

A project focused on designing a platform for active scientific citizenship and literacy would build knowledge about the quality and impact of online knowledge co-creation sites and transform this knowledge into the guidelines for a knowledge management system including artificial intelligence tools. Researchers would motivate the stakeholders to experiment in their natural environments with system prototypes and to deliberatively decide about the socio-technical functionalities.

Such a project would provide a socio-technical system facilitating the inclusive and empowering governance of the knowledge co-creation rather than developing a formal accreditation system. Finally, such a project would have a strong potential for generating long-term societal benefits. The guidelines produced by the project would facilitate more efficient investments when funding online knowledge co-creation platforms. Increase in the number of platforms will enable lay people and scientists to co-create knowledge by means of public forum debates and organize around policy advocacy groups. These groups would, in turn, engage the public to science and increase the quality of policy decisions as they support their arguments with deliberatively constructed, evidence-based knowledge rather than pseudo-scientific claims.

References

Attride-Stirling, J. 2001. Thematic networks: an analytic tool for qualitative research. *Qualitative Research*, *1*(3): 385-405.

Bauer, M., Durant, J., Evans, G. 1994. European public perceptions of science. *International Journal of Public Opinion Research*, 6(2): 163-186.

Bingle, W.H., Gaskell, P.J. 1994. Scientific literacy for decision making and the social construction of scientific knowledge. *Science Education*, 78(2): 185-201.

Brandsen, T., Pestoff, V. 2006. Co-production, the third sector and the delivery of public services: An introduction. *Public Management Review*, 8(4): 493-501.

Brossard, D. 2013. New media landscapes and the science information consumer. *Proceedings of the National Academy of Sciences*, 110(3): 14096-14101.

Dillon, J. 2016. 50 Years of JBE: From Science and Environmental Education to Civic Science. *Journal of Biological Education*, 50(2): 120-122.

Dillon, J., Stevenson, R.B., Wals, A.E. 2016. Special Section: Moving from Citizen to Civic Science to Address Wicked Conservation Problems. *Conservation Biology*, *30*(3): 6.

europa.eu 2017. *European Centre for the Development of Vocational Training Website*. Available at: http://www.cedefop.europa.eu.

Eysenbach, G. 2003. SARS and population health technology. *Journal of Medical Internet Research*, 5(2).

Fischer, G., Herrmann, T. 2011. Socio-technical systems: a meta-design perspective. *IGI Global*: 1-33.

Fraser, N. 2009. Scales of justice: Reimagining political space in a globalizing world. NY: Columbia University Press.

Gerbner, G. 1969. Toward 'cultural indicators': The analysis of mass mediated public message systems. *Educational Technology Research and Development*, 17(2): 137-148.

Gerhards, J., Schäfer, M.S. 2010. Is the internet a better public sphere? Comparing old and new media in the USA and Germany. *New Media & Society*, 12(1): 143-160.

Honey, M. 2014. A descriptive framework for integrated stem education. In Honey, M., Pearson, G., Schweingruber, H. (Eds.) *STEM Integration in K-12 Education*. Washington, DC: The National Academies Press: 31-51.

Hilgartner, S. 1990. The dominant view of popularization: conceptual problems, political uses. *Social Studies of Science*, 20(3): 519-539.

Irwin, A. 2002. *Citizen science: A study of people, expertise and sustainable development*. NY: Psychology Press

Jasanoff, S. 2003. Technologies of humility: citizen participation in governing science. *Minerva*, 41(3): 223-244.

Jenkins, E.W. 1997. *Innovations in Science and Technology Education, Vol. VI*. Paris: Unesco.

Jolley, D., Douglas, K.M. 2014. The effects of anti-vaccine conspiracy theories on vaccination intentions. *PloS one*, 9(2): 89-177.

Lewenstein, B.V. 1992. *When Science Meets the Public*. Washington, DC: American Association for the Advancement of Science.

Miller, J.D. 2004. Public understanding of, and attitudes toward, scientific research: What we know and what we need to know. *Public Understanding of Science*, 13(3): 273-294.

Rapoport, R. N. 1970. Three dilemmas in action research: with special reference to the Tavistock experience. *Human Relations*, 23(6): 499-513.

Rennie, L., Stocklmayer, S. M., 2003. The communication of science and technology: Past, present and future agendas. *International Journal of Science Education*, 25(6): 759-773.

Shermer, M. 2009. Agenticity. Why people believe that invisible agents control the world. *Scientific American*, 300(6): 36-36.

Wagner, W. 2007. Vernacular science knowledge: its role in everyday life communication. *Public Understanding of Science*, 16(1): 7-22.

Citizen Science: The World and China

Shi Lei, Fu Zhenyu

National Academy of Innovation Strategy, CAST, Beijing, China

Abstract: The past decades witnessed the dynamic development of citizen science globally with the emergence of three major citizen science associations, i.e. the U.S. Citizen Science Association, the European Citizen Science Association, and the Citizen Science Network Australia, which reflects the current trends of citizen science growth. This article traces back the origin of citizen science and discusses its contribution to scientific research, science education and science in policy. Then the trends of development in major economies from three perspectives, including practice, policy, publications are reviewed. The three aspects demonstrate that citizen science develop in dual tacks of both bottom-up and top-down. With the momentum of global citizen science, this article focuses on the case in China. This paper introduces the relevant new practice and its effect and the difference between the world and China. The paper argues that Citizen Science in China is upgrading and it gradually turns into a vibrant ingredient in science culture.

1. Introduction

Citizen science is scientific research carried out entirely or in part by amateur (or non-professional) scientists, it is also known as community science, crowd science (Gura, 2018), including scientific exploration, new technology development and data collection and analysis (Abraham, 2012).Different from the traditional science project, in most cases, citizen science is launched by the public, the citizens in collaboration with scientists therefore it showcases the characteristic of crowd participation.

The rise of citizen science might partly because of the more urgent attention to the climate change and environmental protection. As *Science Engineering Indicators* 2018 reveals that from 1989–2017, the percentage of American public who concerns about the quality of the environment took fluctuation between 50% and 70%, and in recent years (since 2015) the percentage is increasingly climbed up over 60% (NSF, 2018). Science is more likely believed as a useful tool to solve the above urgent problems. Pew Survey showed that 79% of American adults say that science has made life easier for most people and a majority is positive about science's impact on the quality of health care, food and the environment (Funk et al., 2015). In China, the 9th Civic Literacy Survey released that over 80% Chinese say that modern S&T can provide more opportunities for next generation and S&T can make life healthier, easier and more comfortable (Xinhua Net, 2015). Additionally, the development of technology especially the Internet could be another force to push the citizen science forward which provides easy access to science observation and scientific findings sharing and as a whole contributes to the increasing growth of citizen science globally.

Citizen science has a rather long history of development in the western world, however, it is still a new practice in China. This article firstly reviews the definition, development and classification of citizen science, secondly it summarizes the trends of citizen science from four levels including practice, policy and publication. Thirdly, it emphasizes citizen science development in China from policy and practice. Finally, it discusses the potential space for citizen science development in China.

2. Definition of Citizen Science

Though the scientific institution like the Royal Society was established in 1600s which provided the frameworks for scientific enterprises, and the term 'scientist' is coined in 1830s, non-professional scientists also endeavored greatly to the development of science. As anthropologist Louis Liebenberg said that the start of modern science could be linked to the practice of hunters-gatherers when they trace the animals and predict weather condition (Liebenberg, 2013) and Mary Anning's findings show that fossils in Lyme Regis in Dorset explored by amateurs pushed forward the development of Paleontology and Geology. The latter of 20th century, science conducted by scientists was promoted tremendously and science outside institutions declined, however, there were still volunteer meteorologists collecting weather observations and non-professional naturalists sending specimens to university and museum led collections. Due to the technological progress(Castells, 2008) and as the general education level rises especially the wide application of Internet, the degree of public engagement in science is overwhelmingly increased and their participation style is far from the previous ones either from quality or quantity, and the new term 'citizen science' emerged (Haklay, 2015). Citizen science is also described as 'public participation in scientific research,' participatory monitoring, and participatory action research (Hand, 2010). The fact that 225 volunteers across the U.S. collected rain samples to assist the Audubon Society in an acid-rain awareness raising campaign was an illustration in point of citizen science that was recorded. Citizen science then could be generalized that the general public engagement in scientific research activities when citizens actively contribute to science either with their intellectual effort or surrounding knowledge or with their tools and resources. Participants provide experimental data and facilities for researchers, raise new questions and co-create a new scientific culture (EU, 2014). Therefore 'citizen science' showed from its definition that it has two features that on the first hand, the doers namely the citizens, in most cases, are the leading actor, implying it demands a large scale of participation, just as illustrated in the previous case that hundreds of amateurs would participate in it. On the other is that the core of it is to improve the entireness of science or provide more evidence for the existing science.

3. Trends of Citizen Science Development

1) Cases of Citizen Science

One of the forces to push citizen science development is to ease the contrast between endless data produced in the scientific research and the relatively limited data processing capacity. In many cases, the data mining speed is much slower than data collecting. However, citizen science in Zooniverse is an illustration to solve the discrepancy. The purpose of this project is to find out the planet or trace the astronomical objects by means of public participation who can classify and identify the astronomical images. The numerous data and images generated in the past decades should be identified and classified by manual work, the 850 thousand volunteers all over the globe were summoned to finish the task. Presently Zooniverse has had quite a few new discoveries (Simmons et al., 2013). Another case in point is launched by Pro. Bruce who wanted to synthesize a compound catalyst. He designed a cheap and easy-to -do chemical kit and attract more than 500 students from 70 high schools and universities to do it. Very soon, a unique metallic oxide is produced and some research result is also published (Lockwood, 2012).

2) Models for Citizen Science

Based on the quite a few citizen science practices, scholars summarizes the prevalent citizen science type as contributory, collaborative and co-created (Bonney et al., 2009), later contractual and collegial is also added to its type (Shirk et al., 2012). The contractual type means that scientists are invited by the public to get involved in the science research, and report the result. In the contributory type, the research is designed by professional scientists and the public are responsible for data collecting and recording. The collaborative type requires the citizens not only conduct the data collecting and recording but also play their roles in information communication. The co-created type suggests that the scientific projects are co-designed both by professional scientists and the public, and the public participate in every step of the project. Additionally, in the collegial type, the public could conduct the research independently and contributing to a certain field (Zhang et al., 2013) According to Bonney et al. the models for citizen science are listed below, and a remarkable difference among Contributory Type, Collaborative Type and Co-created Type relies on whether the citizens participate in the beginning (design the study, raise the hypotheses) and the end of the scientific process (discuss results for future research).

3) Policys for Citizen Science

According to Walter et al. there are two ingredients influencing framework and decisions on how to lead the public participation, and

Table 1 Models for public participation in scientific research.

Step in Scientific Process	Steps Included in Contributory Projects	Steps Included in Collaborative Projects	Steps Included in Co-created Projects
Choose or define question(s) for study			√
Develop explanations (hypotheses)			√
Design data collection methodologies		√	√
Collect samples and/or record data	√	√	√
Analyze samples		√	√
Analyze data	√	√	√
Interpret data and draw conclusions		√	√
Disseminate conclusions/translate results into action	√	√	√
Discuss results and ask new questions			√

Source: Booney, 2009

they are the purpose of the participation and the nature of the issue. The process of citizen science policy includes discovery to define issues and alternatives; education about issues and alternatives; measurement of public opinion; persuasion towards a desired alternative and legitimization of the decisions (Walters et al., 2000). Certainly, citizen science is frequently a bottom-up movement, therefore the policy support for it always from community level which is and driven by a group of amateurs identifying and addressing a problem by means of scientific methods. In the neighborhood scope, community monitoring of noise starts from a local scrapyard to provide evidence of the nuisance from car braking and crashing.

In Europe, after the Arhus Convention of 1998 on public access to environmental information, participation in decision making, and access to justice, a strong legislation in Europe is followed. The European Citizen Science Association was launched in June 2013, the purpose of which is to involve all European Union member states and engage five million people across the EU to promote sustainability through citizen science, build a Think Tank for citizen science, developing participatory methods for cooperation, empowerment and impact (The European Citizen Science Association, 2018).

In UK one of the representative organizations proposing citizen science is Cancer Research UK. There is a policy development team in this center to develops evidence-based policy to inform Government decisions related to cancer and research (The White House, 2009). This organization promote citizen science through crowdsourced cancer data analysis to the public (over 500,000 volunteers worldwide) through apps and games, allowing anyone, anywhere to do real research. Five citizen science projects are put into the public's hands. They also campaign on issues across the patient pathway, from prevention to early diagnosis to the treatment of cancer. They are seeking political support from all parties on key policy calls including prevention, early diagnosis, treatments etc. to appeal for political action to beat cancer.

In the United States, former Obama administration has launched 'Open Government initiatives' focusing on transparency to facilitate public participation in government decision-making (The White House, 2015). The administration is said to be committed to providing more ways for people to participate in government. They firstly started brainstorm phase and suggestion include creating longer legislative review period to increase the exposure for public review, integrating diverse voices and information into the participation process; gathering at the town hall or forum to access to more general citizens to ponder the serious issues and raising recommendations. In 2015, the White House hosted a forum on citizen science and crowdsourcing to accelerate citizen science and crowdsourcing to address societal and scientific challenges. Members of the public are encouraged to actively get involved in the scientific process by means of raising questions, making observations, conducting experiments, collecting data or developing low-cost technologies and open-source code. The White House believed the public can assist to put forwards to the scientific knowledge and benefit society (The White House, 2015). The White House even provide the Citizen Science Toolkit providing information and resources to help federal agencies utilize the power of public participation. Therefore, citizen science has become one of the contents for certain federal department, therefore citizen science in the department level is also developed. The U.S. Geological Survey is led by the Department of the Interior. The department has realized the merits of citizen science, for it can not only push scientific advances in the professional realm, but also can increase the public awareness and engagement in scientific discovery. Cross-country citizen science cooperation can also be established such as Big Bug Hunt between USA and UK conducted by University of Yok, Biotechnology and Biological Sciences Research Council and Innovate UK.

In Australia, Australian Museum Centre for Citizen Science was established in 2015 in Sydney. Citizen science is regarded an exciting and developing field of science, where the public can make a meaningful contribution to our scientific understanding. Kim McKay AO, Director and CEO of the Australian Museum believed that all over the world museums are ideal organizations to foster, excite and engage non-scientists to make this contribution (Vince, 2015). The Australian Citizen Science Association was established in June 2016 after stages of conception in 2014 and formal election in 2015. The Australian Citizen Science Association defines citizen science as public participation and collaboration in scientific research with the aim to increase scientific knowledge. It has its own 10 principles with the adaption of those of European Citizen Science Associations'. The principles emphasize that through active participation to generate new knowledge or understanding, and have a genuine science outcome, provide benefits to both science and society, participate various stages of the scientific process, receive feedbacks, control its limitations and biases, make available data, citizen scientists are suitably acknowledged, citizen science program should be evaluated, legal and ethical consideration should be included in the project (Australian Citizen Science Association, 2018).

In Japan, there is a Safecast, which was founded just days after the massive earthquake, tsunami and nuclear meltdown terribly hitting Japan in March 2011. The organization is supported by foundations, grants and individual donations to collect the radiation data near Fukushima. Especially when the trust towards nuclear power plant operators and the government hasn't recovered, a network of monitors operated by citizens could serve as an early warning system in the event of another disaster by means of giving people the knowledge and the tools to perceive and sense the environment and make more informed decisions based on accurate information. Additionally, when the evacuated towns and villages were reopened, the data from Safecast would verify the safety level of those area. Safecast holds regular trainings for adults to teach them to assemble their own devices and a kids' workshop is also planned. Plans and directions for building the devices are also available online for free. It is said that citizens who build their own monitors are much more motivated to use them (Makinen, 2016).

From Europe, UK, Australia, Japan to the United States, citizen science policy is formed gradually, and the relevant organizations or agencies are also erected. These policies and practice embodied in the formation of these associations would greatly promote the public's capability of devoting and vice versa contribute to policy development.

4) Publication for Citizen Science

With the development of citizen science practice, the research on it is widely boosted. According to a bibliometrics statics, the evolution of it can be sorted to three stages, including early stage (1996–2009), growing-up stage (2010–2013) and development stage (2014–2016). The criteria to classify the stages due to its increase rate, in the early stage the increase rate is rather stable, while from 2010, the output of the citizen science publication gradually climbed up, when it entered into 2014, the number of papers on citizen science is dramatically enlarged and 2014 to 2016 has witnessed the quickest rise since 2009. The hottest subfield of citizen science focuses on environmental sciences and ecology which has never changed from 2009 to 2016 (Table 2). It is quite reasonable that because of the environment deterioration, climate change has become global problems, both professional scientists and citizen scientists would be more concerned with issues in environmental sciences and ecology. As a Pew survey revealed that 79% adults consider that science has made life easier for most people and a majority is positive about science's impact on the quality of health care, food and the environment (Pew Research Center, 2015).

Table 2 The evolution of citizen science research from 2009–2016.

Rank	Early Stage	Frequency	Growing-up Stage	Frequency	Development Stage	Frequency
1	Environmental Sciences & Ecology	150	Environmental Sciences & Ecology	188	Environmental Sciences & Ecology	390
2	Zoology	53	Biodiversity & Conservation	67	Biodiversity & Conservation	167
3	Biodiversity & Conservation	40	Computer Science	55	Science & Technology	123
4	Engineering	13	Astronomy & Astrophysics	47	Computer Science	104
5	Physical Geography	11	Zoology	47	Zoology	82
6	Life Sciences & Biomedicine	10	Engineering	44	Engineering	64
7	Marine & Freshwater Biology	9	Science & Technology	36	Physical Geography	53
8	Water Resources	8	Education & Educational Research	27	Marine & Freshwater Biology	50
9	Public, Environmental & Occupational Health	8	Marine & Freshwater Biology	22	Astronomy & Astrophysics	43
10	Mathematics	7	Life Sciences & Biomedicine	19	Remote Sensing	41

Source: Zhang et al., 2017

4. Citizen Science in China

Citizen science is relatively a novel term in China and it seldom appears in policy context. While with the increase of Civic Scientific Literacy in China, which is released by CAST in 2015 during the 9th National Survey on CSL that in China the percentage of citizens who can be considered to process the civic scientific literacy has up to 8.47% (Cao, 2015),and much higher than 6.2% in 2015. According to the data in 2015, the majority of citizens support the development of science and technology and have full of hope of it (Xinhua Net, 2015). As the dynamic and prevalent movement of citizen science worldwide spread, the citizen science in China has embarked. Hereinafter are the cases of Citizen Science undergoing in China.

The beginning of citizen science might be seen in the practice of China Ecosystem Research Network (CERN). The CERN was established in 1988 to monitor China's ecological environment changes and comprehensively study major issues related to China's resources and ecological environment. At present, the research network consists of 16 experimental station of farmland ecosystem, 11 forest ecosystem experimental stations, three experimental stations, three desert grassland ecosystem experimental stations, a marsh ecosystem experimental stations, two lakes ecosystem experimental stations, three Marine ecosystem experimental stations, a city ecological station, as well as water, soil, air, biological, five disciplines sub-center waters ecosystem and a comprehensive research center. The main objectives of scientific research in CERN at present are as follows: (1) through long-term monitoring of major types of ecological systems in China, to reveal the change rules and causes of ecological system and environmental elements in different periods. (2) To establish the service function and value evaluation, eco-environmental quality evaluation and health diagnosis index system of major types of ecosystem in China. (3) To clarify the functional characteristics of major types of ecosystem in China and the basic rules of biogeochemical cycles such as C, N, P and H_2O. (4) To clarify the impact of global changes on China's major types of ecosystems, and reveal the role and response of different regional ecosystems to global changes. (5) To elucidate the mechanism of degradation and damage process of major types of ecosystem in China, explore the technical approaches for ecosystem restoration and reconstruction, and establish a number of experimental demonstration areas for comprehensive treatment of degraded ecosystem.

CERN has actively cooperated with universities, primary and middle schools to carry out popular science education. CERN has passed the popular science lecture, the popular science exhibition, the summer camp, the visit inspection and so on, many kinds of ways to receive primary and middle school students into the ecological station, to cultivate young people love nature and love science since childhood. The Ailao mountain eco-station actively assists the local government and related units in organizing summer camp activities. The experts went deep into the school for scientific lectures and other forms to expand the field of popular science, which was widely praised and cited at the same time. Dinghu mountain station and Heshan station make use of abundant resources and profound scientific research accumulation, together with education department. Join hands with primary and secondary schools, actively carry out scientific teaching practice, and receive teachers and students in winter and summer camps with amount of 200 students and 300 college students' practice annually.

Another citizen science practice in China is the Birdwatch China. It's a Bulletin Board Station online. In 1997, some environmental activists in Beijing began to take part in wild bird-watching activities. In order to make everyone fully understand birding activities, improve birding skills, and make more people participate in the team of birding enthusiasts. It's a group of professional experts on birding and the amateur bird watchers. It has profound knowledge of ornithology, rigorous introduction of scientific methods, as well as practical evaluation of bird-watching and perception of activities. There are a lot of announcements for citizen science activity. For instance, the work of the bird circle in Cuihu Lake has been under the leadership and guidance of the national bird circle center and Beijing wildlife rescue center (Beijing bird circle station).

The amateurs are recruited to carry out the bird world record training in Cuihu wetland park, to complete the daily patrol, bird picking, environmental record and other work, to take part in the training on time according to the schedule of China bird watching society, and strictly followed the operating procedures and regulations of the recording.

The third case come from the China Bird Watching Record Center (http://www.birdreport.cn/). The online record center is open to anyone who registers as a member. When the member logs in, he/she first fills in the information about the observation place, date, name of the bird, weather condition, environment and route, notes, then he/she can start a new record of bird. When the new record is open, every sort of bird has a number. Then the observer finishes a record. Then the bird observation is complete. Currently, the China bird watching record has accumulated a huge number of information tracing the bird. According to Annual Report of China Bird Watching (2003–2007), and there were 20936 records of the birds, with 17 catalogues, 70 family and 1078 species, accounting for 80% percent of all birds in the world (Li et al., 2012). These data reveal the distribution of birds in China and its change, and bird migrating pattern, providing very accurate and detailed information for the bird study.

From the citizen science cases above we might conclude that compared with the developed countries, citizen science in China is still in a very early stage. The most common practice of citizen science in China is related with environment protection and bird observation. The participation mode of citizens is to receive knowledge and training of science, the purpose of it is to solve the problems by technology. The existed 'citizen science' so to speak inclines to be more pragmatic. Till now few cases of citizen science related to astronomy and physics or medicine with the purpose of responding to the core question of science is found out. Certainly, these questions are more likely approaching the inquiry of original innovation, and needs rather high capability of scientific literacy. The clear contrast might verify that there is some gap between China and S&T advanced countries. Since citizen science is more likely a curiosity-driven and self-motivated public activity, and China always shows its remarkable characteristics in powerful national input, it might be a long way to go to foster a fertile soil for the maturity of citizen science in China.

From the outlook towards the globe, the citizen science is developing swiftly and robust momentum. The reason of its continuous prosperity might be the intense curiosity of human beings. The intense interest in unknown world is stirred to probe into future and uncertainty. Curiosity is the basic nature of human beings and it's the continuous power to motivate citizen science practice. The second reason is the development of the technology and especially the widespread application of Internet which make the electronic data transferring and stored become easy. The third force to motivate citizen science is the post-disaster monitoring and predicting. It is rather urgent and practical citizen science method. With the disaster erupted more frequently the citizen science in post-disaster could be an irreplaceable. With the civic scientific literacy is raised, people are more interested in science globally, some establishment of organization or agency, not only the developed countries but also some developing countries, the future of citizen science can grow in a larger scale.

5. Conclusion

Through the citizen science cases listed above we might conclude that compared with the developed countries, citizen science in China is still in a very early stage. The most common practice of citizen science in China is related with environment protection and bird observation. The participation mode of citizens is to receive knowledge and training of science, the purpose of it is to solve the problems by technology. The existed 'citizen science' so to speak inclines to be more pragmatic. Till now few cases of citizen science related to astronomy and physics or medicine with the purpose of responding to the core question of science is found out. Certainly, these questions are more likely approaching the inquiry of original innovation, and needs rather high capability of scientific literacy. The clear contrast might verify that there is some gap between China and S&T advanced countries. Since citizen science is more likely a curiosity-driven and self-motivated public activity, and China always shows its remarkable characteristics in powerful national input, it might be a long way to go to foster a fertile soil for the maturity of citizen science in China.

From the outlook towards the globe, the citizen science is developing swiftly and robust momentum. The reason of its continuous prosperity might be the intense curiosity of human beings. The intense interest in unknown world is stirred to probe into future and uncertainty. Curiosity is the basic nature of human beings and it's the continuous power to motivate citizen science practice. The second reason is the development of the technology and especially the widespread application of Internet which make the electronic data transferring and stored become easy. The third force to motivate citizen science is the post-disaster monitoring and predicting. It is rather urgent and practical citizen science method. With the disaster erupted more frequently the citizen science in post-disaster could be an irreplaceable. With the civic scientific literacy is raised, people are more interested in science globally, some establishment of organization or agency, not only the developed countries but also some developing countries, the future of citizen science can grow in a larger scale.

References

Abraham Miller-Rushing, Primack, R., Bonney, R. (2012). The history of public participation in ecological research. *Frontiers in Ecology and the Environment*, 10(6), 285-290.

Cao X. (2018). Survey Result Release: Less than 9% Chinese citizen has CSL [online] http://www.stdaily.com/kjrb/kjrbbm/2018-09/07/content_707103.shtml [Access 20th September, 2018].

Castells, M. (2008). Materials for an exploratory theory of the network society. *The British Journal of Sociology*, 51(1), 5-24.

E. U. (2014). Green paper on Citizen Science for Europe: Towards a society of empowered citizens and enhanced research. *European Commission*.

Funk, C., Rainier, L. (2015). Public and Scientist's Views on Science and Society [online] http://www.pewinternet.org/2015/01/29/public-and-scientists-views-on-science-and-society/ [Access 8 September, 2018].

Gura, T. (2013). Citizen science: amateur experts [online] https://www.nature.com/naturejobs/science/articles/10.1038/nj7444-259a [Access 15 July 2018].

Haklay, M. (2015). *Citizen Science and Policy: A European Perspective* (44): Wilson Center & Commons Lab.

Hand, E. (2010). Citizen science: People power. *Nature*, 466(7307), 685-687.

Lockwood (2012). Cowdsourcing Chemistry[online]available at https://cen.acs.org/articles/90/i46/Crowdsourcing-Chemistry.html [Access 20 July 2018].

Li, X,, Lu, L., Peng, G., et al. (2012). Change in Birds Distribution revealed by China Birds Observation Data. *Chinese Science Bulletin*, 57(31), 2956-2963.

Liebenberg, L. (2013). *The Origin of Science*. Cape Town South Africa: Cyber Tracker.

Makinen, J. (2016). Citizen science takes on Japan's nuclear establishment[online] http://www.latimes.com/world/asia/la-fg-japan-safecast-snap-story.html [Access 8th September,2018].

M, C. (2000). Materials for an exploratory of the network society1. *British Journal of Sociology*, 5(1), 5-24.

NSF (2018). Science and Engineering Indicators 2018 [online] https://www.nsf.gov/statistics/2018/nsb20181/[Access 15th June 2018].

Pew Center (2015). *Public and Scientists' Views on Science and Society* (5). Washington D.C.: Pew Research Center.

Shirk, J. L., Ballard, H. L., Wilderman, C. C., et al. (2012). Public participation in scientific research: a framework for deliberate design. *Ecology and Society*, 17(2), 29.

Simmons, et al. (2013). Galaxyzoo: Bulgeless galaxies with growing black holes. *Monthly Notices of the Royal Astromical*, 429(3), 2199-2211.

The European Citizen Science Association [online] https://ecsa.citizen-science.net/about-us [Access 20 August 2018].

The White House (2009). Enhancing Citizen Participation in Decision-Making [online] https://obamawhitehouse.archives.gov/blog/2009/06/10/enhancing-citizen-participation-decision-making [Access 8th September, 2018].

The White House (2015). Accelerating Citizen Science and Crowdsourcing to Address Societal and Scientific Challenges [online] https://obamawhitehouse.archives.gov/blog/2015/09/30/accelerating-use-citizen-science-and-crowdsourcing-address-societal-and-scientific [Access 12th July, 2018].

Vince, C. (2015) Australian museum announces new centre for citizen science[online] https://australianmuseum.net.au/media/new-centre-for-citizen-science [Access 20th June, 2018].

Walters, L. C., Aydelotte, J., Miller, J. (2000). Putting more public in policy analysis. *Public Administration Review*, 60(4), 349-359.

Xinhua Net (2015). China Association for Science and Technology Released the 9th result of China Civic Literacy Survey[online] http://politics.people.com.cn/n/2015/0919/c70731-27608306.html [Access 8th September 2018].

Zhang, J., Chen, S., Chen B., et al. (2013). Citizen science: integrating scientific research, ecological conservation and public participation. *Biodiversity Science*, 21(6), 738-749.

Zhang, X., Zhao Y. (2017). Evolution Path and Hot Topics of Citizen Science Studies. *Data Analysis and Knowledge Discovery from*, 1(7), 22-34.

Communicating Science with Anthropomorphism
——Examples from Japan

Matthew Wood

Faculty of Life and Environmental Sciences, University of Tsukuba, Tsukuba, Japan

Abstract: Anthropomorphism, or the attribution of human characteristics to non-human animals or inanimate objects, is commonplace in many cultures of the world, but is particularly prominent and pervasive in Japan. Talking furniture on children's TV, city government vegetable mascots, an animated letter 'e' to promote online tax returns—there seems to be no limit to what can be anthropomorphized, and no corner of the culture where it is considered out of place. This of course includes efforts to communicate science, where we can find test tube narrators, angry viruses, friendly elements, and a whole lot more. Scientists, on the other hand, are less enthusiastic about anthropomorphism in scientific discussions and tend to consider it as inaccurate and unscientific. In science, thinking or communicating in anthropomorphic terms is generally derided. Where, then, does this leave the talking microbes and smiling proteins of Japanese science communication? The literature has some to say about anthropomorphism, but nothing specifically about anthropomorphism in science communication. This paper draws on examples from Japan to consider the potential roles of anthropomorphism in the communication of science and related issues.

1. Anthropomorphism in Japanese Culture

Anyone with a passing interest in Japanese popular culture will have noticed how industrious the Japanese are at producing endearing characters. Even the most mundane objects are routinely transformed into walking, talking, thinking, feeling entities, and this is ubiquitous across every aspect of the culture from ancient fables to modern manga.

This bestowment of human characters onto non-human animals, or even inanimate objects, is known as anthropomorphism, and while it is commonplace in all cultures, the Japanese have embraced it with vigor. Anthropomorphism has a long history in Japan with ancient folklore rife with a colorful array of talking animals, mischievous monkeys, spooky wailing rocks, and umbrella boogeymen[1]. Contemporary children's media is packed with talking animals and other objects. Of course this is also the case in many other countries (cf. Disney, Sesame Street, Thomas the Tank Engine), but there seem to be no restraints on what can be anthropomorphized as characters of children's television shows, cartoons, and picture books in Japan. Just one popular Japanese television show includes such characters as singing shoes, a chuckling maracas fairy, dancing potatoes, a flying toothbrush, and even a toilet king. Another program features the capers of a whole village of talking chairs.

Anthropomorphized characters are not restricted to children's stories though. They appear in all manner of roles right across Japanese society. Many companies have their own mascot character, which may be modeled on an animal[2] or something else entirely[3]. Many prefectures and municipalities also have a mascot character to promote local products or characteristics of the region, and these can range from historical figures[4] to animals, fruit and agricultural products[5], geographical features[6], or combinations of

---

1 See the scrolls of Chouju-jinbutsu-giga, or peruse the many tsukumogami 'tool gods' for examples.

2 e.g. *Jetta*, the charismatic red panda for economy airline Jetstar, or telecommunications company Softbank's talking dog, who is also inexplicably the father of an otherwise human household.

3 e.g. National broadcaster NHK's friendly television-shaped monster *Domo-kun*.

4 e.g. Ibaraki Prefectural Government's *Hustle Komon* is inspired by a TV dramatization of the exploits of famous daimyo Tokugawa Mitsukuni.

5 e.g. *Sashibanosa-chan*, the hawk mascot of Ichikai Town, or Akaiwa City's peach and grape character *Akaiwamomo-chan*.

6 e.g. Sapporo city Western District's *Sankakuyamabe*, a walking green triangular mountain.

these[1]. These mascot characters, called *Yurukyara*, are extremely popular with the public, and regularly appear in TV news, variety shows, and other media. There is even an annual 'Yurukyara Grand prix' event to decide the most (and least) popular characters.

In addition to promoting companies and regional specialties, anthropomorphic characters are also employed toward public relations by a variety of government departments[2] and are common in government public education campaigns[3]. I will not even begin to list the countless unsung anthropomorphized characters scattered throughout advertising and promotions, events and festivals, tourist maps, instructional and guidance materials, children's textbooks, souvenir shops, warning signs and so on. Suffice to say that anthropomorphized characters are a significant part of the Japanese sociocultural landscape.

2. Anthropomorphism in Japanese Science Communication

Since this culture of anthropomorphism is pervasive through every aspect of Japanese society, it should not be surprising that it is also common in science education and science communication (N.B. Here, I will limit discussion to visual or illustrated instances of anthropomorphism, rather than suggestions of human-like behavior in articles and text). Once again, this is not unique to Japan. Examples of anthropomorphized characters and narrators are not uncommon among Western science communication materials aimed at children[4]. However in Japan, while anthropomorphized science is more common in children's materials (it is much less likely to be found in textbooks for higher grades and more specialized study), it also appears in explanations of science topics for adults, for example in health communications[5] and informational television shows[6]. Furthermore, Japanese examples are more likely to anthropomorphize not only narrators or explainers but also entities which form part of the actual explanation itself. For example, an explanation of influenza may include anthropomorphized virus particles. Or diabetes might be explained using anthropomorphized sugar and insulin molecules. Examples of such integrated anthropomorphism are quite common in children's science texts and supplementary learning materials[7], where air molecules, water vapor, Bunsen burners, and test tubes all take on a life of their own. Even the periodic table of the elements has been entirely anthropomorphized (several times) with each element adopting an individual persona[8].

Comics are also a rich part of Japanese pop-culture, further encouraging anthropomorphism as part of the characterization and storification process which can turn even the most mundane topic into an interesting adventure. Of course this is also applied to science topics, and a few recent noteworthy examples are *Moyashimon - Tales of Agriculture*[9], *Cells at work* (*Hataraku Saibou*)[10], and *Nymphs of the Plant Hormones* (*Shokubutsu Horumon Gijinka*)[11]. *Nymphs of the Plant Hormones* introduces a range of plant hormones personified as female high school students, complete with profiles, nicknames, and key features. *Moyashimon* is the story of an agricultural university student who can see and talk to microorganisms. Through his activities at school the comic introduces a range of microorganisms and their roles in medicine, disease, and fermented food production. *Cells at Work* portrays a range of human body cells (mainly immune system cells) as youthful employees, and through their actions explains functions of those cells in response to injury and disease. *Nymphs of the Plant Hormones* is a science communication project aimed at raising

1 e.g. The strawberry and deer mashup, *Yumezukin-chan*, of the Nagasaki branch of Japan Agricultural Coop.

2 e.g. Tokyo Metropolitan Police Department's *Pipo-kun*, Ministry of Justice's human rights mascot *Jinken Mamoru-kun*.

3 e.g. *Maina-chan* for the Government's personal identification scheme (My Number), and *Saiban-inkoi* for the introduction of a new jury-based legal system.

4 e.g. Professor Jay Hosler's comics explaining Apis ecology, photosynthesis and more use anthropomorphised bees, ants, and other insects as explainers and explainees. (https://jayhosler.com/science-comics.html)

5 e.g. Citizen-Medical Staff Alliance for the Prevention of Lifestyle-related Diseases. (http://www.kozonokai.org/medical-inf/tounyou/what)

6 e.g. Popular TV health programme *Tameshite Gatten* regularly uses graphics and elaborate models containing anthropomorphized elements. (http://www9.nhk.or.jp/gatten)

7 e.g. Visual Science Encyclopedia (ビジュアル理科事典, 2015, Gakken Plus), Learn Middle School Science the Clever Way – 4 Frame Classroom (中学理科がちゃっかり学べる ゆる4コマ教室, 2018, Gakken Plus).

8 e.g. Wonderful Life with the Elements (元素生活, 2015, Kagakudojin), Element Girls (元素周期 萌えて覚える化学の基本, 2008, PHP Institute), and Learn the Periodic Table with Comics (マンガで覚える元素周期, 2012, Seibundo Shinkosha).

9 http://kamosuzo2.tv

10 https://hataraku-saibou.com

11 http://hormone.webcrow.jp

awareness and understanding of plant hormones. However it is interesting to note that although *Moyashimon*, and *Cells at Work* both contain educational content, they were not necessarily motivated by educational or science communication goals, but rather were highly successful commercial projects.

While Japan is not the only culture to produce anthropomorphized science instruction and communication[1], from personal observation and a Western perspective, anthropomorphism in Japan (both in science communication and in general) is conspicuous for its prevalence, diversity, and widespread acceptance. Certainly the commonplace nature of anthropomorphized explanations of science suggests that this is popular with audiences in Japan, or at least is perceived by authors as popular or easy to understand.

3. Anthropomorphism from a Science and Science Education Perspective

Despite this enthusiastic characterization of scientific explanations in Japan, the world of science has traditionally taken a different view. In science, anthropomorphism is bad, and that has long been the dominant view. The argument is that atoms, molecules, cells and so on don't 'feel' or 'want' anything—their behavior is driven by causative relationships rather than human-like motivations—therefore thinking in anthropomorphic terms is simply incorrect and not scientifically accurate. It may seem that considering chemical bonding in terms of atoms 'wanting' an additional electron is just a harmless figure of speech, and in many cases it is, but some scientists worry that this has the potential to influence the direction of scientific research itself. For example, in a criticism of widespread anthropomorphic thinking in the field of microbiology and a warning of the dangers to scientific progress, Davies (2010) describes how teleological thinking associated with anthropomorphism restricted science's understanding of antibiotics. Descriptions of antibiotics having 'militaristic' functions in a 'war' against invaders impeded conceptions of other possible functions, and delayed the current awareness that antibiotics may have a range of roles in transcription modulation, regulating, and signaling.

Since anthropomorphisms are unscientific, they have also been considered unsuitable for science training and education. For decades, anthropomorphic analogies in science education have been thought to be an impediment to learning. Intuitive anthropomorphic reasoning was seen as characteristic of early stages of childhood development (Dorion, 2011), and therefore a childish means of explanation. However, despite normative appeals to eliminate anthropomorphism from science education, is has recently been shown that anthropomorphisms are common among science students, and in fact may play a variety of roles in the development of students' understanding of scientific concepts by allowing students to work with and build on as-yet incomplete conceptualizations of learning goals. Working with anthropomorphisms may offer teachers an opportunity to bolster a tactic that students already use to develop their understanding (Dorion, 2011).

Findings such as this coincide with a recent apparent softening on anthropomorphism in science education. The American Association for the Advancement of Science publishes Benchmarks For Science Literacy (AAAS, 2009) which serve as a guideline for school science education in the US. In 1993 the benchmarks recommended that by second grade children should be made aware that 'stories sometimes give plants and animals attributes they really do not have'. However this has been deleted from the most recent benchmarks in 2009. Instead the current version states that while 'the anthropomorphism embedded in most animal stories causes some worry', developing an interest in reading is more important than 'rigidly correct impressions'. The benchmarks suggest that anthropomorphism can be ignored in the classroom, or students can be guided toward noticing differences in how animals are portrayed (realistically or anthropomorphized) in different books.

4. Potential Influences of Anthropomorphism

Impacts of anthropomorphism specifically on the public communication of science appear to have been largely ignored in the literature.

1 e.g. Swedish/US company Toca Boca produces chemistry and plant biology apps for children which heavily feature anthropomorphized chemical elements and plants (https://tocaboca.com/apps/), while Dan Green and Simon Basher have published a series of science books featuring anthropomorphized organs, cells, chemical molecules, and even qualities such as mass and weight.

However there is some research from cognitive psychology and science education which provides hints of what these impacts might be.

To understand why people anthropomorphize, Epley, Waytz, Akalis, and Cacioppo (2008) examined and provided evidence for two motivational factors which they named *effectance motivation* and *sociality motivation*. Effectance motivation extends on the idea that the human tendency to anthropomorphize agents is an example of induction—the process of reasoning about an unknown or unfamiliar object based on what is known about a more familiar related object. In the case of anthropomorphism, applying a familiar 'human' framework to interpret an unfamiliar concept (e.g. why trees produce fruit) may render a more digestible, albeit imprecise understanding (e.g. to entice birds to disperse its seeds). Epley et al. (2008) were able to show that people with a strong need for control and understanding were more likely to anthropomorphize animals when they had less understanding of the animals' behavior. Therefore it is reasonable to suggest that anthropomorphized explanations of unfamiliar scientific concepts may similarly contribute to a more understandable explanation by using a familiar framework. In fact there is some support for this in the literature. For example, Brossard, Stoos, and Haftel (2017) found that students involved in structured anthropomorphic storytelling as part of a microbiology course performed better on exams than students in a regular class. Marketing research also has shown that an anthropomorphized demonstration of an influenza drug fighting influenza virus led to greater perception of the drug's efficacy among viewers via a better perceived understanding of how the drug worked (Laksmidewi et al., 2017).

In science communication settings, then, a story of an 'angry' virus invading a 'hapless' cell to 'willfully' inflict damage and chaos, followed by the counter-attack and triumph of the body's 'brave and noble' defenses may set up an immediately accessible and understandable framework for explaining influenza infection and immune response, compared to a more 'scientifically acceptable' explanation of random encounters and a complex chain of chemically triggered reactions. The decision of whether or not to use anthropomorphized characters may then come down to the objectives of the communication and the relative necessity of detailed and precise scientific understanding. Of course this must also be weighed against the risk of harmful misunderstandings and misinterpretations originating from teleological explanations.

Sociality motivation, the second motivational factor described by Epley et al. (2008) stems from the need to be socially connected. Being a social animal, humans have a psychological need to connect with other humans. This need is so strong that in the absence of other human contact individuals will construct a proxy in the form of anthropomorphized animals or objects. Epley et al. (2008) demonstrated this by revealing a higher tendency to anthropomorphize pets among individuals with a strong need for social connection (i.e. fewer human social contacts). It might be expected then that these individuals are not only more likely to anthropomorphize their surroundings, but also respond differently to anthropomorphisms when they are presented. In fact, Tam (2015) found that anthropomorphism in conservation messages led to stronger influence on conservation behavior among people with a strong need for social connection.

The literature also describes other apparent influences of anthropomorphism on attitudes and non-cognitive factors. Chartrand, Fitzsimons, and Fitzsimons (2008) noted previous research demonstrating that the perceived attributes or expectations of other people can lead to behavioral change. They considered whether anthropomorphic images might also demonstrate such influence and found that priming people with stereotypical anthropomorphic attributes of animals led to unconscious changes in behavior intentions which conformed with those stereotypes. In this study people primed with images or thoughts of dogs indicated elevated intentions of loyal behavior (an attribute typically associated with dogs in the sample population), while those primed with cats (typically considered not loyal) indicated decreased intentions of loyalty. In another study investigating viewer interpretation of anthropomorphized software interfaces, realistic human representations of the interface agent (compared with 2D representations and stylized caricatures) were interpreted as more capable and intelligent (King et al., 1996). This suggests that a viewer's assumptions or expectations associated with an anthropomorphic representation are applied to qualities of the anthropomorphized object itself.

5. Summing Up

What does this mean for anthropomorphism in science communication? On the one hand describing phenomena in anthropomorphic terms is not scientifically precise, and there are valid reasons for concern about anthropomorphism in scientific contexts. However, anthropomorphic frameworks may offer a familiar and approachable format for more understandable explanations of often complex and unfamiliar science. One must consider then, what degree of detailed scientific precision is required, and how much importance is placed on ease of understanding and approachability. Is the scientific imprecision of a virus's evil grin outweighed by the familiar and understandable metaphor for danger and anomaly? Do the highly figurative portrayals of something like *Cells at Work* present broad themes of cell biology in an immediately recognizable and memorable way that justifies the sacrifice of specific details and the risk of teleological misunderstandings? Furthermore, in addition to conveying accurate information, science communication efforts are often concerned with emotional, attitudinal, and behavioral outcomes, and anthropomorphisms have been shown to influence each of these. In that case, can a narrator/explainer portrayed as a friendly magnifying glass influence interest and motivation, or even induce behaviors that may be associated with magnifying glasses such as earnestness and attention to detail?

Responses to anthropomorphism are likely to be very complex and highly nuanced. They will certainly be influenced by individual traits and cultural factors. For example, an individual raised in the anthropomorphically rich culture of Japan would be expected to have different acceptance, interpretations and understandings of anthropomorphized science than a counterpart from another country. Universal norms or recommendations may not be possible.

In any case, there seems to be a broad range of potential impacts of anthropomorphism in science communication settings. Some of these may be useful, but of course they need to be reconciled with the need for scientific accuracy, and the decision of whether or not to anthropomorphize will ultimately be determined by the goals of a particular setting. Unfortunately, we currently have very limited understanding of what these impacts might be and how they might contribute to our goals. It would seem prudent, then, to attain a better understanding of the influences of anthropomorphism and how they might be useful for the specific purposes of science communication before summarily dismissing it as childish nonsense.

References

AAAS (2009) Project 2061: Benchmarks for science literacy. Available at: http://www.project2061.org/publications/bsl/online/index.php (Accessed: 9 August 2018).

Brossard Stoos, K. A., Haftel, M. (2017) Using anthropomorphism and fictional story development to enhance student learning, Journal of Microbiology & Biology Education, 18(1). doi: 10.1128/jmbe.v18i1.1197.

Chartrand, T. L., Fitzsimons, G. M., Fitzsimons, G. J. (2008) Automatic effects of anthropomorphized objects on behavior, Social Cognition, 26(2), pp. 198-209. doi: 10.1521/soco.2008.26.2.198.

Davies, J. (2010) Anthropomorphism in science, EMBO reports, 11(10), pp. 721-721. doi: 10.1038/embor.2010.143.

Dorion, K. (2011) A learner's tactic: How secondary students' anthropomorphic language may support learning of abstract science concepts, Electronic Journal of Science Education, 12(2), pp. 1-22.

Epley, N., Waytz, A., Akalis, S. et al. (2008) When we need a human: Motivational determinants of anthropomorphism, Social Cognition, 26(2), pp. 143-155. doi:10.1521/soco.2008.26.2.143.

King, W. J., Ohya, J. (1996) The representation of agents: Anthropomorphism, agency, and intelligence, in Conference companion on Human factors in computing systems common ground - CHI '96, pp. 289-290. doi: 10.1145/257089.257326.

Laksmidewi, D., Susianto, H., Afiff, A. Z. (2017) Anthropomorphism in advertising: The effect of anthropomorphic product demonstration on consumer purchase intention, Asian Academy of Management Journal, 22(1), pp. 1-25. doi:10.21315/aamj2017.22.1.1.

Tam, K. P. (2015) Are anthropomorphic persuasive appeals effective? The role of the recipient's motivations, British Journal of Social Psychology, 54(1), pp. 187-200. doi: 10.1111/bjso.12076.

Citizen Science as Participatory Science Communication

Per Hetland

Department of Education, University of Oslo, Norway

Abstract: Over the last 20 years, citizen science has been understood as either democratized citizen science (Irwin, 1995) or contributory citizen science (Bonney, 1996). The present paper will argue for a third understanding—citizen science as participatory science communication (Metcalfe et al., 2008). This paper will shortly presents some results from three case studies studying citizen science (CS). Three conclusions became apparent: (1) Three main science communication styles can be identified when scientists address general, pure, partisan, and affected publics, and the importance of dialogue is common to all three styles. Notably, the scientists perceived these dialogues as 'invisible' to institutions and society at large. (2) Boundary infrastructures like Species Observation's (SO; a digital infrastructure for citizen science) success depends on its ability to reciprocate to its contributors. (3) Validation and quality assurance of contributed data are crucial for successful participatory science communication. This short presentation will only presents some important elements of participatory science communication, while the later full paper will present both the data and the discussion in a more structured manner.

1. Participatory Science Communication

As a path for promoting quality in science communication in a broader sense, participatory science communication embodies aspects of the communicators, their performance and publics, the knowledge itself, and how quality is assured and communicated. Few discuss participatory science communication (Metcalfe et al., 2008). Consequently, this short paper will outline some crucial elements. Rephrasing Guy Bessette (2004) I would like to define participatory science communication as follows (Bessette, 2004: 9):

Participatory science communication is a planned activity, based on the one hand on participatory processes, and on the other hand on media and interpersonal communication, which facilitates a dialogue among different stakeholders, around a common scientific problem or goal, with the objective of developing and implementing a set of activities to contribute to its solution, or its realization, and which supports and accompanies this initiative.

Such defined there is a close relationship between participatory science communication and what is called boundary infrastructures. Boundary objects refer to elements that link various groups and interests together. Star and Griesemer (Star et al., 1989) defined boundary objects as temporary agreements by different actors and groups on how to relate to a given situation. They describe how a standardized method in natural history for collecting, conserving, marking, and describing finds functioned as a boundary object between amateurs and researchers in what was a research subject among researchers and a subject for hobby activity, exercise of an occupation, or nature conservation among groups of the public. In other words, they establish agreement about what are points of contact in common. Boundary objects are negotiated agreements that contain different interests but, at the same time, open up for slightly different practices. The boundary objects create a dialogue between various interests— and handle stability and ambiguity simultaneously (Wells, 1999). In this way, boundary objects permeate borders at the same time as the established practice is continued. Scientific infrastructures represent 'regimes and networks of boundary objects (and not unitary, well-defined objects), boundary infrastructures have sufficient play to allow for local variation together with sufficient consistent structure to allow for the full array of bureaucratic tools (forms, statistics, and so forth) to be applied' (Bowker et al., 1999: 313-314) and thereby facilitate scientists, amateurs, and stakeholders cooperating across disciplines and organizational boundaries.

2. Doing Participatory Science Communication

By building on the three mentioned case

hetland@iped.uio.no.

studies done by the author: (1) interviews with 17 scientists at two natural history research museums concerning their own role in science communication (Hetland, 2018c), (2) a web-based survey about the use of SO with a total of 404 respondents (Hetland, 2018b), and (3) qualitative interviews with 8 volunteers and 4 boundary spanners (Kirby, 2011) who served as liaisons between knowledge production and knowledge infrastructures (Hetland, 2018a). The present study of participatory science communication utilized participant observation of online communities (Hetland et al., 2016) and through interviews about CS knowledge production and knowledge politics. While the three case studies offer a picture of CS as both democratized citizen science and contributory citizen science, biodiversity mapping is also a huge effort toward participatory science communication. Science is communicated across national, institutional, and disciplinary boundaries as well as across both the professional and amateur distinctions.

The study of scientist identified three more general science communication styles towards four major constructions of publics: the general public, the pure public, the affected public, and the partisan public (Braun et al., 2010). The general public speaks as anonymous individuals; the pure public speaks as concrete individuals, often 'naïve citizens' as the subject of education; the affected public speaks as concrete individuals, including the authentic expert with firsthand knowledge of a specific area of life; and the partisan public speaks as interest groups with knowledge of the landscape of possible arguments.

Dissemination activities involving general and pure publics. Although this is a traditional, well-known, and recognized style, it encompasses a series of individual adaptations, including direct communication and communication through different media channels—usually, a mix of both. Creating enthusiasm is perceived as very important, as is the ability to communicate broadly. Additionally, dissemination activities encourage dialogue, whether synchronic or asynchronic. The scientists perceived these dialogues as invisible, hidden from the institutions and from society at large.

Dialogues with partisan publics. These dialogues are perceived as important because partisan publics provide resources for research and implement policies of relevance to biodiversity development. The scientists again perceived these dialogues as invisible, or hidden, from the institutions and from society at large.

Dialogues with affected publics. Amateur naturalists have a long association with natural history research museums, and recent developments have revived that relationship. Because of the limited resources available for fieldwork, professional science is more concerned with general knowledge and less with local knowledge. On the other hand, amateurs develop extensive local knowledge that professional science sometimes needs. Consequently, lay knowledge or amateur knowledge usually conforms with scientific knowledge. However, some amateur naturalists also like to pursue aesthetic values or seek respect as an amateur naturalist by being first, doing the most, and/or being 'best' in one way or another. However, this competitive element is not unfamiliar to science, either (Conniff, 2011). Aesthetic values may simply relate to the enjoyment of natural beauty or to the beauty of a well-designed collection. Finally, although lay knowledge is sometimes associated with ignorance, conspiracy theories, or similar factors, none of the scientists referred to such problems in their interactions with different publics. The one exception was their concern that, at a more general level, evolutionary theory is losing ground among some sections of the general public. In general, affected publics were seen to create a need for extensive dialogue, often around the topics perceived as hidden from institutions and society at large.

The study of amateur naturalist focuses on building knowledge infrastructures for CS (Karasti et al., 2016a; Karasti et al., 2016b) and on the importance of reciprocity (Mauss, 1950 (2002); Sahlins, 1974). Sahlins's typology of reciprocity included generalized reciprocity; balanced, or symmetrical, reciprocity; and negative reciprocity (Sahlins, 1974). All kind of resources, whether tangible or intangible, can be transformed into a gift (Sherry, 1983). Gifts signify a relation that is not solely exchange based but include emotional and social dimensions (Carrier, 1991) e.g. constitute a specific way of playing out environmental citizenship. Reciprocity highlight one crucial element: personal relevance for different publics participating in CS (Frewer et al., 1999). Stocklmayer attempts to map the science communication field by asking three 'basic questions' about communicating any scientific material: 'from whom?', 'to or with whom?', and 'to what end?' (Stocklmayer, 2013: p. 27). Underlying these three basic questions is the key issue of relevance.

Generalized symmetry comes most clearly into play when looking at skewing. Skewing results partly from people following their own interests or going only to favorite places, not taking into account the interests or needs of science; however, these amateur naturalists might be concerned with the environment at their favorite place. This version of generalized symmetry is also encouraged by knowledge politics, giving priority to species on the Red List and the Alien Species List. But SO also provides an opportunity to identify 'white spots' and thereby select new important places to record. We also see that generalized symmetry comes into play when enrolling young amateur naturalists. The more experienced ones see that this is important for sustaining the community of amateur naturalists, but the recruitment is in many respects problematic. One both help the young ones more, and one is more overbearing with their ignorance, however feel rewarded when they start participating in the validation discussions. Personal relevance is high when it comes to generalized symmetry.

Balanced, or symmetrical, reciprocity is the most apparent version of reciprocity when studying biodiversity mapping. On average, each of our interviewees collected 1,575 different species and submitted 58,000 records. This is generally not done solely for the common good. More often, it is done to have order among their own observations or, more specifically, using SO as their own field diary, making them visible to the community of fellow amateur naturalists, achieving status as a knowledgeable and experienced amateur naturalist, and not the least performing environmental citizenship. Personal relevance is crucial. This knowledge may be used to assist others in the validation work. SO practice a form of openness that sometimes is experienced as problematic, consequently, several Facebook-pages have appeared. A comfortable, respectful atmosphere is important to facilitate opportunities for reciprocity (Kramer et al., 2005), which SO not always accomodate.

Sahlins also noted the concept of negative reciprocity, the attempt to get something for nothing with impunity (Sahlins, 1974). Many of those concerned by negative reciprocity think of it as a form of theft. Other people harvest from SO for their own gain, unconcerned with giving anything back. Examples include illegal hunting of protected species, but sometimes legal hunting of the same species, since public authorities allow hunting if a protected species attacks domestic animals or appears outside their zone of protection. In this sense, SO could be misused to identify places for legal hunting, a use that most amateur naturalists find predatory. They therefore invent their own protection scheme by either not recording sightings of such species or by being vague about the exact location of such sightings. Other activities that the participants in this study perceived as negative reciprocity included the form of validation that led to a feeling of being pilloried, open for ridicule.

3. Concluding Discussion

Three main styles can be identified when scientists address general, pure, partisan, and affected publics, and the importance of dialogue is common to all three styles. At the same time, the scientists perceived these dialogues as 'invisible' to institutions and society at large. Consequently, most of what they do under the third assignment is experienced as surprisingly unrecognised. Apparently, this is not a recent phenomenon but has characterised science communication in natural history research museums for perhaps as long as they have existed. Style is not only a question of 'good', 'bad' or 'average' but is also a question of addressing different publics in a relevant way. For that reason, in addressing the quality challenge, greater attention should be paid to what is relevant for different publics.

To conclude, our main claim is that knowledge infrastructures for biodiversity mapping that facilitate generalized and balanced reciprocity encourage longer lasting participation from amateur naturalists in knowledge production and knowledge politics. First, as illustrated in this study, a large variety of actors—be they volunteers, dedicated amateur naturalists, amateur societies, scientists, and administrators—collaborate in Norway's largest CS project. More importantly, anyone can participate, whether a newcomer with only one recorded sighting per year or a veteran with several thousands. Some of the more skilled veterans may of course find the different levels of expertise a challenge; however, most of them recognize the value of this inclusive participatory principle.

As important aspects of studying reciprocity, three subquestions have been asked. First, who are the SO users? Three out of four are men. One interesting question is if infrastructures for natural history appeal differently depending on gender (Brenna, 2016; Rogan, 1998). We are not able to discuss this question based on the available data. The users are mostly well educated; about three out of four have higher education. They represent a steady user group; about one out of two has been using SO for over five years. More than one out of three use SO every day. Consequently, they represent a competent and dedicated pool of free labor, mapping Norwegian biodiversity.

Second, how does SO facilitate engagement and the building of expertise? Three out of four use the resources provided to enhance their own expertise, both interactional and contributory types (Collins et al., 2007; Collins et al., 2015a; Collins et al., 2015b) thereby providing answer to Stocklmayer's question 'to what end?' (Stocklmayer, 2013: p. 27). Many users are members of one or several amateur societies and participate in different activities to enhance their own skills as field naturalists. About 39% contact scientific institutions to obtain help or submit specimens. Validation is a crucial issue that concerns many users. The organized network of validators prioritizes the species included on the Red List and the Alien Species List. For the rest, apomediation is central (Eysenbach, 2008); however, many comments indicate that it works only to a certain degree, and much of the material is never validated. Many users find this issue troublesome; however, an important group also perceives the validation as Sisyphean. It is never ending, and what is done will never meet everyone's satisfaction. One way of rephrasing the quality is illustrated by the claim, 'the quality is in the quantity'. Third, how does the infrastructure facilitate reciprocity? The two most important reasons for using SO are to 'contribute to biodiversity knowledge generally' and 'keep track of my own sightings'. However, summarizing the emphasis on individual outcomes and comparing these with collective outcomes show both as having more or less equal importance.

The emphasis on individual outcome also reflects the importance of personal relevance. As mentioned, Mauss (1950/2002: 50) describes three crucial obligations in a gift economy: 'to give, to receive, to reciprocate'. The users give their sightings, SO receives the sightings and some are validated, and SO reciprocates the gift by facilitating individual projects within the knowledge infrastructure for every user (if they so prefer). Consequently, there is multiple level of reciprocity and gifts. The responses obviously show much appreciation for such reciprocation; some respondents wish for even more tailored reciprocations for each user. Regarding how the respondents use SO, the two most frequent user forms are 'Submit sightings' and 'Look at today's sightings'. However, summarizing the emphasis on either uploading or downloading information reveals that one third focus on uploading, while two third emphasize downloading. Uploading information highlights the act of giving, while downloading information stresses SO's ability to reciprocate in a relevant manner. All in all, SO contain bundles of rights and obligations for those who participate. Furthermore, ranking list may be experienced as an important form of visualizing reputation and consequently as a symbolic return gift in one specific version of credibility cycles

References

Bessette G. (2004) Involving the community: A guide to participatory development communication, Penang: Southbound and International Development Research Centre.

Bonney R. (1996) Citizen science: A lab tradition. Living Bird 15: 7-15.

Bowker G, Star SL. (1999) Sorting Things Out: Classification and its Concequences, Cambridge, Massachusetts: The MIT Press.

Braun K, Schultz S. (2010) '... a certain amount of engineering involved': Constructing the public in participatory goverance arrangements. Public Understanding of Science 19: 403-419.

Brenna B. (2016) Samlingens poesi. Tidsskrift for Kultur-Forskning 15: 37-49.

Carrier J. (1991) Gifts, commodities, and social relations. Sociological Forum 6: 119-136.

Collins H, Evans R. (2007) Rethinking expertise, Chicago: The University of Chicago Press.

Collins H, Evans R. (2015a) Expertise revisited, Part I – Interactional expertise. Studies in History and Philosophy of Science: 113-123.

Collins H, Evans R, Weinel M. (2015b) Expertise revisited, Part II: Contributory expertise. Studies in History and Philosophy of Science.

Conniff R. (2011) The species seekers: Heroes, fools, and the mad pursuit of life on earth, New York: W. W. Norton & Co.

Eysenbach G. (2008) Medicine 2.0: Social networking, collaboration, participation, apomediation, and openness. Journal of Medical Internet Research 10: 1-9.

Frewer LJ, Howard C, Hedderley D, et al. (1999) Reactions to information about genetic engineering: Impact of source characteristics, perceived personal relevance, and persuasiveness. Public Understanding of Science 8: 35-50.

Hetland P. (2018a) The biodiversity mappers: Knowledge production and knowledge politics. Submitted.

Hetland P. (2018b) The quest for reciprocity: Scitizen science as a form of gift exchange. In: Hetland P, Pierroux P and Esborg L (eds) The deep history of participation in museums and archives: Traversing Citizen Science and Citizen Humanities. Submitted.

Hetland P. (2018c) Styles and publics in science communication. Submitted.

Hetland P, Mørch A. (2016) Ethnography for investigating the Internet. seminar.net 12: 1-14.

Irwin A. (1995) Citizen science: A study of people, expertise and sustainable development, London: Routledge.

Karasti H, Millerand F, Hine CM, et al. (2016a) Knowledge infrastructures: Part I. Science & Technology Studies 29: 2-12.

Karasti H, Millerand F, Hine CM, et al. (2016b) Knowledge infrastructures: Part II. Science & Technology Studies 29: 2-6.

Kirby DA. (2011) Lab coats in Hollywood: Science, scientists, and cinema, Cambridge, MA: The MIT Press.

Kramer DM, Wells RP. (2005) Achieving buy-in: building networks to facilitate knowledge transfer. Science Communication 26: 428-444.

Mauss M. (1950 (2002)) The gift: The form and reason for exchange in archaic societies, London: Routledge.

Metcalfe J, Riedlinger M, Pisarski A. (2008) Situating science in the social context by cross-sectoral collaboration. In: Cheng D, Claessens M, Gascoigne T, et al. (eds) Communicating science in social contexts: New models, new practices. New York: Springer Science+Business Media B.V., 181-197.

Rogan B. (1998) On collecting as play, creativity and aesthetic practice. Etnofoor 11: 40-54.

Sahlins M. (1974) Stone-Age economics, London: Tavistock Publications.

Sherry JFJ. (1983) Gift giving in an anthropological perspective. Journal of Consumer Research 10: 157-168.

Star SL, Griesemer JR. (1989) Institutional ecology, 'Translations' and boundary objects: Amateurs and professionals in Berkeley's Museum of Vertebrate Zoology. Social Studies of Science 19: 387-420.

Stocklmayer S. (2013) Engagement with science: Models of science communication. In: Gilbert JK and Stocklmayer S (eds) Communication and engagement with science and technology: Issues and dilemmas. New York: Routledge, 19-38.

Wells G. (1999) Dialogic inquery: Towards a sociocultural practise and theory of learning, Cambridge: Cambridge University Press.

An Analysis of the Science and Culture in the Era of Artificial Intelligence

Li Zhihong

School of Public Policy and Management, University of Chinese Academy of Sciences, Beijing, China

Abstract: As an important technological development in the new era, artificial intelligence has brought a full impact on the social development, and the science and culture of the era of artificial intelligence presents some new features. The age of AI is characterized by intelligence, information and complexity. Science and culture present the characteristics of intelligent culture and new human-machine interaction. Artificial intelligence has a far-reaching impact on science and culture, so we need avoid the ethical and moral hazards which artificial intelligence brought.

Keywords: Artificial Intelligence; Culture of Science; Human-machine Relationship

1. Introduction

As an important technological development in the new era, artificial intelligence has brought a full impact on the social development. In the era of artificial intelligence the science and culture presents some new features. There are some experts study artificial intelligence from different perspectives. They study the social impact of artificial intelligence, the impact on employment, ethical issues, value issues, and so on. Chris Galloway and Lukasz Swiatek (2018) pointed out it is important to realize the multiplicity of roles that AI is beginning to play in public relations, and will play in coming years. Weiping Sun (2017) pointed out the value conflicts and deep challenges caused by AI. Sumei Cheng (2017) explored the future of artificial intelligence from the philosophical perspective by analyzing the paradigm shift of artificial intelligence. Jing Chen (2017) researched the integration of science and technology and ethics, analyzed the human cultural artificial intelligence, and pointed out technology of artificial intelligence will bring about both technological and cultural changes. Tong Wu (2018) analyzed the social problem and the governance of society of artificial intelligence. Xinrong Huang (2018) pointed out human-machine relationship is not strained in the new artificial intelligence era, machine is not completely opposite to human.

Corresponding author: Zhihong Li, No. 19(JIa), Yuquanlu Road, Shijingshan District, Beijing. lizhih@ucas.ac.cn.

2. What Is AI?

Artificial intelligence (AI) is a new technical science which studies and develops theories, methods, techniques and application systems for simulating, extending and expanding human intelligence. The rapid development of AI technology will bring about a new era. In the past 60 years, artificial intelligence has gone through three waves of development. First, at the Dartmouth Conference in 1956, four Turing Prize winners, pioneers of information theory innovation, and a Nobel Prize winner defined the terms of artificial intelligence, including Minsky, Simon, and McCarthy. Later, the first perceptual neural network software and chat software were invented, which proved the mathematical theorem. Second, Hopfield neural network and BT training algorithm were proposed in 1982, and a new upsurge emerged, including speech recognition, speech translation program, and the fifth generation computer proposed by Japan. Third, Artificial intelligence has entered a new stage with Hinton's in-depth learning technology in 2006 and some successes in image, speech recognition and other fields. Now artificial intelligence is rapid developing and bring many new change.

In general, there are three Characteristics of artificial intelligence. First is intelligence. Artificial intelligence simulates the information process of human consciousness and thinking, and can have human intelligence. Based on the characteristics of machine learning, in-depth learning, self-learning, intelligence is obvious, and can complete the promotion of self-development. Second is informatization. AI is the product of rapid development based on informatization, networking and computerization. AI is not only a revolution in technology, but also a comprehensive transformation. Artificial intelligence will bring disruptive innovation. Third is Complexity. In the new era of artificial intelligence, it will be beyond machine development, with human intelligence, but also beyond human development, in a certain way beyond the activities of individual people.

3. The Current Situation of AI Policy Abroad

With the development to AI technology, many countries have made important policy to push the development of AI. In 2016, the United States published three reports on artificial intelligence, namely, Preparing for the Future of Artificial Intelligence, National Strategic Plan for Research and Development of Artificial Intelligence, and Artificial Intelligence, Automation and Economy Report. The major policy initiatives include: Monitor the development of AI technology; Long term R&D investment in AI; Education for all; Manage AI risks; Artificial intelligence research and development personnel, and so on. The policy is helpful to the AI technology development.

In 2016, the UK published two reports, Artificial Intelligence: Opportunities and Impacts for Future Decisions and Robotic Technology and Artificial Intelligence. In UK reports, the revolutionary impact of artificial intelligence and how to use it are elaborated and planned, especially focusing on the legal and ethical risks brought about by the development of artificial intelligence.

In China, in May 2016, the national development and Reform Commission, the Ministry of science and technology, the Ministry of industry and information technology and the central network office jointly formulated the 'Internet +' three year action plan for artificial intelligence. In July 2017, the State Council issued the 'New Generation Artificial Intelligence Development Plan', which not only made a strategic plan for the development of artificial intelligence, but also established the 'three-step' policy objectives, and strived to build China into the world's major artificial intelligence innovation center by 2030. With these policies, AI technology is developing quickly in China.

4. Science and Culture

Science is a culture. Science and culture are symbiosis. Science development is inseparable from cultural atmosphere. Xingmin Li (2007) pointed out the characteristics of science and culture. Dachun Liu (2012) pointed out that science is a mainstream culture. There are some characteristics of Culture of Science. The spirit of science is inherent in scientific culture. Scientific value is an important guide for scientific development. System is the guarantee of scientific culture. Scientific collaboration is the construction of a scientific and cultural atmosphere. The development of science and technology has penetrated into every field of society and has greatly changed the way of human life. With the development of modern society, the division and specialization of the global knowledge system become more and more intense, which leads to the division of science and humanities. Although the future of mankind depends on the interaction of many factors, the combination of science and humanities is undoubtedly an important decisive factor. The combination of science and humanity is an important factor to promote the coordinated development of society. Scientific and technological resources and economic resources have led to the enormous development of productive forces and rapid economic growth. It can also promote the overall progress of society, including ideology, morality, culture, ideas and so on. With the development of AI, new features of science and culture appear.

5. Impact of AI

Artificial intelligence can simulate the information process of human consciousness and thinking, and can think like human beings, possibly exceeding human intelligence. The mode of human-computer interaction has changed, the combination of artificial intelligence and industry is more convenient, and artificial intelligence gradually enters people's life. Deep neural network, large data and massive computation make the development of artificial intelligence more complicated. The age of AI is characterized by intelligence, information and complexity. Artificial intelligence is the extension and enhancement of human function, which is highly anthropomorphic.

There are some series of key technologies, such as in-depth learning of artificial intelligence, image speech recognition and holographic image, have a profound impact on science and culture. According to the results released by NIST in 2018, the best face recognition technology in the world at present has a recognition accuracy of nearly 99%. With this in-depth learning technology, machine vision can not only recognize, but also understand to a certain extent.

First, the change of way of thinking. Artificial intelligence will bring about the revolution of intelligence and the change of ideas. Knowledge updating speed is faster and faster. The future society is a learning society, learning talents, lifelong learning. The acquisition and selection of knowledge and information is more convenient, and how to better select, analyze and process information will become particularly important.

Second, the impact on education. AI pays more attention to deep learning. As a new change in the field of education, online education and interactive learning will bring new challenges to the traditional way of education. There are more ways to learn, more information and more in-depth learning. Through the mass storage of machines, knowledge can be accumulated quickly.

Third, Challenges posed by new human-machine relationship. Whether machines can surpass human beings and whether machines can surpass human intelligence has always been a new problem brought about by the development of artificial intelligence. AI will be a new type of human-machine interaction. Artificial intelligence will bring new ethical problems. It bring new problems to robot work of the protection of privacy, the fairness of the law.

6. Roles for AI in Science and Culture

Artificial intelligence is the extension and enhancement of human function, which is highly anthropomorphic. A series of key technologies, such as in-depth learning of artificial intelligence, image speech recognition and holographic image, have a profound impact on science and culture. The new artificial intelligence theory, such as deep calculation based on large data and self-learning, will have an important influence on the development of science. The characteristics of artificial intelligence personification have developed a new type of human-machine relationship, which makes scientific cooperation in many fields easier to carry out, and may produce important breakthroughs.

AI will bring a series of cultural innovations. First, the mode of knowledge study will change. There will have more opportunities for study through AI. Through artificial intelligence, educational methods and contents are more convenient, information is more abundant, and human-computer interaction makes education more participatory. Through the function of intelligent storage and memory, the content of education is expanded.

Second, cross-border integration is wide. With human brain intelligence and artificial intelligence greatly developed, the relationship between them and their mutual cooperation has become more in-depth. The whole Internet is the connection between human and machine.

Third, the thinking pattern is changing. Artificial intelligence is not only a new technology, but also a way of understanding and thinking about the world, and a way of life for our future. It is not a simple technological change, but a change of the human's thinking pattern. Artificial intelligence brings about an all directional revolution.

The last, a new kind of human-machine collaboration relationship appear. The intelligence of a person to gain dominance over all other natural objects may be surpassed by machines. AI improves the quality of human science and life, and at the same time brings risks to mankind. Artificial intelligence has a quick development now. We need face the challenges of AI. Artificial intelligence has a far-reaching impact on science and culture, so we need avoid the ethical and moral hazards which artificial intelligence brought.

7. Conclusion

This paper discusses the impact of AI on science and culture. In new era, we need strengthen the development of AI. We need have effective use of AI to enhance interaction between human and machine, in that way, human-machine interaction and collaborative will have good governance. We should improve professional talents and information literacy of the whole people. In response to the challenges brought by AI, we must also regulate ethical problems and legal issues of AI.

References

Chen, J. (2017). Integration of technology and ethics—On human culture of AI technology, *Academics in China,* (9), 102-111.

Cheng, S.M. (2017). Paradigm shift and prospect of AI research, *Philosophical Trends*, (12), 15-21.

Galloway, C., Swiatek, L. (2018). Public relations and artificial intelligence: It's not (just) about robots, *Public Relations Review*, 10, pp.1-7. Available at https://doi.org/10.1016/j.pubrev.2018.10.008

Huang, X.R. (2018). Philosophical Reflection on the upsurge of artificial intelligence, *Journal of Shanghai Normal University*, (4), 34-42.

Sun, W. P. (2017). Reflection on the value of artificial intelligence, *Philosophical Researches,* (10), 120-126.

Li, X.M. (2007). On scientific culture and its characteristics, *Science & Culture Review,* (4), 72-87.

Liu, D. C. (2012). Science culture and culture science, *Journal of Dialectics of Nature*, (6): 1-7.

Wu, T. (2018). Some philosophical reflections on the development and governance of AI, *People's Tribune*, (5), 18-25.

Discourse Construction of 'PX' Risk: Public Risk Perception on Weibo and Zhihu

Yue Liyuan[1], Zhang Zengyi[2]

[1] Institute of Science, Technology and Society, School of Social Sciences, Tsinghua University, Beijing, China
[2] School of Humanities and Social Sciences, University of Chinese Academy of Sciences, Beijing, China

Abstract: Since 2007, the PX (p-xylene) project has triggered a series of mass incidents across China, triggering a constant risk controversy. These disputes are directly related with the public scientific cognition of PX risk, and social media is the main space of public discourse expression. Therefore, the public discourse on social media (Sina Weibo and Zhihu community) was chosen to investigate the risk perception and attitude of anti-PX public from seven dimensions by referring to James Paul Gee's discourse analysis method. On this basis, this paper points out the role that public discourse plays in the construction of PX's 'high-risk' image, as well as the the role of social media in promoting public's anti-PX discourse and actions, and the deep reasons for the public's risk construction of PX was revealed.

1. Introduction

At present, the risk problem caused by the development of science and technology has become one of the hot spots in the research of social governance of science and technology in the field of science and technology and society (STS). With the further development of economic globalization and domestic modern industry, Chinese society has shown some characteristics of the Risk society proposed by Ulrich Beck (1986). Compared with developed countries in the world, China, entering the so-called 'social transformation period' and 'environmentally sensitive period', faces more complex and confrontational risks in the application of science and technology: it should not only complete industrial modernization construction and establish modern systems, but also face the unexpected consequences and side effects of reflecting on industrial modernization. In recent years, the domestic PX issue has provided a typical social example of such risks.

In 2007, Xiamen Haicang PX project was moved to the Gulei peninsula due to the public's collective 'walking' opposition. (Public boycott)The PX project first came into public view. Subsequently, public anti-PX 'walking' protests were held in Dalian, Ningbo, Chengdu, Kunming and Maoming. In the past ten years, there has been a constant controversy over the risks of PX projects. At the same time, public resistance to PX has never stopped, making the word 'PX' beyond its original meaning and a synonym for environmental risks in China at the present stage. In previous PX incidents, the public used mobile phones, express appeal, which to some extent formed the 'public opinion field' opposite to the 'official public opinion field' represented by local governments. What is the public's cognition of PX risk? How does public discourse influence controversy and shape this PX risk image?

According to the full text search by CNKI (China National Knowledge Infrastructure), there are 314 humanities and social science papers with the keyword 'PX' in the title or theme (as of March 1, 2017). Statistics show that since the outbreak of PX incident in Xiamen in 2007, domestic scholars have been increasingly studying the PX issue. From the research content of literature, it is mainly about traditional media news report, government governance and crisis public relations, online public opinion analysis, etc., and there are few researches related to the public.

Among them, Tan Shuang (2013) analyzed the psychological process of the public from anxiety to opposition to PX project, pointing out that the public's group anxiety was not resolved in a timely manner, which eventually led to group behavior. Li Xiuqi (2013) pointed out that the public's political will was strengthened to resist the fight against PX for rights and

Corresponding author: Yue Liyuan, Institute of Science, Technology and Society, School of Social Sciences, Tsinghua University, Beijing, 100084, China. yue_liyuan @sina.com interests. Dai Jia et al. (2014) found the promotion effect of rumors on the public's resistance to PX, and described the process of spreading rumors and effects. From the perspective of public risk perception, some researchers propose to strengthen risk communication or risk management through media. 'The public is the customers, consumers, users and investors of governments and scientists, so the outcome of scientific debates depends largely on the public who determine the candidates and should express their views and concerns,' science communication researchers (Li Zhengwei et al., 2004) said.

2. Sample Selection and Research Methods

1) Research Sample–Public Discourse Corpus

During the Xiamen PX incident in 2007, the main new media used by the public were SMS, QQ, blog and BBS. In the following years, Weibo of various portal websites was launched successively. As information spreads faster and more widely and is easy to interact with, it has become the main online space for the public to express their attitudes and opinions. Sina Weibo is one of the most widely used and typical social media in China (Liu Zhensheng, 2013). In addition, as a new emerging online Q&A community, Zhihu has been widely used in user knowledge dissemination and information quality research(Huang Lucheng et al., 2016). It provides materials for the analysis of public PX risk perception Therefore, this paper lists Sina Weibo and Zhihu's PX topic text as the main research objects. In addition, in order to make a more comprehensive analysis of public discourse, the existing research results have also collected second-hand information such as text messages, blogs and interviews in PX incidents in Xiamen and Dalian, which together constitute a public corpus.

Sina Weibo was set as September 1, 2009 (on-line time) — September 1, 2016. With 'PX project' as the key word, 11,260 original microblog posts were retrieved. Computer programming is used to capture relevant data. After screening, 1,415 individual tweets were obtained. After screening comments, the number of retweets and the readability of the content, 279 microblogs were finally obtained as research samples. At the same time, search the text of the PX topic on Zhihu, including questions and answers. The time span is from 2011 to September 2016. 138 questions have been collected and a total of 1,022 replies have been received. According to the comments of the response text and the number of concerns, a total of 125 questions and 242 replies were selected as samples.

2) Methods

The method of combination of quantization and qualitative analysis is adopted. Discourse analysis is carried out on the basis of content analysis of network text. First, the content analysis was carried out for the sample coding statistics of Weibo and Zhihu as the quantitative basis for further discourse analysis. Then, on this basis, the discourse analysis of anti-PX public was carried out in combination with supplementary materials. Discourse analysis draws on the comprehensive method of discourse analysis proposed by James Paul Gee (1999) .Gee proposes that people create seven 'realities' while speaking or writing an article, that is, seven construction tasks significance、activities、identities、relationships、politics、connections、sign system and knowledge (Gee J P., 2000).

Based on these survey tools, this paper examines the risk perception and attitude of anti-PX public from seven dimensions of the construction task. This paper analyzes how they achieve the task of discourse construction, analyzes the construction effect of public discourse on PX's 'high-risk' image, and the role of new media as the public to construct anti-PX discourse and action, and reveals the deep reasons for the public to construct anti-PX discourse and action.

3) Analyze Unit and Category Settings

According to the specific content of sample, combined with the significance of seven dimensions of discourse analysis further to investigate, set the analytical unit of sample survey as five categories: the text type, posting attitude, PX risk cognition, attitude towards other stakeholders and rhetorical strategies.

(1) *Text type and posting attitude*

According to the content attribute of text, Weibo and Zhihu text are divided into information and comment. Posts about social media texts. Xia Yuhe (2011) integrated microblog public opinion into five modes of identification, consultation, questioning, emotion and dissociation (Xia Yuhe, 2011). Based on this, Jin Ming et al. (2013) operated into five categories: objective attention, seeking solutions, rational questioning, emotional catharsis and dissociation mode (Jin Ming et al., 2013). This paper uses this classification for reference and adds a 'rumors spread' in combination with the sample, setting the attitude of Posting to six categories.

(2) *Awareness of the risk of PX*

Refer to the Likert scale for the respondents' cognitive rating Settings Read through the sample content and summarize it. Divide the public's awareness of PX risk into five categories: ① greater harm, ② there are risks, ③ risk uncertainty, ④ there are risks, but they can be controlled effectively, ⑤ effective for the nation and individuals, but they are demonized.

(3) *Attitude towards other stakeholders*

Based on the general reading classification of all contents, the public's attitude towards other stakeholders (government, enterprises, experts, media and other public) is summarized into five categories more specifically: ① criticism/negation (negative), ② questioning/mistrust (negative), ③ measures of objectivity (neutral), ④ purposeful understanding (positive), ⑤ affirmation (positive).

(4) *Rhetorical strategies*

This paper summarizes the sample content and concludes that rhetorical devices can be divided into verbal rhetoric and visual image rhetoric. Among them, there are 7 types of rhetoric in language: citation, case (comparison/analogy), emergence, logical thinking, metaphor, irony (irony), story telling (joke). Visual image rhetoric includes real photographs, editing images, and symbols.

3. Content Analysis Based on Weibo and Zhihu Samples

First, the word frequency statistics of the entire microblog and Zhihu texts were carried out using the Analysis software Rost News Analysis Tool v3.1. It can be seen that the key words are explosion, xylene, Zhangzhou, aromatics, petrochemical, popular science, toxicity and other words related to PX risk.

1) Type of Text and Posting Attitude of Public Discourse

Statistics show that Zhihu users have more comments (95.0%) than Weibo users (56.7%), accounting for an absolute proportion. Perhaps because the chosen Zhihu text is just the answer to a certain question, more attention should be paid to argumentation and evaluation. It's worth noting that in Weibo, 8.3 percent of the comments calling for (anti-PX) were posted, while Zhihu did not appear. On Weibo (9.5%), more people hold the attitude of 'seeking solutions' than on Zhihu (4.6%). Perhaps because the Zhihu text itself is a response to a particular question, The intention of 'seeking solutions' has been embodied in the problem text. Although the proportion of 'objective attention' attitude on Weibo (30.6%) is greater than that of Zhihu (22.9%), They're on a par in 'emotional catharsis', but the proportion of 'rational doubt' in Zhihu (34.6%) is significantly higher than that on Weibo (12.7%). Moreover, the proportion of 'rumour propagation' (6%) on Weibo is much higher than that of Zhihu (0.8%). By comparison, respondents from Zhihu seem to be more rational. Overall, the proportion of emotional catharsis of Internet users is the highest, followed by objective attention and rational questioning.

Table 1 Posting attitude of public discourse.

Attitude	Weibo	Proportion	Zhihu	Proportion	Overall	Proportion
Objective attention	87	30.60%	55	22.90%	142	27.10%
Seeking solutions	27	9.50%	11	4.60%	38	7.30%
Rational questioning	36	12.70%	83	34.60%	119	22.70%
Emotional catharsis	114	40.10%	85	35.40%	199	38.00%
Dissociation mode	3	1.10%	4	1.70%	7	1.30%
Rumors spread	17	6.00%	2	0.80%	19	3.60%
Total	284	100.00%	240	100.00%	524	100.00%

2) Public Perception of PX Risks

On the negative perception of PX risk, More Weibo users (67.4%) than Zhihu (52.6%); With a neutral attitude, Zhihu (22.9%) is much higher than Weibo (7.9%). However, the proportion of the risk of PX being demonized on Weibo is slightly higher than Zhihu. Probably due to the large number of registered users on Weibo, those who support PX will respond in a timely manner and be affected by the deletion of negative information on Weibo. Overall, the perception of PX damage by netizens is still lower than that of Weibo users.

Table 2 Public perception of PX risks.

Public Perception of PX Risks	Weibo	Proportion	Zhihu	Proportion	Overall	Proportion
1. Negative: Greater harm	19	6.80%	3	1.30%	22	4.20%
2. Negative: There are risks	169	60.60%	123	51.30%	292	56.30%
3. Neutral: Risk uncertainty	22	7.90%	55	22.90%	77	14.80%
4. Positive: There are risks, but they can be controlled effectively	34	12.20%	21	8.80%	55	10.60%
5. Positive: Effective for the nation and individuals, but they are demonized.	17	6.10%	8	3.30%	25	4.80%
6. Poncommittal	18	6.50%	30	12.50%	48	9.30%
Total	279	100.00%	240	100.00%	519	100.00%

3) Public Attitude Towards Relevant Stakeholder of PX Project

Overall, the government (39.8%) and the public (31.5%) had higher mentions. In response to the government's role, 17.8 percent of netizens blamed local governments, while 21.7 percent said they did not trust the government. The corresponding negative attitudes were 'blaming the local government for the incident and accident' and 'criticizing the local government for improper procedures and disrespecting civil rights'. Only 0.3% (1 person) 'understood the difficulties faced by the government'. There are obvious differences in online attitudes towards public roles. Most netizens who hold a negative attitude towards PX come from Weibo. 'Called for popular protest' (5.0%) However, the netizens who objectively viewed the PX project mainly came from Zhihu, and 'criticized the public for being irrational, and their knowledge quality needed to be improved'. Other popular expressions include 'NIMBY (not in my back yard) conflict' between people of different regions (4.7%), 'Understanding the plight of the public and the actions of struggle (interests)' (3.3%), 'The opinion leaders were criticized for spreading rumors to incite the crowd' (1.1%). There is also a palpable distrust of experts, businesses and the media. In contrast, Weibo users are more extreme than Zhihu because they show more emotional and irrational content.

Nearly half (48.9%) of the netizens adopted rhetorical devices in their discourse. In general, visual image rhetoric (picture) is the most commonly used, and linguistic rhetoric mainly includes case comparison/analogy (14.5%) and satire (12.5%). The proportion of Zhihu users using citations (scientific data, materials, laws and regulations), cases and stories is higher than that of Weibo users, while the proportion of Weibo users using hyperbole and image rhetoric is higher. From this point, Zhihu netizens are relatively rational and objective.

4. Analysis of Public 'Anti-PX' Discourse: on the Construction of Harmful 'Presence'

1) Building Activities: 'Walking'—Public Discourse of Action

The construction of public discourse on activities is mainly reflected in facilitating the anti-PX action—'group walk', Including making proposals, calling for and organizing the public to fight against the PX project, This is also evident in the text of the blog post, where the type of text in the comments calling for a revolt against PX is more obvious. Anti-PX demonstrations in Dalian, Ningbo, Kunming, Chengdu and Maoming were organized, lobbied and advocated by the initiators. Tracing back to the Xiamen PX event, opinion leader Lian Yue's blog content 'Xiamen people do this' and 'Xiamen people do this 2', the famous 'Lian ten items'. In addition, there is the QQ group of 'give me back Xiamen's clear water and blue sky' established by the initiator and advocate of the parade. (Jin Ming et al., 2013). And the message forwarded by millions of people in Xiamen, which compares the harm of PX to 'atomic bomb', 'leukemia', 'deformity' and incites demonstrations (Zhang Xiaojuan, 2007).

2) Building Identities: 'We' Are 'Citizens' Who Are Guarding Our Home

Starting from the Xiamen incident, which originated from the enlightenment and guidance of opinion leaders and the rendering of some market-oriented mass media (such as Southern

Table 3 Public attitude towards relevant parties of PX project.

Subject Mentioned	Weibo	Proportion	Zhihu	Proportion	Overall	Proportion
1. The government	94	48.20%	49	29.90%	143	39.80%
① criticism/negation (negative)	48	24.62%	16	9.76%	64	17.83%
② questioning/mistrust (negative)	46	23.59%	32	19.51%	78	21.73%
③ understanding (positive)	0	0.00%	1	0.61%	1	0.28%
2. Petrochemical enterprises	19	9.70%	23	14.00%	42	11.70%
① criticism/negation (negative)	1	0.51%	8	4.88%	9	2.51%
② questioning/mistrust (negative)	18	9.23%	15	9.15%	33	9.19%
3. experts	11	5.60%	33	20.10%	44	12.30%
① questioning/mistrust (negative)	10	5.13%	23	14.02%	33	9.19%
② understanding (positive)	1	0.51%	10	6.10%	11	3.06%
4. Media	14	7.20%	3	1.80%	17	4.70%
① criticism/negation (negative)	4	2.05%	0	0.00%	4	1.11%
② questioning/mistrust (negative)	10	5.13%	3	1.83%	13	3.62%
5. public	57	29.20%	56	34.10%	113	31.50%
① A criticism/negation (negative)	18	9.23%	0	0.00%	18	5.01%
② B criticism/negation (negative)	19	9.74%	43	26.22%	62	17.27%
③ questioning/mistrust (negative)	10	5.13%	7	4.27%	17	4.74%
④ understanding (positive)	6	3.08%	6	3.66%	12	3.34%
⑤ affirmation (positive)	4	2.05%	0	0.00%	4	1.11%
Total	195	100%	164	100%	359	100%

weekend), the anti-PX public constructed the identity of 'we' : 'citizens' guarding the homeland. This identity was continued and established in subsequent PX events(Zhou Haiyan, 2012). a) 'We' - highlighting the sense of belonging to 'home'. Lian Yue emphasizes that he is 'a native of Xiamen', and an open letter written by Lu renzi, a journalist from Dalian, to the city leaders regards himself as 'a citizen of Dalian'. This is also clearly reflected in the public discourse of Weibo, which calls for the public to participate in the anti-PX campaign and emphasizes their local identities such as 'Ningbo people' or 'Kunming people'. For example, 'Retweets for beautiful homeland' 'All friends who love XX(city), unite together!' 'Join the resistance team', etc. The public has put the proposed PX project in conflict with the interests of the local ('homeland'), The expression 'we' is a protest to protect the 'homeland'. This expression emphasizes the sense of identity and belonging to the place, and constructs the 'local community' of 'we'. b) 'Citizens'–emphasizing the legitimacy of rights. Weibo and Zhihu netizens have mostly criticized local governments for failing to release information in a timely manner and for failing to respect citizens' legitimate rights and interests. The construction of 'citizen' identity of ordinary people is mostly derived from the guidance of opinion leader Lian Yue. During the PX event in Xiamen, Lian Yue emphasized that environmental protection subjects should have citizenship. He mentioned citizens and civil society more than once in his blog and enlightened the public's citizenship consciousness.

3) Building Relationships: Confrontation Between Subjects that are Contrary to All Interests

In the anti-PX campaign, the main objects of public resistance are local governments (39.7%) and petrochemical companies. They believe that local governments have formed an interest alliance with petrochemical companies in pursuit of GDP, regardless of ecological environment and people's safety. PX experts, who popularize knowledge about PX in the media and justify PX, have also been opposed by the public due to their consistent position with petrochemical enterprises and local governments and have become what the public calls 'brick-makers'. There is also a public antagonistic interpretation of the state media, with only negative attitudes of criticism and distrust. As for the PX popular science articles in People's Daily and focus interviews, the opposition voices of ordinary people on Weibo have been persistent and fierce.

In addition, there are contradictions among the public. From the perspective of attitude, netizens who oppose PX 'call on the public to fight', while those who objectively view the PX project 'criticize the public's knowledge and accomplishment to be improved'. From the content, it is obvious that the two topics around the opposition. One is the scramble for 'Baidu entry battle'. Students of Tsinghua university who defend the low-toxicity scientific conclusion of PX entry have been constantly satirized, and abused by the anti-PX public, especially after another fire accident happened in Gulei PX plant in 2015. Second, the 'NIMBY' conflict between people in different regions, and the discourse even shows obvious regional attacks.

4) Construction of Symbol System and Knowledge: the Symbolization Transmission of 'risk'

(1) *Shaping the 'prototype' of Xiamen PX event*

The 'Xiamen PX event' is the basis for the domestic public to form a fixed image of 'PX'. After 2007, the fact that Xiamen PX project was resisted to relocate was divorced from the background information such as site selection, and gradually shaped into simple causal logic: Because PX is high risk and has great harm, it attracts the collective boycott of Xiamen people. Many Weibo users expressed such attitude, using the Xiamen incident as a reason to question the government 'Why should we accept the project which Xiamen people driving away?' There is also a demand for environmental justice, which, from a moral point of view, demands that this issue be taken seriously.

(2) *Semantic 'symbolization' transmission*

In previous PX events, some words were gradually simplified as tag symbols and slogans in public opinion after being spread by people through word of mouth and widely spread by mobile phones and social media on the Internet. For example, 'survival', 'life', 'descendants', 'homeland', etc. This kind of symbolized semantic expression is highly targeted, and the appeal is very direct. At the same time, it also shows a strong sense of justice for action, which is easy to spread and transform into the 'group memory' of the whole public. This has been widely seen in the series of PX events, which has promoted the public's action against PX. In addition, the combination of visual rhetoric (46.3%) is also quite extensive. Real photos such as collective 'walking', PX plant explosion, massive death of nearby sea horses, and smoke emission of PX in the clear sky are shocking, which can enhance the sense of scene and enhance credibility, thus enhancing the persuasive power and appeal of posts. The editing pictures of the garden city with wreath metaphor, which is about the city to disappear, have a strong contrast, the effect of alerting people and criticizing the status quo, as well as a strong persuasion effect.

(3) *Rumors spread without scientific logic*

The public also spread rumors in its antagonistic discourse, arbitrarily exaggerating the toxicity and harm of PX. For example, in the incendiary text message of the Xiamen incident, the exaggerated expressions such as 'highly toxic', 'atomic bomb', 'leukemia' and 'distorted child' were adopted without any scientific reasoning or explanation. In addition, hearsay stories are also presented as facts, such as a rumor that 'Kuang Li, the person responsible for the karamay fire, is the director of PX project', which has been included in Dalian, Ningbo and Kunming PX projects.

Public anti-PX discourse equates PX with 'high risk' by exaggerating harm and spreading rumors. According to the sample statistics of Weibo and Zhihu, more than half (60.5%) of the netizens think PX is high risk, while 14.8% of them are skeptical about the risk of PX. Most of the images of PX in microblog are highly poisonous, flammable, teratogenic, carcinogenic and polluting. PX, as a core word, has been stigmatized. The public associated PX events with a series of discontented social phenomena such as environmental pollution, food safety accidents, local government corruption and so on. They tended to regard such incidents as a unified social issue, raise questions from the moral level, and define the situation as 'injustice' (Manuel Castells, 2006). This is also the common practice of constructing collective identity in the new social movement of western society. Therefore, some scholars pointed out that 'it is not automatic to transfer social problems into collective action, but the process of social actors, media and social members to jointly interpret and redefine the situation'.

5) Construction Significance: Protect the Environment and Protect the Future Generations

In each outbreak of PX event, the anti-PX public equated PX with high risk by semantic symbolization and rumor propagation, and constructed the anti-PX action as a just movement to protect the homeland and protect future generations through the establishment of the community identity of 'we'. At the same time, the PX by using a simple and strong calls for public slogan and symbol successful capture people's hearts and minds, prompting a broader group of resistance to the foreseeable risks, and implemented in this movement 'self empowerment', namely to feel its own behavior 'justice' and 'public welfare', 'fearless' and 'sacrifice', and form a sense of belonging and identity, sacredness.

6) Build Positions and Strategies: China's Environmental Justice Movement?

In terms of the description of PX, the anti-PX public argued that the potential harm of PX project was 'in the presence' as much as possible, and they scrambled to define the risk of PX and shape PX into a high-risk image. In terms of discourse mode, life language is the main language, especially the semantic symbol features are obvious. The appeal is direct and direct to the object, showing strong emotions. Although they also try to use scientific discourse as the basis of argument, most of them cannot stand the effective verification of science. In terms of the discourse position, the attitude shown by the public against PX mainly includes: ① PX opposition and boycott, ② the risks involved in the PX project are huge, ③ it is difficult to effective regulation (distrust of local governments and petrochemical companies). The public has repeatedly protested that they want the government to decide to abandon the PX project, or to stop or relocate existing PX projects, promising never to take up the project. We can see that the anti-PX public position was firmly opposed to the construction of PX project near their home. In addition, under the guidance of opinion leaders, the public also regards the central (national) government as a supporter of the movement, giving it legitimacy, while pointing criticism at local governments, enterprises, even experts and local media.

In terms of rhetoric strategy, the first is the use of harm rhetoric(John Hannigan,2009). Use exaggerated or sarcastic language and rhetoric to spread rumors and show the great harm of PX project. Second, the risk of semantic symbol, visual (image rhetoric) transmission. Thirdly, it appeals to the moral? morality – environmental justice, showing a kind of just rhetoric, and endowing the anti-PX campaign with a sense of justice to protect the homeland and protect future generations. Fourth, through the 'walking' campaign, the purpose of stopping the PX project was achieved, which then developed into the 'routine practice' against the PX project, forming a kind of Chinese-style NIMBY movement.

5. Conclusions and Discussions

First of all, the PX project is not universally opposed. According to the results of sample content analysis, the proportion of netizens who think PX is risky is about 60%, that is, more than half of the public holds a negative attitude towards PX. However, it is not as we seem to feel, nearly half of the netizens show a neutral attitude towards PX (27.1%) and a reasonable doubt (22.7%), which is more than an emotional catharsis (38%). This conclusion is basically consistent with some previous studies on media. For example, in the online survey conducted by southern weekend in 2011, more than 40% of the 20,000 participants believed that PX was a 'highly toxic and highly carcinogenic substance' (Southern Weekend, 2014). The global public opinion survey center conducted a survey on the general public in Xiamen, Dalian, Kunming, Chengdu and Maoming in 2014. 36.8% of the public held an 'opposition' attitude, while 39.8% expressed a 'neutral' attitude (Cui Mo, 2014). It can be seen that not all people are opposed to PX. Research shows that public discourse and actions against PX are more positive and intense, which to some extent obscures (cover up) other weak voices.

The second is the promoting effect of new media on public construction of anti-PX discourse and actions. From mobile phone messages, QQ groups and blogs in the PX event in Xiamen to social media such as Weibo and Zhihu, new media has become the main channel and way for the public to express their interest demands, condense people's hearts and participate in the PX project. In particular, Internet media enables everyone to participate in public affairs. Some sensitive issues are restricted and excluded by traditional media, and new media can express and disseminate them. Even those unverified rumors, which have not been verified by professional 'gatekeepers', are widely displayed in the views of netizens. In this case, the power of public opinion is more concentrated. In a sense, the Internet has provided the basic conditions for the transformation of 'mass' into 'public'. In the construction of anti-PX discourse practice, new media has played an important role in promoting, which also reflects the Chinese people's awareness of safeguarding their rights and interests and their increasing attention to environmental issues.

Third, the public construction of anti-PX discourse and action of the deep reasons - interests. The analysis shows that the public mainly focuses on the environmental risks and health hazards of PX project, puts forward the appeal to 'environmental justice', demands for equitable distribution of environmental risks, and on this basis puts forward the claim of 'right' for the health and survival of future generations, which belongs to the 'argument of moral condemnation' put forward by Dorothy Nelkin (2004). This is similar to the 'NIMBY effect' in the west. Judging from the development of the environmental movement in the west, 'NIMBY' movement is inevitable in modern society. But things are different in China, where popular science and demonstrations by state media do not dispel fears based on distrust. PX event is different from the traditional environmental NIMBY movement in western countries, and the public's attention varies greatly among different events. For example, compared with the pursuit of local environmental protection by residents in Xiamen and Dalian, local residents in Ningbo environmental problems are just a fuse, and the key problem is that villagers hope to obtain more material benefits through chemical project implementation. At the same time, all PX events in China have been 'single-issue', and the NIMBY facility-PX plant itself is the only issue concerned by the public. In addition, unlike the organized and systematic environmental movements in the West that continue to fight for the protection of the natural environment, the initiators and protesters of PX events in China are independent. Actions between different events are discontinuous. Once the government is forced to make concessions, the independent events ended.

It can be seen that the public mainly seeks and protects personal interests in the PX event, and the public's anti-PX actions and discourse are a kind of radical means forced to take when their own interests (will) are infringed and the normal expression approaches are blocked. As pointed out by Zhou zhijia (2011), 'at the present stage, the resistance participation of Chinese residents is premised on the presupposition of the government's 'rational' response, which is a manifestation of political self-interest of residents. (Zhou Zhijia, 2011). The case study also shows that the anti-PX public actually depends on the government and they are self-interested. Therefore, the real interest demands behind the public discourse should be paid enough attention to. Based on the rational distribution of benefits, this paper proposes a discussion basis for resolving such controversies.

References

Cui Mo. To 'justify' PX requires a breakthrough of three levels . China Petroleum Daily, 21 April, 2014.

Dorothy Neilkin. Scientific controversies,the dynamics of the American public debate.In Sheila Jasanoff et al. eds, Handbook of science and technology studies. Sheng Xiaoming et al, eds. Beijing: Beijing university of science and technology press, 2004: 343-351.

Gee J P. (2000). An introduction to discourse analysis: theory and method[M].

Hannigan. (2009). Environmental sociology (second edition). Hong Dayou (Translator)Renmin University of China Press: 68.

Huang Lucheng, Jiang Linshan, Miao Hong, et al. (2016). Topic identification and analysis based on online Q&A community – Taking the topic of Zhihu 'the aged' as an example. Book Intelligence Work, (5): 93-100.

Jin Ming, Jin tao. (2013). Analysis of public attitudes and cognitive changes from the progress of golden rice event – Content analysis based on sina Weibo. Business Economics and Management, (11): 89-96.

Li Youmei.(2008).From wealth distribution to risk distribution: A new approach to China's social restructuring. Society, 28(6): 1-14.

Li Zhengwei, Liu Bing. (2004).Biotechnology and public understanding of science: An analysis using the UK as an example. Scientific Culture Review, (2): 61-74.

Liu Zhensheng. (2013) Research on social media dependence and media demand – A case study of college students'

Weibo dependence. Journalism University, 117(1): 119-129.

Manuel Castells. (2006).Power of identity. Social Science Literature Press: 174-175.

Southern Weekend. Who made PX universal sensitive words? Chemical industry panic disorder dilemma. 10 November 2014.

Xia Yuhe. (2011). Weibo public opinion in emergencies: An empirical study based on sina Weibo. Journalism and Communication Research, (5): 43-51.

Zhang Xiaojuan. (2007). New media power in xiamen PX crisis. International Public Relations, (5): 48-49.

Zhou Haiyan. (2012). PX and its aftermath: Research on China's network environmental protection action from the perspective of new social movement. Fudan University: 49-51.

Zhou Zhijia. (2011). Environmental protection, group pressure or interest swept – Analysis on the motivation of Xiamen residents' participation in PX environmental movement. Society, (01): 1-34.

Report of Science Communication to Reach out to an Indifferent Public

——Proposal of the two styles of workshop design which makes a participant a re-communicator

Yukiko Muta

Japanese Association for Science Communication, Tokyo, Japan

Abstract: One of the problems of science communication in Japan is that it has not spread sufficiently among citizens. In order to make scientific culture (seeing things in the scientific sense) permeate to the general public, science communication targeting those who are indifferent to science and technology is necessary. MUTA and KATO (2017) proposed and practiced a workshop design to tailor participants to become re-communicators within a structured model to make scientific culture permeate this layer, and verified the outcome. Participants attempted to conduct transmission activities and to promote science culture to their community. In the process, they demonstrated a willingness to be involved in activities that take advantage of scientific culture. As a result, this design showed that there was the possibility of increasing motivation for activities utilizing scientific culture. In this paper, based on the model of 2017, I will report on further experiments carried out by business faculty university students who belonged to the indifferent demographic. At a citizen festival in the Setagaya (Tokyo) area, they conducted re-communicating activities for citizens walking around the stalls.

1. Background (Science Communication in Japan)

S. Stocklmayer et al. define science communication as 'a process by which culture and knowledge of science is absorbed into a larger community' (S.Stocklmayer et al., 2003).

First of all, will Japanese science communication penetrate society? Science communication in Japan began to receive more attention after the 2011 eastern Japan earthquake. Contents are devised such as setting themes for the general public and communication with awareness of interactivity. On the other hand, participants of 'small-scale interactive events such as at science cafes' and 'medium / large events such as lectures' reported that most of the participants both are highly involved and are originally interested in science. (Kano et al., 2013). In other words, despite the fact that everyone is able to participate, the participants are biased towards the high level of science and technology. If only those who are very interested in science and technology are participating repeatedly in science events targeting general citizens that anyone can join, it is hard to say that science communication is making a contribution to spread to the wider general public. Even if it contributes to further deepening their understanding.

In 'The science and technology consciousness survey 2001' conducted by the Japanese Ministry of Education, Culture, Sports, Science and Technology (MEXT), from the viewpoint of attention to science and technology, the public is classified into three layers of 'Attentive public', 'Interested public', and 'Residual public'. Of the respondents of the survey, 4% fall into the 'Attentive Public', and 23% to the 'Interest Public' (NISTEP, 2001), so 27% in the high involvement in science and technology. Ogawa (2006) calls that 73% in the remaining 'Other public', it states that there is a specific layer of 'missing' we unintentionally excluded from the subject of science communication, and this layer is called 'Indifferent Public'. The Indifferent Public is thought to be an inevitable layer in making science communication permeable to society if science communication has not reached the general public.

Regarding this, Ogawa (2015) insists on the importance of intentionally designing interventions using appropriate means tailored to the purpose, who, for whom, what kind of intervention and heading towards a specific the goal. Therefore, in this research I propose a design of a workshop focusing on 'The Indifferent Public'.

2. Significance of Scientific Literacy Included in Science Literacy

1) Science literacy

Among the general public, as mentioned above, there is a percentage who actively seek scientific culture and knowledge, and those who do not. So, is science communication necessary for those who do not wish for it? If so, why?

There are many situations in life in contemporary society which involve science and technology. For example, a housewife in a position to protect the family's life and health, and an adult in a position to educate children, need to select and judge medical care, medicine, food, and how to cook it. There are times when one is needed to answer children's questions regarding nature and the world around then. Also, in community activities, science can also play a role in solving social issues such as environmental issues and disaster prevention related matters by helping us judge the opinions of experts and their advice. What is useful in such situations is science literacy encompassed by Scientific literacy defined by PISA as 'the capacity to use scientific knowledge, to identify questions and to draw evidence-based conclusions in order to understand and help make decisions about the natural world and the changes made to it through human activity (PISA, 2000 international research results)'.

Scientific literacy is valuable if every person including the Indifferent Public can apply it to daily life. A culture utilizing 'scientific literacy' is 'a science culture' and is aimed at a society in which a scientific culture permeates, it is necessary to intervene to make it possible for anyone to solve questions and tasks encountered in everyday life by using scientific thinking. Scientific literacy can be modeled as shown in (Fig. 1) shows in a chronological order the concrete exploration process leading to decision making by deriving appropriate conclusions. In this research I call this 'the scientific process of problem solving skills' and propose it as the ability to use scientific thinking in the process leading to solving everyday problems.

2) The Scientific Process of Problem Solving Skills

Fig. 1 shows a typical model in which 'the scientific process of problem solving skills' is demonstrated in the process of problem solving. The process of problem solving is shown step by step with numbering. (1) When a problem occurs in everyday life and it is recognized that the scientific solution is useful, then a decision is made to solve the problem using scientific thinking. This is the starting point. (2) Observe the problem. (3) Read information from the phenomenon. (4) Make a hypothesis or examine the books and literature on which generalized information is posted. (5) Repeat experiments and observations to obtain information. Identify consistency by referring to information from the laws of nature. (6) Make your own behavior choices and judgments based on the conclusion and use them as material for problem solving. The speech bubbles show typical utterances that are observed along with scientific thinking at each stage. The upward arrows are an index items embodying the scientific literacy exerted at that stage, and cite the following four competency skills and attitude in regards to science and

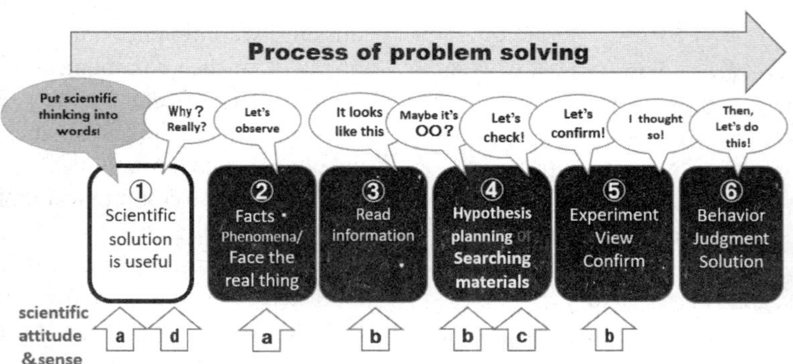

Fig. 1 Scientific process of problem solving skills model.

technology (Nagasaki, 2014). '(a) The ability to think scientifically so that ideas come naturally. (b) The ability to develop a habit of logical thinking based on the fundamental knowledge of science. (c) The ability to acquire correct scientific knowledge easily. (d) The ability to differentiate between fact and antiscience (unscientific) and to have an attitude to study scientifically.' (Science literacy for all Japanese, 2008).

For example, when applying the problem of 'Is it okay to dry laundry outside today?' a situation which anyone can encounter in everyday life, we can apply the Fig.1 model. It can be simulated as follows. (1) (a) (d) When checking the weather forecast on the internet, the information differs depending on the site. Therefore, making a decision which is not dependent on the web is favorable. (2) (a) Let's observe the sky. (3) (b) There are a lot of clouds like sheep floating high in the sky. (4) (b) (c) Upon checking information from (Gakken, 2015) shows that the clouds are 'altostratus' and seems that there is a high chance of rain when the space between the clouds narrows. (5) (b) The distance between the clouds is narrower than before. (6) Let's dry it outside while we have the chance. (The following day) Rain as expected. Drying it outside yesterday was the right choice.

When the problem which needs to be solved develops into a social problem such as medical, environmental, disaster prevention related, etc., the data and information needed varies, and multiple persons from diverse cultural backgrounds including experts become involved. In the cases, 'The scientific process of problem solving skills' may become complicated.

Since explanations by scientists with advanced knowledge should be based on 'the scientific process of problem solving skills' are each person can also utilize 'the scientific process of problem solving skills'. 'Standardized scientific rules exist' (Ogawa, 1998) .Citizens can be involved in science and technology from their respective standpoints, such as being able to judge the legitimacy of the opinions being presented.

3. The Purpose of This Research

In this research, I aim to infuse science culture into the living world of the Indifferent Public and to motivate science and technology involvement among the Indifferent Public, who are less involved in science and technology.

I propose workshop designs of two types (Experiment 1, Experiment 2), that tailor participants to become communicators by utilizating the scientific process of problem solving skills. Experiment 1 establishes guidelines for the creation of transmitters in science communication design that tailors participants to become communicators in closed spaces. Then, workshops are created based on that. Experiment 2 is a workshop design to tailor university students to become communicators who participate as citizen festival staff. They conduct transmission activities by using the scientific process of problem solving skills instantly in the street.

Participants become able to communicate in their own words after understanding the scientific elements. The point at which they actually communicated in their community is the goal of this practical study. Therefore, the possibility of the creation of a communicator from two viewpoints: 'Can this experiment provide skills as a communicator to participants' and 'Participants act enthusiastically as a communicator' was verified and evaluated.

4. Construct Experiment Base

1) Characteristics of Experiment Base

The experiment base of this study was named 'With Science Project (WSP)'. WSP has three main features. The first point is to make the participants only Indifferent Public members. The second point is setting a theme relevant to the life of the Indifferent Public, incorporating the utilization of the scientific process of problem solving skills in the context of that theme. The third point is to tailor participants to become communicators (re-communicators).

Definition of re-communicators Participants who became communicators at the WSP workshop carry out transmission activities but are different from the role of science-communicators, so in this research participants who became communicators are named 're-communicators'. The definition is shown below in three phases. (1) Immediately after WSP activity: A person who participated in the WSP and was able to express the scientific concept contained in the topic presented by the facilitator himself, while actively observing and convincing himself with his own words. (Table 2. Rubric Appraisal B or higher) (2) In daily life (Activity site) after WSP: After participating in the WSP, if you encounter opportunities to communicate scientific topics according to the listener's[1] generation, culture and degree of understanding, you can utilize the scientific process of problem solving skills and instantly communicate topics learned through the WSP. (3) Ideal: People who actively participate in their daily tasks and social discussion with a scientific attitude.

2) 'With Science Project' Concept

Participants Adult Indifferent Public /It is defined as a person not belonging to the science community, not interested in science, or not knowing the word 'science communication'. For recruitment of participants, use bulletin boards of community activities. Do not use websites or bulletin boards used by the science community.

Theme setting Set cultural activities that include elements of natural science, that can be enjoyed by many generations and are relevant to the life of the Indifferent Public.

Activity contents Prepare one impressive scientific main topic that participants want to share with people. While exploiting the context of the theme, incorporate scenes of the scientific process of problem solving skills as appropriate. Prepare a practice menu that allows participants to practice at their activity base.

Role of science communicator Keep in mind that it is an equal position with participants, and following that, they can learn in both directions. In the practical process, pay attention to the language activities accompanying the participants' actions, the participants, and the dialogue between them, and facilitate the participants to actively re-communicate.

5. Experiment 1: Closed Space-style Workshop

1) Practice Summary

Date and time May 24, 2015 (Sun) 10: 00 ~ 14:00

Place Setagaya symbiotic house and its neighboring grass

Theme Let's play with grass

Participants Neighbors, parents with children, mature ages 10 adults. One answered that nine people in the SC (science communication) awareness (Table 1) knew the word is 'Noting= never heard' and one who replied 'very basic' answered that they are not interested in science.

Table 1 Participant information.

ID	Participant	Gender	Age	Age of companion	Job category	Specialty	Science interest	*SC Cognition	Number transferred
1	I.M	F	40's,	2	Architectural Design	Science	a little	Nothing	3
2	Y.Y	F	40's,	-	kindergarten teacher	Humanities	Nothing	Very basic	1
3	M.T	F	70's	-	Housekeeping	Humanities	Yes	Nothing	5
4	K.M	F	40's,	-	clerk	Humanities	Nothing	Nothing	2
5	N.M	F	40's,	5	Housekeeping Child care	Humanities	a little	Nothing	4
6	W.C	F	30's	3	Housekeeping Child care	Humanities	Nothing	Nothing	5
7	N.T	F	30's	4 & 3	Housekeeping Child care	Science	Yes	Nothing	4
8	M.G	F	10's	-	Elementary school 6th grade	-	a little	Nothing	1
9	H.S	M	20's	-	Anime Production	Humanities	Yes	Nothing	0
10	K.R	M	20's	-	A caregiver	Humanities	Yes	Nothing	0

1 In this paper, the subject of the participant's re-communicate is called 'listener'.

So, all the participants were Indifferent Public.

Facilitator Author

Main topic On the leaves of KATABAMI (Creeping yellow wood sorrel/*Oxalis corniculate*) are contents indicating that *Oxalate* is included as a mechanism to protect it from small animals and insects. On the other hand, the mechanism is not a panacea: KATABAMI is edible for some insects. Scientific culture, including scientific concepts (food chain and natural selection) that can be read from the ecology of KATABAMI (Seeing things in the scientific sense), helps us understand and enjoy the depths of nature.

2) Pracitce Program

The practice flow, (1) Explanation of intention of workshop. (2) Participant self-introduction. 〔Enable dialogue among participants〕[1] (3) Introduction of issues / Show off a magic trick to polish money with KATABAMI leaves/Tasting of KATABAMI salad (small amount) / Let's read the picture book 'KATABAMI'. (4) Nearby walks and grass observation / Observation and grass play. 〔By visually confirming the way of life and the brutality of an organism, the possibility of expressiveness of participants themselves is raised. Promote utilization of the scientific process of problem solving skills (Fig. 2). Help the actions they approve and agree with their experimental results, and promote dialogue among the participants〕 (5) Role playing transmission game. (see Ⅴ-3) 〔Do not make participants aware that it is a re-communicator certification tests: provide a fun communication practice place for participants〕 (6) Discussion, impression announcement and questionnaire entry. (7) Activity after practice/Participants practiced transmission activities two weeks after completion of the workshop. (8)About two weeks after the workshop, the author interviewed the participants

Fig. 2 Participants who taste the leaves of KATABAMI.

Fig. 3 Clover leaf looks similar to KATABAMI. But they can't polish money. Participants approving experiment results.

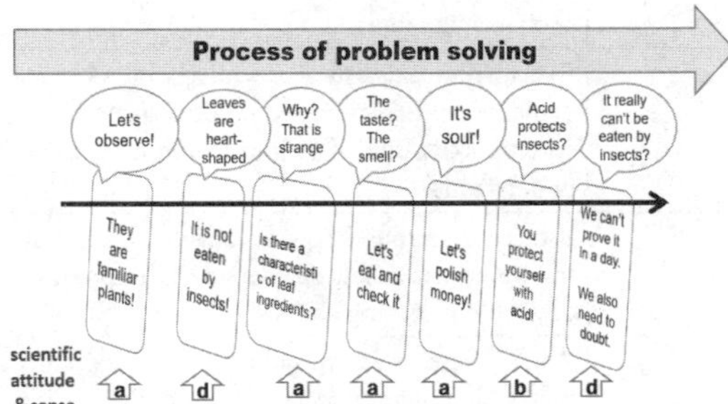

Fig. 4 The scientific process of problem solving skills in Experiment 1.

for the purpose of hearing about the process leading to the transmission activity.

3) Develop Game for Participants' Skill Acquisition and Evaluation

In Experiment 1, I referred to 'Educational Intervention Based on Revoicing' (Tajima, 2008), Proposed to promote active interpretive activities of learners, and practiced communicating with participants by developing an enjoyable 'Role-playing communication game'. There are several purposes to this game, such as reviewing the content of the day, motivation to communicate with others, simulating a place to communicate with others, and evaluating the participants' achievements as re-communicators. In the achievement evaluation video assessment of the state of skill acquisition as a communicator from the performance of the participants was conducted according to the rubric evaluation criteria (Table 2) at a later date.

Rules (1) Alternate the roles of communicator and listener. (2) For the role of communicator, the position of the listener was indicated by a 'To whom?' card (eg. elementary school student, boss, friend, etc) and location by a 'Where?' card (eg. park, in the room etc). (3) The role of the communicator tells the role of the listener of

1 〔〕: Design ingenuity for science communication.

the content of the main topic of the day based on the words appropriate for the opponent, the location and the number of opponents. (4) The role of the listener is to flag the ○, if the contents re-communicated by the role of the communicator can be understood, and ×, if it can't be understood.

Table 2 The rubric evaluation criteria.

Evaluation	Evaluation Criteria
S	Furthermore, it is also possible to communicate using the scientific process of problem solving skills by using expressions such as metaphor.
A	Even without assistance, while being able to demonstrate the scientific process of problem solving skills and it is told to the listener to understand.
B	If there is surrounding assistance, you can tell the opponent based on the scientific process of problem solving skills.
C	When there is surrounding assistance, the minimum content can be transmitted.
D	Even if there are surrounding assistance, contents can't be conveyed.

4) Verification Result of Experiment 1

From the evaluation of the implementation test I examined 'Can this experiment provide skills as a communicator to participants', shown for research purposes. Of the 10 participants, S was 1, A was 3, B was 4, 2 did not participate.

The two people who did not attend the performance were equivalent to Rubric Appraisal B when they demonstrated game tasks at the interview at a later date. Therefore, all participants were recognized as re-communicators and were encouraged to conduct transmission activities.

As for another research objective, I was able to verify that 'Participants act enthusiastically as a communicator' in terms of the amount (number of transmissions) and the quality (scientific correctness of the contents transmitted) of participants.

Verification method At the end of the workshop, I handed out a sealed questionnaire (Document 1) for each participant's estimated number of transmissions. The participant hands the questionnaire to the listener, the listener fills in the answers and sends it to the author. On the other hand, I interviewed participants too. I recorded the number of transmissions and the contents of transmission and verified the consistency with the collected question paper.

Document 1 The main questions that participants ask listeners

Please write if you know the following/ Please do not investigate, just write what you know. (1) KATABAMI leaf shape and flower color? (2)What are the properties of the leaves of KATABAMI? (3)What method can be used to ascertain its nature? (4) What kind of role does KATABAMI play in order to survive? (5)How can you verify that its nature really helps KATABAMI survive? (6)How and when did you learn about(1) (2) (3) (4) (5)?

Eight out of ten participants answered in an interview that they conducted re-communicating activities, and answered that they handed out a questionnaire to 25 people in total. In fact, 25 questionnaires from the listeners were received by the author. Participants were also trying to communicate with SNS etc., in addition to questionnaire collaborators (Table 3). However, in this research I do not assume further transmission from listeners to others. This is because it is impossible for a participant's communication partner to determine whether he/she has acquired the skill of a re-communicator.

Table 3 Results of participants' transmission.

8 out of 10 participants communicated to more than 25 people

listener→ Participant↓	listener 1	listener 2	listener 3	listener 4	listener 5	Paper Recovered	Other destinations
1. I·M	husband	sister	mother			3	Facebook
2. Y·Y	mother					1	Facebook
3. M·T	second son	first son	daughter in-low	neighbor	hobby group	5	Presented @ lifelong learning
4. K·M	daughter (16)	husband				2	father
5. N·M	daughter (5)	husband	friend	friend	friend	5	friend
6. W·C	daughter (3)	husband	nephew (8)	friend		4	Facebook
7. N·T	daughter (4)	husband	nephew (11)	nephew (16)		4	friend
8. A·M	mother					1	father
9. H·S						0	
10. K·R						0	

Regarding the quality of transmission, I confirmed the agreement between the contents of the participants' re-communicating and the contents that the listener understood, in the hope that the scientific contents would be transmitted correctly. However, there were some discrepancies between the content of what the participant communicated and the content that the listener understood. Two listeners incorrectly described the chemical name 'oxalic acid' as 'folic acid' and 'nitric acid'. Interviews convey that the participants who re-communicated to the two persons, correctly reported 'oxalic acid'.

Document 1: Question (5) 'How can you

verify that its nature really helps KATABAMI survive?' This question was asked by the author on the day, and from that I was able to develop scientific perspectives and thoughts on the discussion.

I expected (5) to be a question to develop discussion between communicator and listener, but this item was almost empty.

Also, the style of Experiment 1, is for participants in closed spaces to generate re-communicators. Since re-communication is required after the workshop, I need to rely on participants to choose listeners. As a result, listeners were limited to participants' own families and people within their community.

6. Experiment 2: Open Space-style Workshop

1) Background of Experiment 2

In experiment 1, I needed to rely on participants to choose listeners, so listeners were limited to participants' own families and people within their community. This research has a mission to disseminate science communication to members of an Indifferent public who should exist in various communities. Therefore, it is necessary to conduct re-communicating activities for more diverse Indifferent members.

In the next step (Experiment 2), without depending on participants to select the listener, a re-communicator is created at the open space of a citizen festival where an unspecified numbers of indifferent members gather. In addition, I will design a WSP leaving style for participants to re-communicate with listeners for the first time.

Participants are university students randomly assigned among the festival staff belonging to 'Setagaya Regional Exchange Laboratory'. In other words, scientific communication is not the main purpose, but I considered those who have a willingness to tell listeners something. The main activity on the day of Experiment 2 is re-communication to the listener. Therefore, the author needs to tailor the participants to become re-communicators who can conduct re-communication activities that convince many listeners.

2) Design of Experiment 2

The difference in design from Experiment 1: (1) Participants are already considered to be motivated to communicate with listeners. So, I explain the concrete flow to re-communication along the scientific process of problem solving skills in a short time. And I show examples at the location of re-communication activities in order to refine their skills. (2) The author can confirm the participants' re-communication skills, number, contents and quality during the activity. Therefore, I don't do a questionnaire with the listener or the 'communication game playing a role' which I did in Experiment 1. (3) If the author decides that all participants have acquired the skills of a re - communicator, he / she leaves the activity site so as not to disturb the participants' ingenuity of re-communication or cooperation among them. However, I always check participants' re-communicating on site to see if there are any scientific mistakes. (4) In order to evaluate the purpose of 'Efficiently re-communicating with more listeners' and 'Targeting indifferent people belonging to various communities', the author counts the number of family members of the listener on that day, and asks their age and some personal information.

3) Practice Summary

Date and time April 29, 2018 (Sun) 10: 00 ~ 16: 30

Location Street of the citizen festival (Setagaya Tokyo). The author provides a science workshop stall for citizens

Theme Let's play with grass II

Participants (re-communicators) Five students of SANNO University (School of Management)

＊In the SC awareness (Table 4), since all 5 members answered that they never heard the word 'science communication' is 'nothing', all the participants were Indifferent Public.

Table 4 Participant Information.

	Participant	Gender	Age	Job Category	Specialty	Science Interest	*SC Cognition
1	H.T	F	20	University student	Business	a little	Noting
2	R.K	M	20	University student	Business	Noting	Noting
3	Y.S	F	20	University student	Business	Noting	Noting
4	K.K	M	19	University student	Business	a little	Noting
5	K.T	M	20	University student	Business	Noting	Noting

Facilitator Author

Main topic same as experiment 1. Furthermore, I notice the difference between the white clover (*Trifolium repens L.)* and red clover (*Trifolium pratense L.*) of the legume that lives in this area, which is similar to KATABAMI in leaf shape. For deeply interested listeners, we also assume that the process of scientific process of problem solving skills becomes more complicated than experiment 1 (Fig. 2).

4) Pracitce program

The practice flow, (1) General festival management members greetings and staff introduction. (2) Observe the habitat of 5 participants and KATABAMI. (3) Show the magic that polishes money with the leaves of KATABAMI. (4) Let's read the picture book 'KATABAMI'. (5) Explain the development of re-communicate along the scientific process of problem solving skills. (6) Author shows re-communication model. (7) Skill up participants' re-communicating in a process where participants and authors repeat practice to the listener. 〔For more interesting listeners, the authors propose comparative experiments with other plants, and talk about the ecology of the KATABAMI, and the deep appeal of nature. And show participants the state of the dialogue.〕 (8) If the author confirmed the activities of five participants and determined that they got the skill as a re-communicator, I will leave that activity and facilitation by participants. 〔I am constantly checking whether participants' re-communicate contents are scientifically correct.〕 (9) The author listens to the number of listeners and their age and the community they belong to. (10) End of activity, cleaning. Perform interviews and questionnaires to participants.

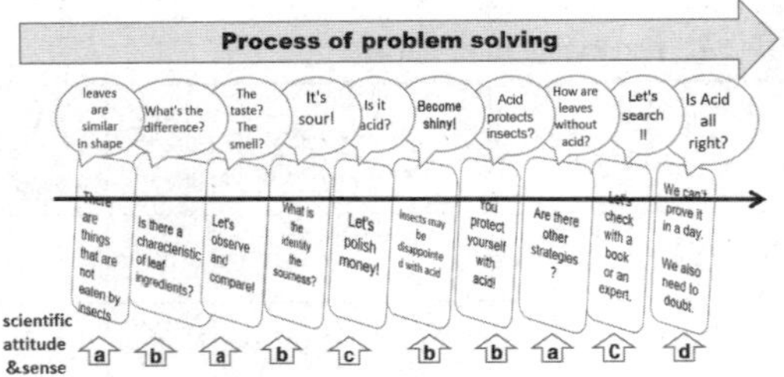

Fig. 5 The scientific process of problem solving skills in Experiment 2.

5) Verification Result of Experiment 2

The purpose of the first object of this research is to verify 'Can this experiment provide the skills required to be a communicator to participants'. It was found by listening to five participants after the completion that they understood the contents of the main topic along with the scientific process of problem solving skills of science at the stage of (5) (6) (7) of the Practice program. The place of implementation is a real-life situation (Fig. 6). In order to cope with situations where many listeners are already waiting, we started working with the author from the participants who acquired the re-communication skills in the (7) process. The rest of the participants observed the activity. As the festival got lively and it became necessary to deal with many listeners, all the participants conducted re-communication activities with the author.

Fig. 6 At regional festival. Open space style workshop.

Fig. 7 Re-communicating activities by university students.

The author proposed to experiment with listeners who showed particular interest, comparing with the plant of Fabaceae to the similarly-shaped leaf of KATABAMI and interacted with the listener while verifying the result. Understood the contents of the dialogue two participants have confirmed the books and materials prepared by the authors themselves, expanded the range of the contents of the re-communicate. And they interacted with listeners according to the age and understanding of the listener. Other participants who observed the situation did the same thing. In situations, their communication skills are like a up as they drew a spiral trajectory.

At this point, I left the activity and left activities to the five re-communicators. However, the author constantly checks the correctness of re-communicates on site, and there was no scientific mistake in the content of the five participants. At first there was a difference in understanding among the participants, but they were working with a relatively loud voice to hear each other's contents. Learning from each other on the spot, eventually the difference in understanding of communication contents and the difference in activity skills diminished. As a result, in Experiment 2, the author was able to provide the skill as a communicator to all five participants.

Regarding the second point 'Participants act enthusiastically as communicators', it is clear that all five participants actively acted as communicators already in the verification contents of the first point, but the concrete amount are shown in Table 5.

Table 5 Transmission results of 5 participants.

Age	Number of Families	Companion
20's	1	Yes
30's	10	Yes
40's	12	Yes
50's	1	No
60's	3	Yes
Over 70	7	Yes
Total	Over 34 families (*)	

* I could not to hear any other things than a scientific dialogue to an independent listener.

In Experiment 2, the intention to propose a design style different from Experiment 1 was in 'Efficiently re-communicate with more listeners', 'Targeting indifferent people belonging to various communities'.

About these, five participants re-communicated to listeners of 37 families or more, and the diversity of the listener's community was shown to some extent. Therefore, it can be said that these two points exceeded Experiment 1.

On that day, I lists the personal information such as the occupation, hobby and interests of the listener I heard: Childcare system / Child care / Early childhood education / Childrearing / Learning cram school / Employee / Part time / Citizen's Orchestra / Flower arrangement / Sports / Soccer World Cup / Area management / Environmental protection group / Juvenile baseball coach / Interest in children's movement · Soccer fan / Nursing care / others

About these, five participants re-communicated to listeners of 37 families or more, and the diversity of the listener's community was shown to some extent. Therefore, it can be said that these two points exceeded Experiment 1.

7. Consideration

As mentioned above, in Experiment 1 and Experiment 2, two objectives of this paper were achieved as a result of verification. I will examine here whether this result leads to the possibility of 'the possibility of infectious scientific culture in the Indifferent Public living world' and 'the motivation for science and technology involved among the Indifferent Public'.

The author assumed the benefits of making participants a re-communicator, with the logic of promoters who wanted to spread scientific communication to indifferent public. On the other hand, I designed two experiments on the premise that participants also feel that they are meaningful to participate in the WSP workshop. In other words, if we can't embed scientific culture in the living world of participants, we thought that science communication would not reach Indifferent Public, which is scarcely conscious of science and technology. The following is an interview statement of the participants who can read that they feel 'carrying out a transmission activity was meaningful'.

According to Experiment 1, 'Not only at the workshop but also on the topic by telling the story, I feel more interested and I feel the content has been acquired (W · C)' 'I want to tell people around me when I learned nature's laws and skillful living strategies I did not notice before (I · M)' 'I have created a material to communicate with my children and colleagues and I am glad that it is always in my head (M · T)'.

According to Experiment 2, 'From the way KATABAMI lives, it was interesting to know that creatures live by their own strategies (K · K)' 'I participated to excite the event, but while I was communicating with the listeners, I felt fun and understanding scientific views. After that, I began to see TV science programs consciously (Y · S)' 'I understood the ecology of KATABAMI, I think that opportunities to conscious of plants growing around us will increase (T · H)'.

Experiment 2 is a design proposed as a possibility of complementing Experiment 1, focusing on the amount of listener and the diversity of the target community, from the viewpoint of 'spreading scientific culture into the living world of Indifferent Public Members'. As mentioned above, it was possible to exceed Experiment 1 in terms of 'Efficient re-communication with more listeners' and 'Targeting indifferent users belonging to various communities'. Meanwhile, from participants in Experiment 2, I didn't hear the intention to participate in activities that utilize scientific culture. Unlike Experiment 1, it may be a major factor that I couldn't spent a lot of time talking about science culture with participants. In the future we think that improvement is necessary for design.

Below, the results of Experiment 1 will be mainly examined. Participants in Experiment 1 had utterances that could be read as willingness of involvement in science and technology. 'I want to see things while thinking about how I can prove it when I feel doubt (I · M)' 'I was doubtful about food, but I would like to think more scientifically from now (N · M)' 'I want to discuss my doubts about foods and additives with friends (W · C)'.

In the interview, 'Were you able to convince the scientific perspective and the significance of the idea presented in the main topic?' among the eight respondents, against the question 'very convinced' is one, 'convinced' is three, 'I think I was convinced' is two. Therefore, it can be said that six participants were almost convinced.

The reason why they said, 'I just wanted to share my persuasive content, so I told my family immediately after I finished my workshop when I got home (K · M)'. There were similar to 'I want you to know' utterances. From this, it can be understood that participants who became communicators trust the facilitation of the promoting entity, and move on to the action of communicating to people through the process of realizing and convincing the usefulness of the transmitted scientific literacy. In other words, the argument of this paper is that science communication, which succeeded in to tailor participants who are Indifferent Public members to become re-communicators, reaches motivates their involvement in science and technology.

However, the information sent by the re-communicator was not necessarily accurately understood scientifically by all the listeners. It is a typical example that the listener misunderstood the name of the chemical substance that the participant said as 'oxalic acid' as 'folic acid' and 'nitric acid'. The design for accurate transmission to the listener is the next task.

More serious for me was the fact that, there was no report that participants and listeners got to the discussion regarding, 'Isn't KATABAMI really eaten by insects?' Ogawa says that when exploring scientific facts, the knowledge that was taught is a fact interpreted among scientists. It is necessary to have a viewpoint that has a revised character, and this way of thinking is important in scientific literacy, and is highlighted in the report of 'Project 2061' (Ogawa, 1998). It is important to accurately memorize the name of the chemical substance in the utilization scenes of the scientific process of problem solving skills.

However, for an Indifferent Public, more important than memorizing the name of the chemical substance as 'leaves of KATABAMI are not eaten by insects' is the ability to show the reasons for doubt. A society in which an Indifferent Public doubts, criticizes, and judges grounds for expert policy and judgment, will function more soundly.

8. For the Future

The long-term vision of this research is the formation of a local community in which each citizen believes 'science is meaningful within our culture', and to connect each other by science culture. In other words, by embedding science as a culture in various communities, we can grow a society where each citizen makes decisions with sound judgment. In order to spread science communication to society, it is not enough for only the author to develop a workshop. Therefore, by asking the scientific community to recruit science communicators who agree with WSP and by developing WSP in each region, science culture can be spread to each community.

The author proposes this as a model of a new science communication 'Embedded model'. Fig. 8 is a conceptual diagram of the embedded model. A science communicator connected to the science community will spread WSP in their own region and community. As new communicators, the participants will expand a network in which small science communication is re-communicated in their own communities. It may be possible to expand it not only in the Setagaya area but also throughout the world.

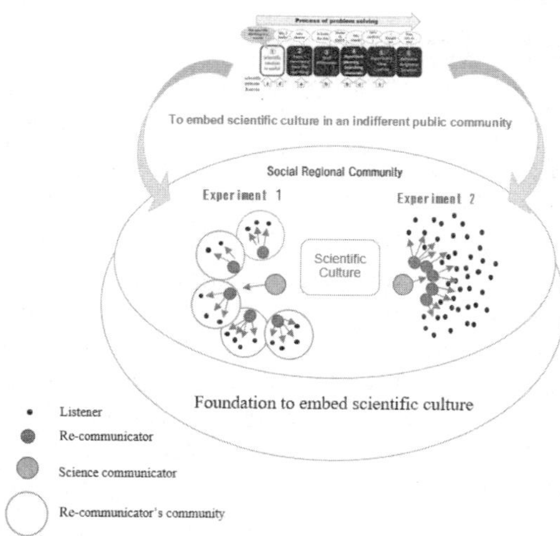

Fig. 8 Deployment vision of embedded model.

In order to develop and carry out a workshop in reality, the science communicator needs to have the ability as a workshop designer too. The workshop is 'The program is made and practiced, the performance is evaluated and analyzed, and the program is being improved based on the result (Mizukoshi, 2014)'. It is not only about the program at that time, it is necessary to consider regionality, environment, conditions, cultural background of participants, etc.

I think that the process of realizing this long-term vision can be considered as one form of science communication defined by Stocklmayer et al.

Acknowledgments

This research was completed with the advice of Professor Hiroshi Kato (Open University of Japan) and CSCL seminar, and Ms. Taeko Tada, a plant ecologist who instructed specialized fields. I sincerely thank you. Also, Ms. Miwako Iioka of 'Regional Symbiosis / IIOKA san-chi de asobo' in Setagaya Ward, Professor Tomoko Nakamura (SANNO University) promoting 'Setagaya Regional Exchange Laboratory' and Ms. Rie Suzuki (FUTAKOTAMAGAWA Merchant Street Director), I was indebted to them for my workshop. I deeply appreciate that.

References

American Association for the Advancement of Science (1989). Science for All Americans/Project 2061.

DOWNS. Cal. W. (1999). Communication Audits, CAP Publications.

Gakken's picture book (2015). Clouds·weather, Supervised by Masamitsu Morita, Gakken.

KANO, et al. (2013). Segmentation and targeting of participants in science cafes - From the Viewpoint of the Extent of Engagement in Science and Technology, Japanese Journal of Science Communication,13: 3-16.

MIZUKOSHI S. (2014). 21st century media theory, The Society for the Promotion of the Open University of Japan.

MUTA. KATO (2017). Report of science communication to reach out to an indifferent public – Proposal of the workshop design which makes a participant a re-communicator – 41(1) 23-35.

NAGASAKI, E (2014). Background, Science and technology spirit Beginning and now. Research report on science and technology literacy, JST Center for Science Communication, 5-17.

NISTEP (2001). The survey for public attitudes towards and understanding of science & technology in Japan. Ministry of Education, Culture. Sports, Science and Technology, Japan (In Japanese).

OECD (2000). Program for International Student Assessment (PISA), Survey international summary of results.

OGAWA. M (1998). Rediscovering science – Western science as a different culture, Rural Culture Association Japan.

OGAWA. M (2006). Exploring possibility of developing indifferent public-driven science communication activities. 30(4), 201-209.

OGAWA. M (2013). Towards a'design approach' to science communication Edited by John K, Gilbert and S, Stocklmayer, Communication and engagememt with science and technology, Routledge, 1-18.

Science and technology spirit project (2008). Science Literacy for all Japanese.

STOCKLMAYER, et al. (2003). Science communication in theory and practice, Maruzen Planet Co., Ltd.

TADA. T (2007). Katabami, Fukuinkan Shoten Publisher inc.

TAJIMA. A (2008). Educational intervention based on 'revoicing' and the understanding of concepts: From the Standpoint of Bakhtin's Theory.

Scientists and Non-scientists Compete for Authority Over Science in West Africa

Bankole Falade

South African Research Chair in Science Communication, Centre for Research on Evaluation, Science and Technology, Stellenbosch University, South Africa

Abstract: The uptake of science communication in Africa is often hampered by religious and traditional belief systems, previous experiences with science and conspiracy theories about Western interventions in African affairs. This paper examines the oral polio vaccine revolt in Northern Nigeria and historical boundary disputes in vaccination controversies in Europe and the United States. The paper also examines the tetanus toxoid controversy in Cameroun, the Ebola crisis in West Africa and draws similarities from the Zika virus disease crisis in Brazil. The contestation over boundaries with science in religiously conservative Africa is made more complicated by the dual role of some scientists as religious leaders and as previous studies show, the public may have similar levels of trust in both. Religious intervention can also be double-edged, acting as a hindrance and partner in the propagation of scientific ideas.

1. Introduction

1) The Progression of Truths

Truth in pre-colonial and pre-slavery Africa was informed by the African traditional belief systems, the gods: Orisa, Orunmila, Esu, Yemoja, Ifa, Osun goddess, etc., enforced by traditional practices and taboos. Divination by Ifa and the word of the King represented the truth. Truth in colonial Africa was informed by the Abrahamic religions, the one God: Christianity, Islam, and the scientific method enforced by the English law and legal system. Post-colonial era presents a mixture of pre and colonial practices, the old and the new coexisting: Abrahamic religions, African traditional belief systems, atheism, the scientific method all enforced by English laws and legal system, taboos and the traditional practices of appeasing the gods. PEW research (2010) shows that side by side with their high levels of commitment to Christianity and Islam, many people in Africa retain beliefs and rituals characteristic of traditional African religions. In four countries, over 50% believe that sacrifices to ancestors or spirits can protect them from harm. In 14 countries, more than three-in-ten people say they sometimes consult traditional healers. The authors noted that while the recourse to traditional healers may be motivated in part by economic reasons and an absence of health care alternatives, it may also be rooted in religious beliefs about the efficacy of this approach.

2) The Two Statements of Truth

Both science and religion (Gaskell et al., 2010) are used as the basis of statements about the truth. From the scientific perspective, scientific truth trumps any other form of knowing, and some scientific authorities argue that religion is at best wishful thinking, at worst a pernicious force in society. Some authors however argue that the confidence and trust (see Luhmann, 1998) people have in science is similar what they have in religion. For Durkheim (1912), faith in science is not necessarily different from religious faith. Einstein (1940) believes science can only be created by those who are thoroughly imbued with the aspiration toward truth and understanding, a desire which springs from religion. Einstein (1950) who separates his notion of religion from that of a God said he could not conceive of a genuine scientist without that profound faith: 'science without religion is lame, religion without science is blind'. Einstein's assertion here has been subject of great debate. While some authors use this to back up their claims of his religious views, others argue this is not the case. Einstein (1954) however made clear his religion does not include the notion of a God, which he describes as a product of human weakness. But can scientists, driven by a desire which originates from religion, believe in God? While Professor Francis Collins (2006), a geneticist, argues that this is the case in his book titled 'The language of God: A scientist presents evidence for belief', Professor Richard Dawkins (2006), an evolutionary biologist, argues in the opposite that God almost certainly does not exist in his book titled 'God Delusion'.

Science according to Thomas Kuhn (see Knight, 2004) is not some disinterested and isolated search for truth, free of metaphysics and manifestly a good thing in a naughty world. The history of science, Knight (2004) argues, is like the history of France, prone to revolutions and each turn was associated with new beliefs about the world, and with new language as the scientific community was converted to new ways of seeing and believing.

For Habermas (2003, 2006), the post secular society continues the work for religion that religion did for myth and secularisation is not a zero-sum game as a democratic common sense remains osmotically open to both sides, without relinquishing its independence. Openness to both sides indicates coexistence in what Moscovici describes as cognitive polyphasia, the coexistence of different forms of thought (Moscovici, 1984).

3) Science Coexisting with Religion

In Nigeria, Pastor (Dr) E A Adeboye, former University lecturer, with a PhD in mathematics heads the Redeemed Christian Church of God. Pastor William Kumuyi, former University lecturer, with a First Class in Mathematics, heads Deeper Life Bible Church and Dr Daniel Adekoya, PhD molecular genetics, former medical researcher, heads the Mountain of Fire and Miracles Ministry. These three churches have the largest congregations of Pentecostal Christians in Nigeria with branches all over world. For these pastors, being a scientist does not lead to rejection of religion, both coexist.

A study of PhD's written in a South African University by the author shows that 43% of doctorates acknowledged faith in an omnipotent God and this was across all faculties. Some of the acknowledgements said:

> '... I would like to thank the almighty God for ...'
>
> 'GOD *Unknlunkulu* (Zulu language) arose from beneath and created in the beginning men, animals, and all things.'
>
> *'My dank aan my Skepper vir sy genade aan my'* (Afrikaaans language) Interpretation: My gratitude to my Creator for his mercy to me.

Another study of some communities in South Africa reveals the same belief systems, coexistence of science and religion in some respondents.

> 'When you take what is written in the Bible, and that which the scientist tells you, you must be able to weigh the one against the other ...'
>
> '... He didn't want to breast feed, and he was becoming weaker ... We had to take him back to hospital ... He was given nine injections ... I could take all that feeling to me, to me personally. I was like if anything happens with this child, I prefer God to do it to me, not to him.'

A study of trust in cultural authorities in Nigeria (Falade and Bauer, 2018) found high levels of trust in scientists (M=3.65, SD=1.58) and religious leaders (M=3.61, SD = 1.58) when compared with the military (M=1.93, SD=1.70), politicians (M=0.7, SD=1.24), judiciary (M=2.45, SD=1.63) foreign NGO's (M=3.85, SD=1.82) and local NGO's (M=3.08, SD=1.76). The correlation between scientists and religious leaders was also significant and a factor analysis groups them together in one factor. Whoever trusts a religious leader in Nigeria is likely to also trust a scientific expert, a strong evidence of cognitive polyphasia (see Moscovici, 1984).

Factors	Public sector	Indepen-dents	Cultural authorities
Trust in military leaders	0.829		
Trust in Judiciary	0.788		
Trust in politicians	0.605		
Trust in foreign NGO's		0.913	
Trust in local NGO's	0.45	0.865	
Trust in religious leader			0.845
Trust in scientists and professors		0.428	0.842
Variance explained	36.2%	19.9%	14.5%

Structure matrix for trust items (source Falade and Bauer, 2018)

2. Case Studies

1) The Polio Vaccine Controversy in Northern Nigeria

In line with the goals of the Global Polio Eradication Initiative (GPEI), African leaders in 1996 launched the polio eradication campaign tagged 'Kick Polio out of Africa'. That year, polio was rampant in 41 African countries, but by 2002 most countries, including several states in Southern Nigeria, were declared free of the disease. As part of the global effort, national immunisation days were set aside in Nigeria by the federal government commencing in the last quarter of the year 2000. This campaign was resisted from the onset by some religious leaders in Northern Nigeria who described the exercise as being against Islamic injunctions and rumours of contamination with the AIDS virus were also widespread (Falade, 2015).

Public revolt The crisis was aggravated in July 2003, when in the midst of a nationwide campaign, two very influential Islamic groups: the Supreme Council for Shari'ah in Nigeria (SCSN) and the Kaduna State Council of Imams and Ulamas declared that the vaccine contained anti-fertility substances and was part of a western conspiracy to reduce the population of the developing world. Given the revered status of the two groups among Muslims, the stage was set for a major revolt. The revolt peaked when some states in Northern Nigeria, led by Kano State, banned the use of the OPV citing its 'contamination by sterilizing substances'. Kano governor, Ibrahim Shekarau described the ban as 'the lesser of two evils ... to sacrifice two, three, four, five even ten children to polio than allow hundreds of thousands or possibly millions of girl children likely to be rendered infertile …'

Notably, many of the scientists supporting and opposing the vaccination exercise were from different disciplines (see Jasanoff, 1987) and some double as Islamic scholars. Among those in opposition were a medical doctor, a pharmacist, a medical biochemist and a professor of science education. In favour were a professor of medicine and immunology, a professor of virology and a pathologist.

Religious perspectives The West was seen as having a secret Muslim depopulation programme following wars in Bosnia, Afghanistan and Iraq, which appear to be against Muslim countries. An Islamic cleric said: 'If they really love our children, why did they watch Bosnian children killed and 500,000 Iraqi children die of starvation and disease under an economic embargo?' Also, in 1996 there was the infamous Pfizer Trovan drug trial in Kano State Northern Nigeria during which some children died. Kano, the most populous state in Northern Nigeria, was the first to ban the vaccine. An Islamic sect also declared immunisation un-Islamic and called on Muslims to shun the exercise and a cleric told his followers that some verses in the Holy Koran can be used to cure diseases.

Scientific perspectives The debate over scientific risk was largely avoided. One in 200 infections leads to irreversible paralysis. One in every 2.7 million first doses leads to vaccine-associated paralytic poliomyelitis (VAPP). Circulating vaccine-derived polioviruses, cVDPV is when the vaccine changes genetically (rare) and circulates in the population. Immunodeficient vaccine-derived poliovirus (iVDPV) also arises when some individuals become chronic long-term excretors of VDPV. Kano's chief health officer said given the low level of education, it may lead to a mass rejection and the UNICEF chief also argued that communicating medical risk to illiterate and remote populations isn't always possible.

Science and religion working together Kano State eventually accepted the potency of the vaccine produced in Indonesia. Speaking on the acceptance, an Islamic preacher said, 'from what we were told at the meeting, the polio vaccine to infertility ratio had been exaggerated.' A Kano business man, who was also at the meeting said 'even though I do not understand most of the medical jargon of the committee, I am convinced the polio vaccination should go on.' The Kana State Governor, Ibrahim Shekarau, who banned the vaccine later administered it on several babies. Also, following a five day immunisation tour of Egypt in 2007, the Emir and spiritual leader of Gombe, Abubakar Alhaji Usman Shehu admitted to journalists that although he went on the trip as a 'doubting Thomas' he has since been convinced of the vaccines compatibility with Islam.

Vaccination: an international debate McKinnon and Orthia (2017) compared 19th- and 21st-centuries Australia and found that government campaigns have not changed much and have been based on scientific facts which are however, likely to get lost in the plethora of information sources on the internet. For Leask, Willaby and Kaufman (2014) societal circumstances may contribute to a growing parental hesitancy and include increasingly 'crowded' vaccination schedules; lower prevalence of vaccine-preventable diseases; hyper-vigilance of parents in relation to children and risk; and

Table 1 Challenges to the scientific authority of vaccines in history and across continents (Falade, 2014).

Country	Year	Communication themes/suspected effects	Opposing groups	Vaccine
United States	1721–1722	Transmits syphilis, black plague, leprosy	Clergy and scientists	Smallpox
England	1840–1871	'Mark of the beast', against 'civil liberties' transmits syphilis	Working class, liberal reformers, church, scientists, etc.	Smallpox
United States	1890–1922	'Medical tyranny', coercion, un-Godly	Scientists, anti-vaccination League, Christian Scientists	Smallpox
Brazil	1904	'Torture code', 'vile secretions expelled from sick animals'	Middle class, elite, church, press, congress members, etc.	Smallpox
United Kingdom	1974	'Neurological complications'	Scientists, public	DTP
Japan	1975	'Neurological complications'	(Fallout of UK episode)	DTP
United Kingdom	1998	Autism	Scientists, public	MMR
USA, Australia, New Zealand	1998+	Autism	(Fallout of UK episode)	MMR
Cameroon	1990	Sterilising vaccine	Catholic priests and opposition politicians	Tetanus Toxoid
Tanzania	1990	Sterilising vaccine (anti-hCG)	Scientists, Islamic preacher	Tetanus Toxoid
Nigeria	2001–2009	Western conspiracy, contaminated with HIV and anti-fertility substances	Islamic groups, politicians, scientists	OPV
United States	post 2000	Asthma, diabetes, Gullian-Barre syndrome, encephalopathy, autism and inflammatory bowel disease, mercury exposure, intussusceptions, Gulf war illness	Scientists, the public	DT, DTaP, DTP, Hepatitis B, Measles, MMR, OPV, Rubella

an increasingly consumerist orientation to healthcare (see also Pereira et al., 2013). The debate over the appropriateness of vaccination campaigns have been with us for centuries and cuts across continents and given recent events in the United States (Song, 2013), it may continue in the foreseeable future. Scientific facts are not enough but must be nested within other authorities.

The Tetanus Toxoid: controversy driven by religion In Cameroon, girls ran away from schools to avoid vaccination teams fearing the vaccines would sterilize them (Feldman-Savelsberg et al., 2000). It was during a period of public disagreement between the pro-life Catholic group and government over the safety of the Tetanus Toxoid (TT) vaccine being administered on girls of childbearing age only. The aftermath of the controversy was a sharp rise in teenage pregnancies and abortion in the northwest province as the vaccinated girls sought to confirm their ability to have offspring.

There were indeed trials using anti-hCG vaccines partly funded by the WHO (Jones, et al., 1988; Jones, 1996) which involved two intramuscular injections and a promise of a contraceptive effective for six months. The basic principle of a contraceptive (or anti-fertility) vaccine is to use the body's own immune defence mechanisms to provide protection against an unplanned pregnancy. The development of this contraceptive however, became a weapon of choice for the pro-life campaigners and it transmuted from an 'anti-fertility vaccine' to a 'sterilising vaccine'.

2) The Ebola Virus Disease: Science, Religion and Common Sense

The Ebola virus disease (EVD) outbreak in West Africa which claimed over 10,000 lives also shows the roles of non-scientific authorities in the uptake of scientific information (Falade et al., 2017). Prior to the outbreak, the disease had been restricted to Central and Eastern Africa. The only case in West Africa was in Cote d'Ivoire in 1994 and the patient survived. Thus, the West African publics had no previous experience with the disease which ravaged Sierra Leone, Liberia, Guinea and Mali, all sharing land borders and was transmitted by air travel to Nigeria.

Scientific perspectives Ebola causes fever, vomiting, headache, muscle pain, diarrhoea, nausea, haemorrhagic fever, etc. and shuts down the immune system. To prevent the transmission of the disease, the public were told to avoid shaking hands, wash hands frequently and avoid bush meat such as bats and monkeys – the vectors. The symptoms were however very similar to other known diseases such as Lassa, Dengue and Malaria fever. Malaria also causes high fevers and vomiting, cholera causes vomiting and diarrhoea, diseases which are common in Africa. This similarity initially confused western medical practitioners and traditional healers leading to several new infections from contact and misdiagnosis.

Religious and traditional perspectives Pastors were laying hands on the sick to cure them of 'spiritual attacks'. A leader told pastors that all those who fasted for 100 days should have no fear of Ebola but should avoid laying hands the sick. A Muslim opinion quoted the Quran: 'the Prophet said: There is no "Adwa" i.e. transfer of a disease by itself, but with the permission of Allah.' Bassonians, Kru and Grebo people spend time on mats mourning their dead ones and wash the bodies as part of traditional rites. Many traditional secret societies believe a dead member's ghost will torment others if they failed to observe the traditional rites of passage of washing the corpse. Both traditional and religious perspectives provided a steady stream of new hosts for the disease.

Accommodation of science by religion and tradition Many churches ordered the suspension of the practice of shaking hands as a sign of peace and the serving of Holy Communion in the mouth. A religious leader in Liberia called on fellow men of God to stop laying hands as a means of healing persons and imams to stop bathing dead bodies:

> *I beg you in the Name of Jesus ... please stop laying hands on people in order to cure them of Ebola. To our venerable imams, I respectfully appeal to you in the name of Allah ... please stop bathing dead bodies.*

There were also calls for the suspension of traditional practices of mourning the dead. The church was also a source of medical aid. Several religious institutions in Sierra Leone, including the United Brethren in Christ Church Conference, received funds from abroad for medical equipment and the Abundant Life Chapel Home in Liberia served as an orphanage for children who lost their parents to the disease.

3) Zika: A benign African Virus Turns Invasive in the Americas (Falade, 2018, in Press)

Zika was discovered in monkeys in 1947 in Uganda and was later isolated in the Aedes africanus mosquito. The virus had no known adverse effect in Africa and it is thought that the population has pre-existing immunity. The disease moved from Africa to Asia and by 2015, it was spreading in Brazil, in a population with no immunity. The Aedes species transmitting Zika in the Americas are the Aedes aegypti and Aedes albopictus. which also transmit yellow fever, dengue fever and Chikungunya. The symptoms of infection, fever, rash and joint pain are similar. But while the other diseases were previously established in the Americas, Zika was an unknown. Also, the known diseases did not affect the foetus. The Aedes mosquito is also found in Florida and Hawaii, and in hot weather, in northern parts of the United States. It is also found in warm parts of Europe including France, Portugal, Spain and Italy. Given its movement from Africa to Asia and the Americas, anxiety over its possible spread to the United States and Europe from returning travellers raised a global alarm at a time the spread of the Ebola virus disease from Africa to the United States and Europe was still fresh in memory. Just like Ebola, the symptoms initially confused western medical practitioners.

Zika virus infection during pregnancy can cause infants to be born with microcephaly. Microcephaly also has other causes, including genetic defects, alcohol exposure in pregnancy and infection by rubella. Zika is not related to rubella or cytomegalovirus which harm the foetus, but is related to yellow fever, dengue and West Nile virus, which are not known to harm the foetus. Also there is an increased risk of neurologic complications including Guillain-Barré syndrome following infection with Zika.

The Church, Politics and abortion The Catholic Church is opposed to abortion and all forms of contraceptives but these are part of the options for women against the disease, reigniting the debate over church, state and choice in Brazil. A Cardinal in Brazil said mothers must accept babies born with microcephaly 'as a mission,' and that abortion was out of the question but Pope Francis was more open to contraceptives arguing that avoiding pregnancy is not an absolute evil like abortion which is a crime and an absolute evil. Some Brazilians sought solace in God. A woman, when confronted with a diagnosis of possible microcephaly for her pregnancy said, 'It's God's will: he wanted us to have a baby like this'.

In the United States, a report noted that while the feud on Capitol Hill over response to the virus appears to be a fight over how much money is needed, beneath the surface are issues that have long stirred partisan mistrust, including Republicans' fears about the use of taxpayer money for abortion and possible increased use of contraception, and Democratic worries about protecting the environment from potentially dangerous pesticides (see also Mooney, 2005).

3. Conclusion

The religious leaders of modern day Africa are highly educated, some are senior scientists, and research has shown that the public have similar levels of trust in both authorities. This raises the dilemma of who to trust when both disagree on an issue. The polio vaccine controversy and the Ebola crises have shown that though there may be disagreement among scientists and between scientists and religious leaders, these are resolved in the long term in favour of science. Notably, the causes of scientific uncertainties were different in the Ebola and OPV data. In the OPV data, it was in disagreement among scientists in the interpretation of findings while in the Ebola data, symptoms mimic previously established diseases and there was no cure. In the OPV case, scientific authority eventually overcame religious concerns, in the Ebola case, while there were differences at the early stages, scientific, religious and traditional authorities eventually cooperated to contain the virus.

References

Collins, F.S., 2006. The language of God: A scientist presents evidence for belief (No. 111). Simon and Schuster.

Dawkins, R., 2006. *The God delusion*. Boston: Houghton Mifflin Co.

Durkheim, E., 1912/2001. *Elementary forms of religious life* (C. Cosman, Trans.). New York: Oxford University Press.

Einstein, A., 1940. Science and religion. No 3706 (605).

Einstein, A., 1950. My position concerning God is that of an agnostic. Letter dated October 25, 1950 addressed to Mr Morton Berkowitz. Available from http://www.lettersofnote.com/search?q=+einstein downloaded 06/09/2018.

Einstein, A., 1950. My position concerning God is that of an agnostic. Letter dated October 25, 1950 addressed to Mr Morton Berkowitz. http://www.lettersofnote.com/search?q=+einstein (Accessed 06 September 2018).

Einstein, A., 1954. The word God is the product of human weakness. Letter dated January 1954 addressed to philosopher Erik Gutkind http://www.lettersofnote.com/search?q=+einstein (Accessed 06 September 2018).

Falade, B. A., 2014. Vaccination resistance, religion and attitudes to science in Nigeria PhD thesis London school of Economics and Political science available at http://etheses.lse.ac.uk/911/.

Falade, B. A., 2015. Familiarizing science: A western conspiracy and the vaccination revolt in Northern Nigeria. *Papers in Social Representations 24*(1), 3.1-3.24 [http://www.psych.lse.ac.uk/psr/].

Falade, B, A. and Coultas, C. J., 2017. Scientific and non-scientific information in the uptake of health information: The case of Ebola. *South African Journal of* Science 113 (7/8), 1-8. http://dx.doi.org/10.17159/sajs.2017/20160359.

Falade, B. A., and Bauer, M. W., 2018. 'I have faith in science and in God': Common sense, cognitive polyphasia and attitudes to science in Nigeria. *Public Understanding of Science, 27*(1), 29-46.

Falade, B. A., 2018. Leveraging media informatics for the surveillance and understanding of disease outbreaks: the case of the Zika virus disease spread in the Americas. *South Africa Journal of Science*, in press.

Feldman-Savelsberg, P., Ndonko, F. and Schmidt-Ehry, B., 2000. Sterilizing vaccines or the politics of the womb: Retrospective study of a rumour in Cameroon. *Medical Anthropology Quarterly 14*(2), 159-179.

Gaskell, G., Stares, S., Allansdottir, A., et al., 2010. Europeans and biotechnology in 2010: Winds of change? Luxembourg: European Union.

Habermas, J., 2003. *The Future of Human Nature*. Cambridge: Polity Press.

Habermas, J., 2006. Religion in the public sphere. *European Journal of philosophy, 14*(1), 1-25.

Jasanoff, S. S., 1987. Contested boundaries in policy-relevant science *Social Studies of Science,* 17, 195-230.

Jenkins, H. E., Aylward, R. B., Gasasira, A., et al., 2010. Implications of a circulating vaccine-derived poliovirus in Nigeria. *New England Journal of Medicine, 362*(25), 2360-2369. https://doi:10.1056/NEJMoa0910074.

Jones, W., 1996. 5 contraceptive vaccines. *Baillieres Clinical Obstetrics and Gynaecology, 10*(1), 69-86.

Jones, W., Judd, S., Ing, R., et al., 1988. Phase I Clinical trial of a World Health Organization birth control vaccine. *The Lancet, 331* (8598), 1295-1298.

Knight, D. M., 2004. *Science and spirituality: the volatile connection*. London; New York: Routledge.

Leask, J., Willaby, H.W. and Kaufman, J., 2014. 'The big picture in addressing vaccine hesitancy' *Human Vaccines & Immunotherapeutics*. 10(9), 2600-2602. http://dx.doi.org/10.4161/ hv.29725.

Luhmann, N., 1998. Familiarity, confidence and trust: Problems and alternatives. In D. Gambetta (Ed.), *Trust: Making and breaking of cooperative relations*. Oxford: Blackwell.

McKinnon, M. and Orthia, L. A., 2017. Vaccination communication strategies: What have we learned, and lost, in 200 years. *Journal of Science Communication*, (03), AO8 1-16.

Mooney, C., 2005. *The republican war on science.* New York: Basic Books.

Moscovici, S., 1984. The phenomenon of Social representations. In R. M. Farr & S. Moscovici (Eds.), *Social representations*. Cambridge: Cambridge University Press.

Pereira, J. A., Quach, S., Dao, H. H., et al., 2013. 'Contagious comments: What was the online buzz about the 2011 Quebec measles outbreak?' *PLoS One* 8(5): e64072. Ed. By L. G. Filion.. https://doi.org/10.1371/journal.pone.0064072.

PEW., 2010. Pew Forum on Religion & Public Life / Islam and Christianity in Sub-Saharan Africa. Chapter 3: Traditional African Religious Beliefs and Practices. Accessed 09/03/2018 from http://www.pewforum.org/2010/04/15/traditional-african-religious-beliefs-and-practices-islam-and-christianity-in-sub-saharan-africa/.

Song, G., 2013. Understanding Public Perceptions of Benefits and Risks of Childhood Vaccinations in the United States. *Risk Analysis*, n/a-n/a. doi: 10.1111/risa.12114.

From Knowledge Is Power to Thou Shalt Communicate

Wang Dapeng, Li Honglin, Chen Ling

China Research Institute for Science Popularization (CRISP), Beijing, China

Abstract: Scientists are the creator of science knowledge and they have the power to influence people's behavior through science communication, that is, as the source of science communication, scientists should bear the responsibility to communicate to the public, only through this way, can we build a solid foundation for innovation and prosperity of the whole nation. Generally, our narrative or discourse mainly focus on the frame of responsibility of scientists to communicate their science to the public, and emphasize that they should carry out this responsibility to the whole society, but neglect the gain and loss frame of this endeavor, which means we should help scientists weight or know what they could get from science communication, this article would draw frame analysis to discuss how to promote their science communication behavior, thus enhance the well-being of human.

Keywords: Science Communication; Responsibility; Frame

1. Introduction

Science and technology are embedded in virtually every aspect of modern life, and communicating science is much more important than ever before, just because only the public understand science, could they make reasonable decisions involving science-relevant issues, such as GMOs, vaccination, food safety, etc. With the institutionalization and professionalization of science and scientists, science gradually becomes a discipline that needs more training and learning, even some scientists could not understand other practitioners work which is out of their own field.

Especially, when we enter into the post-modern era, science gradually becomes a field complicated by economy, politics, culture and ethics, among other things. Therefore, scientists, as one important participants, should carry out science communication to the public through various channels. As the creator of science knowledge, science method, scientists stand at the front of science research, they could avoid the mistake of science communication that some journalists or other science communicator could make during their communication work.

We should admit that nowadays science communication is mainly carried out through various channels and platforms by or with scientists, such as newspaper, TV, magazines, radio, internet, and finally social media. Based on this recognition, the relationship between scientists and journalists is a hot issue which needs more attention from the science research. Many scholars examined the theme for a long time and produced many helpful outcomes for the practice of science communication, such as U.S., Germany, Japan, U.K., and other countries (Besley et al., 2013; Gascoigne et al., 1997; Kreimer et al., 2011; Martin-Sempere et al., 2008; Peters et al., 2008). According to some scholars, a gap and barrier exists between scientists and journalist (Peters, 2013), even there is oil and water, so they should overcome the creative tension if the science communication would be successful.

2. Carry on Responsiblity

Many scholarship and documents emphasis the science community should carry on their responsibility of communication science to the public, Brown (1971) stated that scientists had a responsibility to inform the public, and the Bodmer (1985) report emphatically stated that scientists' personal duty was '…learn to communicate with the public, be willing to do so, indeed consider it your duty to do so'. And research shows that most scientists do indeed feel that that they have a responsibility or duty to communicate their research findings with the general public (MORI, 2001b; People Science and Policy Ltd., 2006b; Searle, 2011). Such responsibilities are being articulated and incorporated in some national and international guidelines for science communication, such as the International Council for Science Committee on Freedom and Responsibility in the Conduct of Science (2010), the European Commission (2005).

China, like other countries, paid much attention to science communication these years. It issued the Law of the People's Republic of China on Popularization of Science, which is the first law dedicated to science communication on the world, then it issued and implemented the Outline of the National Scheme for Scientific Literacy (2006–2010–2020), all the documents attached more importance to science communication in China, and also ask scientists communicate their science to the public. In 2016, China President addressed a speech to researchers and scientists, which emphasized that science communication is of equally important to science research. And many scientists here believed communicating science to the public is their own responsibility, which could demonstrated from their communication activities and their comments on some popular media.

However, scientists still face some dilemma in communicating science, Sagan effect still needs to be overcome inside the science community, and the overall climate of science communication for scientists also requires some evolutional change.

As one director of CAST mentioned during a conference for science popularizers in 2017, scientists still face 4 tough situations in science communication, that is, they are unwilling to do science communication, they disdain to do science communication, they are unskilled to do science communication and they are afraid of doing science communication. All in all, the above mentioned situation could be attributed to the issue of responsibility. On the one hand, scientists themselves do not aware that they must bear on this kind of responsibility, or else they could improve their communication skills through different channels. On the other, the social culture does not provide a benign climate for scientists to carry their responsibility, such as communicating science could be blamed or looked down upon by their peers, and the journalist also could sensationalize some controversial issues, even hurt the passion of some scientists. What is more, the public could not be tolerant about the inaccuracy of science news (actually the inaccuracy mainly originate from journalists' agenda setting's purpose), and scientists would become the target of public blaming.

In a survey dealing with the relationship between scientists and journalists in 2016 (548 scientists answered all the questions), authors of this article found that nearly all the surveyed scientists (95.64%) thought science communication was very important and most of them (69.58%) would like to participate in science communication. We could interpret this results as majority of the interviewed scientists take science communication as their responsibility, however, when comes to their actual behavior or the frequency of their media contact, the survey exposed some mixed results, which means some gaps exist between the potential and actual effect. So what is the problem.

3. Gain & Loss Framing

Framing is cast information in a certain light to influence what people think, believe, or do. It is often used to communicate that an issue is a priority or a problem, who or what might be responsible for it, and what should be done about it (Iyengar, 1996). Hundreds of studies across a range of fields have tested the ability of gain/loss framing strategies to influence specific types of behaviors (National Academies of Sciences, Engineering and Medicine. 2017). And the framing also could be applied to analyze the gap between scientists' awareness of their responsibility of communicating science and their action.

In some cultures, regarding science communication by the scientists for the public, duty could be used instead of responsibility, such as UK, which means a commitment or expectation to perform some action in general or if certain circumstances arise. A duty may arise from a system of ethics or morality, especially in an honor culture. Many duties are based created by law, sometimes including a codified punishment or liability for non-performance. By nature, this kind of duty would have some guarantees to make it happen. However, when come to science communication, even various documents or scholarship emphasis the duty or responsibility of communicating science to the public by scientists through different platforms, there is little mandatory measure to push them forward, of course, only some grants would require applicants should explicit narrate their outreach activities in proposal.

On the macro level, the whole society and science community have reach the common understanding that communicating science have many benefits, not only for the individuals, but for science itself, the whole society (Gregory et al., 1998). This kind of benefit is also applied by institution, government, and science community to persuade scientists engage in science communication. In 2017, authors of the articles interviews nearly 30 media-savvy scientists, who are very active in using social media, such as blog, Wechat and microblog, to communicate their science to the public, all of whom believed that communicate science would improve the public scientific literacy, thus lay a solid foundation for the innovative-orientated nation in future.

However, we could find that if we apply gain and loss framing to their communication behaviors, we stress too much gain of their actions for the whole society, but less gain for the single person (here the scientists themselves), and the loss side of that.

As we mentioned above, communication science also needs some new-acquired skills which do not include in their professional training. And on the other hand, time is forever equal to anyone, each one of us have 24 hours a day, if they appropriate some time or energy to communicate their science to the public, which would mean that he or she should borrow some time from their research work. And carrying scientific research and publish their findings still is a priority for many science researchers. Just based on this judgment, the UNESCO-Kaling prize winner for science popularization Prof. Jean Audouze, a French astrophysicist, said in an interview with Indian Journal of Science Communication that he came into science communication profession only after his retirement (Patairiya, 2013). We tend to believe his practice would avoid Sagan-effect on one hand, and could exert more time and energy to the field on the other.

According to the above analyze, we could conclude that even scientists know they have the responsibility to communicate science to the public, but if they have other more important issues, such as publishing research findings, applying grant, to deal with, this kind of communication work would be put aside. The reason rests with our society and science community do not explain what kind of benefits or gains they could get from science communication for themselves.

Some research found that scientists are also motived to communicate with the general public by feelings of accountability to the taxpayers; or a need to promote their area of research to maintain or increase public funding (Searle, 2013). And other motivations include a desire to educate and inform the public and share knowledge and learn, as well as gain public approval or recruit new scientists and science students (European Communities, 2007; Gascoigne, 1997; Gregory et al, 1998; Pearson et al., 1997). However, after analyzing all the mentioned benefits in these literatures, we could find that they to a certain degree mixed the benefits for individual scientists with benefits for the whole community. Hence, we here try to discuss the benefits for individual scientist in three aspects.

Communication is a must skill for scientists, not only in science communication field, but in science research area. By engaging in science communication activities, scientists could learn and grasp necessary skills of communication, which could finally benefit their professional development. Equal to communicating science to the general public, publishing research findings is also some kinds of story-telling. The essence of this could be generalized and be universal.

Crowdsourcing and citizen science is much popular nowadays, which in nature is to bring layperson's knowledge or contribution to science research, we could find some research articles in prominent journals have more than hundreds even thousands authors, this revolutionized the traditional of science research. Communicating science to the general public also bring some new perspectives for scientists, for the public member could raise various questions that scientists did not pay attention to before communication, and some questions could be incorporate into their research or open new areas for their future work.

Some research found that communicating to the public through media could increase the citation of the original source by peers or scientists (Phillips et al., 1991), in 2018, new study also suggests good research pushed through social media gets more citations (Lamb et al., 2018). For scientist, citation is a key factor of their influence in science community, especially in their own specialized field, and according to the above mentioned research, communication could contribute to their research findings' citation. Hence, what we should do is to promote the recognition of this phenomenon.

4. We Need a Paradigm Shift

As we analyzed before, the conventional and existing practice of persuading scientists get involved in science communication start from claiming that they have the responsibility, but we do not clearly articulate what kind of benefits this kind of engagement could bring to them. So we suggest that a paradigm shift should be put on the agenda of carrying out science communication research, from stressing the responsibility to explaining what kind of gain that scientists could get from this practice, only by experiencing the benefits that science communication offers to individual scientist, could they enjoy communicating science to the general public, and finally their responsibility could be fulfilled.

And we should admit that even some literatures did try to explore the true benefit underneath this practice, and conclude some reasonable achievements, however, we still need to do some quantified research to testify the suggestions provided by this article, and that would be our future research focus.

References

Besley, J. C., Nisbet, M. (2013). How scientists view the public, the media and the political process. *Public Understanding of Science, 22*(6), 644-659.

Bodmer, W. (1985). *The Public Understanding of Science*. Retrieved from http://royalsociety.org/displaypagedoc.asp?id=26406 (accessed 14/7/2009).

Brown, W. T. (1971). The scientist's responsibility to the public. *Psychiatric Quarterly*, 45(2), 227-233.

European Commission (2005). The European Charter for Researchers and the Code of Conduct for the Recruitment of Researchers. From http://ec.europa.eu/euraxess/index.cfm/rights/index.

European Communities (2007). *European research in the media: the researchers point of view*. Report December 2007.

Gascoigne, T., Metcalfe, J. (1997). Incentives and impediments to scientists communicating through the media. *Science Communication, 18*(3), 265-282.

Gregory, J., Miller, S. (1998). *Science in Public: Communication, Culture, and Credibility*. New York: Perseus Books.

International Council for Science Committee on Freedom and Responsibility in the Conduct of Science. (2010). Advisory note on science communication. Retrived 28 January 2011, from http://www.icsu.org/publications/cfrs-statements/science-communication/.

Iyengar, S. (1996). Framing responsibility for political issues. *Annals of the American Academy of Political and Social Science*, 546, 59-70.

Kreimer, P., Levin, L., Jensen, P. (2011). Popularization by Argentine researchers: the activities and motivations of CONICET scientists. *Public Understanding of Science, 20*(1), 37-47.

Lamb, C. T., Gilbert, S. L., Ford, A. T. (2018). Tweet success? Scientific communication correlates with increased citations in Ecology and Conservation. *PeerJ*, 6, e4564.

M. K. Patairiya. (2013). 'Science-communication' journals: navigating through uncertainties. *JCOM* 12(01), C06.

Martin-Sempere, M. J., Garzon-Garcia, B., Rey- Rocha, J. (2008). Scientists' motivation to communicate science and technology to the public: surveying participants at the Madrid Science Fair. *Public Understanding of Science, 17*(3), 349-367.

MORI.(2001b). The Role of Scientists in Public Debate. Retrived 9 Aprial 2009, from http://www.wellcome.ac.uk/About-us/Publications/Reprots/Public-engagement/wtd003429.htm.

National Academies of Sciences, Engineering and Medicine. (2017). Communicating science effectively: a research agenda. National Academies Press.

Pearson, G., Pringle, S. M., Thmos, J. N. (1997). Scientists and the public understanding of science. *Public Undersatnding of Science*, 6(3), 279-289.

People Science and Policy Ltd. (2006b). Factors Affecting Science Communication: A survey of scientists and engineers. Retrived 12 February 2011, from http://www.peoplescienceandpolicy.com/projects/survey_scientists.php

Peters, H. P. (2013). Gap between science and media revisited: Scientists as public communicators. *Proceedings of the National Academy of Sciences, 110*(Supplement 3), 14102-14109.

Peters, H. P., Brossard, D., De Cheveigné, S., et al. (2008). Interactions with the mass media. *Science, 321*(5886), 204-205.

Phillips, D. P., Kanter, E. J., Bednarczyk, B. et al. (1991). Importance of the lay press in the transmission of medical knowledge to the scientific community. *The New England Journal of Medicine*, 325(16): 1180-1183.

Searle, S. D. (2013). Scientists' Engagement with the Public. In Gilbert, J. K., Stocklmayer, S., Stocklmayer, S. M. eds. *Communication and engagement with science and technology: Issues and dilemmas: A reader in science communication*. Routledge, 43-44.

Searle, S.D. (2011). *Scientists' communicaiton with the gengeral publci-an Asutralian survey*. Canbeera: The Australian National University.

Between Collectivism and Individualism

——Chinese scientists' choice

Wang Huibin[1,2]

[1] National Academy of Innovative Strategy, CAST, Beijing, China
[2] Institutes of Science and Development, CAS, Beijing, China

Abstract: The rise of modern science was accompanied with the spread of individualism. While the so-called 'big science' has encouraged scientists to be cooperative, it is individualism, instead of collectivism, which has played the basis of western science. Nevertheless, for Nineteenth-century Chinese literati who received western science and the early Twentieth-century Chinese scientific community, it was easy to accept the epistemological individualism but difficult to appreciate the political, partly due to traditional Chinese patriotism and national demand for power and wealth. Especially in the era of the PRC, the dominant collectivism has strongly disciplined Chinese scientists, no matter educated in capitalism or socialism countries.

Since modern science was introduced to China in the 19th century, Chinese scientists, as well as literati accepting science, have been dealt with the tension between individualism and collectivism. Compared with western peers, Chinese scientists do have been more collectivism when faced with the nation and their partners. In this paper, I would defend the central place of individualism in western science, and then historically describe and analyze Chinese scientists' choice which seems more collectivism than individualism and discuss how collectivism has effected Chinese scientific practice.

1. The Descent of Individualism in Science—Or Is It?

Philosophers usually regard empiricism epistemology, which is essential for modern science and highlights the importance of individual sensation in the production of scientific knowledge, as a typical individualism (Arblaster, 1985; Lukes, 1973). The standpoint could be exemplified by the motto of the Royal Society: 'Take Nobody's Word for It'. Nevertheless, a growing number of scholars have noted the social dimension of scientific research, and discussed issues of order and authority in the production and dissemination of knowledge, in the effect of Merton's sociology of science, Thomas Kuhn's argument on 'scientific community' in the *Structure of Scientific Revolution* and then the sociology of scientific knowledge (SSK). Especially after Price suggested the famous concept of 'big science', individualism seems fading and descending in the scope of science. But, is it?

Undoubtedly, 'big science' has been an ideal concept showing the expanding of scientific research organization and fund and, naturally, reminding scientists meeting the nation's requests and being more cooperative than the period of 'small science'. However, it could not be inferred that collectivism has replaced the priority of individualism in the field of science. Among the world, research papers are not signed by research teams. Similarly, scientific awards are still mainly presented to individual scientists.

The stabilization of individualism in scientific practice lies on the stabilization of its metaphysical presupposition. According to Mario Bunge, there are ten modes of individualism, such as metaphysical, logical, semantic, epistemological, methodological, ethical, and political and so on, but they are interrelated and parts of individualism as a whole (Bunge, 2000). Clearly, all of these modes share an atomic basis and conform to the pattern from particular to general. Especially for the epistemological and the political, their correlation has been confirmed by the common form of British restored polity and experimental science in the 17th century. From a cognitive perspective, 17th century British experimental philosophers warned against dogmatism in their work: 'No isolated powerful individual authority should impose belief.' What's more, the working way of experimental community

Supported by the China Postdoctoral Science Foundation funded project (2018M631550).

also provided an ideal polity model at that time (Shapin et al., 1985). In Yaron Ezrahi's opinion, one of the most central beliefs in modern Western culture is that 'seeing is knowing', and western science provided liberal democracy ideological resource named 'attestive visual culture', namely, the mode of attesting, recording, accounting, analyzing, confirming, disconfirming, explaining or demonstrating by showing and observing examples in a world of public facts (Ezrahi, 1990). Individualism is still the hardcore of western science.

2. In Search of Power and Wealth: Pioneers of Chinese Scientists and the Nation

In Chinese context, after Jesuits' translating natural philosophy at the turn of the Ming and Qing Dynasties, Protestants as well as Jesuits translated modern science into Chinese during Late Qing. Nevertheless, translating has never simply 'moved' into another context but also generating new meanings from understandings of translators and, what's may be more important, readers (Liu, 1995). For Chinese scientists, before the foundation of P.R.C, they were calling for Mr. Democracy and Mr. Science but developed a tradition of 'saving China through science' and a variation of liberalism 'rejecting individualism and emphasizing the balance between individuals and the group' (Huang, 2006).

Faced with successive military and diplomatic failures, Chinese literati who grew up in the tradition of patriotism and the system of civil examination recognized the necessity of studying western science and shifted from the speculative paradigm to the empirical (Shang, 2001). Some even explored scientific methodology and introduce inductive method to strengthen the imperial China. Wang Tao (王韬), the first local translator, shaped inductive logic in three dimensions: rejecting authority, recommending empiricism and emphasizing the position of Francis Bacon in the history of science. On the contrary, Wang did not plan to challenge existed political framework consisting of the empire and people, but made suggestions to improve the empire's govern (Wang, 2002). His different attitudes on the epistemological and political could be explained by his central target which was achieving China's prosperity with the help of science and its inductive method, instead of maximizing individual freedom (Cohen, 1974). Yan Fu (严复) was also frustrated in the civil examinations and spoke more highly of induction. Basing on John Stuart Mill's thoughts, Yan had begun to discuss inductive logic in his translations, articles and lectures on strengthening China since 1890s. Unlike Mill's inductive logic could be seen as a tool that allows each individual to test the authority (Staley, 1999), Yan regarded individual liberty as a means to achieve national goals of promoting civil intellectual and morality. Therefore, Yan's esteem of inductive logic was correcting logical errors in traditional thinking which hindered the wealth and power for the country (Schwarz, 1964).

Although Wang and Yan may be limited by their misunderstandings of foreign thoughts, Republican Chinese scientists carried forward the collective character. Founded in 1915, the Chinese Science Society has been regarded the milestone of the institution of science in China. Nevertheless, Ren Hongjun (任鸿隽), one of its founders, has been still the most typical of 'saving China through science' until now. As Zuoyue Wang demonstrated, on the one hand, the society and its members kept trying to seek autonomy from the government; on the other hand, their activities seemed so-called 'scientific nationalism' which tried saving China through science by operating with the government (Wang, 2002). Among the large groups of patriotism scientists, we could exemplify He Zehui (何泽慧), a famous female scientist because of her marriage with Qian Sanqiang (钱三强) and their research on the phenomena of nuclear fission. She represented the 'scientific nationalism' in one letter to her sister in 1937:

Studying ballistics, I may be asked to return to China to service for the Military Industry Department. Armed with my calculation, Chinese soldiers could shoot their guns accurately. If they asked me earlier, Japanese soldiers have already retreated. (as cited in Wang et al., 2017, p.33)

3. The Rise of Red Scientists

After the foundation of the People's Republic of China, hundreds of abroad scientists keeping the ambition of 'saving China through science' returned and participated in careers of the 'new China'. When Liang Sili (梁思礼) and his accompaniers knew the foundation of the 'new China' on the ship sailing to China, they warmly celebrated and even made a national flag according to the description on the radio (Liang,

1999). Nowadays, Chinese scientific and technological elites, compared with the western counterpart, do represent stronger patriotism (Cao, 2004).

Besides insisting the tradition of collectivism on the issue of individuals and the nation, scientists have been disciplined to unite. In 1950, the Representative Meeting of All-China Natural Scientific Workers (中华全国自然科学工作者代表会议) was held and one slogan hung on the meeting platform was 'All-China natural scientific workers should be united to fight for a better financial condition' (Yonglong, 1950). Also at the meeting, the Premier Zhou Enlai (周恩来) gave an address in the title of 'Building and Uniting'. He admitted that the scientific community tended to be single but it was time when scientists should unite (Zhou, 1990). Moreover, Vice-Premier Li Jishen (李济深) asked Chinese scientists to learn Soviet scientists to use collective working method to study and solve problems, and abandon working alone and denying communicating as scientists in capitalist countries (Li, 1990).

The meeting decided to establish the All-China Federation of Natural Science Societies (中华全国自然科学专门学会联合会) and the All-China Association for Science Popularization (中华全国科学技术普及协会). It was partly via these organizations that the newly governing Chinese Communist Party united and utilized scientists to a great extent. Scientists, or the new expression 'scientific workers', were requested to constrain their individualism tradition which was absolutely wrong and learn to operate and, on the other hand, dedicate for the people. It is worthy of pointing out that this kind of collectivism, including patriotism for the nation and cooperative with partners, has played a key role in the high-speed development of Chinese science and technology. Especially in the middle of 20th century, a large number of scientists and engineers worked for the nation and the people selflessly. Several Chinese scientists have mentioned the factor in their oral history, including but not limited to the Noble Prize winner Tu Youyou (屠呦呦). Indeed, the first published paper reporting the antimalarial function of artemisinin was in the name of the research team. When accepting the Nobel Prize, Tu Youyou acknowledged all of teams that advanced their research and spoke highly of the cooperative spirit of Chinese scientists. Song Jiashu (宋家树), one academician of Chinese Academy of Sciences mainly due to his contribution for the project of 'two bombs, one satellite', suggested the first atom bomb was a model of the cooperation among science, technology and industry and also the cooperation among cadres, intellectuals and workers.

In recent years, several kinds of the spirit of Chinese scientists in specific fields have been continually suggested and established as models of Chinese scientists. The most famous may be the spirit of 'two bombs, one satellite': patriotism (热爱祖国), dedication (无私奉献), self-reliance (自力更生), struggle (艰苦奋斗), cooperation (大力协同) and courage (勇于攀登). In addition, the spirit of Manned Space is abstracted as struggle (特别能吃苦), fight (特别能奋斗), concentration (特别能攻关) and dedication (特别能贡献); and the spirit of manned deep submergence is rigorous and realism (严谨求实), united and cooperation (团结协作), struggle and dedication (拼搏奉献) and courage (勇攀高峰).

4. Conclusion

While individualism has played one of the basis of modern science, collectivism has charactered Chinese scientists in two aspects. Coping with the relation between individuals and the nation, Chinese scientists emphasize supporting and dedicating for the nation. Relatedly, they would be more cooperative and even selfless in the process of scientific research. In a diachronic view, Chinese scientists' choice between collectivism and individualism was shaped upon the national desire to be strong and the ideological discipline. It could be forecast that collectivism would be insisted and keep advancing Chinese careers of science and technology.

References

Arblaster, A. (1985). *The rise and decline of western liberalism.* London: Basil Blackwell.

Bunge, M. (2000). Ten modes of individualism—none of which works—and their alternatives. *Philosophy of the Social Sciences*, 30(3), 384-406.

Cao, C. (2004). *China's scientific elite*. London: Routledge Curzon.

Cohen, P. A. (1974). *Between tradition and modernity: Wang T'ao and reform in Late Ch'ing China.* Cambridge (Mass.): Harvard University Press.

Ezrahi, Y. (1990). *The descent of Icarus: Science and the transformation of contemporary democracy.* Cambridge (Mass.): Harvard University Press.

Huang, K. (2006). Western liberalism in modern China. In J. Huang, ed. *The interaction and integration of Chinese and extraterritorial cultures*, Taipei: Himalayan Foundation, pp.341-378.

Li, J. (1990). Address of Vice-premier Li Jishen. In. Z. He, G. Yin, X. Zhang, eds. *Chinese societies for science and technology*. Shanghai: Press for Science Popularization in Shanghai, pp.475-476.

Liang, S. (1999). The Reddest sun, the dearest nation. In the Committees of Cultural and Historic Data of the Chinese People's Political Consultative Conference, ed. *Notes on international students' return*, Beijing: China Press for Cultural and History, pp. 101-107.

Liu, L. (1995). *Translingual practice: Literature, national culture, and translated modernity, China, 1900—1937*. Stanford: Stanford University Press.

Lukes, S. (1973). *Individualism*. Colchester: ECPR Press.

Schwarz, B. (1964). *In search of wealth and power: Yen Fu and the west*. Cambridge (Mass.): Harvard University Press.

Shang, Z. (2001). The Impact of modern science on Late Qing intellectuals from 1886 to 1894: A case study on the natural science section of the essay contests of Shanghai Polytechnic Institute. *Studies in Qing History*, (3), 72-82.

Shapin, S., Schaffer, S. (1985). *Leviathan and the air-pump: Hobbes, Boyle and the politics of experiment*. Princeton, NJ: Princeton University Press.

Staley, K. W. (1999). Logic, liberty, and anarchy: Mill and Feyerabend on scientific method. *The Social Science Journal*, 36(4), 603-614.

Wang, C., Zhang, L. (2017) *Scientific dream, Chinese dream: Album of the exhibition on Chinese modern scientists*. Beijing: China Press for Science and Technology.

Wang, T. (2002). *Tao's garden series*, Shanghai: Shanghai Bookstore Publishing House.

Wang, Z. (2002). Saving China through science: The Science Society of China, scientific nationalism, and civil society in Republican China. *Osiris*, 17, 291-322.

Yonglong. (1950). An introduction of the Representative Meeting of All-China Natural Scientific Workers. *Science Bulletin*, 8, 363-366.

Zhou, E. (1990) Building and uniting. In. Z. He, G. Yin, X. Zhang, eds. *Chinese societies for science and technology*. Shanghai: Press for Science Popularization in Shanghai, pp.489-494.

Science and Technology Innovation Team Engagement with Popular Science: Model and Implementation

——Case study of national key laboratories participating in popular science

Liu Xuan[1], Li Xinyu[2]

[1] National Academy of Innovation Strategy, CAST, Beijing, China
[2] Northeastern University, Shenyang, China

Abstract: A complete science communication work is a communication chain that involves different subjects of communication. There are differences in the mode and path of different science communication guidance. In this study, case study is used to sort out and analyze the science popularization work carried out by the State Key Laboratories, and to explore their synergistic relationship and transformation paths. The research shows that the current model of scientific research teams participating in science popularization is shifting from task-based passive communication to responsibility-based active communication. Meanwhile, due to the significant differences in the core values, cost structure, collaborative subject and IT application of the scientific research teams, it is possible to adopt the reconstructed science popularization transformation path and the progressive popular science transformation path in the process of transformation. The research results provide a clear model and reference for the government to guide scientific research teams in conducting science popularization works in an orderly manner.

1. Introduction

Science communication, as a new academic field emerging from philosophy of science and history of science, has a close relationship with both traditional science popularization and communication. Science communication in contemporary China has three designations: science popularization, public communication of science & technology, and science communication, representing three groups and three models of science communication respectively. Science popularization, featured by nationalism, utilitarianism and scientism, is a mainstream and orthodox model of communication dominated by China Association for Science and Technology (CAST). Researchers for public communication of science & technology are mostly communication scholars who mainly focus on means and efficiency of communication. Advocates of science communication are mainly scholars in the field of history of science and science philosophy, who are most likely challenging the three major ideologies concerning science popularization. The three models are still interacting and integrating intensively.

Twenty-five years ago, for the first time, the Royal Society organized and compiled the report Public Understanding of Science, in which it was appealed that promoting public understanding of science is a part of the professional duties of every scientist and that scientists must learn to communicate with the public [1]. In 2007, 'Several Opinions on Strengthening National Science Popularization Competence' promulgated by eight departments of China such as the Ministry of Science and Technology clearly put forward to 'gradually establish and improve the public-oriented science and technology information release mechanism in the process of implementing major engineering projects, science and technology programs and science and technology major projects of the nation, so as to enable the public to learn and master associated scientific and technological knowledge and information'.

The impact of science and technology on society is exerted through science communication and the power of science and technology lies in its popularity [2-3]. Situated at the head end of the communication chain of science, scientists must assume the primary responsibility for science popularization [4]. Scientific workers should combine scientific research with science popularization [5]. In 2004, the Ministry of Science and Technology issued relevant documents calling for the implementation of the policy 'State Key Laboratories Open to the Public'

Corresponding author: Liu Xuan, No.3 Fuxing Rd., Beijing, P.R. China (100863).

and pointed out the necessity of opening up scientific research laboratories to the public [6]. Consequently, some scientific research teams have assumed the responsibility of scientific communication. However, a complete science communication work is a long chain of communication involving different communication subjects. Wu Guosheng proposed that there are five subjects in the current scientific communication system, namely scientific community, government, media, public and non-governmental organizations [7-8]. Public participation and influence are highlighted, the European Consensus Conference offers this possibility [9]. As an important tool for the practice of the consultation between the public and the scientific community, the consensus conference breaks through the public understanding of the deficit model [10], becomes a hallmark of the democratic model at the practical level. It is of great significance in exploring new ways for the public participation in science [11]. Undoubtedly, posing higher expectations and requirements for the science popularization duties of scientists today. In particular, Facing the new model of scientific knowledge production [12], which broke through the age of elites and authoritative science. The scientists, as 'producer of scientific knowledge', still play the role of 'the primary subject'; therefore, it is advisable to guide a batch of scientific workers to be engaged in research activities for life [13]. The interaction between the public and science spans the 'communication gap' [14], promoting public understanding of science is still the call of the times in contemporary society and the development of modern science and technology. More complex relationship between public and science, the public is heterogeneous group of people, and publics act in social contexts and shift their attention and knowledge [15-16]. In some ways, the public is also dynamic group. Engagement of scientific research teams in scientific communication is a research subject reflecting the common demand of the government and the public.

The ways in which universities and scientific institutions launch science popularization work are increasingly diverse. For example, the 'Public Science Day' has become a branding activity for the engagement of scientific research institutions in science popularization [17]. Science popularization workers unfold popularization exhibition of scientific research resources jointly with scientific research personnel and technology enthusiasts to exhibit high-end scientific research resources in a straightforward way popular with the public, which is not only helpful to encourage scientific research personnel to carry out basic research and cutting-edge exploration, but also favourable to establish an effective interaction between scientific research and the public [18]. In the previous studies, the scholars have conducted targeted analysis on the work patterns and paths that the scientific research teams take part in science popularization, whereas there is a lack of systematicity on the whole to a certain degree. Due to differences in types of teams and forms of engagement in science popularization work, it is difficult to use questionnaires. Therefore, it is a good research idea to apply case study method in allusion to such problems, aiming at the creation of concepts, propositions and theories [19]. Discovering the characteristics of the study object through penetrating its superficiality based on the various primary and secondary information about the case and associated analysis and exploration [20], this method is intrinsically explanatory, exploratory or descriptive. Therefore, case study method is adopted for the purpose of this paper to analyze the scientific research teams of state key laboratories, sort out the models and modes of engagement in science communication, find out the problems of the scientific research teams in science communication, and explore their synergetic relationship and the transformation path.

2. Study Method

1) Introduction of the Case

Case study method is adopted mainly for the following reasons. First, this study focuses on the discussion about current situation of science popularization engagement of scientific research teams undertaking major scientific and technological innovation projects, which belongs to the scope of 'what' and 'how'. By using the case study method, the current phenomenon can be reasonably summed up to effectively tap into the potential law behind it and find its theoretical basis [21-22]. Second, the reason to choose certain a state key laboratory as research object is that the laboratory has undertaken all the major national science and technology innovation projects and also a large amount of work in science popularization, which contribute to its certain representativeness.

The laboratory focuses on basic research and

applied basic research, extending to application development at the same time. It has a sound talent team with 77 fixed scientific research personnel. The laboratory has been awarded as an outstanding innovative group by the National Natural Science Foundation of China, an innovation team by the Ministry of Education, a research team for national defense and one of the first batch innovation teams in the 'Giant Project' of Hebei Province.

2) Data Collection

Data are collected for the study by in-depth interviews with some scientists and scientific research personnel, consulting the archived files about scientific communication as well as on site observation of the science popularization exhibition hall and the laboratory. Diversified sources of data can ensure that data are mutually supplementary and cross-validated to increase the validity of the case [23] and meanwhile provide proper guarantee for the adequacy, accuracy and pertinence of the data.

From December 2016 to March 2017, field investigation of the laboratory was conducted successively for a total of 5 times and interviews with 13 middle and senior scientific research personnel were made for 20 men times in a total of

21.5 hours, with 103 thousand words of interview text sorted out. The whole interview process is scientific and standard. Each interview involves at least three researchers in a semi-structured way, interviewing the interviewees according to the interview outline designed for the study on one hand, and questioning closely based on the answers and ideas of the interviewee on the other hand to obtain information as detailed as possible, with attention paid particularly to penetrating discussion about the examples of the phenomena or opinions put forward. Detailed information of the interviews and on-site observations were recorded; the interview material was discussed thoroughly and information was sorted out timely. Upon finishing of the interview, the collected information was subject to internal discussion in time so that key points of the interview are sorted out and the inconsistency of information described in the same interview and that among different interviewees are recorded, with defective information marked to be supplemented and perfected in the next interview and field observation. In addition to formal interviews, other means are adopted for the study to further collect and master the laboratory-related data, mainly including: ① Secondary information, such as the information on its official website, books, documents, newspaper and on-line reports regarding the laboratory. ② Archived files, with relevant information obtained from the laboratory such as handbook of development, organization structure, operation process, promotion PPT and video. ③ Field observation, e.g. visiting the key laboratory. ④Informal communication: The author communicates and exchanges idea with the scientific research personnel to strengthen intuitive understanding (Table 1).

Table 1 Table of code sources.

Source of Data	Data Classification
Primary information	Information obtained through in-depth interviews
	Information obtained through informal interviews
	Information obtained through field observation
Secondary information	Team information obtained from the Internet
	Information reported by the social media
	Information obtained from the internal documents and brochures

3) Data Coding and Analyzing

The researchers summarize and sort the information obtained through multi channels and the author reviews and collates all the documents before coding. In this paper, the author follows the coding idea of exploratory research method and analyzes the data of the laboratory in the form of open coding. The three coding personnel read and compare the information summarized and extract main contents therein, aiming to code the logical relations among major concepts and contrast the coding results, which will pass if

the three opinions are the same or be subject to discussion for the inconsistency in case of disagreement until the research members arrive at an agreement. Such method diminishes the onesidedness of the conclusion resulting from personal prejudice and subjectivity so as to ensure the integrity of information acquired [24].

4) Reliability and Validity

To improve the quality of current case study, processing was made for this study from the aspects of construction validity, internal validity, external validity and reliability, as shown in Table 2.

Table 2 Reliability and validity.

Verification	Strategy	Application Stage	Specific Practice
Construction Validity	Multiple data sources	Data collection	Ensure consistency in opinions by multi-channel sources of data such as in-depth interviews, archived files, field observation and secondary information inquiry
	Evidence chain formation	Data collection	Raw data - statement identification - relevant construction concept extraction - preliminary theory construction - further verification and modification between theory and data information - theoretical model formation
	Report validation	Data collection	Submit the information to the interviewees for review and verification to ensure correct understanding of the contents
Internal Validity	Model matching	Data analysis	Discover matching model for the study; the concept model is in reasonable agreement with the study conclusions
	Interpretation establishment	Data analysis	Interpret and explain layer by layer according to the logical frame
	Analyzing opposite competitive interpretations	Data analysis	Make interpretations by a number of researchers and prepare opposite data for the existing interpretation so as to review and revise the original interpretation
External Validity	Theoretical guidance	Study design	Establish theoretical framework of this paper through document review and analysis, which is an in-depth dialogue between the study and the documents
	Case study	Study design	Precise analysis
Reliability	Detailed and reliable case study plan	Data design	Prior to formal study, a number of researchers discuss about the study design thoroughly, revise and form a detailed study plan
	Information bank establishment	Data collection	Establish information data base and proceed with detailed classification according to the content, acquisition channel and acquisition time
	Repeated implementation	Data analysis	Different researchers analyze the data information separately, then compare the analysis results, with unanimous opinion arrived finally after discussion and debate
	Presentation by diversified evidence types	Data analysis	The case analysis presents the evidences obtained from multi-channels such as internal documents, secondary information, interview contents and field observation

3. Study Findings

1) Characteristics of Task-based Passive Communication

The science popularization work of traditional scientific research teams is mainly intra-industry oriented aiming at the scientific opinions that the scientific community has uttered, which is no more than an announcement to outside, without intention to consider whether the public understands it in the process of science popularization. At the same time, scientists are not so clear about how to communicate science to outsiders. Sometimes they do not realize this problem and more often they realize it but can do nothing to change. Therefore, in the process of science communication, many people come only for the scientists' reputation, which cannot achieve sound effect. See Fig. 1 for details.

(1) *Task-based passive communication*

The task-based passive communication initiated by administrative orders is to cope with the temporary requirements of government departments at all levels, such as Science and Technology Week (held by the Ministry of Science and Technology in May annually), National Science Popularization Day (held by China Association for Science and Technology in September annually), visiting and opening as required by education management departments or routinely. Either providing relevant information and content or opening the laboratory to the public for visiting is completed passively, without work planning for science popularization before project implementation, thus they are contingent science popularization tasks.

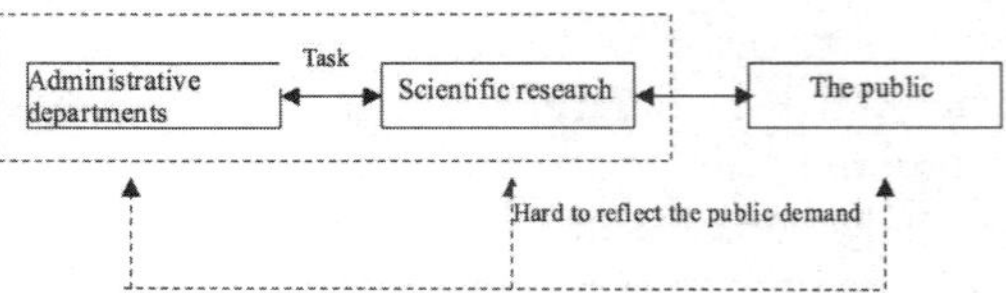

Fig. 1 Characteristics of Task-based Passive Communication of Popular Science.

(2) *Subjects of science popularization work*

'We just do what the administrative division

requires and do not care much about how the public feeds back.' said the office director of the innovation team. In previous science popularization process, scientific research teams were required by the administrative departments to carry out science popularization work, so they had little direct contact with the public. To put it in another way, the sector proposing to conduct science popularization work is administrative departments at all levels while the sector receiving science popularization is the public. This results in the disjunction between supply and demand in the process of science communication.

(3) *Cost of science popularization*

'We also hope that more people can understand what we do. In the task-based passive communication process, we were asked to prepare a variety of publicity materials for distribution. On one hand, the materials are so professional in content that they are not popular with the public; on the other hand, we spend lots of effort and time doing the work but they can't understand at all, which is a waste of manpower and material resources. Putting the material cost aside, it is much too time-consuming.' said the office director of innovation team.

(4) *Application of new media for science popularization*

'In the past, new media technology was hardly used by scientific research teams for science popularization, whether as a way of demonstration or a means of communication, since it is not the mainstream of our work. Although animation or other forms are well-received, we do not have the professional production staff, and what's more, the production cycle will be long for the need of professional review. In this process, we have no creative team for science popularization or special unit for the organization of science popularization activities. Rather than the economic cost, no one is able to use it or operate it, nor a professional third-party outsourcing operation team which not only understands what we do, but also, more importantly, able to make science-based compilation for the public to understand.' said the office director of the innovation team.

2) Characteristics of Responsibility-based Active Communication

Currently, scientific research teams take the demand of the public as origin and destination of communication, in which the public put forward its scientific demand to be responded scientifically by the scientific research teams. The responsibility-based active communication process is turned into the communication model of 'scientific research teams—creative team—the public' which, as a major revolution itself, subverts the traditional path of scientific communication. With the introduction of the new requirement for transformation of science popularization, it is also necessary for the scientific research teams to make transformation and adjust the process of science communication in order to adapt better to the present era [25]. The innovation lies in the shift of task-based passive communication led by administrative orders to the public-centered responsibility-based active communication.

Task-based passive communication, as passive collaboration between administrative orders and scientific research teams, is not social responsibility but orders. The public-centered responsibility-based active communication is the collaboration among various units in the science communication chain where scientists interact with the public via the link of knowledge and the public becomes the core of science communication.

Responsibility-based active communication is composed of the following three parts (as shown in Fig. 2), in which the first one is collaboration between the front end of science communication and the public. On the one hand, the public affects the purchase, production, storage and logistics of creative models of science popularization service through orders, feedback and evaluation; on the other hand, the back end of science communication effectively meet the rapid and diversified demand of the public through deep collaboration (the synergistic relationship is shown in Fig. 2).

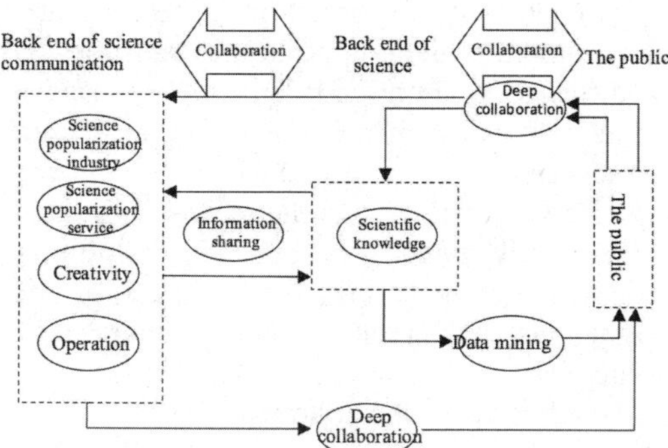

Fig. 2 Characteristics of Responsibility-based Active Communication.

(1) *Collaboration between front end of science communication and the public*

At the front end of science popularization, the public can influence the popular science products and services provided by scientific research teams through demand posting, online evaluation and feedback with the support of Internet technologies. At the same time, the scientific research teams can promote the innovation of public services of science popularization by analyzing public demand, behavioral characteristics and popular science demand through data mining based on all kinds of public information. The following aspects are mainly included: ① Online science publicity. Place stress on the dynamic data mining of real-time information of public behaviors and realize rapid response and satisfaction to the public demand; catch the public data periodically through major online platforms; and guide the releasing time and form of science popularization through intelligent analysis. ② Off-line science experience. Unfold public-oriented new product development and design based on dynamic data mining to achieve accurate personalized service. Different receivers have different demand for the same scientific knowledge, so potential demand of the public may be explored through the interaction in the scientific community to provide popular science products and services that satisfy different public needs. ③ Realizing in-depth scientific guidance. Enhance the overall service value at the ports where research teams contact with the public by constructing an O2O service system from 'merely online' or 'merely offline'.

(2) *Collaboration between front and back ends of science communication*

In the supply chain of science popularization service provided to the public, Internet technology enables the information to be fully and timely shared which supports the transmission and diffusion of front-end demand to the back end, as well as the response and feedback from the back end to the front end, not only realizing the rapid response to popular science demand of the public effectively but also helpful to realize rapid update and transmission of the information which predicts science popularization product market.

It is reflected in the following aspects: ① As for public demand, information sharing should be realized through the Internet so as to achieve rapid response of science popularization to the public demand on science. ② As for knowledge updating, the scientific research teams make demand prediction through mining public information and publish the predicted demand scientifically and reasonably. ③ Support the personalization of popular science products through research & development and production technology improvements. Promote the publication of science popularization industry and achieve specialization and premiumisation of popular science products.

(3) *Collaboration between back end of science communication and the public*

In the context of new media, the science experience of the public, i.e. science popularization effect, is one of the most important issues for the government administration sections and scientific research teams. On one hand, the public affects the types, forms, creativity and operation works of science popularization provided by science teams through the order, feedback, evaluation and the like; on the other hand, the back end of science popularization supports the diversified demands of the public effectively through deep collaboration and promotes the science experience of the public comprehensively.

They are reflected in the case from the following aspects:

① To improve constantly by learning about the public opinions. The scientific research teams have access to the public evaluations on the Internet in terms of science popularization they provided so that they may improve the problems in science popularization process. ② To support the diversity and difference of the public through the update of popular science knowledge and science popularization exhibition technology. ③ In-depth collaboration brought about by information sharing between science and technology R&D team and creative team constantly improves the experience of science popularization and promotes science popularization effect.

3) Transformation Path from Task-based Passive Communication to Responsibility-based Active Communication

Based on the analysis towards the scientific research teams, they have undergone the transformation from task-based passive communication to responsibility-based active communication. The subsections above have analyzed the operation characteristics of traditional and public-oriented science popularization works and the following part will expound the way to realize transformation. There are two project teams under the state key laboratory, presenting different methods and paths in the process of science popularization transformation.

Interview surveys show that the paths for science popularization transformation of the two project teams follow the realization mechanisms from resources to capabilities and then to transformation, but the two apply different use methods of resources and capabilities. Project team I adopts a reconstructed science popularization transformation path while Project Team II chooses a progressive path.

Specifically, Project Team I chooses a reconstructed transformation path mainly for the following reasons. First is resource base. The initial science resources and processes of Project Team I only suit the traditional market and can't adapt to current informatization conditions, so it has to resort to fundamental revolution to adapt science popularization to the characteristics of networking. Second is the organization inertia which is formed in the process of traditional science communication. New science communication work can't be unfolded without essential adjustment.

Project team II chooses progressive science popularization transformation path mainly because it is dominated by young people with intersection of multi-disciplines, it has ever adopted new media technology for communication and the content of the achievements are also suitable for the adopting of new media. Therefore, Project Team II makes use of its own resources and advantages to realize science popularization transformation from task-based passive communication to responsibility-based active communication (as shown in Fig. 3).

(1) *Case 1: reconstructed science popularization transformation path*

Project team I undertook task-based passive communication work by making some display panels, brochures and professional contents or organizing visiting to the laboratory for equipment show. When their project director saw the change in science popularization habits of the public, they decided to adopt new media technology for transformation.

For Project Team I, the primary task for transformation is to acquire new resources, including talents, brand, IT, creative industry resources, etc. ① Talent resources. In order to understand and get familiar with the operation of new media, professional creative personnel for science popularization should be introduced, multidisciplinary intersection and integration should be made and the internal organization conflicts and pressures should be reduced. ② Brand resources. Create some branded scientific activities to enhance the popularity of scientific research. For instance, the first top-notch original technology show 'I am the future' co-produced by Hunan Satellite TV and Vivid Media and jointly launched by Bureau of Science Communication, Chinese Academy of Sciences Science recently. ③ IT resources. Cooperate with data companies to capture online public

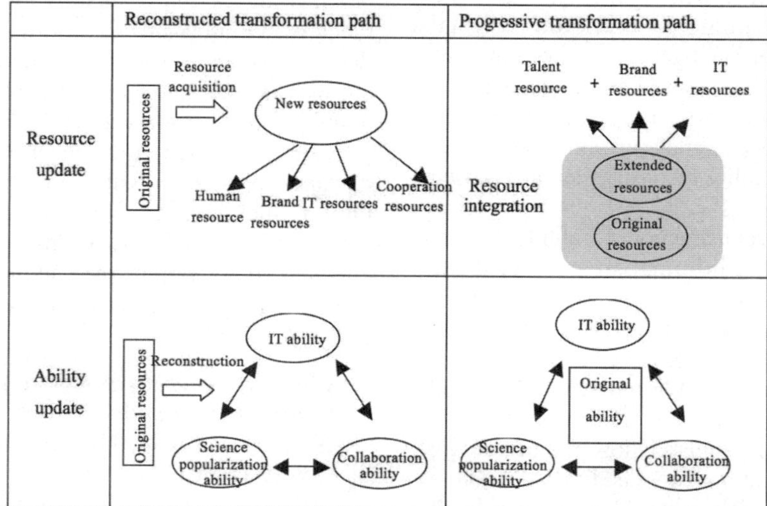

Fig. 3 Transformation path from task-based passive communication to responsibility-based active communication.

information and unfold intelligent analysis by applying the data analysis system to promote research & development and service design for science popularization; integrate the public data resources to enhance public satisfaction. ④ Supplier resources. Project team I assists science popularization creative production units in providing and reviewing scientific knowledge, forming public-oriented popular science resources.

With the science popularization model dominated by scientific research teams, some science popularization creative teams transit in the direction of higher level and specialization. Faced with brand new market characteristics, science popularization creative teams need new resources, including equipment, talents, IT and so on. ① Science resources. The science popularization work is closely linked with scientific contents, so the Project Team I of the scientific research teams applies current scientific equipment to carry out science popularization work and the creative teams do not need to purchase relevant resources. ② Talent resources. The project team has a large number of scientific research personnel who, however, do not know so well about science popularization work and can't fulfill some science popularization tasks, affecting the social image of the entire R&D team. Later, teams with new media communication and operation experience are introduced to conduct collaborative development. ③ IT resources. In response to the rapid and personalized demands of the public, project team I and creative companies realize real-time sharing of information.

On the basis of obtaining new resources, the two sides reconstruct their abilities through rapid organizational learning to break the original organization inertia, overcome the resource fogyism and procedure stereotype and restructure the business processes and mode of operation for science popularization. To this end, three aspects of ability reconstruction are carried out.

(a) IT ability reconstruction. IT ability, aiming at data mining and information sharing, is the basis of public-orientation. Project team I introduces IT resources to effectively analyze public behavior data, explore the behavioral characteristics of the public on science demand as well as the different types of the public and carry out science popularization product design and planning accordingly. In addition, IT ability reconstruction which is based on data mining has changed the content and direction of information sharing between scientific research teams and creative units, providing technical support for the collaboration necessary for the public-orientation of all links of science communication.

(b) Science popularization ability reconstruction. In terms of science popularization product and service supply, they cooperate with creative companies to supply popular science products rapidly to the public so as to adapt to the heavy demand of the public on science knowledge in the context of new media. In the process of continuous collaboration, creative companies understood the science content and adopted suitable ways to interpret and express it. Under the guidance and help of the project team, a specialized science popularization creative company comes into being which drives the development of science popularization industry.

(c) Reconstruction of collaboration ability for science popularization. Project team I reconstructs the key points for collaboration during science communication. The original key point is the collaboration between the government and scientific research teams, research institutions neither contact with nor care about the public but their own research content and whether they have completed the tasks assigned by the authorities. Such model result in the difficulty in completing the science popularization task that Project Team I undertakes, and evaluations from all parties bring about a good deal of troubles to the project team. Accordingly, Project Team I changes the key point of science popularization collaboration into the public and make science popularization product and service designs that adapt to the new media era, and share the public order, feedback and evaluation information with the creative units on a real-time basis so that the creative units update its understanding of changes in public demand, and process science popularization resources and design science popularization activities according to public demand. In order to realize the deep collaboration of the science popularization chain, the project team holds occasional seminars and conducts deliberations on public suggestions and other issues. The transition process is shown in Fig. 4.

(2) *Case II: progressive science popularization transformation path*

As compared with the reconstructed model of Project Team I, Project Team II, as a team undertaking science popularization creation voluntarily, has creation basis for science

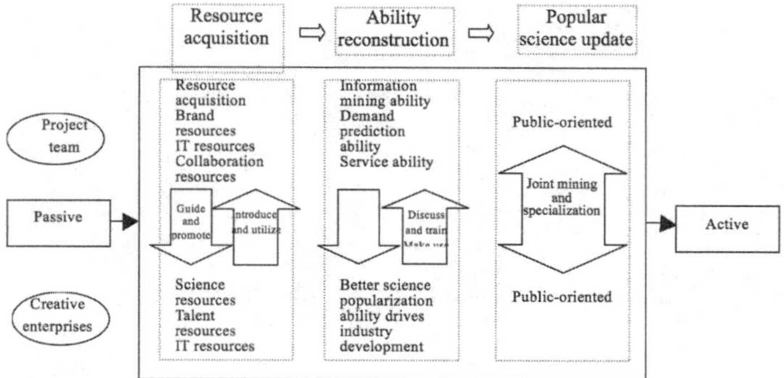

Fig. 4 Reconstructed science popularization transformation path.

popularization to a certain degree. Therefore, Project Team II unfolds science transformation in a forward-looking way based on the utilization and integration of existing resources, and gradually expands its original ability by phases to achieve progressive innovation in science popularization. The focus of resource integration of Project Team II lies in the reorganization of existing resources to make up for the current disadvantages of science popularization resource and strengthen resource advantages, including: ① Talent resources. Turn every project member into a high-end science popularization talent who can carry on popular science explanation and creation. ② Brand resources. ③ IT resources. Plan and supplement IT resources to strengthen public data mining. Readjusting the available resources and balancing the relationship between science popularization informatization and traditional science communication is an important task in the preliminary stage of transformation.

(a) IT ability development. Project Team II strengthens collaboration of Wechat and APP new media, and by platform data exchange, realizes transmission and sharing of information between the public science demand and science knowledge provided by scientific research teams, enhance data mining and analysis ability, guide scientific comments of the public and improve scientific literacy of the public.

(b) Business ability development. The project team combines the public welfare undertakings organically with the profit-making science popularization industry and makes industrialization attempts while conducting scientific communication.

(c) Collaboration ability development. Influenced by the public orientation, Project Team II increased the frequency of communication with various supporting enterprises to enhance synergetic effect. With the support of Internet technologies and information systems, resources sharing, interconnection and intercommunication is realized. Some supporting enterprises start to pay attention to the development trend of science. Some supporting enterprises also proposed design ideas and participated in popular science research and development, which consequently improves the overall quality of science popularization and realizes the matching and collaboration between back end and front end.

To sum up, in order to cope with the impact of new media on traditional science popularization, maintain the attention to major science and technology innovation projects and undertake the social responsibilities of scientists, Project Team II makes full use of its resource base and integrates relevant science popularization resources such as human resources, brand, IT and supporting enterprises, and gradually develops scientific communication abilities and collaborative creation abilities based on such resources, realizing the transformation from task-based passive communication to responsibility-based active communication. The transformation process is shown in Fig. 5.

4. Conclusions and Discussions

Scientific research teams start to consciously participate in popular science work and assume the responsibility of scientists even better [26]. The process of science communication is unfolded at all levels in this study which explores the major factors of science popularization work for major science and technology innovation projects, the path of science communication, and transformation of task-based passive science communication to responsibility-based active science communication, with different realization paths offered. The research findings provide theoretical support for governmental authorities

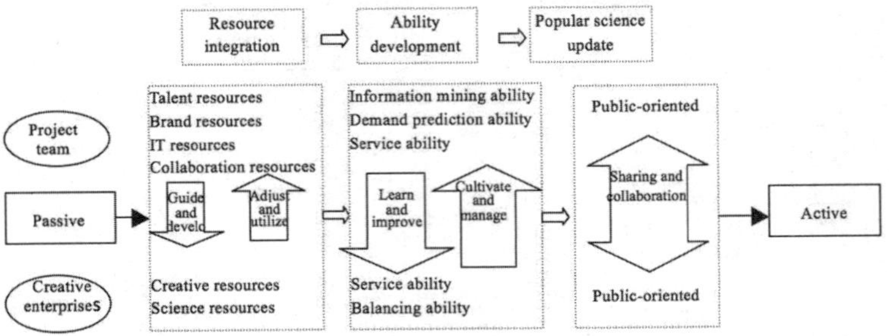

Fig. 5 Progressive science popularization transformation path.

and scientific research teams to carry out science popularization work effectively.

1) Changing Engagement Model of Scientific Research Teams in Science Popularization

The emergence of the Internet and the increasing emphasize on science popularization work have changed the status of the public in science communication. The development of the Internet has further strengthened the status of the public in science communication, making the public the origin and destination of science communication and forming the empowerment effect of the public. It also enables scientific research teams to realize transformation with lower information and channel costs and higher collaboration costs and thus constitute the public-centered basic conditions. Therefore, the structure of science communication shifts from the original cooperative operation between scientific research teams and the government to that between scientific research teams and the public. Task-based passive communication is shifted to responsibility-based active communication.

2) Realization Path for Engagement of Scientific Research Teams in Science Popularization

In the context of network, science communication of scientific research teams shifts from the original task-based passive communication to responsibility-based active communication.

The impact of new media technology and the changes in public demand have invalidated the original science popularization work and science popularization resources of Project Team I and forced the team to find new resources and contents to complete science communication works. Project Team I acquires new science popularization resources in various ways. In order to make full use of these resources, the team overcomes the original organization inertia mainly through IT ability reconstruction, business ability reconstruction and collaborative ability reconstruction so as to achieve the reconstructed science popularization transformation.

Science popularization work of Project Team II has already involved some new market resources which can effectively buffer the popular science demand for scientific research teams caused by the changes in the external environment so that it chooses a progressive transformation mode. Project Team II integrates internal and external resources and optimizes the processes to give full play to its advantages. Meanwhile, it improves and promotes the ability of collaborative creative units in phases. On one hand, it analyzes the public preferences more accurately through data mining and guides the creative units to improve efficiency and promote the professionalism of science popularization creative units at the same time. It realizes progressive science popularization transformation by overcoming organization inertia through development of IT ability, business ability and collaboration ability. Due to differences in resources in the process of transformation, scientific research teams may achieve active communication through either reconstructed science popularization transformation path or progressive science popularization transformation path. Thereinto, the reconstructed science popularization communication path acquires resources to form ability reconstruction; progressive realization mechanism realizes ability development through resource integration and then achieves science popularization transformation further. The research in this paper shows that the resource base and organization inertia of scientific research teams are the main reasons for the realization of science popularization transformation path. However, in general, both of them follow the realization mechanism from resources to abilities and then to science popularization communication in terms of transformation path (See Fig. 6).

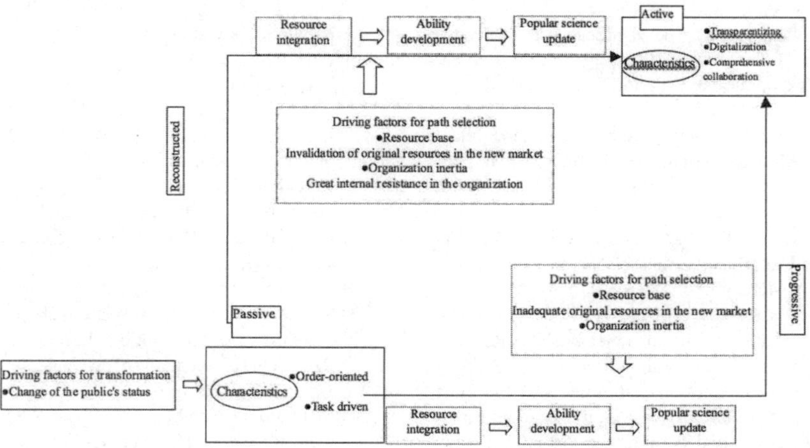

Fig. 6 Theoretical frame for transformation and realization path of science communication.

3) Differences between Task-based Passive Communication and Responsibility-based Active Communication

It is the collaboration between scientific research teams, the public and creative units that task-based passive communication and responsibility-based active communication emphasize essentially. The fundamental goal and basic operation structure of the two type of communication are similar, but there aresignificant differences in terms of core values, cost structure, collaborative subjects and IT application which are shown down below.

First, in the process of science popularization, the science popularization effect of scientific research teams mainly lies in improving the public experience to form a word-of-mouth effect, so the core values of public-oriented responsibility-based active communication is to create value by innovating popular science products and services for the public. While for task-based passive communication, scientific research teams are mainly concerned about the completion of political tasks and pay little attention to the effect of communication and demand of the public.

Second, task-based passive science communication path is completely different from the responsibility-based active communication path. Different structures just support different operation modes of science popularization. In traditional markets, the cost of the public for information acquisition and channel construction are high. In the network environment, each science popularization node may win large amount of popular science market and public information and the public power has been continuously strengthened, thus forming the public-oriented science communication.

Third, in the traditional market, the public is at the end of science communication process and lack of contact with the innovation team, thus having little influence. The collaborative subject of traditional science communication is the government. However, in the Internet environment, the public is at the initial and tail ends of science communication where science customization is submitted by the public, various science popularization services also directly contact with the public and the public becomes the core of science communication. Therefore, the collaborative subject of public-oriented science popularization is scientific research teams and the public.

Finally, in traditional market, there is little contact between research teams. The contact between scientific research teams and creative units could be even less. IT application only provides appropriate support for science popularization. However, things are different in the internet environment. On one hand, the science popularization work of scientific research teams are to be conducted based on the Internet; on the other hand, scientific research teams need to respond to personalized popular science demand of the public through data mining, intelligent prediction and data marketing. Therefore, IT application results in a comprehensive sharing of popular science information and provides support for accurate decision-making.

References

[1] Royal Society. Public Understanding of Science [M]. Translated by Tang Yingying. Beijing: Beijing Institute

of Technology Press. 2004: 43-45.

[2] Li Yunqing, Wang Huilan. The Significance and Effective Approaches of Universities Participating in Popular Science Work in the New Period [J]. Science Focus. 2008, (6): 69-70.

[3] Xu Zhifeng, Chen Zhimin, Wang Pengjuan. Introduction to Modern Science and Technology [M]. Changchun: Northeast Normal University Press, 2006: 200-225.

[4] Bian Yulin. Science Popularization is so Important that it can not be Undertaken Solely by Science Writers [J]. Science, 1993, 45 (2):4.

[5] Zhu Guangya. Carrying Forward Glorious Traditions, Shouldering Historical Missions and Pushing Forward the Cause of Science Popularization [R]. Speech by leaders of the CPC Central Committee and the State Council at the National Working Conference on Popularization of Science and Technology, 1996: 22.

[6] Ministry of Science and Technology. Notification on Providing Public Access to State Key Laboratories [Z]. 2004.

[7] Wu Guosheng. From Science Popularization to Science Communication [N]. Science and Technology Daily, 2000-09-22 (3).

[8] Liu Huajie. Integrating Two Traditions: A Discussion on the Scientific Communication [J]. Science News, 2002, (18): 5-7.

[9] Joss S., Durant J. (1995). Public Participation in Science. The Role of Consensus Conferences in Europe. London: Science Museum.

[10] Weigold M. F. Communicating Science: A Review of the Literature. [J]. Science Communication, 2001, 23(2): 164-193.

[11] Joss S., Durant J. The UK National Consensus Conference on Plant Biotechnology [J]. Public Understanding of Science, 1995, 4(2): 195-204.

[12] Gibbons, Michael, et al. The New Production of Knowledge. SAGE Publications, 1994.

[13] Cai Guojun, Li Tianbin, Feng Wenkai, et al. Research Laboratories Carrying on Science Popularization Activities to Improve the Scientific Quality of the Public [J]. Research and Exploration in Laboratory, 2015, 34 (08): 131-134.

[14] Logan R. A. Science Mass Communication: Its Conceptual History[J]. Science Communication 23.2(2001): 135-163.

[15] Einsiedel E. F. (2000). Understanding 'Publics' in the Public Understanding of Science. In M. V. G. Dierkes, Claudia (Ed.), Between Understanding and Trust. The Public Science and Technology (pp. 205-215). Amsterdam: Harwood Publishers.

[16] Van Dijck J. After the 'Two Cultures': Toward a '(Multi)cultural' Practice of Science Communication[J]. Science Communication 25.2(2003):177-190.

[17] Yang Jing, Wang Nan. Research on Current Status of Science Popularization Activities in Universities and Research Institutions in China [J]. Study on Science Popularization, 2015, 10 (6): 92-101.

[18] Ao Nihua, Gong Huiling, Ju Siting, et al. The Opportunities and Challenges of the Frontier Scientific Popularization: A Case Study of Popular Science Exhibitions [J]. Science Management Research, 2016, (3): 1-4.

[19] Eisenhardt K. M. Building Theories from Cases Study Research[J]. Academy of Management Review, 1989, 14: 532-550.

[20] Li Jianming. Application of Case Analysis in Management Research [J]. Shanghai Economic Review, 2004, (02): 76-79.

[21] PARÉ G. Investigating Information Systems with Positivist Case Study Research [J]. Communications of the Association for Information Systems. 2004(13): 233-264.

[22] Huang Jiangming, Li Liang, Wang Wei. Case Study: from Good Stories to Good Theory: A Review of the Research Forum on China's Enterprise Management Cases and Theory Construction[J]. Management World, 2011 (02): 118-126.

[23] Yin R. K. Case Study Research: Design and Methods [M]. CA: Sage Publications Inc. 2008.

[24] Mao Jiye, Zhang Xia. Normativity and Current Situation Assessment of Case Study Method: A Review of China Enterprise Management Cases Forum[J]. Management World, 2008 (04): 115-121.

[25] Tang Shukun, Zheng Jiuliang. Reflection on the Transformation of Science Popularization Work under the Context of National Development [J], Study on Science Popularization, 2016, 11(1):10-15.

[26] Xie Qihui. Research on the Science Popularization Resource Application Model of National Science Popularization & Education Bases: A Case Study of State Key Laboratory on Fire Science of University of Science and Technology of China [C]. Anhui First Science and Technology Doctor Forum of Science Popularization Industry - Proceedings of Academic Exchange Forum on Science and Technology Communication System and Platform Construction. Wuhu, Anhui Province: Society Department of the Association for Science and Technology of Anhui Province, 2016: 267-270.

The Project of Collecting Historic Data of Old Scientists' Academic Life

——Brief history, status, and outlook

Yang Zhihong[1], Zhang Li[2]

[1] Institute of Innovative Talents, National Academy of Innovation Strategy, Beijing, China
[2] School of Humanities, University of Chinese Academy of Science, Beijing, China

Abstract: The project of Collecting Historic Data of Old Scientists' Academic life (the collecting project, for short) has been carried out since 2010 in China. By the end of June, 2018, there have been 512 distinguished scientists or scientist groups that were initiated by the collecting projects, with a collection of more than 100,000 original materials, 250,000 digitized materials, and 13,000 hours of audio and video data, which outcomes remarkable achievements, including the accumulation of scientific and technological historical data of China, the research on the growth of scientific and technological talents, and the publicity of scientific and technological figures. It also takes the lead in establishing a set of standards in China, which has led the community to pay more attention to the standard collection and preservation of data of scientists, engineers, and other professionals. In this paper the origin, development status, practical experience of project management and development prospect of the collecting project are analyzed and introduced, some of which may be useful in the organization and management of large projects of social science and history program.

Keywords: Historic Data; Scientist; Academic Life; Collecting Project; Project Management

1. Introduction

Old scientists are the living archives and living history of the development of science and technology for the People's Republic of China. Their history of academic growth reflects the progress of science and technology development and the education in the two contexts before and after the establishment of the new China. Applied by the China Association for Science and Technology(CAST), in November 2009, the 'Implementation Plan for the project of Collecting Historic Data of Old Scientists' Academic life' was approved by the State Council. In May 2010, the collecting project was officially launched and implemented by the CAST, together with 11 ministries and commissions.

By the end of June 2018, 512 old scientists and groups had been collected for academic growth, and more than 3,000 collectors across the country had participated in the project, accumulating abundant data. This project is of significant value, whether from the perspective of preserving the historical literature of the development of science and technology in the new China, the perspective of carrying forward the scientific tradition, the accumulation of scientific and technological publicity materials, or from the perspective of the study of the growth law of talents, and from the perspective of caring for old scientists.

2. Working Mechanism and Academic Norms

1) Management Mechanism

Since the approval of the implementation plan of the collecting project by the state council, a leading group composed of the principal leaders of the CAST and 12 leading members of the ministries and commissions has been formed, responsible for the macro-guidance of the collecting project and the formulation of major policy measures.

At present, the project office is set up at the National Academy of Innovation Strategy, with a five-party management mechanism, in which the collecting project leading group office (Department of Research and Publicity, CAST) totally coordinates, the National Academy of Innovation Strategy is responsible for project

Corresponding author: YangZhihong, No. 3 Fuxing Rd., Beijing, P. R. China (100863). yangzhihong@ cnais.org.cn.

management and database & web site construction, the collection base (library, Beijing University of Technology) is responsible for historical material collection and sorting, the collection team management units in Association for Science and Technology at provincial level cooperate in their jurisdiction, which is also referred to the joint management mechanism of five organizations, as shown in Fig. 1.

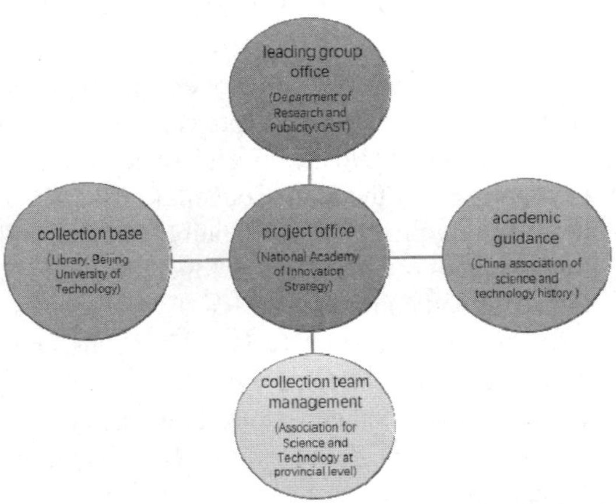

Fig. 1 Joint management mechanism of five organizations.

A regular '5+1' working meeting system has been established, in which the 5 means the above five organizations, the 1 means the China Science and Technology Publishing House which takes charge of scientists biographies publication, to ensure that all parties perform their respective duties, communicate in a timely manner, supervise each other, and improve work efficiency and work quality

2) Basic Documents

The collecting project is a groundbreaking work and there are no available materials for reference in the collecting process, methods, standards, specifications, training, and management. The Department of Research and Publicity, CAST coordinated with the China Association of Science and Technology History and the library of Beijing university of technology to jointly study and formulate the supporting rules for the implementation of the collecting project.

At the beginning of 2010, after more than four months of meticulous and in-depth work, they developed eight basic documents, including the Working Rules of the Leading Group, the Working Rules of the Office of the Leading Group, the Measures for Project Management, the Standards for Acceptance of Collection Work, the Standards for the Qualification of Collectors, the Measures for the Management of Collection Base, the Provisions on the Confidentiality of Collectors and the Agreement on Intellectual Property, and nine other technical documents, including the Collecting Workflow, the Collecting Work Specification, the Collecting Work Training Program, the Training Manual, the Collection Base Data Transfer Acceptance Standard, the Data Digitization Program, the Database Construction Plan, the Data Storage Program, the Data Display Program, which form a complete guidance document system and ensure the quality of the collection.

3) Collecting Cycle

Collecting cycle is shown as Table 1. The management mode of major phases, including the determination of solicitation lists, collecting list review, training of collecting personnel, signing of project appointment documents, inspect review, mid-term review, concluding review, has been iteratively refined.

Table 1 Project schedule of collecting team for the collect ing project.

No.	phase work	Time Schedule																																			
		Approval Year(Month)												Execution Year(Month)												Conluding Year(Month)											
		1	2	3	4	5	6	7	8	9	10	11	12	1	2	3	4	5	6	7	8	9	10	11	12	1	2	3	4	5	6	7	8	9	10	11	12
1	Solicitation list	■	■																																		
2	Collecting list review			■																																	
3	Collecting training				■																																
4	Project approvel					■																															
5	The 1st working section						■	■	■	■	■	■	■																								
6	Inspect review											■	■																								
7	The 2nd working section												■	■	■	■	■	■	■																		
8	Midterm review																	■																			
9	The 3rd working section																		■	■	■	■	■	■	■	■	■	■	■	■	■	■	■	■	■		
10	Concluding review																																			■	■

In the nine years since the implementation of the collecting project, a set of quality assurance measures has been formed, for example, taking the training as the necessary conditions for the establishment of the collecting team, undertaking the collecting project and signing the project appointment letter, taking the mid-term review as the important grasp of process management, taking the handover of data as the precondition for the final acceptance, taking the final acceptance as an effective means of the inspection in the concluding review. Administrative mobilization and academic norms are the biggest characteristics of the collecting project.

3. Progress to Date

1) Historical Material Outcomes

The collecting project has so far collected academic historical data of 512 old scientists and groups, as shown in the Fig. 2. The average age of the collection in 2010 was 88.5 years old, followed by a decrease in the age of collection. In 2018, scientists over 83 years of age are collected. A large number of precious historical materials of academic value and display value are accumulated, including 361,518 minutes of videos, 419, 341 minutes of audios, 105,228 pieces of physical materials (including manuscripts, letters, photos, etc.) and 256058 pieces of digital data, as shown in the Table 2.

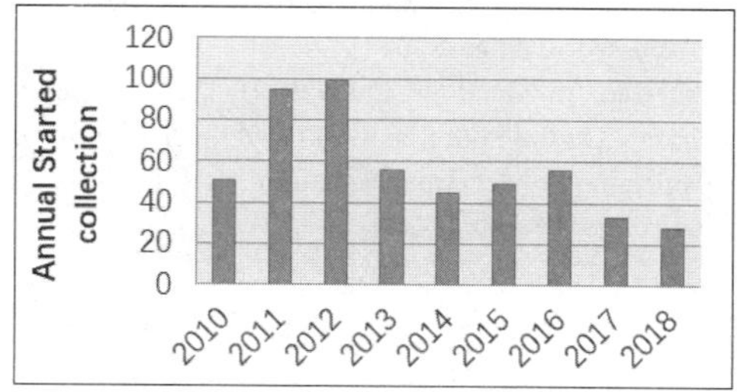

Fig. 2 Annual started collection.

Table 2 Collected historical materials.

Year	Original Materials	Digital Materials	Video(min)	Audio(min)
2010	7391	33423	48973	39833
2011	17523	59382	73874	78120
2012	28767	70515	89364	124183
2013	21894	42222	59486	71666
2014	15074	24919	44287	51844
2015	14579	25597	45534	53695
In Total	105228	256058	361518	419341

2) Research and Publicity

(1) *Biographies*

120 biographies are published so far based on the research of the collected data of the collecting project. These biographies are defined as 'academic biographies', which distinguish them from other biographies of literariness and propaganda. The term 'academic' has a dual meaning. One refers to the subject of the academic career of the individual scientist, the other emphasizes the 'academic nature' of biographies, written based on historical materials that have been carefully examined and debated. The research results and methods of the history of science, especially the modern history of China, have been fully absorbed. These biographies has become important literatures for Chinese scientists study.

(2) *Exhibitions*

In December 2013 'the dream of science and technology, the Chinese dream—modern Chinese scientists exhibition' was held successfully in National Museum, which presented the struggle of science and technology staff in more than 100 years and displayed the characters and spirits of Chinese scientists. Since then its itinerant exhibition has been to 31 cities of 26 provinces and autonomous regions of the country, with the scene audience of more than 870000 people and high attention from all walks of life. As the first large-scale exhibition with the theme of a group of scientists since the founding of New China, it has become an image textbook that condenses the spirit of Chinese scientists. In addition, a donation ceremony and a selection exhibition of donation materials are also held at the end of the annual concluding review of the collecting project.

(3) *Database and the Museum of Scientists in China (Web version)*

Since 2013, in order to support the processing, research, and publicity of the collected historical data of scientists, the informatization project has been started by the National Academy of Innovation Strategy. The website is the first full-directional, visualized, multi-faceted portal to show the academic growth of modern Chinese scientists. It has set up 'public edition' and 'academic edition' to show the track of academic growth of old scientists from many angles, propagandizing and carrying forward the lofty moral character of scientists undefined patriotic dedication and the scientific spirit of seeking truth and pragmatism. It also provides accurate and reliable research literature query for professional researchers.

4. Outlook

Through continuous efforts from 2010, the rescue task focusing on collecting and sorting out the data of the older generation of scientists has been basically completed. The scale and coverage of the collection will be further expanded, and the collection and research of academic data of young and middle-aged science and technology experts will be launched. Centering on 'collecting, storage, research and publicity', the collecting project will be a long-term work, intensifying data collation and research, promoting resource sharing, and establishing a first-class platform for Chinese scientists research.

A museum of Chinese scientists (physical museum) is looked forward to opening in 2020, which will be a part of the national science and technology communication center that was officially approved by the state council with a construction area of over 60,000 square meters. It will be the only national museum in the world with the theme of scientists, and it will be the collecting management center of Chinese scientists' literature and material objects, academic thought research center, scientific spirit and scientific culture publicity and exhibition center, as well as the spiritual palace and emotional home of Chinese scientists.

5. Conclusion

The Collecting project is a public welfare academic project with perfect combination of administrative leadership system and academic norms under the theoretical guidance of multi-discipline including history, archival science, management science, etc. It takes the lead in establishing a set of standards in China, which has led the community to pay more attention to the standard collection and preservation of data of scientists, engineers, and other professionals. Key reasons for the success of the project are summarized as following:

(1) Leadership: A big success project needs strong leadership and top-level design.

(2) Rules first: Before the start-up, rules were made and are respected during the implementation process.

(3) Personnel: There is a group of people engage in it who share the same ideas and goals.

References

Chunfa Wang. (2011). Origin, progress and significance of 'the Collecting Project'. *The Chinese Journal for the History of Science and Technology*, 32(2), 139-148.

Li Zhang. (2015). Biography series on the Project of Collecting Historic Data of Old Scientists' Academic Life: A new exploration into the biography of scientists. *The Chinese Journal for the History of Science and Technology,* 36(1), 107-109.

(2011). Implementation plan for the Project of Collecting Historic Data of Old Scientists'Academic Life. *The Chinese Journal for the History of Science and Technology*, 32(2), 305-308.

National Museum for Modern Chinese Scientists [on-line]. Available at http://www.mmcs.org.cn/1228/ index.shtml [Accessed 12 August 2018].

Wenjing Gao, Li Ma, Lijie Liu. (2019). For our common cause—On-the-Spot Report of the Project Office of the Collecting Project. *Modern Science*, (21), 80-83.

Long Live the Scientists: Tracking the Scientific Fame of Great Minds in Physics

Wang Guoyan

University of Science and Technology of China, Hefei, China

Abstract: This study utilizes global digitalized books and articles to examine the scientific fame of the most influential physicists. Our research reveals that the greatest minds are gone but not forgotten. Their scientific impacts on human history have persisted for centuries. We also find evidence in support of own-group fame preference, i.e., that the scientists have greater reputations in their home countries or among scholars sharing the same languages. We argue that, when applied appropriately, Google Books and Ngram Viewer can serve as promising tools for altmetrics, providing a more comprehensive picture of the impacts scholars and their achievements have made beyond academia.

Keywords: Scientific Fame; Own-group Preference; Google Corpus; Altmetrics

Some say that a man dies three times. The first time is when his heart stops beating and he dies physically. The second is when people come to his funeral and his identity is erased from society. The third time is when nobody on the earth remembers him anymore. Then he is really dead.

—'Dragon Raja' by Lee Yeongdo

1. Introduction

Books are the stepping stones to human progress. According to UNESCO (United Nations Educational, Scientific and Cultural Organization), the number of estimated published books in 2017 alone is up to 2.2 million. [1] Such a large collection is undoubtedly a rich archive of human history and civilization. Yet, as one of the most telling embodiments of knowledge stock and advancement, books have not captured sufficient attention in quantitative research evaluation.

Fortunately, with access to Google Books and the Google Books tool Ngram Viewer, scholars are now able to trace cultural evolution on a long time scale based on digitalized texts and trillions of words. This application of high-throughput data collection to study human culture can be traced back to Michel et al. (2011). In this pioneering study, the authors utilized the Google Books corpus and conducted text-based statistical analysis to trace cultural trends. That innovative research method soon captured academia's attention and was adopted in the arenas of digital history (Sternfeld, 2011), the history of science (Laubichler et al., 2013), economics (Roth, 2013), social psychology (Greenfield, 2013; Acerbi, 2013; Zeng et al., 2015), and cultural psychology (Pettit, 2016).

Google Books has also been utilized to assess the fame of great scientists throughout history. Based on the word frequency of people's full names mentioned in books, Bohannon's Science Hall of Fame was built a as an objective evaluation of scientific fame over centuries (Bohannon, 2011a; 2011b). Moving beyond previous work, this paper utilizes both Google Books, which covers 36 million global digital books, and Google Scholar, which indexes 91 million academic items,[2] to examine the scientific fame of top physicists. We particularly focus on and compare two of the greatest physicists, Isaac Newton and Albert Einstein, depicting their fame evolution over centuries and exploring what they are famous for.

Our research reveals that the great minds are gone but not forgotten. Early scientists are still on the public's lips in modern society. Their scientific impacts on human history have persisted for centuries. This holds true for other prominent physicists as well. We also found that while Einstein's scientific fame has exceeded that of Newton among intellectuals since the mid-20th

1 Data source: http://www.worldometers.info/books/. Accessed on January 18, 2018.

2 The types of academic items Google Scholar includes are research articles, books, patents, case laws, and citations. In this research our search excludes patents and citations.

century worldwide, there is a different pattern of fame when the own-group preference is differentiated by the language of digitalized corpus. The computational analysis confirms that the influence of Einstein is largely related to his two contributions on general relativity and quantum theory, while the most frequently mentioned scientific achievements of Newton are the law of universal gravitation and the laws of motion.

This paper makes the following contributions to the literature. To begin with, this is the first attempt, within our best knowledge, to explore scientific fame based on the combination of both books and articles indexed by Google. In addition to depicting the evolution of fame of great minds, we also explore their most accredited achievements based on co-occurence analysis. Second, our study contributes to the discussion on the expected role of scientists in science promotion. By comparing the indicator for scientific impact (i.e., work being cited by scholarly publications) and the indicator for scientific fame (i.e., name being mentioned in books), our study sheds some light on how to gauge scientists' contributions beyond academia. We argue that, when applied appropriately, Google Books and Google Ngram Viewer can serve as promising tools for altmetrics, providing a more comprehensive picture of the impacts scholars and their achievements have made on society.

The rest of the paper is structured as follows. In the next section, we delineate our method and case selection justifications. Section 3 presents our analysis. In Section 4, following a summary of main findings, we conclude our paper discussing limitations and future research venues.

2. Method and Data

1) Notion and Measurement

There is no agreed-upon definition of scientific fame or its measurement. The term can be traced back to the book *The Life of Sir Charles Linnæus*, a biography of a Swedish botanist, physician, and zoologist whose fame is centered on his enduring achievement of binomial nomenclature (Stöver, 1794). *Yet* many scientists are ordinary folks and are little known to the public (Astin, 1957; Menard, 1971; Merton, 1970). Some scholars have argued that the fame of scientists should be confined to professional achievements, while others believe scientific fame goes beyond academia (Menard, 1971; Bohannon, 2011a). Feist (2016) noted that regardless of either intrinsic or extrinsic research, the assessment of fame in science ultimately rests on productivity and its impact on advancing the research front. Previous studies often used being elected to prestigious societies or winning research awards or prizes as proxy indicators of scientific recognition or reputation (Bronk, 1976; Youtie et al., 2013). Instead of relying on the subjective judgment of panel experts, Bohannon's Science Hall of Fame (2011a) innovatively uses the appearance of people's names in books to capture scientists' influence across different domains throughout history. This is the approach we adopt in tracking and recording the fame of great scientific minds and their achievements.

2) Case Selection

The focus domain in this work is physics. Among the myriad scientists, we purposely choose Isaac Newton and Albert Einstein for illustration based on the following considerations.

To begin with, both Newton and Einstein are in the field of physics, which makes their comparison relatively free from discipline differences of publication distribution. And given that both names consist of two words with similar length, the quality of their retrievals is fairly comparative. As the two most influential physicists in the history of science (Whittaker, 1943; Baker, 1984), Newton and Einstein are appropriate candidates for evaluating the historical fame of individuals. Finally, who is more influential has been a topic attracting much attention in the global scientific community. The debate has not been settled for more than half a century (Gribbin, 1987; Graneau, 1993). Successor scientists have commemorated them on special anniversaries, such as 'Science 1943: Aristotle, Newton, Einstein, the three-hundredth anniversary of Newton's birth' and 'New Scientist 1987: Newton vs. Einstein, the three-hundredth anniversary of Newton's theory of gravity'. In 2005, the Year of World Physics and also the Centenary of Einstein's Special Relativity Theory, the UK Royal Society conducted polls of both academia and the public. The results showed that both Royal Society members and the British netizens surveyed considered Isaac Newton, a British scientist, to have greater influence on both science and humankind than Albert Einstein (*The Royal Society News*, 2005). It would be interesting to find out whether that also holds true beyond the geography of the UK.

3) Data

Our main datasets are Google digital books and scholarly articles. As shown in Table 1, the Google corpus used in this study consists of over 36 million global digital books and 91 million articles published since the 16th century.

Table 1 Coverage of Google Books and Google Scholar.

Google Corpus	Total Returned Hits	After 2000	20th Century	19th Century	18th Century	17th Century	Before 1600
Books	36,540,000	14,600,000	14,700,000	5,680,000	808,000	319,000	433,000
Scholar	91,258,000	43,061,000	46,935,600	1,180,630	80,300	2	0

Note: This search was conducted through the library at London University in July 2015. For Google Books, the command we used is 'allinurl: books' at https://books.google.com. For Google Scholar we set up the custom range of years in the advanced search.[1]

To gain a complete picture of these physicists' fame worldwide, we took a variety of languages into consideration. We first tested our search terms in 57 different languages using the Google Books search engine based on trials and errors.[2] The volumes and shares of the written languages of these books are displayed in Table 2. As shown, the dataset we analyzed covers more than 126 thousand books mentioning Newton and 149 thousand books mentioning Einstein since the 20th century. This finding seems opposite to that of the UK polls in 2005. Table 2 also reveals that, similar to other publication datasets, the coverage of Google Books is also biased in language (Lin 2012; Liu et al., 2018; Liu et al., 2015). Books written in Chinese, Russian, Korean, and other non-English languages are highly underrepresented in the Google Books corpus. This caveat needs to be borne in mind when using this search engine for analysis.

3. Analysis

1) Global Fame over Time

Figure 1 depicts the quantity of digitalized books and academic items mentioning Newton's and Einstein's full names. It needs to be pointed out that our returned hits consist of two scenarios: books mentioning Newton or Einstein and books written by either of them. But given the number of returned hits and the number of books even the most prolific scholar can write, it is reasonable to believe that the majority of the retrieved books are those mentioning them. Section 4 discusses this in more detail. Blue denotes Newton while red represents Einstein. The dotted lines mark the deceased years of the two great scientists, respectively. Please note that any key word search is only effective for accessible content. In our case, only Google Books with a preview available can be searched for names in the text.

In Figure 1A, the number in the upper left ($1e^4=10^4$) is the unit of the absolute number of books indexed in the Google Books search. For Newton, the most glorious period of his historical influence is between 1680 and 1880, when records mentioning Newton amount to more than 60% of all records mentioning him. As demonstrated, the deaths of these great scientists do not dim their glory. On the contrary, the popularity of both physicists has risen dramatically since the 1980s. One may wonder what happened in the period of 1880 to 1980, when relatively fewer books talked about Newton and Einstein. One speculation is that two dark clouds appeared in the world of physics at the end of the 19th century: the Michelson–Morley experiment and the ultraviolet catastrophe in

1 It needs to be pointed out that depending on the search date and location, the returned hits vary. Take Google Books, for example. We repeated the same search in July 2018, and the returned hits are not 36,540,000 as in July 2015 in the UK but are 38,400,000 in China and 39,300,000 in South Korea. This pattern also holds for Google Scholar. As shown in Table 1, the coverage of Google Scholar was 43,061,000 articles published over the period of 2000–2015 when the search was conducted at London University in July 2015. In August 2018, its coverage of the same period extends to 46,840,000 and 48,720,000 articles when searched in China and South Korea, respectively. Our experiences echo previous findings on the coverage scope, growth rate, and indexing quality of Google Books and Google Scholar (Abdullah et al., 2014; Fagan, 2017; Halevi et al., 2017). For an extensive review of the pros and cons of Google Books and Google Scholar as well as their utilities in research assessment, please refer to Google Scholar Digest at http://googlescholardigest.blogspot.com/p/bibliography.html.

2 The variations of name order (e.g., 'Isaac Newton' and 'Newton, Isaac') were taken into consideration.

Table 2 Retrieved hits in Google Books by different languages.

Scientist	Search Word	Retrieved Hits in 57 Languages	2000–2015 (prop.)	1900–1999 (prop.)
Isaac Newton (1643–1727)	Isaac Newton	Records in 49 languages with the same name spelling as in English	95.78%	98.18%
	Исаак Ньютон	Russian	2.01%	1.28%
	아이작 뉴턴	Korean	1.17%	0.12%
	アイザック・ニュートン	Japanese	0.51%	0.31%
	‘艾萨克·牛顿’OR‘艾薩克·牛頓’	Chinese (simplified or traditional)	0.21%	0.06%
	إسحاق نيوتن	Arabic	0.18%	0.04%
	אייזק ניוטון	Hebrew	0.15%	0.02%
	আইজাক নিউটন	Bengali	0.01%	0.00%
	आइजैक न्यूटन	Hindi	0.00%	0.00%
	Total retrieved hits of Newton		125593	36434
Albert Einstein (1879–1955)	Albert Einstein	Records in 49 languages with the same name spelling as in English	96.25%	96.78%
	Альберт Эйнштейн	Russian	2.35%	2.08%
	אלברט איינשטיין	Hebrew	0.42%	0.62%
	アルバート・アインシュタイン	Japanese	0.35%	0.22%
	알버트 아인슈타인	Korean	0.26%	0.08%
	‘阿尔伯特・爱因斯坦’OR‘阿爾伯特・愛因斯坦’	Chinese (simplified or traditional)	0.21%	0.17%
	البرت اينشتاين	Arabic	0.05%	0.06%
	अल्बर्ट आइंस्टीन	Hindi	0.01%	0.00%
	আলবার্ট আইনস্টাইন	Bengali	0.00%	0.00%
	Total retrieved hits of Einstein		149169	38154

Data source: Google Books search. Data accessed in January 2016.

Note: In the other 49 languages such as German, French, Italian, Spanish, and Portuguese, both name spellings are exactly the same as in English.

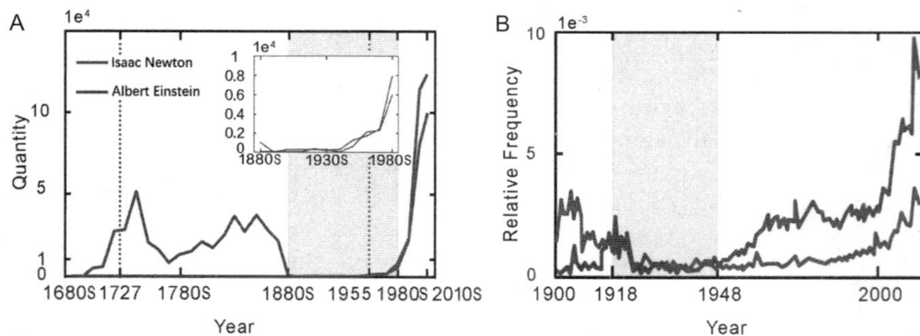

Fig. 1 Global fame of Newton and Einstein.

Data source: Panel A is Google Books search. The time coverage of the main graph is from 1680s to 2010s with an inset zooming in on the period of 1880s—1980s; Panel B is Google Scholar search. The time coverage is from 1900 to 2012.

Accessed in January 2016.

Forty-nine languages with the same name spelling as in English are included.

Planck's law (Kelvin, 1901). They overturned the framework of the old theory system, leading to the birth of quantum and relativity theories (Hoover, 1977; Sanghera, 2011). This, when combined with the breakout and aftermath of World Wars Ⅰ and Ⅱ, partially explains the silence of Newton's voice from 1880 to 1980.

Figure 1B maps the scientists' influences in academia via a Google Scholar search engine. The y values, i.e. relative counts, were computed by dividing the number of Google Scholar indexed articles mentioning the scientist in a given year by the total number of Google Scholar indexed articles in that year. As shown, Newton leads in fame until 1918, then both scientists parallel for 30 years. The year of 1948 is the watershed. Since then, Einstein has gained more popularity than Newton worldwide.

One key message conveyed by Fig. 1 is that globally Einstein seems to enjoy greater fame than Newton, which is contrary to the results of the 2005 UK polls mentioned previously. If the findings of both the surveys and the computational analysis are valid, one possible explanation for this discrepancy is that scientific fame varies by location. People are more recognized within their own group (Egghe et al., 2004; Tang et al., 2015), be it by country or by shared language. To test this speculation, we utilized the language option of Google Books Ngram Viewer and examined the physicists' historical influence by various languages.[1]

2) Global Fame Moderated by Language

Figure 2 was generated with Google Books Ngram Viewer with a default smoothing of three.[2] The four panels illustrate the scientific fame of Newton and Einstein in books written in British and American English together, British English alone, American English alone, and German, respectively. Both scientists seem to be preferred in their own groups. In the British English case, Newton has consistently enjoyed more popularity than Einstein throughout the last 100 years (Panel B). This finding is consistent with 2005 UK polls results. In sharp

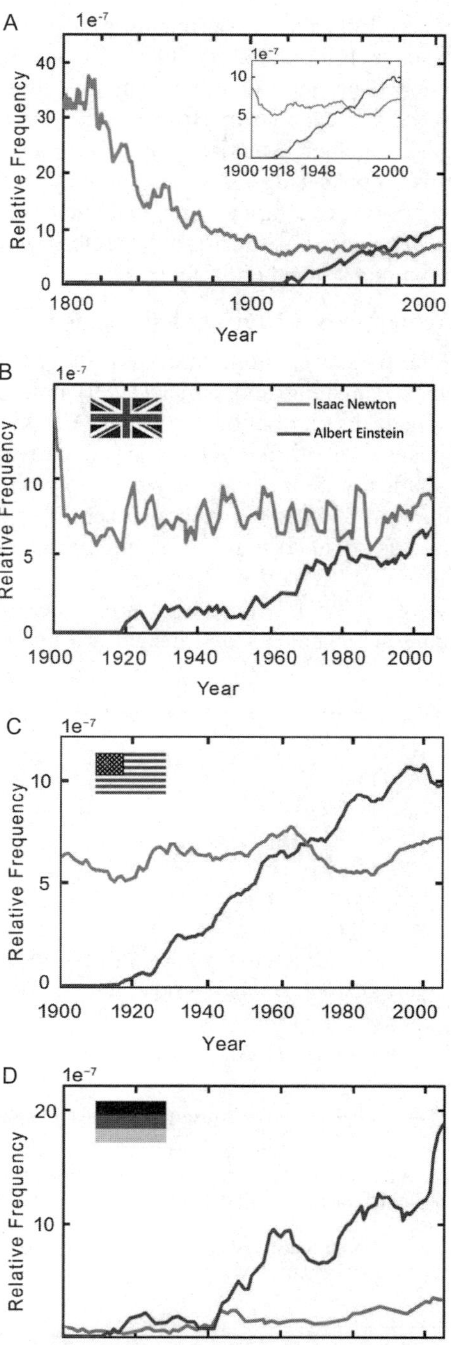

Fig. 2 Scientific fame by language.

Data source: Google Books Ngram Viewer. Accessed in January 2016.

Panel A: Only books in English (both British English and American English) are considered.

Panel B: Only books in British English are considered.

Panel C: Only books in American English are considered.

Panel D: Only books in German are considered.

The time coverage is from 1800 to 2008 for Panel A and from 1900 to 2008 for Panels B~D.

1 Since 2010 the Google Books Ngram Viewer has been available at http://books.google.com/ngrams. Its source, i.e. Google Books, was generated through Partner Program and the Library Project (please refer to https://www.google.com/intl/en/googlebooks/about/). These millions of books published between 1500 and 2008 were digitally scanned and the corpus was winnowed. For more details please see Michel et al. (2011) and Twenge et al. (2012). The searchable corpora with the Google Books Ngram Viewer were developed by the Google Team and can be downloaded at http://storage.googleapis.com/books/ngrams/books/datasetsv2.html. Please note that Google Books Ngram Viewer cannot be used to determine how often a search term occurs in a book.

2 The default smoothing parameter of Google Books Ngram Viewer is set at 3. Taking the year of 2000 as an example, a smoothing of 3 means that the data shown for 2000 are a mathematical average of original data for 2000 plus three values before and three after 2000 (i.e., a moving average of seven-year data, including 1997, 1998, 1999, 2000, 2001, 2002, and 2003). This setting avoids spikes and makes trends more apparent (Michel et al., 2011; Greenfield, 2013). More information on graphs generated by Google Books Ngram Viewer is available at https://books.google.com/ngrams/info.

contrast, Einstein, a German-origin physicist who later immigrated to the US, has been mentioned far more than Newton since the mid or late 20th century in British and American English together, American English alone, and German books (Figs. 2A, 2C, and 2D). This comparison of citing book language uncovers a high relevance of own-group preference, or approximately location of fame.

3) Achievements They are Famous for

Scientists are commemorated for various reasons. Hawking and Israel (1989) note that the greatest contributions made by Newton in the history of science are the law of universal gravitation and the three laws of motion. In addition to physics, Newton also contributed to the fields of mathematics, astronomy, and the philosophy of nature (Newton et al., 1959; Newton, 1999). In comparison, the most notable achievements of Einstein are quantum mechanics and general relativity (Whittaker, 1943; Torre, et al., 2000). Then, a question arises: what are they famous for as evidenced by the Google corpus?

As demonstrated in Fig. 3A, the contributions of Newton surpass the Laws of Motion and the Law of Gravity. This is particularly true prior to mid-1800. As shown the name of Isaac Newton is mentioned in Google Books far more than his most accredited physics theories are.

One may notice the close intertwining or convergence of lines representing scientific achievements and the scientists' names in later periods. But the approximately same frequency of mentioning the achievements and the scientists' names does not mean that the mentions appear simultaneously within the same corpus books. So, we next conducted co-occurrence analysis to examine to what extent these scientists' most accredited achievements contribute to their enduring reputations (Table 3).

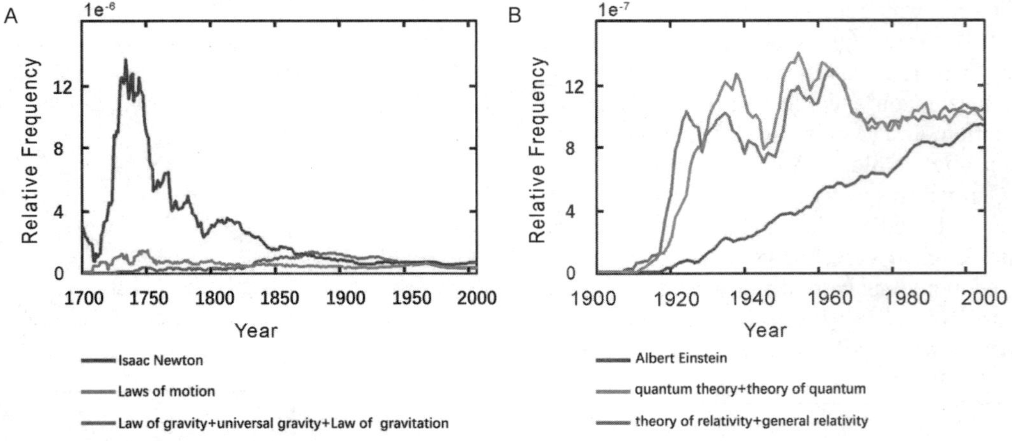

Fig. 3 The major scientific contributions of Newton and Einstein.
Data source: Google Books Ngram Viewer. Accessed in December 2017.
The time coverage is from 1700 to 2008 for Panel A and from 1900 to 2008 for Panel B.

Table 3 Co-occurrence analysis of contributions and physicists.

Key Words	1900–2015	
	Records in Books	% of Total
Isaac Newton	156,100	/
Newton and gravity (including law of gravity, universal gravity, universal gravitation, law of gravitation)	23,700	15.2%
Newton and calculus	12,880	7.9%
Newton and laws of motion	10,440	6.7%
Albert Einstein	178,200	/
Einstein and relativity (including general relativity and special relativity)	50,130	28.1%
Einstein and quantum	30,020	16.9%
Einstein and photoelectric	4,200	2.4%
Einstein and mass-energy	2,420	1.4%

Data source: Google Books search. Data accessed in January 2016.

As depicted in Table 3, books mentioning Isaac Newton often deal with the law of universal gravity (15%) and calculus (8%). The influence of Albert Einstein is highly correlated to the theory of relativity (about 28%) and quantum theory (almost 17%). This, on one hand, reveals what achievements contribute to the glory of Newton and Einstein; on the other hand, the residue (i.e., the uncorrelated proportion) suggests other elements (such as other achievements or even anecdotes) exist.[1]

4) Other Great Minds in Physics

We extended our analysis to other great physicists in history. We first compiled a list of 234 of the most influential scientists based on the following four books:

The 100 Most Influential Scientists of All Time (Rodgers, 2009)

The 100 Most Influential Scientists (Simmons, 2009)

100 Scientists Who Changed the World (Tiner, 2002)

The Most Influential Scientists in Human Civilization (Luo, 2012)

After removing scientists born before the year 1 BC, such as Aristotle and Plato, we next searched each name in Wikipedia and allocated the scientists to different research fields. We then retrieved their name appearance frequency in all Google Books published after 2000. The top 20 physicists listed in Table 4 show that in the field of physics, Albert Einstein, Max Planck, Isaac Newton, Blaise Pascal, and Galileo Galilei are the top five most influential and best-known physicists in the 21st century.

Table 4 lends further support that great scientists are gone but not forgotten. [2] As shown, the early scientists such as Newton, Pascal, and Galilei are still on the public's lips in contemporary society. Undoubtedly, the achievements of these scientific pioneers have become the cornerstones of human scientific knowledge. But it is also true that besides scientists' contributions to scientific advancement, their scientific spirit, anecdotes, and other special attributes may remain

Table 4 Top 20 physicists ranked by scientific fame.

Rank	Name of Physicist	Nationality of Origin	Year of Birth	Rec. of Being Mentioned in Books	Rank	Name of Physicist	Nationality of Origin	Year of Birth	Rec. of Being Mentioned in Books
1	Albert Einstein	German	1879	145,000	11	Michael Faraday	UK	1791	49,500
2	Max Planck	German	1858	129,000	12	James Watt	UK	1736	47,900
3	Isaac Newton	UK	1643	121,000	13	Clerk Maxwell	UK	1831	43,100
4	Blaise Pascal	France	1623	118,000	14	Enrico Fermi	Italy	1901	41,400
5	Galileo Galilei	Italy	1564	96,100	15	William Thomson	UK	1824	40,700
6	Stephen Hawking	UK	1942	82,800	16	Robert Hooke	UK	1635	40,200
7	Niels Bohr	Denmark	1885	80,100	17	Max Born	German	1882	38,000
8	Joseph Henry	USA	1797	79,100	18	Wolfgang Pauli	Austria	1900	37,200
9	Richard Feynman	German	1918	65,300	19	Robert Oppenheimer	USA	1904	36,400
10	Werner Heisenberg	German	1901	55,400	20	Thomas Young	UK	1773	36,400

Data source: Google Books search. Data accessed in January 2016.
Time coverage: 2000–2015

1 Our manual examination of the remaining books mentioning Newton but not gravity, gravitation, calculus, or motion suggests that many are related to natural philosophy, alchemy, and theology. In addition to mentions of relativity, quantum, photoelectric and mass-energy, books mentioning Einstein also discuss his ethical and philosophical views on cosmic religion, his educational philosophy, and even his private life.

2 The scientific fame of these 20 physicists persists after they pass away.

as well. Students and the public learn about the great scientists via scholarly publications, textbooks, and popular science books. When a scientist becomes a public figure, his life experiences and achievements are often adapted into various biographies and even enlightening stories. In stories such as 'Newton and His Apple', the scientific spirit has been used to encourage and to inspire each succeeding generation for hundreds, if not thousands, of years.

This also holds for Stephen Hawking. In the list of top 20 physicists in the 21st century shown in Table 4, Hawking ranks sixth and far ahead of some classic scientists including James Watt, Michael Faraday, and so on. In addition to Hawking's great scientific achievements, his passion in promoting science given his special physical condition has been seen as inspiring to the public and especially younger generations.

4. Conclusion and Discussion

1) Summary

Big data is an important theme in natural and social sciences today. Adopting and expanding the approaches developed by Bohannon (2011a, 2011b), Michel et al. (2011), and Lin et al. (2012), this work exemplifies how to track the historical achievements and influence of great scientific minds. Utilizing the language option of Google Books Ngram Viewer, we find evidence in support of own-group preference, which explains the discrepancy of the UK survey and computational analysis results based on the Google corpus. The co-occurrence analysis demonstrates the main contributions that scientists are known for by later generations

The wheel of history is always moving forward, with new developments emerging and old theories gradually fading away. This is an inexorable law of history, with each scientist's face and influence having a certain vitality cycle. Yet our study suggests the great minds live long intellectually. A scientist can only live several decades in physical form, but his contributions can live on in human history for hundreds or even thousands of years. Today when we talk about pioneers of science before the Christian era such as Aristotle, Euclid, and Archimedes, we find that these names have been written into history, respected and eulogized repeatedly by generations of newcomers. This may be the highest level of the meaning of life that scientists are pursuing.

2) Future Research

It is an accepted practice to use citations to measure the scientific impact of researchers or their research output. Yet for an academic paper, the citation life cycle is usually about 3 to 5 years, and the count of citations often declines drastically afterwards (Glänzel et al., 1995; Tang 2013; Rogers 2010). For a long time bibiometricians have used or abused this impact factor while also acknowledging the difficulties in quantifying long-term scientific impact (Wang et al., 2013; Waltman, 2016). With the development of Google digitalized texts and computational analytical methods, we can track names being mentioned and assess the evolution of scientists' fame on a global scale over centuries. Given the variety of book types indexed in Google, this opens another possible research venue of altmetrics: measuring the influence of scientists beyond academia (Costas et al., 2015).

In his critical overview of altmetrics, Bornmann (2014) pointed out that no accepted framework existed to measure societal impact. Different from the existing altmetrics, which assess the impact of scientific work mostly on social media, we argue that when applied appropriately, Google Books can serve as a promising source of data for altmetrics, and the appearance frequency of scientists' names in books can serve as alternative metrics, which are still in great flux, to capture scientists' intellectual contributions to a broader public (Bornmann, 2014; Fenner, 2013; Piwowar, 2013). In the future it would be interesting to differentiate the fame of a scientist within academia and among the public with the categories provided in Google Books search (textbook, fiction, scholarly book, etc.). This paper focuses on physics; research looking into scientific fame of figures in other disciplines or even across disciplines is also worthy of further examination.

3) Limitations

In spite of their potential value for evaluation research, we would like to raise several notes of caution regarding the data and the method.

There are two sources of false positives. Searching individuals' names in books or scholarly publications can yield two types of results:

works written by them and their achievements cited or mentioned by others. Given that Google Books and Ngram Viewer cannot differentiate the specific locations of search terms within books, false positives are inevitable. The good news is that even for an extremely prolific scientist, the actual number of articles one can write and publish is only in the hundreds. As an illustration of authorship versus citation, we searched Alexandre Dumas (1802–1870) who, though not a scientist, was a prolific French writer who dedicated his life to writing and published 357 books. We searched his full name in Google Books on December 20, 2017, and retrieved 23,400 hits. This suggests that the number of times mentioned can serve as a proxy indicator of fame of prominent individuals and is not a reflection of their authorship.

The other source of false positives is name disambiguation. For instance, searches for just 'Newton' could result in mentions of other people or even the International System of Units (SI) unit of force. The common name problem, translation, and transliteration can also lead to an overestimation of scientists' fame. On the other hand, name changes over time associated with marriage and other reasons can produce false negatives, or an underestimation of reputation. In this sense the measurement of scientific fame works best for great scientists or scientists with unique names. For more details, please refer to MacRoberts and MacRoberts (1989) and Tang and Walsh (2010).

We also need to be aware of the issues of eponymy and obliteration, which have been discussed rather comprehensively by McCain (2011). When a scientist's achievements are so important as to become common knowledge, it is possible that only the achievements themselves rather than the scientist's full name are referenced. For instance, when people talk about the law of gravity, they may not mention Newton's name at all. In this case, using full name as a proxy causes an underestimation of fame. As McCain noted, 'it is often observed that very well- established concepts, experimental methods, empirical findings, and models are mentioned in the literature without being linked to their originators' (p. 1413). This in our case will incur false negatives of recording scientists' fame. However, we argue that a scientist's full name never being mentioned once throughout a whole book is rare and thus has minimal impact on our findings.

It should be noted that for any computational analysis the amount of retrieved data should be sufficiently large. Scientific fame is no exception. Despite its potential, researchers should be aware of the caveats of the Google corpus, such as language bias against non-English languages, the quality heterogeneity of indexed books, and whether it is appropriate only for highly influential scientists rather than general scientists, when conducting fame analysis. It needs to be pointed out that though Google Books and Ngram Viewer make it possible for us to analyze already famous scientists' fame, it would not be possible to identify those who are not yet famous scientists.[1]

Another caveat of this novel measurement is that it may not separate mentions of a famous figure for infamy rather than mentions of the figure for positive glory. Fortunately, in our case of two well-acknowledged great minds in science, the majority of the books on Newton and Einstein are based on a positive or at least a neutral evaluation. This saves us from the concern about evaluating controversial figures.

Acknowledgments

The research is supported by the National Social Science Foundation of China (#14CXW011), Korea Foundation for Advanced Studies (International Scholar Exchange Fellowship 2017-2018), and the Humanities and Social Sciences Project of the Ministry of Education of China (#WKH3056005).We would like to express our sincere gratitude to two anonymous reviewers and Professor Ludo Waltman for their insightful suggestions which substantially improved this manuscript. We thank Professor Fei Xu for the inspiring discussion, Jiafei Shen and Feifei Han for their help in handling the figures. Author contributions: G.W., C.L., and L.T. designed research; G.W. analyzed data; and L.T and G.H. wrote the paper. The authors declare no conflict of interest.

References

Abdullah, A., Thelwall, M. (2014). Can the Impact of non-Western Academic Books be Measured? An investigation of Google Books and Google Scholar for Malaysia. Journal of the Association for Information Science and Technology, *65*(12), 2498-2508.

1 We would like to thank one anonymous reviewer for directing us to this point.

Acerbi, A., Lampos, V., Garnett, P. et al. (2013). The Expression of Emotions in 20th Century Books. PLOS ONE, 8(3): e59030.

Astin, A.V. (1957). Scientists and public responsibility. *Physics Today* 10(11), 23-27.

Baker, E.P. (1984). Einstein versus Newton: the gravity of the situation. Ohio: Online Computer Library Center.

Bohannon, J. (2011a). The science hall of fame. Science, 331(6014), 143-143.

Bohannon, J. (2011b), Google Books, Wikipedia, and the Future of Culturomics. Science, 331(6014), 135-135.

Bornmann, L. (2014).Do altmetrics point to the broader impact of research?An overview of benefits and disadvantages of altmetrics. Journal of Informetrics, 8(4), 895-903.

Bronk, D. W. (1976). Alfred Newton Richards (1876–1966). Perspectives in Biology & Medicine, 19(3), 413.

Costas, R., Zahedi, Z., Wouters, P. (2015). Do 'altmetrics' correlate with citations? Extensive comparison of altmetric indicators with citations from a multidisciplinary perspective. Journal of the Association for Information Science and Technology, 66(10), 2003-2019.

Egghe, L., R. Rousseau (2004). How to measure own-group preference? A novel approach to a sociometric problem. Scientometrics, *59*(2), 233-252.

Fagan, J.C. (2017). An evidence-based review of academic web search engines, 2014–2016: Implications for librarians' practice and research agenda. Information Technology and Libraries, 36, 2.

Fenner, M. (2013). Letter from the guest content editor: Altmetrics have come of age. Information Standards Quarterly, 25(2), 3-3.

Feist, G. J. (2016). Intrinsic and extrinsic science: A dialectic of scientific fame. Perspectives on Psychological Science, 11(6), 893-898.

Glänzel, W., Schoepflin, U., (1995). A bibliometric study on aging and reception processes of scientific literature, Journal of Information Science, 21, 37 – 53.

Graneau, P., Graneau, N. (1993) Newton versus Einstein: How matter interacts with matter. Sheffield: Carlton Press.

Greenfield, P. M. (2013).The changing psychology of culture from 1800 through 2000. Psychological Science 24(9), 1722-1731.

Gribbin, J. (1987). Newton vs. Einstein. New Scientist 116(1586):45-45.

Hoover, E.R.(1977). Cradle of greatness: National and world achievements of Ohio's Western Reserve. Ohio: Cleveland.

Hawking, S. W., Israel, W. (Eds.). (1989). Three hundred years of gravitation. Cambridge University Press.

Halevi, G., H. Moed, J. Bar-Ilan (2017). Suitability of Google Scholar as a source of scientific information and as a source of data for scientific evaluation—Review of the Literature.Journal of Informetrics 11(3), 823-834.

Laubichler, M.D., Maienschein, J., Renn, J. (2013). Computational Perspectives in the History of Science: To the Memory of Peter Damerow. ISIS 104(1): 119-130.

Liu, W., et al., (2015).China's global growth in social science research. Journal of Informetrics, 9(3), 555-569.

Liu, W., et al., (2018). Author address information omission in Web of Science—An explorative study. Journal of Informetrics, forthcoming.

Luo, J.W. (2012). The most influential 100 scientists in human civilization. DAyi Publishing, Xiamen (in Chinese).

Macroberts, M. H., Macroberts, B. R. (1989). Problems of citation analysis: A critical review. Journal of the American Society for Information Science, 40, 342-349.

McCain, K. W. (2011) Eponymy and obliteration by incorporation: The case of the 'Nash Equilibrium'. Journal of the American Society for Information Science and Technology, 62(7), 1412-1424.

Menard, H. W. (1971). Science: Growth and change. Harvard University Press, Cambridge, MA.

Merton, R. K. (1970). Behavior patterns of scientists. American Scientist, 3(2), 213-220.

Michel, J. B., Shen, Y. K., Aiden, A. P., et al. (2011). Quantitative analysis of culture using millions of digitized books. Science 331(6014), 176-182.

Newton, I. (1999). The Principia: mathematical principles of natural philosophy. University of California Press.

Newton, I., Cohen, I. B., Schofield, R. E. (1959). Isaac Newton's papers and letters on natural philosophy.

Pettit, M. (2016).Historical time in the age of big data: cultural psychology, historical change, and the google books ngram viewer. History of Psychology 19(2), 141-153.

Piwowar, H. (2013). Altmetrics: Value all research products. Nature, 493(7431), 159-159.

Rogers, J.D. (2010).Citation analysis of nanotechnology at the field level: implications of R&D evaluation. Research Evaluation 19(4), 281-290.

Roth, S. (2013).The fairly good economy: Testing the economization of society hypothesis against a Google Ngram View of trends in functional differentiation (1800—2000). Journal of Applied Business Research 29(5), 1495-1500.

Sanghera, P. (2011). Quantum physics for scientists and technologists: Fundamental principles and applications for biologists, chemists, computer scientists, and nano-technologists. Wiley-Blackwell.

Simmons, J. G. (2009). The 100 most influential scientists paperback. New York: Fall River Press.

Sternfeld, J. (2011). Archival theory and digital historiography: Selection, search, and metadata as archival processes for assessing historical contextualization. The American Archivist 74(2), 544-575.

Stoever, D. H., Trapp, J., Linné, C. V. (1794). The Life of Sir Charles Linnæus. London, E. Hobson. Library of Congress. ISBN 978-0-19-850122-0.

Tang, L. (2013). Does 'birds of a feather flock together' matter-Evidence from a longitudinal study on US-China scientific collaboration. Journal of Informetrics 7(2), 330-344.

Tang, L., Walsh, J.P. (2010). Bibliometric fingerprints: name disambiguation based on approximate structure equivalence of cognitive maps. Scientometrics 84(3), 763-794.

Tang, L., Shapira, P., J. Youtie (2015). Is there a clubbing effect underlying Chinese research citation increases? Journal of the Association for Information Science and Technology 66(9): 1923-1932.

Tiner, J.H. (2002).100 scientists who changed the world. New York: Gareth Stevens Publishing.

Torre, D. L., Daleo, A., I. García-Mata (2000). The photon-box Bohr-Einstein debate demythologized. European Journal of Physics 21 (3): 253-260.

Twenge, J., Campbell, K. W. Gentile, B . (2012). Increases in individualistic words and phrases in American books: I960–2008. PLoS ONE 7 (7): e40181-e40181.

Waltman, L. (2016). A review of the literature on citation impact indicators. Journal of Informetrics 10(2), 365-391.

Wang, D., Song, C., Barabási A.L., (2013). Quantifying long-term scientific impact. Science 342(6154): 127-132.

Whittaker, E.T. (1943). Aristotle, Newton, Einstein, Science 98(2542), 250-254.Youtie, J., Rogers, J., Heinze, T., et al. (2013). Career-based influences on Scientific Recognition in the United States and Europe: Longitudinal Evidence from Curriculum Vitae Data. Research Policy, 42(8), 1341-1355.

Zeng, R. Greenfield, P. M. (2015). Cultural evolution over the last 40 years in China: Using the Google Ngram Viewer to study implications of social and political change for cultural values. International Journal of Psychology 50(1), 47-55.

Difference Between the Educational Patterns of the USA and the Soviet Union

——A study on Chinese scientist Jiang Yiyuan

Wang Liyuan

Department of Innovation Talents, National Academy of Innovation Strategy, CAST, Beijing, China

Abstract: The educational experience of the Chinese scientist Jiang Yiyuan reflected the difference between the Agricultural Machinery Education of the USA and the Soviet Union during the 1940s and the 1950s. The former educational system emphasized practice, and the other paid more attention to theory. Jiang Yiyuan absorbed both sides of those two patterns in his study.

1. Introduction

The history of Chinese modern education of science and technology reflected the process that China learned from the other countries. Japan, the USA and the Soviet Union all have been the example of China. The results of the learning were not limited in the field of education and academic. It produced labor resources of science and technology for China and influenced the pace of its modernization. How did a Chinese student become a scientist through the modern education of science and technology? How did different education patterns of different countries work on the learners? The case of Jiang Yiyuan（蒋亦元）, who is the Academician of the Chinese Academy of Engineering, can give some answers.

Jiang Yiyuan is a Chinese specialist of Agricultural Machinery. He was born in 1928 in China and received education of Agricultural Machinery of America and the Soviet Union during the 1940s and the 1950s. According to his oral interview, educations of Agricultural Machinery of America and the Soviet Union were very different, but they both deeply influenced his research and teaching career. It is necessary to discuss the difference between those two educational patterns and the effect of the education to a scientist through the experience of Jiang Yiyuan.

This study is mainly based on the history of Chinese modern higher education of science and technology, as well as the personal experience of the scientist Jiang Yiyuan. The research data came from the history materials of Nanking University, the description of the subject of Agricultural Engineering in China, and the manuscripts, memorials and interviews of Jiang Yiyuan. Literature analysis, oral history interview and comparison methods are used in the study process.

2. Educational Experience of Jiang Yiyuan

Jiang Yiyuan spent his child time in a rural countryside of China. Although he left the countryside when he went into school, he never forgot peasants of his hometown. It is one of the reasons that Jiang Yiyuan chose agricultural engineering in the university. In fact, he entered the electrical machinery department of the science and engineering college of Nanking University at beginning in 1946. But the US advanced farming machines and American professors in the agriculture college attracted him so much, so he decided to join the major of agricultural engineering to combine his care about agriculture of China with the research into electrical machinery. Through an entrance exam, Jiang Yiyuan began the new life of agricultural engineering study in the college of Agriculture and Forestry of Nanking University.

After graduation Jiang Yiyuan was invited to be an assistant by the dean of department. But he thought the vast black soil in north-east China could provide a suitable condition for the development of farm mechanization, so he went to the northeast plain. As soon as arriving Harbin, he found the soviet textbook of agricultural machinery discussed more theories than that of USA. He started to learn Russian by teaching himself and then mastered it through an accelerated training. (Wang et al., 2008)

In 1957, Jiang Yiyuan was sent to the Soviet

Union for a further academical training of agricultural machinery under the guidance of M. H. Летошнев. M. H. Летошнев was one of the distinguished experts of agricultural machinery. He not only gave some academical directions to Jiang Yiyuan in the Agricultural College of Leningrad, but also introduced Jiang to visit some other Soviet leading research institutes of agricultural machinery, which gave more inspiration in theory for Jiang.

3. American Education Pattern: Emphasizing Practice

Nanking University is a private academical institution supported by American missionary interests. Its general education resource and staff came from the US. The College of Agriculture and Forestry of this university was founded in 1914. It was the oldest and one of the largest agricultural colleges in China. In agricultural education, Nanking University tried to improve contacts between China and America, as well as to provide well-trained personnel to extend modern scientific agriculture throughout the nation. At the beginning, Agricultural engineering was a division in the Department of Agronomy of Nanking university. C. W. Riggs, who received his master degree of science from Cornell University in US, served as the head of the division. In 1948, Nanking University received an approval of full department status for agricultural engineering. (Hansen et al., 1949)

In the fall of 1947, Jiang Yiyuan became a freshman of the agricultural engineering major in the Department of Agronomy of Nanking University. He found that the teaching of the major put practice and application in the first place. It could be reflected in two main aspects. The first was the courses and textbooks. The second was the activities students attended. Here is the curriculum served as a sample and guide to Chinese colleges to establish departments of agricultural engineering in 1947. The subjects included were thought suitable in the United States, and required further change and modification to meet the problems of China. (Hansen et al., 1949)

From Table 1, it can be found that some basic abilities of agricultural engineering and management were required by the curriculum. According to the interview of Jiang Yiyuan, some handbooks of engineering were used in the courses instead of textbooks, because they were more practical. For example, Mr. Riggs taught students to look up some parameters in the handbooks of machinery or architecture, then to use them into the machine design and building practice. (Jiang, 2014)

Table 1 Curriculum in agricultural engineering department.

Junior			
First Semester	Credit	Second Semester	Credit
Heat Power Engineering	3	Machine Design	2
Machine Design	2	Heat Power Engineering Lab	1
Testing of Materials	1	Farm Machinery	3
Hydraulics	3	Electrical Engineering	3
General Horticulture	4	Advanced Farm Shop	3
Electrical Engineering	3	Agricultural Economics	3
Farm Shop	3	Animal Husbandry or Forestry	3
Crop Production	3	Crop Production	3
Physical Education		Physical Education	
Total	22	Total	21

Senior			
First Semester	Credit	Second Semester	Credit
Farm Structures	3	Farm Power II	3
Electrical Engineering Lab	1	Rural Industry	2
Soil and Water Conservation	3	Dairy Machinery	2
Repair of Farm Machinery	3	General Plant Pathology or Entomology	3
Farm Power.	3	Agricultural Engineering Seminar	1
Soil and Fertilizers	3	Thesis	1
Agricultural Engineering Seminar	1	Physical Education	
Thesis	1		
Physical Education			
Total	18	Total	15

Prof. Hansen was another teacher of agricultural engineering of Nanking University. He led students out of classrooms to attend some activities, such as remodeling the agricultural machinery and equipment building. The Agricultural Engineering Division was located on the main campus of the university. Its shops and classroom buildings were adjacent to each other and near a small field for primary farming practice and machinery operation and adjustment. Moreover, Prof. Hansen redesigned a shed-type building and fitted it for a farm machinery laboratory with student participation. They put in a suitable concrete floor, repaired the roof, and improved the lighting so that the building became well adapted for the exhibit and study of the new farm equipment. (Hansen et al., 1949)

The case of the division of agricultural engineering reflects a typical American educational pattern of agriculture in 1940s. It emphasized practice and application because of two main reasons. The one is the service object of the agricultural engineering of American was the commerce, the products stood at the most important place. The other reason was that they would like to developed farming management and machinery in China where mostly depended on the traditional agriculture. The extending of the modern agriculture needed practice more than theory obviously.

4. Educational Pattern of the Soviet Union: Stressing on Theory

During the 1950s, China began to learn from the Soviet Union at every aspect. The Chinese agricultural machinery education was strongly influenced. Chinese agricultural university copied the educational pattern of agricultural machinery of the Soviet Union. They brought in the courses and books from the Soviet Union and use the same education projects, methods and exams. Experts from the Soviet Union were invited to Chinese agricultural universities to guide the teaching and research. During this period, Jiang Yiyuan was a teaching assistant in the Northeast Agricultural College. It was the beginning that he touched the Soviet educational system of agricultural machinery. He found the Soviet textbooks of agricultural machinery showed many complex theories. It was so different from the American education ways which only provided general concept and structure.

In 1957, Jiang Yiyuan got a chance to study in the Soviet Union. He chose to go to Agricultural College of Leningrad, because he wanted to learn more theories of agricultural machinery. M. H. Летошнев in Agricultural College of Leningrad was one of the most influential academician of the Soviet Academy of Agriculture. His theory on agricultural machinery was famous for its meticulous analysis and thorough explain. (Jiang, 2014)

Jiang Yiyuan wrote more than 100 study articles and notes from 1956 to 1959, when he was learning the Soviet agricultural machinery theories. For example, Research on the Cleaning Components After Picking Ears of Corn, Mechanical Analysis of Farm Machinery Wheels, Study on the Hanging Planter of Grain and Flax, Research on the Application Device of Special Fertilizer and so on. (Jiang, 2014)

Except learning in the Agricultural College of Leningrad, Jiang visited the main Institutes, colleges and factories of agricultural machinery in Soviet Union. He stayed in each place for two weeks and absorbed the experience from both theory and practice. He found that some academical views could not be realized in the production.

In the 1950s, the education of agricultural machinery of the Soviet Union provided a different pattern from that of the US. It was decided by the origin of the discipline in Soviet Union. Soviet scholars pointed out that the creation of the knowledge system should be based on the theory. Some researchers even thought that theory must instruct the practice at first. But another had realized that sometimes there was long distance between hypothesis and reality. However, professors and scholars in the academies paid more attention to the theoretical research and formed the tradition in the major of agricultural machinery in the Soviet Union. This affected Chinese education of the major in the particular historical period.

5. Conclusion

Jiang Yiyuan received different education of agricultural machinery from the US and the Soviet Union. The educational patterns of agricultural engineering and agricultural machinery of American and the Soviet Union had their own features. The former educational system emphasize practice, and the other paid more attention to theory. Jiang Yiyuan absorbed both sides of those two patterns in his studies, and then used them in his teaching and invention. His attitude had changed. In the early time of his teaching and research during the 1950s, theory attracted him more. But with the experience of research increasing, he thought practice were important as well as practice

References

Hansen, Edwin L., Mc Colly, Howard F., Stone, Archie A., et al. (1949). *Introducing Agricultural Engineering in China.* Agricultural and Biosystems Engineering Technical Reports and White Papers.

Wang Tingmao, Yu Shuai (2008). Splendid Achievement Comes from His love to Farm Machinery: A Record of the Famous Specialist of Agricultural Machinery, the Academician of the Chinese Academy of Engineering, Mr. Jiang Yiyuan. *Farm Machinery,* No. 3C, 2008, P39-41.

Jiang Yiyuan, Shi Yan, *Oral interview,* 27 November, 2014, The Database of Collection Project of Historical Materials of Chinese Modern Scientists.

Jiang Yiyuan, Shi Yan, *Oral interview,* 2 December, 2014, The Database of Collection Project of Historical Materials of Chinese Modern Scientists.

Jiang Yiyuan, *Manuscripts*, The Database of Collection Project of Historical Materials of Chinese Modern Scientists, 2014.

Formation and Development of Space Engineering Culture with Chinese Characteristics Based on Scientific Spirit

Cai Wenyi, Yang Xuejiao, Rao Chenglong

Research and Development Center, China Academy of Launch Vehicle Technology, Beijing, China

Abstract: Combined with the history of China's successful aerospace development, this paper discriminates the definitions of engineering and space engineering, engineering culture and space engineering culture, studies the significance and role of scientific spirit in China's space engineering culture and engineering practice from the logical relationship among scientific spirit, engineering practice and engineering culture. And finally it considers several construction measures of space engineering culture with Chinese characteristics based on scientific spirit

1. Introduction

Engineering culture takes engineering as the object of cultural research. The core of engineering is spirit of science, the essence of culture is human spirit, and the engineering culture is the unity of scientific spirit of engineering and humanistic spirit of culture. Great cause creates advanced culture, and advanced culture promotes great cause. It is of great significance to deeply understand and understand the basic connotation and spiritual essence of space engineering culture based on scientific spirit, and cultivate, inherit and popularize excellent space engineering culture.

2. The Origin and Connotation of Space Engineering Culture with Chinese Characteristics

1) Engineering and Space Engineering

(1) *Engineering*

Engineering is a general term for various disciplines formed by applying the theory of natural science to specific industrial and agricultural production department. It is a work with a certain period carried out with large and complex equipment, such as water conservancy engineering, chemical engineering, construction engineering, etc. Engineering is both a process and a result.

(2) *System engineering*

In a narrow sense, system engineering refers to the technical process, that is, the technology of analyzing and designing the system's constituent elements, organizational structure, information flow and control mechanism in order to better achieve the system's goals. System engineering in a broad sense refers to the scientific method of planning, researching, designing, manufacturing, testing and using complex systems for organization and management. It is a scientific method of universal significance to all systems.

(3) *Space engineering*

Space system is an engineering system composed of spacecraft, space transportation system, spacecraft launch facilities, space measurement and control system, user equipment and other support equipment to complete a specific space mission. Space system engineering refers to the scientific system and method of organizing and managing space model planning, research, test, production, personnel, materials, guarantee conditions and funds, which is covered the whole process and rich in content.

The space engineering mentioned in this paper refers to a comprehensive project to develop, explore and utilize the space and celestial bodies, especially the design, manufacture, test, launch, operation, return, control, management and use of spacecraft and space transportation systems. The key development areas of China's space industry include four major application businesses and four major systems engineering. The four major applications are satellite remote sensing, satellite navigation and positioning, satellite live TV and satellite communication. The four

Corresponding author: Cai Wenyi, No.1 Nandahongmen Road, Fengtai District, Beijing, China, P.O. Box 9200-38. caiwenyi0803@163.com.

major system projects are: high-resolution earth observation project, Beidou navigation project, manned spaceflight project and moon exploration project.

2) Engineering Culture and Space Engineering Culture

(1) *Engineering culture*

The formation and evolution of any kind of culture comes from people's specific labor production practices. Engineering culture takes engineering as the object of cultural analysis and research, is the culture of engineering, and is produced in the concrete production practice of engineering project construction. The core of the project is scientific spirit, and the essence of cultural is humanistic spirit. Engineering culture is the sum of the universally recognized and followed values formed in the process of engineering, the institution and material carrier reflecting these values, and material culture reflected by the engineering results.

According to difference forms, the content of engineering culture is divided into four levels: spirit, institution, material and behavior. The whole content of engineering culture is an organic entirety. The core of engineering culture is the spiritual value system, which guides the nature and direction of engineering culture. Institutional culture includes all kinds of rules, concepts and regimes, which have dual functions of guiding and restricting engineering practice. Material culture is the carrier of spiritual culture and institutional culture. The material level of engineering culture includes various engineering signs, slogans, billboards, books and other elements. Behavior culture is the dynamic embodiment of the working style, spiritual style and interpersonal relationship produced by all employees in the concrete engineering practice, and is the reflection of engineering team spirit and values.

(2) *Space engineering culture*

Space engineering is a national strategic project with high technology, high risk and high investment. Space industry is characterized by national will, advanced science and technology, huge system and comprehensive integration. After more than 60 years of development, China's space industry has made brilliant achievements represented by Two Bombs and One Satellite, manned space flight and lunar exploration, etc. China's space engineering technology has reached the world's advanced level and has had great benefits and positive effects on economic construction, national defense construction and scientific and technological progress.

China's space industry has been implanted with the gene of Chinese excellent traditional culture from its initial period. Generations of personnel have continuously strived, based on scientific guidance, insisted on seeking truth from facts, walked out of a development path of self-reliance and independent innovation, and formed a space engineering culture with Chinese characteristics. China's space engineering culture refers to the sum of the ideology, consciousness, concept and materialized ideology Chinese space characteristics that have been cultivated and gradually formed by the space system based on scientific spirit in the long-term space engineering practice. Space engineering culture is an industry culture, an integral part and a subsystem of the whole social culture. According to the theory of connotation level of engineering culture and the concrete practice of China's space engineering, this paper divides the components of China's space engineering culture into four parts, namely, space spirit culture, space institutional culture, space behavior culture and space material culture.

3. Analysis of the Logical System of Scientific Spirit, Engineering Practice and Engineering Culture

Since 1956, China's space industry has gone through more than 60 years, from small to large, from weak to strong, from imitation to independent innovation, creating remarkable achievements in the fields of missiles, rockets, satellites, manned spaceflight, deep space exploration and so on, stepping out of a unique way of China's space development and forming a unique Chinese space engineering culture on the basis of inheriting the excellent traditional culture of the Chinese nation. The following will explore the scientific spirit contained in the formation and development of space engineering culture, as well as the logical system relationship among scientific spirit, engineering practice and engineering culture from the spiritual, institutional, behavioral and material levels.

1) The Elements of Scientific Spirit

The scientific spirit originates from people's

exploration and practice of the natural law, is the scientific subject's values, modes of thinking and codes of conduct, which are determined by the nature of science and run through scientific activities, and is the method and thought embodied in scientific knowledge. Scientific spirit is a diversified concept. From the viewpoint of system theory, scientific spirit can be regarded as a system composed of several elements that are interrelated and mutually restricted.

The spirit of science includes the spirit of exploration, truth-seeking, innovation and cooperation, of which the core is truth-seeking, which emphasizes thinking, exploration, pragmatism and innovation, and is the inner spiritual element of the subject of scientific knowledge activities. Pursuing the truth of knowledge, insisting on the objectivity and dialectic of knowledge are the primary features of scientific spirit, taking practice as the foundation and innovation as the soul are the internal requirements of scientific spirit.

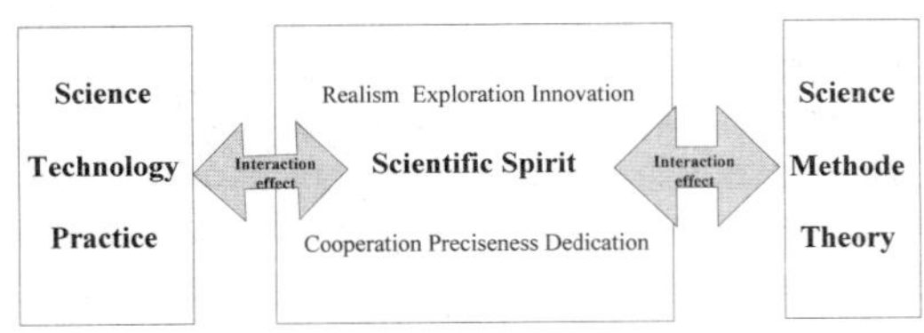

Fig. 1 The elements of scientific spirit.

2) Scientific Spirit in Space Engineering Culture

(1) *Space spirit level: dedication and cooperation with space spirit as cultural leadership*

Space spirit culture is the core of space engineering culture, influencing and guiding the construction and development of system culture and material culture. Space spirit culture refers to the ideology formed by personnel in the practice of space engineering. It is a guide for people to conduct research and production, reflects the common pursuit and common understanding, and is a spiritual power to promote the development of space industry. Its core essence is the *three spirits*, namely, the traditional spirit of space, the spirit of *two bombs and one satellite* and the spirit of manned spaceflight. The *Three Spirits* of are the concrete embodiment and inheritance of space engineering culture in different historical periods, and are the combination of Chinese national spirit and space engineering practice.

The traditional spirit of space originated in the early period of China's space career in the 1950s. After the arduous pioneering and scientific exploration of the older generation of space science and technology workers represented by Qian Xuesen, the traditional spirit of *self-reliance, hard work, strong coordination, selfless dedication, rigorous pragmatism and courage to climb* was formed and formally solidified. The spirit of *Two Bombs and One Satellite* is another refinement and sublimation of the traditional spirit of space. It was formally put forward in 1999, of which the connotation is *selfless dedication, self-reliance, hard work, strong coordination and courage to climb.* The spirit of manned space flight was conceived and formed by astronauts in the manned space project from the 1990s to the beginning of the 21st century. It was formally put forward in 2003, and its connotation is special ability to endure, to combat, to tackle, and to devote.

(2) *Space institution level: rigorous norms with standard regulations as framework guidelines*

Space system culture is a series of organization forms, work norms, management regulations and rules with the characteristics of the space industry formed by space enterprises and departments in their long-term practice of space engineering in order to achieve the overall goal of space product development, production and testing. For example, aerospace engineering specifications include quality standards, working procedures, operating guidelines, production regulations, safety measures, etc. Space management institutions includes legal institution, personnel institution, management institution, production management institution, personnel incentive institution etc.

Space system culture is in the middle layer of space engineering culture system, which is the guarantee of space spiritual culture, reflects the value orientation and behavior norms of space engineering practice subject, and has the characteristics of mandatory, normative and guiding. In the early days of the start-up, the older generation started and practiced the *three strictness* style of *serious attitude, strict requirements and strict methods*. In the course of advancing the space project, generations of personnel have continuously summed up, formed higher standards and more rigorous working principles, and restricted and guided the practice of space engineering, which is a summary of the practical experience and lessons of space engineering and provides an important guarantee for the sustained and rapid development of China's space industry.

(3) *Space behavior level: be realistic and pragmatic and guiding by seeking truth from facts*

Space behavior culture refers to the organization system, management institution, moral code, conduct code and work habits with space characteristics, which is a behavior pattern formed by space talents who have been engaged in space engineering for a long time and regulates the behavior of space personnel. It is a goal around which the entire space team is moving in one direction. The space behavior culture takes the system engineering method as the operation rule and the organization management institution and system institution as the basic framework, which is a summary of the space engineering practice and lessons.

Space engineering is different from ordinary engineering projects. It is a complex system engineering, belongs to the field of cutting-edge science and technology, is at the forefront of scientific and technological development, integrates many new disciplines and technologies, is universally recognized as *high*, *fine*, *sharp* and *new* engineering projects. In the process of space engineering practice, work is done with a serious attitude, rigorous style, strict organization and discipline, in strict accordance with the rules, regulations, requirements and with high standards. Respect objective laws, make prudent decisions with a scientific attitude, and ensure safety, reliability and safety. Start with the details, pay attention to the details, and achieve zero defects.

(4) *space material level: innovative development with cultural symbols as propaganda carrier*

Space material culture refers to the products and various material facilities created by space personnel and the culture they represent. Space material culture is the material basis and activity achievement for space personnel to engage in space engineering practice, as well as the external manifestation and carrier of space spirit culture, space institutional culture and space behavior culture. On the one hand, material culture is restricted by spiritual culture, institutional culture and behavioral culture, and has the characteristics of subordination and passivity. On the other hand, material culture is also the external form that people feel the existence of space engineering culture, which is vivid.

The material culture of space is mainly manifested in contribution to society of space industry, leading scientific and technological progress with high and new technology, safeguarding national security with national defense construction, promoting economic development, shaping spiritual models with advanced images etc. Space material culture includes not only specific space products, but also the environment, conditions, production tools and other implements, propaganda slogans and logos, such as the Long March rocket, the Dongfanghong satellite, the Chang'e satellite and the Shenzhou manned spacecraft. Space material culture is the material basis and achievement of space engineering practice on which space personnel rely. Each generation of space personnel, under certain objective material conditions and in accordance with the law of scientific and technological development, constantly innovates space material culture, spreads space engineering culture to the public and promotes the outstanding achievements of space engineering practice.

3) The Relationship Between Space Science Spirit, Engineering Practice and Engineering Culture

The spirit of space science, engineering practice and engineering culture are mutually logical, mutually supportive and mutually reinforcing, and constantly promote their own progress in the process of blending. For more than 60 years, space people have always acted according to scientific laws, continuously improved the technical level of space engineering, sought key technological breakthroughs, aimed at the forefront of the development of space science and technology in the world, and relying on science and unremitting exploration, they have acquired a number of core technologies with advanced level and independent intellectual property rights. With the development of the times and the progress of science and technology, the space engineering culture based on scientific spirit will play a more prominent role in promoting the practice of space engineering.

(1) *Space scientific spirit directs engineering practice which practices scientific spirit*

Everything needs scientific development and must respect its development law and operation

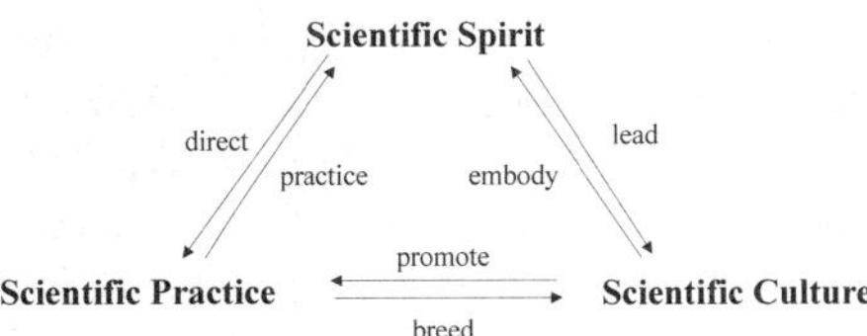

Fig.2 The logical relationship diagram of scientific spirit, engineering practice and engineering culture.

law, which is the inherent requirement of scientific development. The spiritual level is the most profound core, pulling the exploration of the practical level. Space engineering culture is formed based on scientific spirit, driven by the sustainable development of space engineering, which in turn provides deep impetus for the development of space engineering. Space engineering is a high-tech system engineering with high technology density and high technology concentration. China's aerospace engineering culture advocates science and technology to strengthen military and aerospace service.

In the 60 years of engineering practice, China space has always respected science, respected knowledge, been meticulous, practical and realistic, followed the law of scientific and technological development, scientifically formulated the development path and key points in combination with the actual level of national economic and technological development, successfully developed two bombs and one satellite through long-term rigorous exploration, realized major projects such as manned spaceflight and Chang'e moon exploration, solved technical problems, and continuously innovated and expanded.

(2) *Space engineering practice breeds engineering culture which promotes engineering practice*

Space engineering is one of the most important engineering in the modern science era. On the basis of inheriting the existing scientific and technological achievements, in the concrete engineering practice, the Chinese traditional culture and the development of modern science and technology are fused and condensed, forming the engineering culture with space characteristics, and thus promoting the engineering practice. The continuous development of space engineering practice has laid the foundation and conditions for the formation of space engineering culture in terms of spirit, institution, material and behavior, and on the other hand, it has continuously generated new demands for promoting the development of space engineering culture.

Manned spaceflight project is a national key project with the largest scale, the most complex system, the highest technical difficulty, the highest requirements for quality reliability and safety, and extremely risky in China's spaceflight field. It has formed a scientific and rigorous decision-making system, organizational system, technical system, planning and coordination system, quality management system and human resource system, and has bred an engineering culture with the characteristics of the times represented by the manned spaceflight spirit, fully verifying the important significance and guarantee role of China's spaceflight engineering culture for spaceflight engineering practice.

(3) *Engineering culture embodies scientific spirit, which leads engineering culture*

Space engineering culture reflects the common pursuit, common values and common interests of space personnel and has the functions of guiding, condensing, encouraging, regulating, restraining and supervising their thoughts and behaviors. These functions and functions play an important and irreplaceable role in the development of space industry. The scientific spirit is deeply rooted in the culture of space engineering, and the world-renowned achievements made by China's space industry are closely linked with the persistent scientific exploration spirit of all previous generations of personnel in space.

As discussed earlier, the typical elements of scientific spirit are reflected in the spiritual, institutional, material and behavioral levels of space engineering culture, including exploration and innovation, concerted efforts, selfless dedication, preciseness and realism, etc. In the process of space engineering practice, the space personnel knows the mission and responsibility of strengthening the army by science and technology and serving the country by space, and always insists on scientific preciseness and seeking truth from facts in all aspects of scientific research, production and operation management, to ensure the successful completion of all scientific research and production tasks.

4. Thoughts on Measures to Strengthen China's Space Engineering Culture Construction

The development of China's aerospace industry in the past 60 years has proved that the aerospace engineering culture has always been guided by the scientific spirit, developed with the continuous development of aerospace engineering practice, promoting innovative development of aerospace engineering practice with the development of times. With the acceleration of the transformation and application of space science and technology, in order to support the transformation and development of space science and technology industry, the space engineering culture should also further expand its depth and breadth.

1) Internalize the Space Spirit and Fully Inherit and Develop It in Practice

The development of China's aerospace industry in the past 60 years has proved that the aerospace spirit was conceived in the aerospace engineering practice, tested by the engineering practice and enriched and developed with the practice. Space spirit is an important part of space engineering culture, an embodiment of the Chinese national spirit, an important support for the sustained development of China's space industry, and has had a wide and far-reaching impact on all aspects in society. The relationship between inheritance and development should be properly handled in accordance with the law of continuous development of things and stage-by-stage leaps. Strengthen ideological guidance, set up lofty ideals, internalize the space spirit and actively practice in various tasks of the space project to contribute to the great rejuvenation of the Chinese nation at an early date.

2) Adhere to the Spirit of Scientific Exploration and Respect Science, Knowledge and Talents

With the spirit of scientific exploration, generations of space personnel have implemented the strategy of *powerful country* in science and technology and *powerful country in talent* into the practice of space engineering, and have always insisted on ensuring success as the highest principle, quality construction as the life project, with improving the safety and reliability of the project as the center, following the law of scientific and technological development, relying on science, seeking truth and pragmatism, and solving a large number of technical problems through long-term rigorous and meticulous exploration. In order to realize the goal of building a space power at an early date, we should further increase the investment in science and technology resources, continue to improve the science and technology evaluation system, and create a harmonious and enterprising academic atmosphere. At the same time, we should create an atmosphere of respect for science, talents, labor and creation, fully respect scientific and technological personnel, improve the treatment level of space researchers, and ensure the steady development and continuous growth of the talent team.

3) Vigorously Cultivate and Advocate the Spirit of Innovation, Improve the Capability of Space Research

The characteristics of space engineering practice determine that space engineering culture and continuous innovation of science and technology support and promote each other. Every scientific research experiment carried out in space engineering is a high-tech practice and an arduous innovation practice. Space engineering culture not only embodies in the implementation of space engineering projects, but also embodies in technological innovation, business model innovation and management innovation. To adapt to the new situation and seek new development, it is necessary to deeply plant innovation factors in the space engineering culture, cultivate and advocate innovation spirit in the space field, keep up with the development trend of world science and technology, meet the demand of rapid development of space science and technology, attach importance to basic, frontier and strategic technology research, apply the latest achievements of science and technology development, break through key technologies, continuously improve the capability and level of space research.

4) Promote Integration of Defense and Civilian and the Application of Space Engineering Culture

Space technology plays an increasingly prominent role in the new technological revolution, is one of the important symbols of advanced productive forces in society. At present, the development trend of aerospace commercialization is rapid, science and technology and industry are deeply integrated, and a new industrial form represented by Internet thinking is accelerating, pushing the efficient combination of technology, industry and capital with new rules and extending to various industries. China's space industry has begun to explore commercial cooperation to promote the deep integration and development of the space industry and other industries. The construction of space integration of defense and civilian demonstration base and projects for innovation and development will promote the transformation and application of space technology in various fields of the national economy, accelerate the expansion of space engineering culture to more fields and provide support and services to the whole society.

5) Expand Communication Channels and Ways to Enhance the Influence of Space Engineering Culture

The development of space engineering culture and creative industries will drive the spread of culture with industrial development. Make full use of the special connotation and characteristics of space engineering culture, make space elements and Chinese traditional culture deeply integrated, expand the form and communication channels of space engineering material culture, build the space engineering culture communication capacity and system, utilize digitize technology and emerging media, explore the establishment of the *internet + Space Engineering Culture* platform, innovate communication methods, make full use of existing mature technologies, platforms and channels, show the influence of space engineering culture through various channels, form a good environment for the development of space engineering culture, promote the development of space engineering culture in depth and breadth, and realize the socialization and universality of space engineering culture.

5. Conclusion

Culture is connected, and space engineering culture is space and Chinese. After more than 60 years of engineering practice, China's aerospace industry has always adhered to scientific exploration, seeking truth from facts, and continuously innovated and developed, bringing China's aerospace engineering technical capability to the world's advanced level. China's space flight has been implanted with excellent cultural genes with Chinese characteristics from its initial stage. Space engineering culture is the internal driving force for inheriting the space spirit, training space talents, shaping the space image, boosting the space engineering practice and promoting the sustainable development of the space industry, and it is also the indispensable spiritual pillar for China to achieve the goal of building a space power in the new era.

References

Feng Baojing. Analysis of Space Spirit and Socialist Core Values, *Journal of Guilin University of Aerospace Technology*, 2016(3), 417-420.

Hu Chengguang. Analysis on the Essence of Engineering Culture, *Heilongjiang Researches on Higher Education*, 2012(4), 29-35.

Hu Xianfei. Review on Study of Scientific Spirit in China, *Journal of Changsha University of Science& Technology (Social science)*, 2011(11), 19-23.

Li Cunjin, Chen Yonggang. Analysis of Innovation Characteristics of China's Aerospace Engineering, *Science and Technology Entrepreneurship Monthly*, 2014(12), 8-10.

Li Huinan. Innovation Promotes Space Culture Development. *Research on Ideological and Political Work*, 2015(12), 30-32.

Liu Hongbo. Current Situation,Questions and Prospect of Engineering Philosophy. *Science & Technology Progress and Policy*, 2007(11), 62-65.

Louis L. Buccirell. *Engineering Philosophy*. Delft University Press, 2003.

Ma Laiping. Analysis on the Core and Contents of Scientific Spirit. *Journal of literature, History and Philosophy*, 2001(4), 51-54.

Zhang Chunying. Research on Characteristics and Connotation of Chinese Space Culture. *Journal of north china institute of aerospace engineering*, 2016(2), 52-55.

Bibliometric Analysis of Culture of Science Research

Cao Xuewei, Gao Xiaowei, Chen Rui

National Academy of Innovation Strategy, CAST, Beijing, China

Abstract: The concept of culture of science is characterised by complexity. It is a multi-dimensional and cross-disciplinary research domain. In this paper, bibliometric analysis has been applied to the field of culture of science to obtain a structured overview of the characteristics and the developments in this research domain. In total, 1038 publications published between 1993 and 2018 related to 'culture of science' or 'scientific culture' or 'science culture' were identified in Scopus. The 1038 publications were mainly belong to Social science (460) and Arts and humanities (324), others distributed in journal of natural science like Medicine, Engineering, and Computer science and so on. The main research countries were concentrated in regions with science origin, such as United States (248), United Kingdom (112), Spain (53), Germany (51) and Italy (51). The mainly authors were Bauer, M.W. and Elliott, P. The key words focused on Human, Science, Culture, Article, Education, History and Cultural Anthropology. The details were shown below.

1. Introduction

The amount of scientific literature available on a specific research discipline or research topic is often overwhelming, which makes it challenging for researchers and practitioners to have a structured overview of relevant information (Zhou et al., 2015). Bibliometric analysis is a technique which makes it possible to provide a macroscopic overview of large amounts of academic literature. Through a quantitative analysis of information on the publication history, the characteristics and the development of scientific output within a specific field of research can be mapped (Li and Hale, 2016). The number of different journals publishing on a specific topic and the subject categories allocated to publications can give an indication on the variety of research themes, and the multidisciplinary character of a research domain. Bibliometrics can reveal the latest advances, research directions and leading topics in a particular field of research (Wang et al., 2014). It is widely used to rank applications for academic positions, and to evaluate the performance of journals, countries and institutions.

In this paper, bibliometric analysis has been applied to the field of culture of science research. The research topic of culture of science is a relative new one in terms of academic research. Our culture is a scientific one, science and technology' progress may cause public concerns. Cultural studies analysis the role of expertise throughout society. Many journals address the history, philosophy and social studies of science, its popularisation, and the public understanding of society. There is an increasing worldwide consensus on the vital importance of science personal, social, economic, and political development. This has spurred many countries to increase their investments in science and technology. Nations must also promote cultures that celebrate science and its values of reasoning, openness, tolerance, and respect for evidence. This has led to a continuously growing publication rate regarding culture of science, which makes it difficult to obtain a comprehensive overview on the topic. This difficulty is enhanced by the complexity of the concept of culture of science.

The goal of this descriptive paper is to provide a macroscopic overview on the main characteristics of culture of science publications based on a bibliometric analysis. The information presented in this paper provides a clear picture on the research progress achieved in the domain of culture of science research, and it can assist researchers and practitioners in identifying fundamental influences from authors, journals, countries, institutions, references and research topics.

2. Data and Methods

We use the method of bibliometrics to analyze the international research progress of culture of science from 1993 to 2018. The data for this part were mainly retrieved from Scopus on July

11, 2018. 'culture of science' or 'scientific culture' or 'science culture' was used as search topic. In total, 1038 publications related to culture of science were identified, but 28 publications didn't show the publish year clearly. All types of publications were included in the search. Looking at the document types, the majority is article (n = 587) and review (n = 139), others are conference paper (n = 89), book chapter (n = 84) and book (n = 50), the other document types are limited in numbers. The freely available software program VOSviewer was used to analyze and visualize relationships between authors, countries, co-citations and terms.

3. Results and Discussion

1) Publication Output and Growth Trend

The number of peer-reviewed publications is an important indicator to measure the development trend of a scientific research discipline or subject. Figure 1 showed the number of science of culture publications increased since 1993 (the number of publications every year before 1993 were less than 10, we ignored them). There were less than 20 publications per year on science of culture from 1993 to 2000. Since 2001, an increasing number of publications could be observed every year, with exceptions in 2005 and 2015, where a decline could be observed. A peak of publications is reached in 2017 (n = 96), for the number of publications in 2018 is still incomplete up to now. In general, the cumulative number of publications on science of culture keep increasing from 1993 to 2018.

According to Price's law, which evaluates the overall growth of scientific publications in a specific research domain (Price, 1963), the growth of a research domain goes through four phases: (1) a precursors'phase, where a small body of scientists begins to publish on a new field, (2) the proper exponential growth, where an increasing number of scientists is attracted by the many aspects of the subject that still have to be explored, (3) a consolidation of the body of knowledge and (4) a decrease in the number of publications (Dabi et al., 2016). The data showed that science of culture research was still in the exponential growth stage, the number of publications on science of culture will increase in the future.

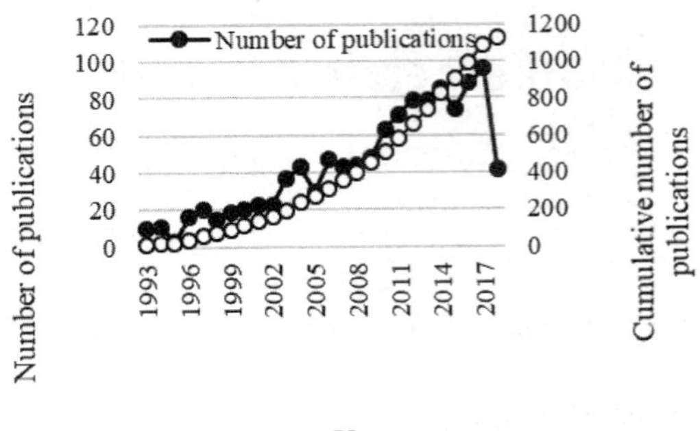

Fig. 1 Number of culture of science publications and cumulative number of culture of science publications by year.

2) Country or Territory Distribution

Table 1 shows the top-10 of most productive countries and territories on culture of science research. The USA produced the most publications (n = 248), followed by United Kingdom (n = 112), Spain (n=53), Germany (n=51) and Italy (n=51). In the extension of the information on countries and territories, a geographical inequality can also be seen when looking at the continents. Europe and North America were the main research area on this field.

Table 1 Top-10 of most productive countries or territories publishing on culture of science.

No.	Countries or Territories	Number of Publications
1	United States (North America)	248
2	United Kingdom (Europe)	112
3	Spain (Europe)	53
4	Germany (Europe)	51
5	Italy (Europe)	51
6	Canada (North America)	44
7	Brazil (South America)	35
8	France (Europe)	35
9	China (Asia)	32
10	Australia (Oceania)	25

3) Authors and Their Cooperation

The authors who published two publications (84) were the largest proportion, and the second proportion of authors was only credited in one publication (58). There were 17 authors who published more publications ($\geqslant$3), it was consistent with the law that only a small group of productive authors contributes to a significant share of publications on a specific topic (Liu et al., 2012).

Table 2 showed the top-17 of most productive authors publishing on science of culture topic.

The ranking was based on the author's publication numbers who as the first author or corresponding author. Bauer, M.W. and Elliott, P. were the most two productive authors on science of culture topic with 5 publications, followed by Roth, W.M. and Inkster, I. with both 4 publications. Another 14 authors all published three publications. For the 17 productive authors, there were 5 persons belong to United Kingdom, followed by United States and Spain with both 3 persons, respectively.

A high number of co-authored publications indicates a closer relationship among the authors within the same domain and a greater opportunity for future collaboration (Wang et al., 2014). The cooperation pattern (i.e. co-authorship) of the authors publishing on science of culture was analyzed with VOSviewer (See figure 2 below). Firstly, we analysis all the 2016 authors (including only one document author) for a global perspective. Figure 2A showed the authors network for the 500 authors with the greatest total link strength clustering into 35 clusters. The size of the circles represents the amount of publications, and the line between two authors represents the cooperation between them. Cluster 1 with the mainly authors Fara G.M., Azara A., Cecchini A. and Giammanco G. was the biggest and latest one which including 79 authors. Then we further showed the network of cluster 1 (Fig. 2B). It was a complex and centrally radiating network.

Secondly, we analysis the authors in the network published at least two papers on the topic. Authors who are not connected with other authors in the network are not included. In the cooperation network, 19 major cluster of 43 authors can be distinguished. This results showed 19 isolated clusters, each cluster only including 2-3 authors. Then we showed the 5 clusters which including 3 authors, they formed a close partnership on culture of science research. Regarding the authorship, a possible bias should be noted. Authors with non-English name could not be distinguished very well by VOSviewer, we had to manual merger for this problem.

Table 2 Top-17 of most productive authors publishing on science of culture.

No.	Author name	Country of Author	Number of Publications as First Author or Corresponding Author	*h*-Index
1	Bauer, M.W.	London School of Economics and Political Science, London, United Kingdom	5	19
2	Elliott, P.	University of Derby, School of Humanities, Derby, United Kingdom	5	7
3	Roth, W.M.	University of Victoria, Faculty of Education, Victoria, Canada	4	49
4	Inkster, I.	SOAS University of London, London, United Kingdom	4	7
5	Livingstone, D.N.	Queen's University Belfast, School of Geography, Belfast, United Kingdom	3	19
6	Solomon, J.	Open University, Milton Keynes, United Kingdom University of Plymouth, Plymouth, United Kingdom	3	19
7	Daïan, J.F.	Universite Grenoble Alpes, Grenoble, France	3	11
8	Subramaniam, R.	National Institute of Education, Singapore, Singapore City, Singapore	4	10
9	Finzen, A.	Stephanstr. 61, Berlin, Germany	3	10
10	Carson, R.N.	Montana State University – Bozeman, Department of Education, Bozeman, United States	3	6
11	Nieto-Galan, A.	Universitat Autònoma de Barcelona, Barcelona, Spain	3	6
12	Lessl, T.M.	The University of Georgia, Athens, United States	3	6
13	Montgomery, S.L.	University of Washington, Seattle, Seattle, United States	3	5
14	Olvera-Lobo, M.D.	Universidad de Granada, Granada, Spain	3	4
15	López-Pérez, L.	Universidad de Granada, Colegio Máximo, Granada, Spain	3	2
16	Dvořák, T.	Filosofický Ústav AV ČR, Prague, Czech Republic	3	1
17	Greifer, I.	—	3	1

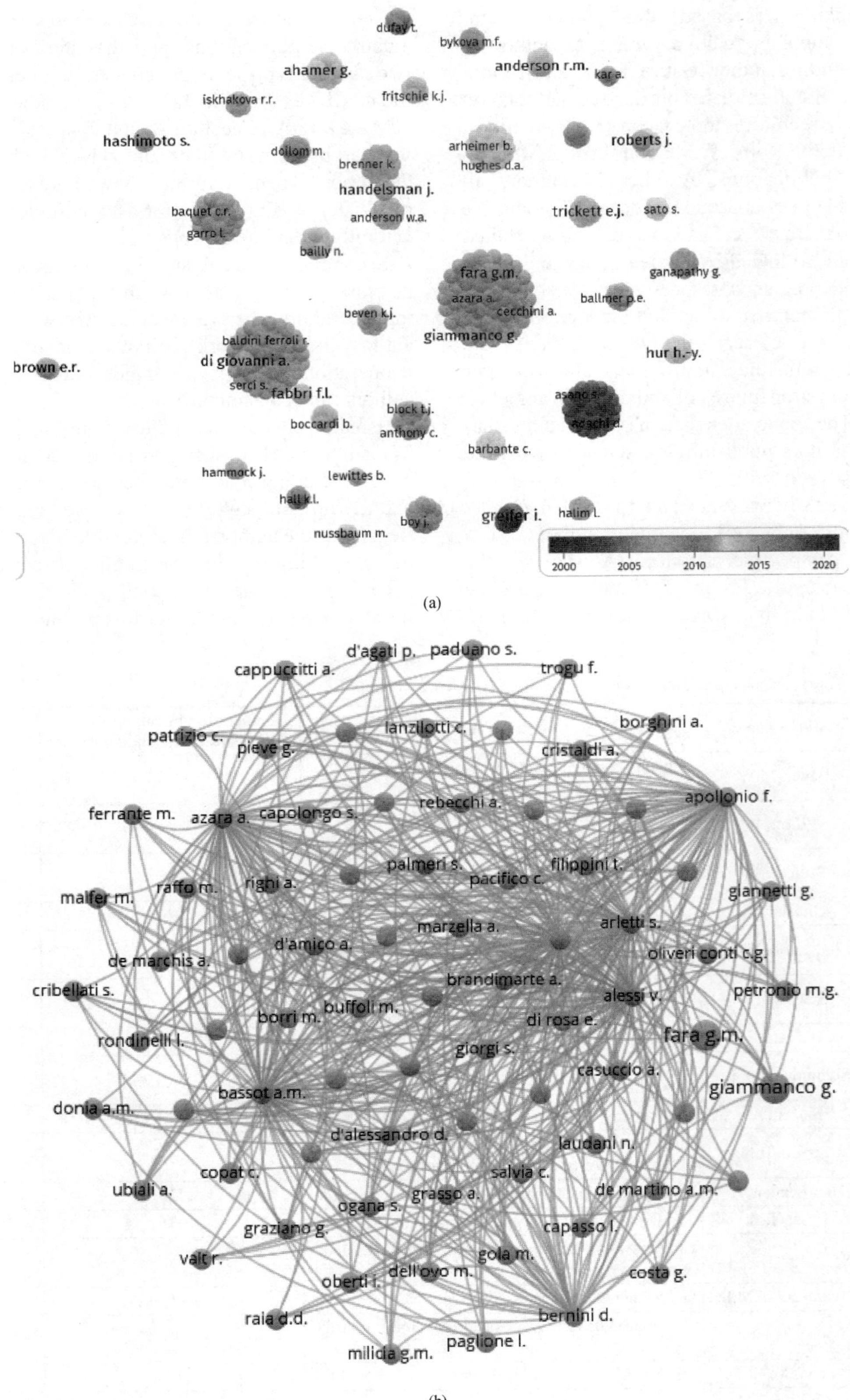

Fig. 2 Authors cooperation network in science of culture research.

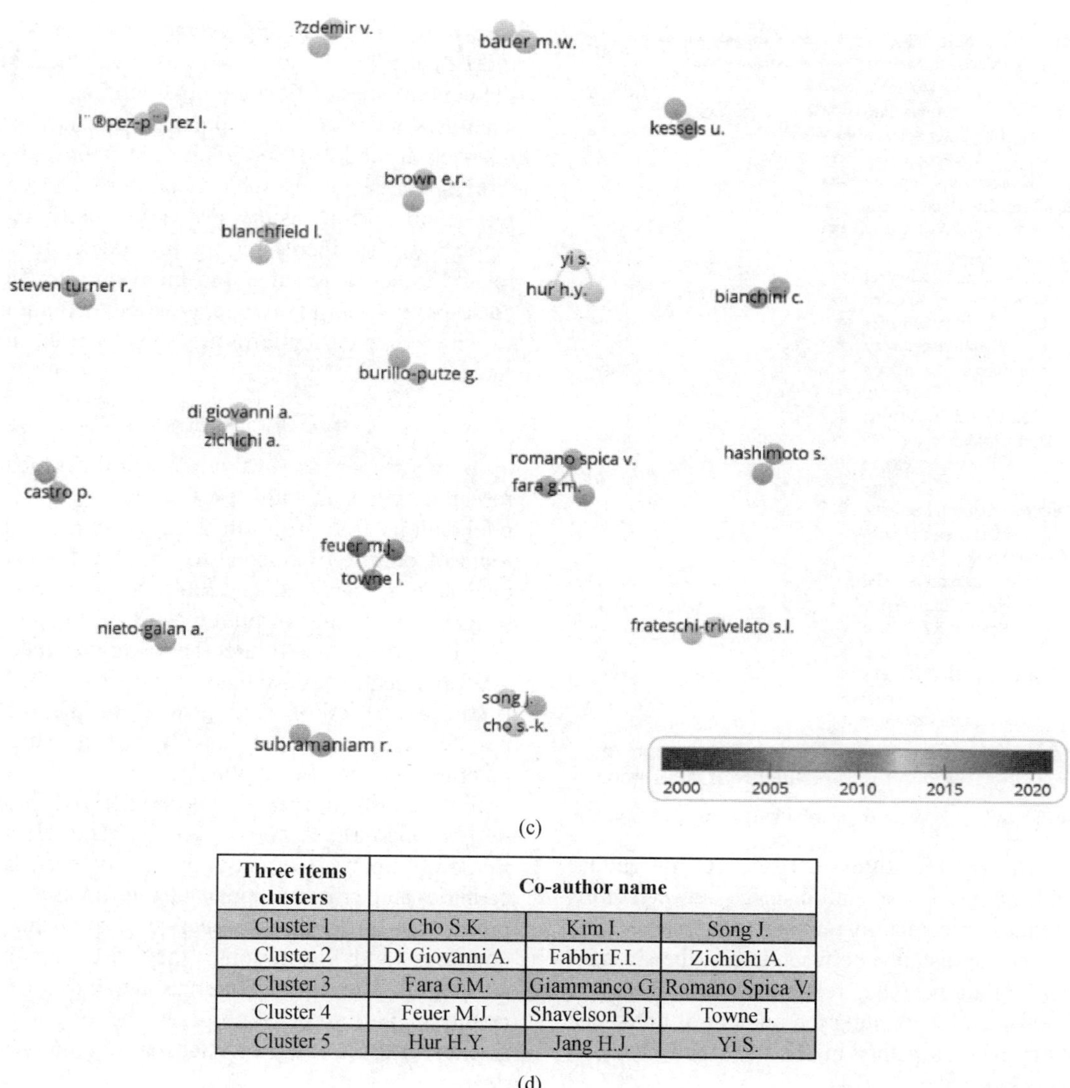

(c)

Three items clusters	Co-author name		
Cluster 1	Cho S.K.	Kim I.	Song J.
Cluster 2	Di Giovanni A.	Fabbri F.I.	Zichichi A.
Cluster 3	Fara G.M.`	Giammanco G.	Romano Spica V.
Cluster 4	Feuer M.J.	Shavelson R.J.	Towne I.
Cluster 5	Hur H.Y.	Jang H.J.	Yi S.

(d)

Fig. 2(Continued)

4) Subject Categories

The subject categories reflect a particular field of research. The 1038 publications were divided into 30 subject categories mainly. A great diversity in subject categories can be seen, it indicated a wide variety of research themes, and the multidisciplinary character of science of culture research. Figure 3 showed the mainly 30 subject categories on science of culture. Most of the publications on science of culture belong to Social Science and Arts and Humanities, followed by natural science subject, like Medicine, Engineering, Computer science, Psychology and so on. This results showed that the theme of science of culture was not only focus on philosophical research, but also distributed in various science and technology fields.

5) Terms Analysis

An analysis of the terms that are used in the titles and abstracts of publications can provide insight in main topics and research trends in the domain of science of culture. VOSviewer was used to analyze and visualize the terms. First, all noun phrases were extracted from the titles of the 1038 publications. Terms with a general meaning, such as'article' and 'conclusion', were not included. Only terms that occur in at least four publications were considered. 120 terms met this threshold. The result of the terms analysis is presented in Fig. 4. The size of the circles represents the occurrence of a term, i.e. the higher the size, the higher the occurrence of

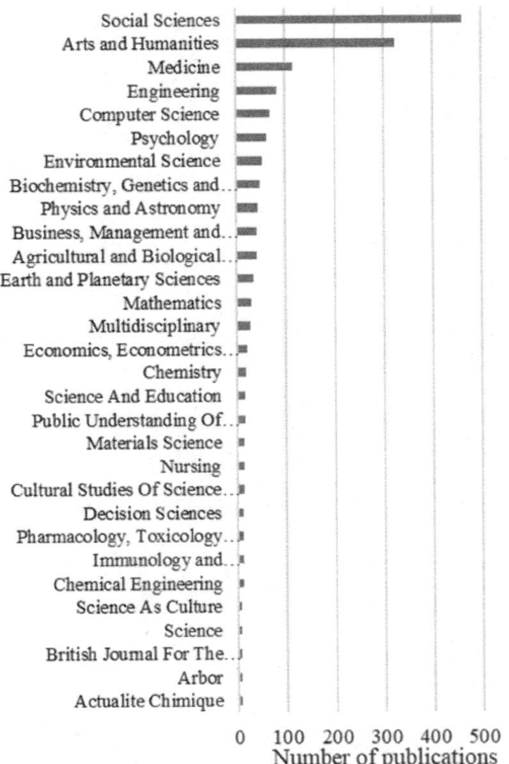

Fig. 3 Subject categories distribution of science of culture.

a term in the titles of science of culture publications. The overall distance between terms provides information on their relatedness. The shorter the distance between terms, the stronger their relation. The relatedness of terms is determined by counting the number of times that terms occur together in the titles and abstracts (Rodrigues et al., 2014).

The terms map shows how the terms of the publications cluster together, and illustrates 11 clusters (See Fig. 4(a)), the colours are used to distinguish different clusters. The most common keywords in the largest cluster were: art, creativity, culture, ethic, gender, humanity and innovation. This cluster seems to entail publications on philosophical research, terms suggest a more theoretical focus. Figure 4B showed the terms analysis of the safety culture publications, but with time information. The colour of a term indicates the term's average publication year. The average publication year of a term is calculated by taking the average of the publication years of all publications that have the term in their title. Terms that are used more towards 2014 are shown in red, while terms that are used more towards 2006 are shown in blue. Looking at the time periods, most research around 2006 was conducted in the content area of Ethics (corresponding terms such as science culture, women, gender). Most research around 201008 focused on philosophy (corresponding terms such as culture, science, reflection, doctor, history, physic). Finally, the map showed an increasing trend in publications related to social practice and innovation as the corresponding terms (image, practice, evolution, experience, contribution) are mostly used in recent years.

4. Conclusions

In this paper, an evaluation on the global research trends in culture of science research publications from 1993 to 2018 is given. The topic of culture of science has been a field of extensive research since 2000. Latest trends suggest the number of publications in this field will increase in the future. The study includes 1038 publications on culture of science, which is still in a period of rapid growth. But the total number of papers is relatively small, which belongs to a 'small discipline'.

Some positive aspects could be derived from the bibliometric analysis. Firstly, The USA, England, Spain, Germany and Italy are the countries and territories dominating the publication production. From the perspective of publishing countries and institutions, they are mainly concentrated in North America and developed countries in Europe. Europe is the origin of science, and the study of science and culture is deeper.

Secondly, it can be concluded that there is much collaborative re-search in the safety culture domain, as multi-authored publications make up about three quarters of all publications. Bauer, M.W. and Elliott, P. are the most productive authors, both of them are affiliated to the UK. Because of the discipline nature, the research of culture of science tends to be more personal, the common collaborators in publications are 2–3 people.

Finally, some limitations of this bibliometric study should be addressed. First of all, the search was limited to publications listed in Scopus. Although Scopus is among the largest global data-bases, it does of course not contain all publications in the field of safety culture research. Secondly, bibliometric analysis uses quantitative methods. Hence, the content or the quality of publications cannot be interpreted (Dunk and

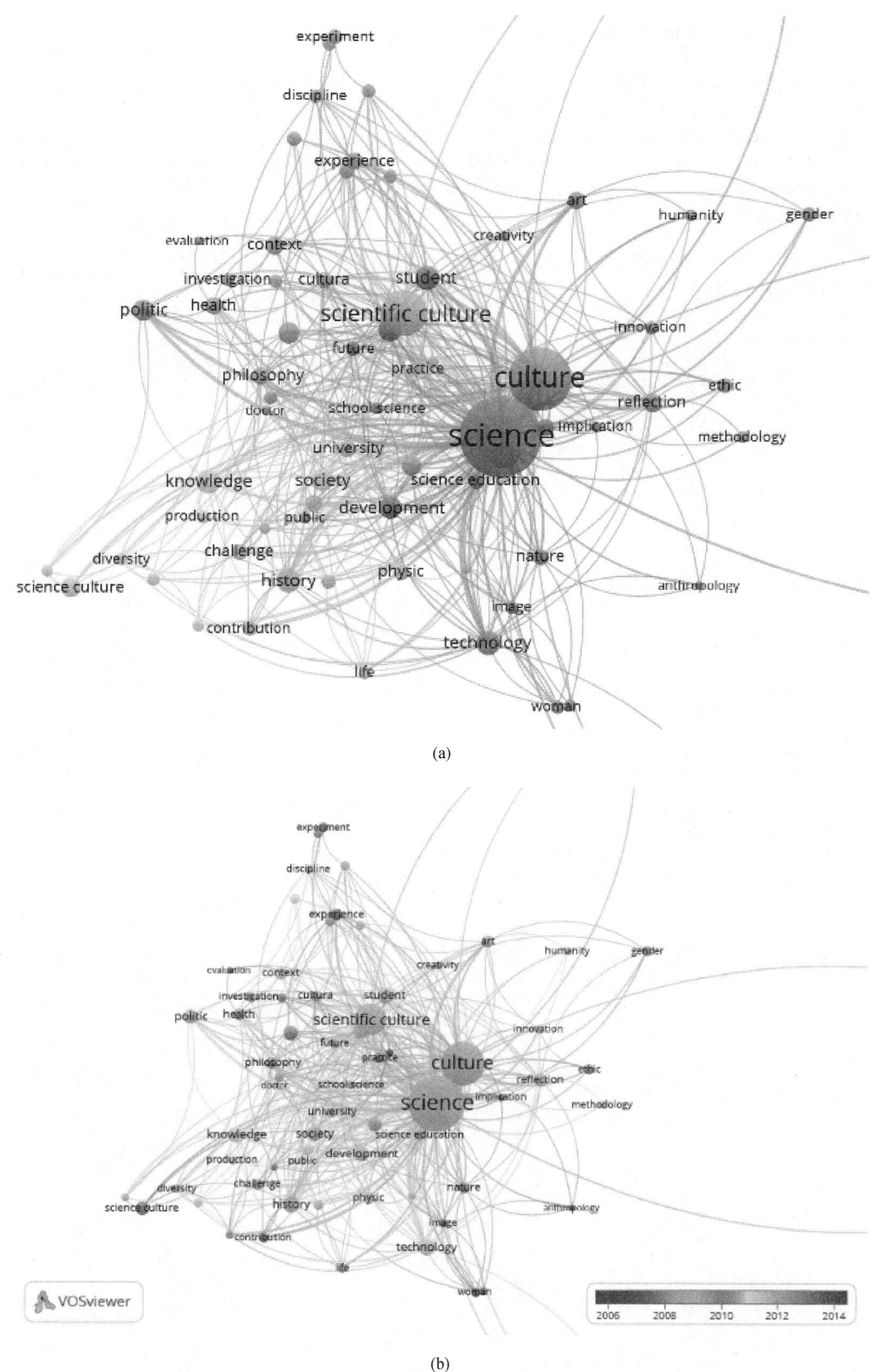

Fig. 4 Terms analysis on the theme.

Arbon, 2009). This can imply that some of the publications were included in the analyses notwithstanding they address a different topic than safety culture; they can address for example the topic of safety climate. Based on these limitations characterising bibliometric analysis, a deeper content analysis is re-commended for further research.

References

Dunk, A.M., Arbon, P., 2009. Is it time for a new descriptor 'pressure injury': A biblio-metric analysis. Wound Pract. Res. 17 (4), 201-207.

Li, J., Hale, A., 2016. Output distributions and topic maps of safety related journals. Saf. Sci. 82, 236-244.

Liu, X., Zhan, F.B., Hong, S., et al., 2012. A bibliometric study of earthquake research: 1900–2010. Scientometrics 92, 747-765.

Price, D.J.S., 1963. Little science, big science. Columbia University Press, New York.

Rodrigues, S.P., van Eck, N.J., Waltman, L., et al., 2014. Mapping patient safety: A large-scale literature review using bibliometric visualisation techniques. BMJ Open4 (3).

Wang, B., Pan, S.-Y., Ke, R.-Y., et al., 2014. An overview of climate change vulnerability: A bibliometric analysis based on Web of Science database. Nat. Hazards 74, 1649-1666.

Zhou, Z., Goh, Y.M., Li, Q., 2015. Overview and analysis of safety management studies in the construction industry. Saf. Sci. 72, 337-350.

The Responsibility of Science and Technology Workers to Taxpayers: A Theoretical Framework

Xu Jie, Liu Xinyang, Deng Dasheng

National Academy of Innovation Strategy, CAST, Beijing, China

Abstract: The social responsibility of scientists or science and technology (S&T) workers has been widely discussed and reached a consensus, while few literatures have noticed their responsibility to taxpayers. Based on the background, this paper mainly focuses on the responsibility of S&T workers to taxpayers. It firstly discusses relationships between S&T workers and taxpayers. Then, three theoretical models, which are the Concentric-circle Model, the Corresponding Decomposition Model, and the Process Analysis Model, are established to reveal special responsibilities of S&T workers to taxpayers. Finally, it comes to a summary that three models give us a theoretical framework for further exploring this issue in the future.

Keywords: Science and Technology Workers; Taxpayers; Responsibility

1. Introduction

With the continued and steady growth of Chinese economy, the research and development (R&D) funding in science and technology (S&T) in China has been steadily increasing. The funding source of S&T activities in China is mainly composed of four parts: government funds, self-raised funds by enterprises, foreign funds and other funds. About one third of R&D expenditure of the country was from government funds. The government S&T appropriation increased from 168.85 billion yuan in 2006 to about 776.07 billion yuan in 2016, accounting for 4.13 percent of the government financial expenditure in 2016. The central government S&T appropriation have increased faster than government financial revenue in the same period. Driven by government funding, China spent 1567.67 billion yuan on R&D in 2016, accounting for about 2.11 percent of GDP. Meanwhile, 81.7% of government revenue was contributed by taxation. It means that tax is the main source of funds for S&T activities in China. Most of the government S&T appropriation allocates to colleges, universities, research institutions and other nonprofit institutions, in order to support them for basic research, applied research, promotion and application of S&T achievements S&T service etc., and the enterprises, colleges and research institutes are clusters of S&T workers. In other words, it is taxpayers who support the S&T workers to carry out R&D activities in China.

The social responsibility of S&T workers is not a new topic. In the 1930s, British scientists led by Bernard firstly raise the issue of the social responsibility of scientists[1], many domestic and foreign scholars have hotly debated the issue of 'what social responsibility S&T workers should undertake' and have formed a consensus in many aspects. However, few studies paid attention to the social responsibility of S&T workers to taxpayers. It means few studies discuss the responsibility and obligations of S&T workers from the perspective of being responsible to taxpayers, and even S&T workers themselves haven't considered relationship between their work and taxpayers.

Therefore, in a legal society with the growing awareness of citizenship in the context of market economy, it is necessary to discuss 'the responsibility to taxpayers' of Chinese S&T workers. Based on these situations, the study discusses the issue that what social responsibility that S&T workers should take when using financial funds to carry out S&T activities, and put forward three kinds of analysis framework. It theoretically discusses and proposes the indirect rights and obligations between S&T workers and taxpayers.

2. The Relationship Between S&T Workers And Taxpayers

1) The Financial Appropriation is an Important Source for S&T Workers to Take R&D Activities

In 2016, the total government budgetary expen-

Corresponding author: Xu Jie. xujaja@163.com.

diture for S&T was 776.07 billion yuan, accounting for 4.13 percent of the financial expenditure in that year. The government funds for S&T are composed of two parts: the central government funds and the local government funds. The central government funds for S&T are mainly used for the basic operation of research institutions, the construction of infrastructures, scientific research projects and scientific popularization. In 2015, the Ministry of Science and Technology and the Ministry of Finance integrated and optimized original S&T projects (special projects, funds, etc.) into five kinds of S&T projects (special projects, funds, etc.), including the National Natural Science Foundation, Major National Science and Technology Special Projects, Main National Research Projects, Special Projects (Foundation) for Technology Innovation, and Special Projects for Research Base and Talents. Financial S&T funding is the pillar of the routine operation of government-funded research institutions, universities and other institutions. The financial S&T appropriation is directly allocated to various national S&T projects, which is also the direct funding source for S&T workers to carry out financial research projects.

2) The Indirect Rights and Obligations Between Taxpayers and S&T Workers

Tax revenue is the most important source of financial revenue in China. In 2017, the tax revenue was accounted for 83.7% of the general public budgetary revenue [2]. The state relies on official power to levy taxes on taxpayers based on laws. Taxpayers pay taxes in accordance with laws to meet the demand for public goods and services. It means the S&T activities funded by the government are supported by taxpayers. In some sense, the research achievements produced from scientific research activities are supported by taxpayers as well. From the perspective of the social contract theory, tax revenue is indeed for the purchase of public goods and services by citizens in the form of taxation. Taxpayers provide financial revenue for the government in the form of taxation. The government appropriates financial revenue to S&T workers to carry out various S&T activities. Therefore, an indirect relationship of responsibility and right is established between taxpayers and S&T workers through a third party that is government(see Fig. 1). Taxpayers have the obligation to pay taxes, and they also have claims to the S&T workers indirectly through the government to guarantee their rights.

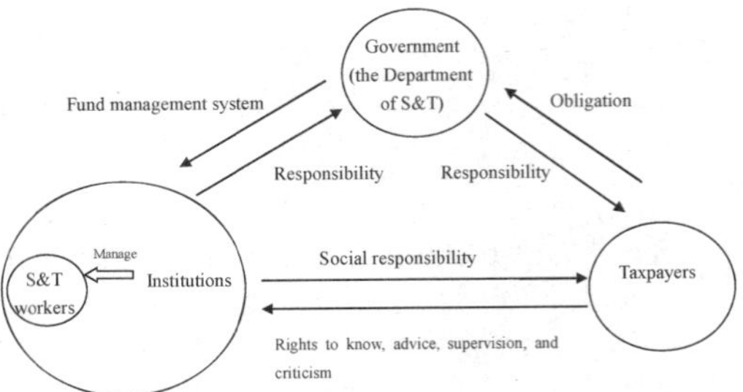

Fig. 1 Indirect relationship between S&T workers and taxpayers.

3. Analytical Models for the Responsibility of S&T Workers to Taxpayers

1) The Concentric-circle (CON) Model

The book *Social Responsibility of Science Workers* proposed two analytical models which are layered model and development model for social responsibility of S&T workers[3]. The other useful reference is from the research on corporate social responsibility (CSR). Gerva (2008) focused on the conceptual structure of CSR and the relations between its elements[4], it proposed three different schematic descriptions of CSR: pyramid, intersecting circles, and concentric circles. The CSR pyramid was framed to embrace the entire spectrum of society's expectations of business responsibilities and was constituted of 4 categories: economic (make profit), legal (obey the law), ethical (be ethical) and philanthropic (be a good corporate citizen). The model categorizes the different responsibilities hierarchically in order of decreasing importance. The most fundamental is economic responsibility. The intersecting circles (IC) model of CSR recognized the possibility of interrelationships among CSR domains, and rejected the hierarchical order of importance. The concentric-circle (CON) model (see Fig. 2) is similar to the pyramid in that it views the economic role of business as its core social responsibility, and it outlined the noneconomic social responsibilities as embracing and permeating the core economic responsibilities.

On the basis of models above of social responsibility of S&T workers and the corporate, this study firstly puts forward analytic framework of the CON Model for S&T workers' social responsibility to taxpayers, as shown in Fig. 2.

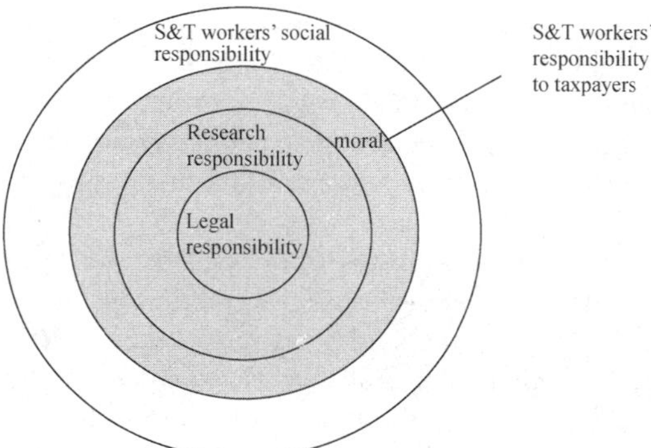

Fig. 2 The CON Model S&T workers' responsibility to taxpayers.

The CON Model regards legal responsibility when S&T workers take S&T activities using of financial funding as the core and basic responsibility, emphasizing the interaction between different levels of social responsibility and taking social responsibility, which is extended from inner to outer, as an extension of the legal responsibility. The CON Model regards the responsibility that S&T workers should take to taxpayers as a form of S&T workers' social responsibility, which contains legal responsibility, moral responsibility and other responsibility.

The responsibility of complying with laws and regulations is taken as the core responsibility. The reason is S&T workers are social citizens in a legal society, and abiding by the law is the most important and basic requirement for every citizen. Within the center of the circle, it lays the basic requirements for S&T workers when using financial funding to carry out S&T activities, which is similar to other social citizens, the S&T workers should abide by the relevant laws and regulations of scientific research expenditure management, including legally using S&T funds, never devouring the money, and respecting for intellectual property rules, etc.

In addition to abiding by laws and regulations, S&T workers should follow the general scientific research ethics and social morality, and assume corresponding moral responsibility. The general scientific research ethics require S&T workers to be honest, objective, serious, realistic, democratic, and ensure academic equality when conducting S&T research. From the perspective of scientific research ethics, the S&T responsibility to taxpayers is emphasized moral responsibility other than the responsibility of abiding by laws. For example, S&T workers should respect the science spirit and pursue innovation when applying for financial funds for research. In the use of financial funds, it requires using the limited funds to create more achievements, value, that is improving the efficiency and efficiency of the expenditure of funds.

On the basis of abiding by laws and academic ethics, S&T workers should be responsible for their most fundamental 'contributor' — the taxpayers, when carrying out S&T activities with financial funds. Some behaviors of S&T workers are beyond the scope of legal responsibility and moral responsibility in general. It can be considered as the responsible behaviors to taxpayers took by S&T workers. Such as publicizing the usage details of research findings, explaining the significance of the research achievements in a more general language, participating in science popularization activities, promoting public understanding of science, and improving the public scientific literacy. When making suggestion to the government, S&T workers should represent the public interest. Under the current laws and social norms in China, there are no binding regulations of the duty to taxpayers that S&T workers must comply with. Therefore, the responsibility that S&T workers should be responsible to taxpayers is an extension of legal and scientific research moral responsibility which lays within the scope of social responsibility of S&T workers.

The most peripheral circle of social responsibility indicates that S&T workers should devote themselves to the sustainable development of human society, protect the ecological environment with S&T, and have the heart of maintaining world peace and promoting social progress. All the above responsibility is within the social responsibility of S&T workers.

2) The Corresponding Decomposition Model

Tthe second model is 'the Corresponding Decomposition Model' from the perspective of taxpayers' right. This model analyzes the relationship between the right of taxpayers and the social responsibility of S&T workers, and defines the responsibility of S&T workers to taxpayers. In addition to the procedural rights of taxpayers in the process of taxation, the logical extension of basic rights of taxpayers in legislation includes the right of consent, the right to know, the right of participation in tax legislation, and the right of supervising financial expenditure. From the perspective of social contract theory, taxation is essentially the purchase of public goods and public services by citizens through taxation [5]. The government budgetary expenditures are used to support S&T workers to take scientific theory research, experimental development, technology application, academic exchange, knowledge diffusion, and other related activities of S&T. It means S&T workers are ultimate grantee of taxpayers when they are conducting S&T related activities. S&T workers should disclose the possible influence of their behaviors, either positive or negative, to the public. According to this logic , the taxpayers rights to know, the right of consent, supervision, option and sharing are most closely related to the responsibility of S&T workers, as shown in Table 1.

Table 1 The taxpayers' obligation and S&T workers responsibility.

Rights of Taxpayers	Responsibilities of S&T Workers
The right of consent	Applying for S&T funds
The right to know	Usage details of funds, achievements released, social impact (science popularization)
The right of supervision	Usage details of funds, achievements application
The right of option	Policy-making consulting, offering advice and suggestions
The right of sharing	Achievements application

The right of consent is extended from the right of taxpayers' consent on taxation, not only for taxation but also for financial expenditure. In 1954, the Constitution of China established the power of deliberation and supervision of the budget by the National People's Congress of the People's Republic of China [6]. Governmental S&T appropriations, through a variety of distributions, finally become scientific research funds for S&T workers. Therefore, in accordance with the consent right of taxpayers, government departments such as the Department of Science and Technology which manage S&T funds should bear corresponding social responsibility, as well as S&T workers. In order to obtain the consent of taxpayers, government departments have the responsibility to explain the composition of budgeting, S&T workers also have the responsibility to explain the significance of research projects and the composition of project budget, as well as the S&T workers should be responsible for ensuring scientific research project to be beneficial to the national development, the public interests, and a fair and just in project reviews etc.

The right to know is developed by the right of informing about tax legislation. For many rights of taxpayers, the right to know is the most fundamental and can be regarded as the primary right. Furthermore, the right to know is the basis of the public participation right. Therefore, S&T workers are responsible for disclosing the details of the usage of research funds to taxpayers. It is legal right of taxpayers to know what they have achieved with their tax. It is proper for taxpayers to take part in scientific popularization activities and promote the improvement of citizens scientific literacy. It is important for taxpayers to understand the significance and social value of S&T activities so that they have sufficient information and knowledge to understand the role of S&T in social development.

The right of supervision is the derivative right of taxpayers in legislation and the derivation of the right to know as well. The right of supervision for financial expenditure has required establishing a supervision system under the present regulations and laws. In such an institutional environment, it is the responsibility for S&T workers to voluntarily disclose the details of S&T funds usage to the public, accept research project audits and public supervision. In addition, S&T workers should avoid improper S&T achievement application which may be harmful to human beings sustainable development or be contrary to social ethics. It means their behaviors in the application of scientific research achievements need to accept public supervision because S&T workers have the responsibility to ensure reasonable application of scientific research achievements.

For example, in the Standing Committee meeting of the twelve National People's Congress, the minister of the Department of Finance entrusted by the State Council reported the allocation and usage of financial S&T funds, which is a response to taxpayers' rights to know and the right of supervision. Besides taxpayers' right to know, the right of consent and supervision, the broad sense of taxpayers' rights also includes the right of sharing, the right of participating public affairs, the right of option and so on. For example, S&T workers who receive public financial support for their research are encouraged to give back to the public so that the public can benefit from their research. Through this way, the public can share the achievements of the S&T research.

3) The Process Analysis Model

The third analysis model of the social responsibility of S&T workers to taxpayers is the Process Analysis Model. This model analyzes whether they have the responsibility to taxpayers at a certain stage in the whole process of scientific research projects. In a narrow sense, scientific research project can be divided into the early stage that is project approval application, the middle stage that is systematic research, and the later stage includes project review, achievement release, and achievement transformation etc. From the perspective of social contract theory, as the contributor of S&T workers, taxpayers have the right to claim these rights. According to the time line, it is reasonable to analyze responsibility that S&T workers should take at each stage of scientific research projects.

In the early stage that is project application, S&T workers need to explain the reasons of the research meaning in both technical aspects and the social aspect, and accept peer review and supervision in order to get sponsors for S&T funds. At present, open project bidding has become a competitive way of S&T project application. Due to the high specialization of S&T development, the public actually do not have the capability to evaluate it. It is more likely to be reviewed by the science community. Therefore, it is necessary for S&T workers, both as individuals and as members of the science community, to assume social responsibility, and to explain to taxpayers why S&T workers need financial support.

In following process of researching, S&T workers use the funds to carry out research. During the period, S&T workers have the responsibility to disclose how to use project funds sponsored by finance. S&T workers should comply with the requirements of regulations and supervisions. It means they should use funds in reasonable and efficient ways.

After the study is over, S&T workers must go through such as project review, achievement appraisal, evaluation of awarding, etc. S&T workers could publicize the research achievements in the form of papers or patents. At this stage, S&T workers mainly need to explain to taxpayers what achievements have been made by research funds. They are responsible to taxpayers to let them know or understand about their research achievements. They have the obligation to accept project audit and supervision from both the government and the public.

When the research project is completed, some S&T achievements may be further transformed to business or industry applications. In a certain sense, the transformation of S&T project achievements to explore their socio-economic values is responsible for taxpayers. After taxpayers 'invest' in the research activities of S&T workers, they benefit from the advanced application of modern S&T. It conforms to the spirit of contract in modern society. Even for basic research, even though the research achievements may not be transformed into applicable products in the near future, the majority of the public has understood the great contribution and value of scientific research to the long-term development of human beings and the world. Therefore, S&T workers are also required to further explain the value and the influence of the achievements to society. In short, it is popularization of science.

4. Main Experience in Building Association for Science and Technology System

The taxpayers pay taxes and form the state or local government revenues which then are allocated to S&T workers to conduct S&T activities. From the most fundamental relationship between these two aspects, the right of the taxpayer's consent, the right to know and the right of supervision are most closely related to S&T workers' responsibility to taxpayers. This paper puts forward three analytical models, trying to deconstruct the responsibility of S&T workers to taxpayers. Each analytical model has its own characteristics, and the responsibility of S&T workers to taxpayers also change and develop with the times. Neither the three models of the CON Model, the Corresponding Decomposition Model, and the Process Analysis Model can cover all the responsibility of S&T workers to taxpayers. The CON Model points out the relative status of the S&T workers' responsibility to taxpayers in a wider concept of the S&T workers' social responsibility. However, it has some disadvantages, such as the boundary of legal responsibility, moral responsibility and social responsibility are unclear. Some moral responsibility may be incorporated into the legal responsibility along with the social development. The Corresponding Decomposition Model tries to establish correspondences between the right of taxpayers and the responsibility S&T workers. The Process Analysis Model reveals the details of S&T workers' responsibility in each period of a research project. Therefore, there is no a analytical model can fully cover all kinds of responsibilities. S&T workers' responsibility to taxpayers is changed as well, and the purpose of the study is not to exhaust all, but to put forward a theoretical framework and a research perspective to analyze the issue of S&T workers' responsibility to taxpayers.

References

[1] J. D. Bernal, *Science and Human Welfare*[J]. Science and Society, 1956,20(2): 97-110.

[2] Ministry of Finance of the People's Republic of China, *2017 fiscal revenue and expenditure situation*[N/OL]. http://gks.mof.gov.cn/zhengfuxinxi/tongjishuju/201801/t20180125_2800116.html. [2018-08-30]

[3] Development Research Center of China Association for Science and Technology. *Social Responsibility of Science Workers*[M]. Bejing: The Science Publishing Company, 2009.

[4] Geva, A. Three models of corporate social responsibility: Interrelationships between theory, research and practice [J] .Business and Society Review, 2008 , 113(1):1 -41.

[5] Shuying LI. *Corporate Social Responsibility in the Perspective of Social Contract Theory* [J]. Journal of Renmin University of China, 2007, 21 (2): 51-57. (in Chinese)

[6] Wang wenyan. Preliminary exploration of the operating mechanism of government scientific research funds [M]. Beijing: China Economic Press, 2005. (in Chinese)

Climate Change Communication as Political Agenda and Voters' Behavior

Muhammad Azfar Anwar[1], Rongting Zhou[2], Fahad Asmi[2], Aqsa Sajjad[3]

[1] School of Public Affairs, University of Science and Technology of China, Hefei, China
[2] Dept. of Science and Technology Communication and Policy, University of Science and Technology of China, Hefei, China
[3] History of Science and Technology, University of Science and Technology of China, Hefei, China

Abstract: 'Climate Change Communication' is taking the strategic position in the international and national politics around the globe. In the recent decade, different developing nations have started considering 'Climate Change Communication' as an integral part of the political campaigns and social development. Specifically, the current document comprised of two sections. In the first section of the document, authors briefly compared the attributes related to 'Climate Change Communication' in the top leading political parties' manifesto for the general election 2018 – Pakistan, in a qualitative manner. In the second part, the difference of opinion among voters of different political parties towards 'Climate Change' examined. In a birds eye view, the perceived seriousness of 'Climate Change' as a real challenge among voters mapped by the independent factors of 'Urbanization' 'Industrialization' 'Transportation' and 'Waste management' through the primary quantitative survey of 732 voters in the country (Pakistan). The finding highlight (1) youths' understanding of 'socio-scientific issues,' i.e., Climate Change is easy to communicate, (2) the smart use of media by political parties in framing and communication about 'Climate Change' plays a significant role in Climate Change Communication (Public Understanding of Science). The current study concludes that 'Public Understanding of Science' holds a critical role in developing regions' future political dynamics.

Keywords: Politics; Climate Change Communication; Media; Youth; Pakistan; Socio-Scientific Issue

1. Introduction

People deal with risks all the time, and the perception of those risks do not rely on facts or their values of a worldview but framing a risk indicates the ultimate actions. Climate change affects the efforts to achieve developmental goals and future sustainable objectives (Biagini et al., 2014). Climate change action consists of mitigation and adaptation. Mitigation indicates causes of global warming where adaptation indicates the impacts of climate change on society and the attached activities (Sussex, 2006). Corner et al. (2010) outline various recommendations about public communication to increase climate change concerns and to stimulate required behavioral changes. Environmental and cultural settings play an important role in shaping public perception towards global warming risks (Akerlof et al. 2013). While it is inevitable to exclude socio-economic changes while assessing climate change impacts. These changes can magnify or lessen the climate change consequences (Abildtrup et al., 2006).

Brulle, Carmichael and Jenkins (2012) Indicate that the effect of economic conditions and elite cues is much larger than the scientific risk communication when it comes to policy preferences. The study concludes that the information based science advocacy have little impact on public climate concern where political mobilization led by elites can generate high concerns about climate change. Different political systems process the same scientific data on climate risks and opt mitigation options differently. There is a need to know more about the conditions which influence the political processing of scientific data and the behavioral outcomes (Bernauer, 2013). Weber and Stern (2011) explain the public and scientific understanding of climate change. Understanding can influence the concerns and risk management decisions, whereas the effect of education and information on other various actions is limited. Lack of understanding or concern is not a constraint but the politics of climate change.

Ockwell et al. (2009) indicate that climate communication to influence only attitudinal change seems to be ineffective. The individual attitude and resulting behavior can be arbitrated by societal norms and free riding. The study suggests that communication strategies with political and psychological factors can trigger the demand for climate regulations by strengthening the grass root acceptance. Strong leadership along with public acceptance of regulations could solve various restrains attached to possible actions. Carvalho, van Wessel and Maeseele (2017) Investigate the public engagement on climate change. They indicate that communication practices not only help create the conditions for political engagement, but they also comprise the modes of such engagement. Ryan and Ramirez (2016) Find that climate agenda can be developed by linking it to local issues. Political and social support for climate change policies can help to sustainable policy implementation even beyond the electoral cycles and government changes. Climate change has not yet got salience in electoral competition between political parties.

The study will qualitatively analyze the manifestos of main political parties in Pakistan to investigate the climate change issue on their political agenda. The investigation will provide the real insight into the politicization of climate change and its implications for industry, urbanization, waste and transportation by these political parties. The study will strive to explore the strategic position these parties are taking on climate change and how these efforts are sensitized by political parties and their leadership.

Pakistan is the world's sixth largest populous country, and the population has been projected to 100 million by 2050 which will cause great stress on its resources. With ambitious economic goals, carbon emission for Pakistan is expected to increase by 300 percent in the next 15 years with the high demand for energy and transportation. By Paris climate accord, Pakistan has pledged to reduce emissions by 2030, but the government has still to balance this target with its economic development goals. Asian Development Bank has alarmed the situation by indicating temperature rise particularly in northern areas with more water stress as less per capita water will be available in coming years. This would indicate great challenge as high temperature cause more melting of glaciers and high evaporation rates leaving the less irrigated water and low yield of certain crops (Qamar, 2017).

At a global level, more than 524,000 people have died as a direct result of over 11,000 extreme weather events and economic losses of 3.16 trillion dollars between 1997 and 2016. Pakistan has been ranked at 7th position, with 523.1 people died per year a total of 10,462 deaths during 20 years and economic losses of 3.8 billion dollars (0.605 percent of the GDP in the 20-years period). During this time slice, Pakistan had suffered from 141 extreme weather events like cyclones, storms, floods, Glacial Lake Outburst Floods (GLOFs) and heatwaves. Among these floods and heavy rains have badly affected the livelihood of people by compromising the growth targets. Pakistan is ranked 4th regarding property damage, and the major impact to these damages come from the 2010 floods. Along with this, Pakistan has suffered from prolonged droughts (1998–2002, 2014–17), heat waves (2011, 2014), the 2014 cyclone Nilofar, and GLOF events (Sönke et al., 2015).

Addressing to these grave climate change challenges Pakistan has entered in the list of countries which have passed legislation on climate change with its Pakistan climate change act 2017. Despite investing more than 8 percent of its GDP on climate change mitigation and adaptation, Pakistan suffered from the most severe implications related to it. These limitations call for the research on the political will to implement the climate policies, and along with the politics, it is necessary to analyze the behavior of the voters to whom these policies are beneficial. Pakistan is the only nation to feel that life has worsened in seven Asian countries surveyed by Climate Asia. Pakistanis are also the people in South Asia to feel most strongly that climate change includes extreme weather conditions, variation in rainfall and rising temperatures which have a huge impact on their wellbeing. Health worries are common across communities while other primary concerns differ according to location. People in Pakistanis have the lowest confidence in the government to tackle climate change and perceive that socioeconomic differences are increasing due to government failure on this issue. They are taking actions according to the information they receive on climate change by working with communities and get support from NGOs. Many people feel helpless with no access to the necessary resources. 65% of people do not know that what 'climate change' means (Qualitative, n.d.). The climate change communication in Pakistan should focus on motivating those who are taking actions to respond more and encourage those are struggling to act. These communication practices will mediate the government and citizens to interact more on this issue. Media and interpersonal communication are both necessary as people get more information from cable in an urban setting and in the rural setting they trust their communities.

The literature on Pakistan's climate shows the impact of transportation, industry, urbanization and waste at provincial and area level with limited information of on their politicization. To enhance public engagement for adaptation against the climate change requires the strong role of government in promoting and supporting of these activities. The local leadership is effective in implementing and monitoring of climate policies with the participation of locals in the process. Deep understanding of local climate knowledge of public is necessary to form effective mitigation and adaptation, which could be achieved through political and social surveys at the local level. This study is field based research that will provide quantitative insight of understanding of climate change, trust in politics, attitude and pro-environmental behavior of voters. The study finally discussed the qualitative analysis of manifestos and quantitative analysis of the pro-environmental behavior of the voters to provide a comprehensive outline of climate change communication as political agenda and voters' perception.

2. Literature Review

The existing literature informs us that public engagement shares a vital part in national climate mitigation policies (Cara, 2010). Changing public behavior to engage them in emission control and other preparedness activities is important to address climate change. The tools developed by scholars and other research institutions get legitimation with their integration in policy, and this competition does not imitate the difficulties in design and communication as knowledge brokerage strategies are not often tested empirically (Adelle, 2015). Knowledge co-production induces social learning in adaptation and co-management reproduces governance arrangements at various levels to address the particular problem (Armitage et al., 2011). Scientific knowledge presents limitations when it comes to addressing climate change as it involves public ideology, so communication strategies should assume societal approaches to motivate pro-climate behaviors (Bain et al., 2012). The different characteristics, i.e. what motivates national climate policy; how it is supported by technical and scientific knowledge; communicating the climate change messages, integration and co-ordination at multi-level for implementation are challenges for climate change adaptation strategies (Biesbroek et al., 2010). It is thus required to inform the public with detailed knowledge for climate change to understand the problems and consequences (Bord et al., 2000). Communicating the risks associated with the real problem can enhance public awareness which is necessary for their engagement and effective policy implementation to address the risk (Boudet et al., 2014).

'Media coverage of climate change matters' as it is convenient to motivate public action in the politicized environment (Boykoff & Boykoff, 2007). The cultural context of journalists can influence media coverage of climate change problems and thus demands cross-cultural research in media studies (Brossard et al., 2004). Climate change communication is evolving as values, personal priorities and relationships are being investigated by research along with their practical implication (Environment, 2012). Climate change messages from science educators, communicators and journalists have failed to convey the seriousness of the problem and should find innovative methods for better communication (Corbett, 2004). Addressing Climate change needs specific behavioral changes and support for environmental policies and social marketing which has shown limitations with public engagement (Corner and Randall, 2011). The climate change communication can encourage public attitudes, intentions and behaviors for related climate change consequences (Dickinson et al., 2013). Media coverage, information access, and education are strong predictors of climate change knowledge where expectations and norms are inversely associated (Kahlor and Rosenthal, 2009). Media and public communication can play a vital role to find sustainable solutions to the societal phenomenon (Nisbet & Scheufele, 2009). Boykoff and Boykoff (2004) express that press reporting in the US contributes to deviation of popular discourse from scientific discourse on global warming by highlighting the discrepancies between press reporting and scientific discourse. Antilla (2005) explores the US newspaper representation of scientific knowledge on climate change and indicates that newspaper reports have been succeeded in maintaining confusion among the public by enduring the myth of a lack of scientific consensus on human causes of climate change.

Public polarization along with different news cables indicate major implications about how they comprehend climate change (Feldman et al., 2012). Climate change appears to be negatively associated with anxieties and policies to mitigate these changes, but effective climate change communication can change public perception (Jang, 2013). Public engagement at cognitive, affective and behavioral levels is important factors for effective climate change mitigation. Targeted and tailored information provision to initiate a systematic shift to get public out of their carbon consumption comfort zone (Lorenzoni et al., 2007). There should be a change in policies to achieve public engagement by reframing the climate change communication. These decisions are too important to leave just in the hands of elected representatives as it has a huge effect on life style of citizens and thus demands active public participation (Nisbet, 2009). Fear induced depiction of climate change could be counterproductive as far as public engagement is concerned. The later could be achieved by incorporating daily life concerns in climate change communication rather than just realizing them helpless with fearful messages and images (O'Neill & Nicholson-Cole, 2009). Climate change is more than a technical issue. The difference over climate change reveals the deeper levels of various attitude towards risk; different political believes, ethical concerns, ideological fronts, and technology along with their effects on human wellbeing (Hulme, 2009). Spence et al. (2011) indicate that low public participation to mitigate climate impact is due to lack of personal experience. Risk perception about environmental changes must be linked with individuals' surroundings and real-life experience. Carvalho and Burgess (2005) call for a cultural view of assessing risk perception associated with climate change. There is a dire need to handle the social dynamics involved in communication research for current and future actions to moderate the risks of climate change. Whitmarsh (2009) explores the public understanding of climate change focusing on unprompted knowledge and believes. Public understanding of climate change and global warming is positively associated with their concerns, experience and prevailing knowledge.

The core factors in climate change communication are identified as 'purpose and scope of the communication, audience, framing, messages, messengers, modes and channels of communication, and assessing the outcomes and effectiveness of a communication' to yield important insight and engagement (Moser, 2010). Political division in American public on the issues on global warming is caused by the politicization of climate change. Furthermore, the political messages addressing the concerns over global warming are contributing to growing this divide (Mccright & Dunlap, 2011). Public risk perception is an important determinant to address any political, economic or social action in response to particular risk. Public risk perception in America is strongly associated with experiential factors, and the same is true for policy support (Leiserowitz, 2006). Lewandowsky et al. (2013) argue that the American public's attitudes have become more polarized towards science particularly climate change. Results indicate that conservatism and free-market worldview strongly predict denial of climate science. The public indifferences over climate change can be ascribed to comprehension deficit. There are indications that climate change has become more politicized and public differences on climate change do not originate from their disbelief on science (Kahan et al., 2012). Hart and Nisbet (2012) incorporate motivated reasoning theory, the theory of social identity and persuasion theory to detect increased public polarization on climate change and indicate political partisanship as a boosting factor for positive political polarization on climate change. Carvalho (2007) investigates informal depictions of scientific knowledge about climate change to illustrate a set of ideas and values necessary to legitimate a program of action. The representation of science influences political programs and has consequences for assessing the responsibility of governments and the public in addressing climate change.

The Table 1 presents the top ten highly cited articles in the domain of climate change communication, and they have accumulated the insight from perception to framing; from fear to climate scepticism; from communication to polarization; from risk analysis to polarized science literacy and representation of scientific knowledge. Which will pave the way for the empirical analysis of the present study to analyze the effect of knowledge, social values and political trust on environmental behavior mediated through attitude and source credibility of information.

Table 1 Top cited papers in 'Climate Change Communication'.

Author	Frequency	Title
Lorenzoni et al. (2007)	87	Barriers perceived to engaging with climate change among the UK public and their policy implications
Nisbet (2009)	81	Communicating climate change: why frames matter for public engagement
O'Neill & Nicholson-Cole (2009)	77	Fear won't do it: promoting positive engagement with climate change through visual and iconic representation
Hulme (2009)	76	Why we disagree about climate change
Moser (2010)	68	Communicating climate change: history, challenges, processes, and future directions
Mccright & Dunlap (2011)	64	The politicization of climate change and polarization in the American public's views of global warming, 2001–2010
Leiserowitz (2006)	57	Climate change risk perception and policy preferences: the role of effect, imagery, and values
Kahan et al. (2012)	55	The polarizing impact of science literacy and numeracy on perceived climate change risks

3. Methodology

To examine the prime objective, the quantitative data in the form of 'Manifesto' for the general election 2018 in Pakistan taken into consideration. Specifically, the manifesto of the country's leading three parties taken to examine 'keywords' trends. In other words, the first sub-section of analysis comprises the descriptive analysis by keywords used related to 'Climate Change' / 'Climate Change Communication'.

In the quantitative section, the structured questionnaire survey conducted across the rural and urban regions of Lahore, Karachi, and Peshawar. The study was purposefully driven to understand the perceived behavior of citizens towards 'Climate Change' / 'Climate Change Communication.' This sub-section of the research elaborates the instrument used sample-related issues and analysis procedure.

1) Instrument

To make the findings credible and valid for further study, the quantitative survey adapted from the existing pool of literature. Specifically, the list of each of the items adopted in the current study is shown in table 2 below. The three items scale for 'Knowledge about Climate Change' (KCC) adapted from Vainio and Paloniemi (2013), whereas the three items scale for 'Trust in Politics' (TP) adapted from Pavlou, Tan and Gefen (2003). 'Social Value' (SV) as a phenomenon, adapted different flavors as it's been studied in different social and behavioral environment, in the current study the three items scale adapted by Hong, Tam and Hong (2006) to understand the 'Social Value' of Climate Change as SSI. Moreover, the instruments for 'Source Credibility' (SC) and 'Attitude towards Climate Change' (ACC) adapted from Bhattacherjee (2016) and Venkatesh (2000) respectively. All constructs measured o the five Likert scales with the response ranged from the strongly Disagree to Agree.

Table 2 The instrument of scale adapted in the current study.

Construct	Instrument of Scale	Adapted Source
Trust in Politics (TP)	1. Politicians and their views are generally reliable. 2. Politicians and their views are generally honest. 3. Politicians and their views are generally trustworthy.	Pavlou et al., 2003
Knowledge about 'Climate Change' (KCC)	1. I think that I am well informed about the causes of 'Climate Change.' 2. I think that I am well informed about the consequences of 'Climate Change.' 3. I think that I am well informed about ways to fight with 'Climate Change.'	Vainio & Paloniemi, 2013
Social Value (SV)	1. People who are important to me would want me to understand Climate Change. 2. People who influence my behavior would think I should understand about Climate Change. 3. People whose opinions I value would prefer me to understand about Climate Change.	Hong et al., 2006

(Continued)

Construct	Instrument of Scale	Adapted source
Source Credibility (SC)	1. The political party's representatives providing the information was knowledgeable about Climate Change. 2. The political party's representatives providing the information was trustworthy about Climate Change. 3. The political party's representatives providing the information was credible about Climate Change.	Bhattacherjee & Clive Sanford, 2016
Attitude towards 'Climate Change' (ACC)	1. To understand climate change is a good idea. 2. To understand 'Climate Change' as Socio Scientific Issue (SSI) is interesting. 3. I like to understand 'Climate Change' as Socio Scientific Issue (SSI).	Venkatesh, 2000
Pro-Environment Behavior (PEB)	1. I do prefer to reduce 'Energy Consumption' and to increase its efficient use. 2. I do prefer to use an efficient way of transport, instead of individual transport. 3. I prefer to separate the waste for recycling	Vainio & Paloniemi, 2013

2) Data Sampling and Collection

During the first phase (manifesto analysis), the web resource of each of the national political party used to get access to the manifesto. To assure the data readability of the files 'manifesto', the format conversion is also performed to make the raw data available for 'text analysis' in a descriptive manner.

However, to achieve the objectives from the quantitative survey, the traditional and online data collection source adopted. In the pre-test examination of the construct and the questionnaire, the 25 individuals (university students) participated to review the constructs' items. Moreover, the construct circulated in the local language as most of the targeted population feel convenient to respond in a familiar language. The modified and revised questionnaire circulated during the first half of 2018 among the political and social workers in the focused political parties. Only 217 responses in total collected and considered for the quantitative survey in the current study. However, the considerable 515 respondents (registered voters) collected from the social networking sites, and online survey support. In other words, only 57.2% online response observed in the current research as initially 900 respondents (registered voters) circulated through an electronic medium.

The demographic profile of the respondents (registered voters) can be seen in Table 3 below. Specifically, 95% of the respondents (registered voters) are younger than the age of 35. However, the almost equal proportion of both genders can be observed. Nearly 57% of the respondents (registered voters) in the collected survey are from the urban region of the country. Interestingly, the elected party in the general election of 2018 observed as dominating in the collected sample

Table 3 Respondents (registered voters) profile from the collected survey.

Demographics		Political Affiliation			Total
		PTI (357)	*PML-N (177)*	*PPPP (198)*	732
Gender	Male	141	104	125	370
	Female	216	73	73	362
Age group	Under 25	322	131	137	590
	25 ~ 35	31	35	44	110
	Above 35	4	11	7	32
Geographic	Urban	208	95	117	420
	Rural	149	82	81	312
Most preferred medium to follow political campaigns (among mentioned four channels)	SNS (electronic media)	209	93	105	407
	Television	82	54	44	180
	Print Media	39	19	35	93
	Local Representative	27	11	14	52
Most important issue related to Climate Change (among mentioned four factors)	Urbanization	154	97	117	368
	Transportation	162	55	54	271
	Industry	09	11	16	36
	Waste	32	14	11	57
Which institution(s) is/are the most influential while dealing with Climate Change as SSI. (among mentioned four bodies)	International Agreements	67	52	54	173
	Environment Organizations	71	27	22	120
	Government	56	14	30	100
	Citizens	15	13	12	40
	All of Above	148	71	80	299
What kind of impact of 'Climate Change' is observable in your town? (Response type: Yes, or No)	Increase in Temperature	356	172	195	723/732
	Seasonal / Rain Shift	334	159	182	675/732
	Human Health Issue	347	171	189	707/732
	Life Style Change	340	170	187	697/732

as well. Moreover, it can be concluded that the youth's behavior can be predicted as the strong determinant of the public response in the country towards 'Climate Change' / 'Climate Change Communication'.

Through the descriptive findings from the current study, a few of the interesting patterns are observed. (1) Social Networking Sites are getting serious attention by the citizens while creating any perception about the performance of any political campaign, as 55% of the respondents (registered voters) believe the significant role of Social media while image building. It is further followed by 'Television,' Print Media' and 'Local Representatives respectively. (2) The current study contribute unique value. For example the respondents (registered voters) believe that the 'Urbanization' and 'Transportation' are a far more responsible factor for 'Climate Change' as compare to the other factors (Industrialization and Waste Management). It provides an interesting and potentially valued implication for the current study. (3) The citizens' collective response about the responsibility and the need of initiative expects the involvement of all resourceful institutions to collaborate and provide best possible facilitating conditions for citizens to cope with 'Climate Change' as 40% of the respondents believe the collective initiative as the possible solution of 'Climate Change' in the current era of time. (4) Alarmingly, almost all of the respondents believe that the climate change is real as it accounting for the increase in the overall temperature, the seasonal shift in rain, life style, and the increase complex health issues in the society. In the tabulation form, the descriptive results from the respondents are shown in table 3.

4. Analysis

Conceptually, analysis part comprises the qualitative analysis of the manifesto of top three national political parties where specifically, the primary and secondary category of words examined separately in the analysis as shown in Table 4 below. In the quantitative section, citizens' (voters') opinion about climate change communication is presented.

1) Qualitative Analysis (Manifesto Analysis)

The manifesto of political parties allows the voters to position themselves in certain parties by exploring the issues on their political agenda. The issues communicating through the manifesto give the opportunity to comprehend the political strategies a particular party opts to address the issues on board merely the politicization of the certain issue and how political parties address those issues. The underlying analysis indicates the politicization of climate change and the strategic positions taken by political parties. The analysis will focus on three mainstream national political parties of Pakistan namely Pakistan Tehreek e Insaf (PTI), Pakistan Muslim League-Nawaz (PML-N) and Pakistan People's Party Parliamentarian (PPPP). The study will analyze their manifestos for the 2018 general elections, held on 25th July. The analysis is performed at three stages with Keywords, primary words and secondary words respectively. The three-stage analysis counts the numbers by which the keywords, primary words, and secondary words are being repeated in the manifesto. The assumption behind this frequency count shows the issue salience in manifesto and indicates the seriousness a certain party devotes to treat that issue (N. Carter, 2013). The prominence of climate change in manifesto by political parties of Pakistan reflects its importance as political agenda to contest and compete for the election.

Table 4 Type of keywords analyzed qualitatively.

Category	Type of Keywords	Keywords (Items)
Category 1	Keywords	Climate Change, Environment, Global Warming, Carbon
Category 2	Primary Keywords	Urban, Industry, Transportation, Waste
Category 3	Secondary Keywords	Pollution, Rural, Agriculture, Green, Forestation

(1) *Keywords (Category 1)*

The analysis reveals that the term 'climate change' has been used 15 times in the PTI manifesto, 7 times in PMLN manifesto and 14 times in the PPP manifesto. The term 'environment' has been used 9 times by PTI, 8 times by PMLN and 4 times by PPP. 'global warming' appears one time in PMLN, two times in PPP and not mentioned by PTI. The term 'carbon' is used 0 times by PTI, 5 times by PMLN and one time by PPP as shown in Fig. 1 below. The party view as mentioned in the manifesto is discussed party wise to understand their relevance as political agenda. PPPP has claimed to sensitize its members to climate change and environmental sustainability as it affects urban and rural settings of Pakistan. PPPP envision climate change as an instrument for its foreign policy to build South-South coalition for better policy stance at global climate change regime. It puts climate change and environmental projects at top priority agenda for its government. They are aware of the current trends research as the manifesto says that Pakistan emits less than 1 percent of global greenhouse gas but still ranks at 7th position in climate risk index. They also indicate that South Asia is facing many climate change challenges like water scarcity, flash floods, and smog and they believe that Pakistan and India can cooperate to improve environmental conditions in the region. PPPP sees ecological conditions as serious as other national security issues like protection from violence, poverty, hunger, and unemployment. The manifesto indicates the party's comprehension over the growing climate change-related threats to Pakistan's economy and society and call for comprehensive plans to safeguard the socioeconomic infrastructure for sustainability.

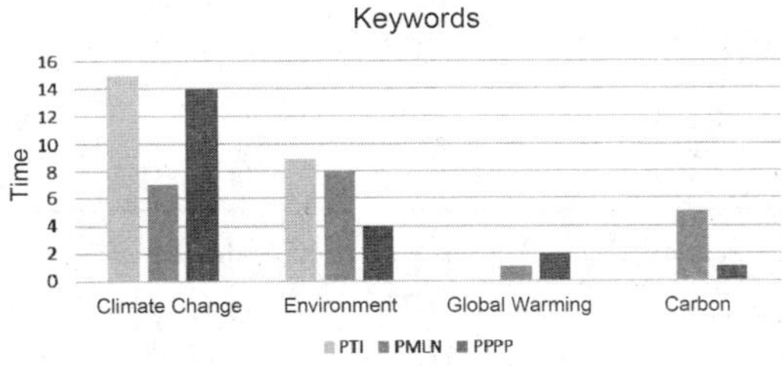

Fig. 1 Keywords (Category 1) in each manifesto.

PTI looks the priority issues facing Pakistan are increasing farm input costs, monopolized agriculture markets, low access to finance, ill-working of institutions, degrading ecosystem by climate change and low export potential of agriculture. PTI acknowledge Pakistan as one of the most water-intensive economies in the world, and the situation is getting worse with climate change, population growth and water losses in agriculture along with regional water disputes. PTI is determined to invest more in research related to climate change adaptation and disaster relief management. PTI is also aware of the fact that Pakistan ranks 7th among the most climate change vulnerable countries and its economy bears a loss of 3 billion dollars annually by air and water pollution thus calls for a concrete action plan to safeguard the economy and livelihood of the people by reversing the environmental degradation.

Nations all over the globe are engaged in a race against time to cope with climate change consequences, to improve environmental degradation and to achieve their long-term economic development. PMLN envisions a greener Pakistan secure from adverse effects of climate change with a commitment to protect not only its industry, agriculture and financial goals in the short term but a sustainable environment to our generations to come. PMLN is also aware of the fact that Pakistan being the least contributor to global warming yet ranks among the most vulnerable countries to climate change impacts. PMLN strives to present a balanced portfolio of green energy and local resources to assure sustainability and reduced economic burden caused by environmental degradation. PMLN will plan to encourage mass participation in environmental conservation programs through data-based advocacy.

(2) *Primary Keywords (Category 2)*

The term 'Urban' has been used 11 times by PTI, 8 times by PMLN and 24 times by PPPP. The term 'Industry' has been used 11 times by PTI, 19 times by PMLN and 11 times by PPP. 'Transportation' has been used 3 times by PTI, 13 times by PMLN and 1 time by PPPP. The term 'Waste' has been used 4 times by PTI, 2 times by PMLN and 2 times by PPP as shown in Fig. 2 below.

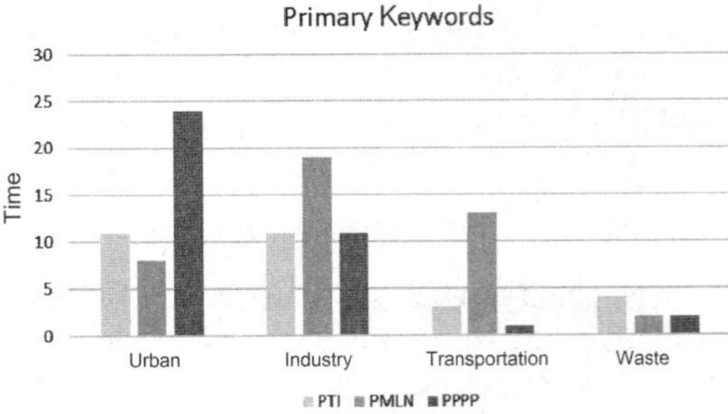

Fig. 2 Keywords (Category 2) in each manifesto.

PPPP will be Introducing energy standards and incentives to improve energy efficiency in buildings, industries, and transportation sector and protect them through legislation. PPPP is ready to adopt the best practices for sewage treatment to reduce the demand for potable water by industry and agriculture. It will also develop standards and regulations for sewage disposal. PPPP will encourage major investments in sanitation infrastructure in urban and rural communities. PPPP will introduce deeper reforms in agriculture, trade, and industry, natural resource management, and energy which will form the basis for rebuilding our economy. PPPP is concerned for wastage of energy will be curtailed at all levels by increasing public awareness and encouraging the private sector to manufacture energy efficient appliances. PPPP envisages strong potential in this emerging industry and believes a significant economic contribution by creating employment in the Country. PPPP will regulate hospital waste management and will stretch waste disposal services in rural and urban communities.

PTI plans to integrate true environmental costs of projects in economic decisions and to develop an 'Eco budget.' PTI would look for activities protected by legislation to initiate mandatory ecological education, green building codes, zero waste policy and reducing plastic use in Pakistan. PTI will ensure that Pakistan as a nation takes charge to improve ecological conditions and manage climate change to protect its citizens, natural endowments and economic investments. Affordable and sustainable clean energy provision is a priority, and PTI will promote clean transport, support green infrastructure, effective waste management and build efficient agriculture sector to protect biodiversity. PTI will lift local economies with strategic investments in tourism, minerals, and renewable energy. There is a dire need to shift towards greener and sustainable energy alternatives such solar, wind or hydropower. PTI will provide safe drinking water in urban slums and rural communities. PTI will expand rainwater catchment capacity in urban and rural areas. PTI will promote efficient and clean transport with integrated warehousing facilities, a brief view of extended keywords is shown in Fig. 3.

PMLN has formulated effective policies to mitigate environmental degradation like 1,500 MWs renewable energy projects, national Forests Policy, EURO II fuel standards, less duty on electric vehicles, Green Pakistan Program in 2016. PMLN aims to build resilience against climate change through initiatives like afforestation and conservation of biodiversity. PMLN will expand the scope of environmental conservation efforts by promoting green industry growth; agriculture is resilient against climate change, green energy, cutting carbon emission by 10 percent, knowledge sharing for innovation, enforcement of emission targets and development of climate change-related technologies. PMLN will establish clean energy fund to provide incentives for clean technologies. PMLN will build and connect renewable energy zones to grid nodes. PMLN will strengthen the linkages between industry, research, and academia with the provision of grants for applied research projects. PMLN aims to reduce the carbon footprint by promoting green energy (wind, solar, biogas) from 0.5% to 5%. PMLN will provide safe public transport schemes for women, free transport in ICT federal schools, built international standard urban transport systems throughout the country. Transportation plans further include the provision of finest public transport in 25 major cities and ensure no citizen is 500 meters away from a bus stop and does not wait for more than 10 minutes. PMLN will strive to employ the best available technology to recycle municipal and industrial waste and water sewage. As the energy has been discussed along with primary words, it is necessary to present their frequency as well to show the concern of parties to use clean and renewable energy for these sectors.

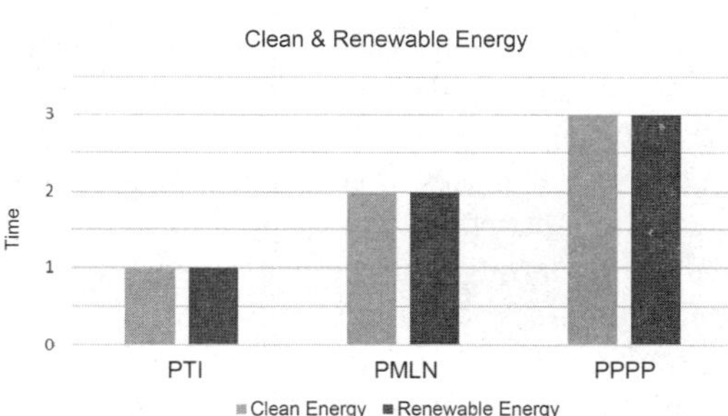

Fig. 3 Keywords (Category 2-extended) in each manifesto.

(3) *Secondary keywords (Category 3)*

The term 'Rural' has been used 8 times by PTI, eight times by PMLN and 25 times by PPPP. The term 'agriculture' has been used 13 times by PTI, 26 times by PMLN and 14 times by PPPP. PTI is the only party to mention pollution in the manifesto for 5 times. 'Green' has been used 11 times by PTI and 14 times by PMLN. 'Forestation' has been used 2 times by PTI, 5 times by PMLN and 2 times by PPPP as shown in Fig. 4 below. PPPP sees price support system as the most significant policy intervention in agriculture and aiming to provide assured income to wheat farmers. Recent trends in

Climate change indicate that Pakistan may run dry by 2025 as there exists a 50 percent gap between demand and supply. Climate change, water scarcity, and periodic floods have critically challenged the development and population wellbeing. Focuses afforestation including social forestry on more than 500,000 has to act as carbon sinks. PPPP supports efficient irrigation systems in the agriculture sector, but also indicate them as a valuable resource for the functioning of industry and markets and promises to deploy drip/sprinkler irrigation systems on more than 4 million hectares of agriculture farmlands by 2023 if elected. PPPP will focus on promoting climate SMART agriculture approach in agricultural policy, planning, and research.

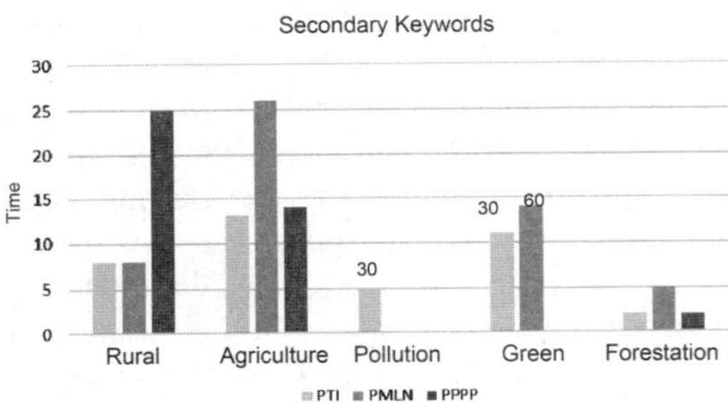

Fig. 4 Keywords (Category 3) in each manifesto.

PTI aims to take serious actions by setting a green growth agenda to tackle climate change and investing in long-term climate change plans to address real causes of environmental pollution for better adaptation and mitigation. This is achievable through strengthening institutions, 10 billion tree plantation in 5 years, and by improving our disaster relief management and risk reduction. PTI aims to provide decent 'green' jobs and equip our country to face challenges of climate change and environmental pollution. PTI endeavors for rural electrification through renewable and off-grid solutions. PTI will build 1.5 to 2 million urban and 3 to 3.5 million rural housing units for adequate living. PTI will reduce water losses in Agriculture by endorsing best practices, smart interventions and better monitoring of farm lands. This includes incentives for farmers to conserve water, adoption of regenerative agriculture and market-driven crop mix. PTI will promote the use of wasteland for fish farming.

PMLN with its Green Pakistan initiative will promote urban forestry by ensuring the provision of state land for plantation. It will be accompanied by strengthening the Forests Department to monitor better implementation of afforestation efforts. Pakistan needs a speedy and effective response towards a green economy is PMLN manifesto for 2018–2023 to manage the dangers posed by climate change and to reverse the environmental degradation in the previous decades. PMLN intends to invest in rainwater harvesting and to encourage the use of drainage water in agriculture. Improved rural transportation, ambulance services, financial sector assessments and complete electrification of rural areas. Promote the use of smart technologies by staff to supervise health, education, agriculture, livestock, and irrigation fields. PMLN focuses on farmer education in ICT and credit access to them to modernize agriculture.

2) Quantitative Analysis (Survey Analysis)

During the statistical analysis of the quantified statistical survey, the Structural Equation Modeling adopted with the support of SPSS-AMOS (v.21). AMOS helps to perform Confirmatory Factor Analysis (CFA) and the path analysis of the proposed model, to understand the significance and importance of the proposed hypothetical relations in the current study. In the following subsections, the measurement model (EFA) and structural model will be observed.

(1) *Measurement model*

In the first stage of the measurement model, the validity of the data and the reliability of the relationships proposed in the current study taken under the examination. Initially, the KMO (Kaiser Meyer Olkin) observed as 0.866 which is considered to claim sample adequacy as suggested value supposed to be higher than 0.70 (Hair, Black, Babin & Anderson, 2014). To examine the internal reliability of each of the construct, the Cronbach Alpha (CA), Composite Reliability (CR) and Average variance extracted (AVE) also measured as shown in table 5 below. Statistically, the computed value of CR and CA supposed to be above 0.70 (Hair, Black, Babin & Anderson, 2010) as followed in the current study. Moreover, the recommended value of AVE is observed in the recommended value of 0.50 (Hair et al., 2014). The factor loadings of each of the item of all constructs are also noted above 0.70 as recommended by the previous pool of quantitative behavioral studies (L. Carter & Bélanger, 2005).

Table 5 Factor analysis and internal reliability examination.

Construct	Items	FL	CA	AVE	CR
Trust in Politics (TP)	TP1	0.842	0.836	0.681	0.884
	TP2	0.831			
	TP3	0.802			
Knowledge about Climate Change (KCC)	KCC1	0.825	0.787	0.660	0.853
	KCC2	0.815			
	KCC3	0.797			
Social Value (SV)	SV1	0.864	0.913	0.710	0.880
	SV2	0.834			
	SV3	0.830			
Source Credibility (SC)	SC1	0.846	0.872	0.680	0.864
	SC2	0.838			
	SC3	0.789			
Attitude towards Climate Change (ACC)	ACC1	0.806	0.828	0.606	0.822
	ACC2	0.798			
	ACC3	0.729			
Pro-Environment Behavior (PEB)	PEB1	0.870	0.853	0.666	0.856
	PEB2	0.829			
	PEB3	0.745			

FL= Factor Loading, CA= Cronbach Alpha, CR= Composite Reliability, AVE= Average Variance Extracted

For the discriminate validity of the current study, the square root of each of the construct's 'Average Variance Extracted' (AVE) computed, and compared with the correlation value of each of the construct. To claim the discriminant reliability of the collected sample, the value of each of the construct's square root of AVE observed higher than the correlation with all other constructs as advised previously (Chin, 1998). The tabulation manner, the discriminant validity is shown in table 6 below.

Table 6 Discriminant validity of the constructs.

Construct	TP	KCC	SV	SC	ACC	PEB
TP	0.825					
KCC	0.334**	0.812				
SV	0.343**	0.274**	0.843			
SC	0.442**	0.320**	0.473**	0.824		
ACC	0.339**	0.275**	0.494**	0.454**	0.778	
PEB	0.248**	0.197**	0.500**	0.355**	0.564**	0.816

** $p<0.01$; * $p<0.05$

TP=Trust in Politics, KCC=Knowledge about Climate Change, SV=Social Value, SC=Source Credibility, ACC, Attitude towards Climate Change, PEB=Pro-Environment Behavior

Note: the square root of AVE's mentioned in the diagonal (bold and underlined).

Moreover, the VIF scores between the range of 1.33 to 1.49 observed to analyses the possibility of multicollinearity among the variables used in the study. The findings conclude that the no multicollinearity issue exists in the conducted survey. Furthermore, the common method biases measured by Harman's one-factor test, which provided significant results, as no single factor observed while holding the variance higher than 50% (Hair, William C. Black, Barry J. Babin & Rolph E. Anderson, 2010). Specifically, the highest score of variance noted as 37.74% by the single construct in the current study.

(2) *Structural model*

The confirmatory factor Analysis performed before examining the structural model, the model fitness indices noted and examined against each of the recommended value in the case of absolute, relative and non-centrality based indices. Specifically, the GFI and AGFI are recommended to be about 0.90 (Hooper, Mullen et al., 2008). Moreover, the NFI and FLI are advised to be above 0.95 (Hooper et al., 2008). Furthermore, the value of RSMEA and CFI are advised to be below 0.80 and above 0.95 respectively. In the current study, the measurement model satisfied the all measurement indices recommended thresholds as mentioned in Table 7. However, the structural model represents the slight variation in the case of NFI measurement only (as advised

Table 7 Measurement and structural model fitness indices.

Fitness Indices	Recommended	Measurement Model	Structural Model
Chi-square (X^2)		360.006	434.753
the degree of freedom (df)		120	124
CMIN / df	<5.0	3.00	3.51
GFI	0.90 (Hooper et al., 2008)	0.951	0.941
AGFI	0.80 (Bollen, 1990)	0.930	0.918
TLI	0.95 (Hu & Bentler, 1999)	0.959	0.950
NFI	0.90 (Malaquias & Hwang, 2016)	0.953	0.943
CFI	0.95 (Hu & Bentler, 1999)	0.968	0.958
IFI	0.95 (Hu & Bentler, 1999)	0.968	0.959
RMSEA	<1 (<0.08) (Hooper et al., 2008)	0.052	0.059

by Hooper (2008)). However, NFI satisfied the lower limit of 0.90 as advised by Malaquias and Hwang (2016) the supporting results from measurement and structural model are followed by the hypotheses testing in the next subsection of the document.

During the structural path analysis of three exogenous variables (Trust in Politics (TP), 'Knowledge about Climate Change' (KCC), and 'Social Value' (SV)) and three endogenous variables ('Source Credibility' (SC), 'Attitude towards Climate Change' (ACC) and 'Pro Environment Behavior' (PEB)). The strong positive effect of 'Trust in Politics' (TP) over 'Source Credibility' (SC) (H1(a) b=0.287) and 'Attitude towards Climate Change' (ACC) (H1(b) b=0.176) noted. Among all three exogenous factors, 'Knowledge about Climate Change' (KCC) observed as the weakest influencer over the dependent constructs 'Source Credibility' (SC) (H2(a) b=0.147) and 'Attitude towards Climate Change' (ACC)(H2(b) b=0.117).

Interestingly, the dominating role of 'Social Value' (SV) is observed as its strongly affected the perceived 'Source Credibility' (SC)(H3(a) b=0.373) and individual's Attitude towards Climate Change (ACC) (H3(b) b=0.488). In other words, it can be concluded that the 'Social Value' (SV) has the strongest influencing power while defining individuals' 'Attitude towards Climate Change' and their 'Pro-Environment Behavior.' As shown in Table 8. Moreover, the graphical explanation of the proposed model is shown in Fig. 5. During the further analysis, the study concludes that the individual's 'Attitude towards Climate Change' dominates the perceived 'Source Credibility' wile deciding and thinking about 'Pro-Environment Behavior, (PEB) among citizens of Pakistan (H4 b=0.100) and (H5 b=0.609).

Table 8 Hypotheses evaluation according to the proposed model.

Hypotheses	Statement	Overall
H1(a)	TP+→SC+	0.287***
H1(b)	TP+→ACC+	0.176***
H2(a)	KCC+→SC+	0.147***
H2(b)	KCC+→ACC+	0.117**
H3(a)	SV+→SC+	0.373***
H3(b)	SV+→ACC+	0.488***
H4	SC+→PEB+	0.100**
H5	ACC+→PEB+	0.609***

*** $p<0.001$; ** $p<0.01$

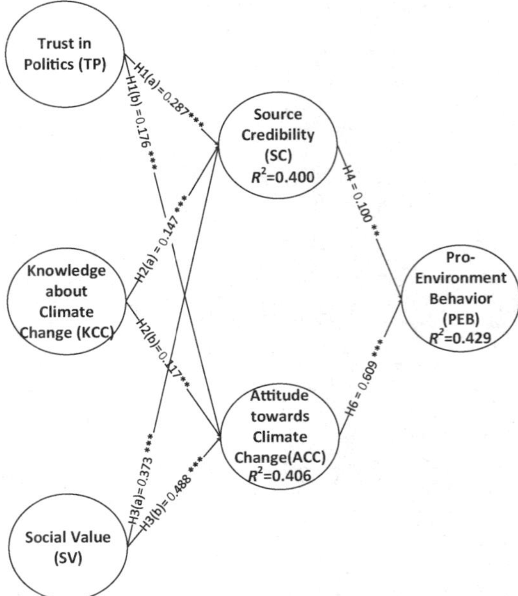

Fig. 5 Graphical path analysis of the proposed model of the current study.

5. Discussion and Conclusion

Conceptually, the discussion and conclusion comprise the possible lead extracted from the quantitative and qualitative analysis performed in the analysis. For a strategic position to respond to climate change, the parties can adopt one of the three strategies 'adversarial strategy' to openly reject the climate change; 'dismissive strategy' to ignore climate change; or 'accommodative strategy' to integrate climate change into their political discourses and policy programs (N. Carter, 2013). Climate change denial did not have any political expression as there is no opposition to the advancement of climate-related policies among political parties and revealed by their manifestos too. If not opted adversarial strategy the other option is to be indifferent to climate change by choosing a dismissive strategy. This strategy implies that political parties do not take clear positions on climate change with no policies or discourses to address climate change and leaving it as no issue for electoral competition. This strategy is also not applicable to the scenario as clearly defined policies showing the explicit position of political parties is available. The strategic position taken by political parties in Pakistan is an accommodative strategy with the gradual inclusion of climate change into their political discourse. They recognize the issue and pledge to take certain actions, but there would be great variation on the ambition of stated policies that these political parties are willing to support or promote. This is sometimes done to control the extent of policy differentiation and electoral completion on climate change. Political parties in Pakistan show low intensity integration of climate change with gradual inclusion of climate change in political discourses and policies but remain focused on valence issues like corruption, health and economic growth as compared to positional issue like climate change which does not reflect much demand from voters or when it's difficult to distinguish the policy stance of other parties (Ryan & Ryan, 2017).

This is certain that political, economic, sociocultural, scientific and technological factors have great effect on coverage of climate change issues in Pakistan (Sharif & Medvecky, 2018). Michael Kugelman indicates that there are voices coming from political parties on the climate change issue, but these issues were not resonating on the campaign trail (Ebrahim, 2018). Despite all the climate change posed challenges, it does not acquire major space in the debates during the campaign for new parliament to be elected. Social service issues and corruption have dominated the campaigns by candidates (Salam, 2018). The economic cost of climate change mitigation and adaptation is going up with wide its multi-level implications. Pakistan ranks among the most climate vulnerable countries instead of spending more than 8 percent of the national budget on climate change. Lawmakers in national parliament should be sensitized about the seriousness of the issue, and political parties should inform this through their party manifestos for upcoming elections. This can be seen for example in the NDC submitted by Pakistan which focuses on energy, agriculture, industry, land use, and forestry, and waste, closely linked with SDGs (Tauqeer Sheikh, n.d.).

The 18th Amendment of constitution empowers provinces with environmental management in an unusual way which leaves minimum space for the federal government to chalk out minimum standards to protect its citizen's health from climate change consequences. The federal government has left with certain policy options which include both governance and economic perspectives. Putting a carbon tax on high sulfur fuel to create an environment fund that could be invested to improve health-related issues caused by their emission in the air. The government could even help private sector for waste management by providing special grants or providing funding of such schemes. The federal government could compel provincial governments for compliance with climate change mitigation and adaptation goals by blocking the funds. Politicians at both federal and provincial level would start to work on climate change consequences especially water, air and waste management (Ahmad, 2018). Climate change is considered a national security issue, but politicians are not ready to understand the gravity of its consequences. They are mostly focused on roads and mega transport projects, which will further complicate the problem (Junaidi, 2018). Deforestation is serious environmental issue with social and economic implications from ecosystem to health and from high poverty to high temperature. The temperature rise in Karachi during the last five years has indicated a five degrees' centigrade change, and deforestation is a major contributor. More than 19,000 trees have been chopped down for government transport project in the last 18 months, while 5,000 more trees are cut down for other development projects (Dawn, n.d.).

Rising temperature, droughts, agricultural loss, melting of glaciers, variation in rain patterns, health issues and water problems are directly linked to climate change, and the problem is growing with every passing day. The world's highest temperature was recorded in Nawabshah this year. Climate change will severely affect agriculture, forestry, biodiversity, industry, water, and health. Urban lands are getting more congested with high-rise building at the expense of urban green cover/forest (Ilyas, 2018). PMLN is determined to strengthen the 'Pakistan Climate Change Council,' the 'Pakistan Climate Change Authority' and the 'Pakistan Climate Change Fund' envisaged in the 'Pakistan Climate Change Act of 2017'. PTI's chairman Imran Khan recognizes the 'foremost challenge' facing Pakistan is climate change. According to his views, climate change is no more a distant threat but is currently affecting Pakistan. Deforestation, water scarcity, and urban pollution are the most alarming factors. He seems determined to initiate 'Ten Billion Tree Tsunami' program particularly in urban areas within first 100 days if elected (Ebrahim, 2018).

PPPP chairman Bilawal Bhutto Zardari set his party's focus on water shortage, solid waste management, and urban sewage system. He mentioned that terrorism and climate change as two major problems and required regional and global dialogue for a sustainable solution. The manifesto does not seem to be so expressive on climate change as no separate chapter devoted to this issue. Imran Khan and Bilawal Bhutto Zardari are both in support of small to medium dams and water storage plans in all over the Pakistan (Ebrahim, 2018). Climate change communication has a very important role to play to enhance the climate change adaptation and to realize the threats which may not be seen or experienced directly by many people. People engagement in climate change communication with provision of more relevant information can reduce the vulnerability and can change the adaptive behaviour and could be helpful in framing the case in scientific, political, economic and social spheres.

The study concludes that the citizens take active supportive role in 'Climate Change Communication' and coping strategies if the synergized behavior is observed from the law enforcement agencies, legislative authorities and NGOs. The collective grand strategic framework is immediately required. The high positive affect of 'Social Value' demands smart 'Climate Change Communication' framing strategy so the practical change can be created. On the present moment, the credibility of source is climate change communication is comparatively weak as compare to attitude of individuals towards pro-environment behavior. This represents the positive sense of responsibility among citizens towards Climate Change. In terms of implications of the study, the effective and smart framing strategies are required with the support of electronic media as may communication channel. The comprehensive legislative structure can be reviewed further to map the current understanding of citizen's behavior towards 'climate change' and their readiness to adopt future 'environment friendly' initiatives.

References

Abildtrup, J., Audsley, E., Fekete-Farkas, M., et al., (2006). Socio- economic scenario development for the assessment of climate change impacts on agricultural land use: a pairwise comparison approach. *Environmental Science & Policy*, *9*(2), 101-115. https://doi.org/10.1016/j.envsci.2005.11.002.

Adelle, C. (2015). Contexualising the tool development process through a knowledge brokering approach: The case of climate change adaptation and agriculture. *Environmental Science & Policy*, *51*, 316-324. https://doi.org/10.1016/j.envsci.2014.08.010.

Ahmad, K. (n.d.). An environmental manifesto - Newspaper - DAWN.COM. Retrieved September 8, 2018, from https://www.dawn.com/news/1404779.

Akerlof, K., Maibach, E. W., Fitzgerald, D., et al. (2013). Do people 'personally experience' global warming, and if so how, and does it matter? *Global Environmental Change*, *23*(1), 81-91. https://doi.org/10.1016/j.gloenvcha.2012.07.006.

Antilla, L. (2005). Climate of scepticism: US newspaper coverage of the science of climate change. *Global Environmental Change*, *15*(4), 338-352. https://doi.org/10.1016/j.gloenvcha.2005.08.003.

Armitage, D., Berkes, F., Dale, A. et al. (2011). Co-management and the co-production of knowledge: Learning to adapt in Canada's Arctic. *Global Environmental Change*, *21*(3), 995-1004. https://doi.org/10.1016/j.gloenvcha.2011.04.006.

Bain, P. G., Hornsey, M. J., Bongiorno, R., et al. (2012). Promoting pro-environmental action in climate change deniers. *Nature Climate Change*, *2*(8), 603-603. https://doi.org/10.1038/nclimate1636.

Bernauer, T. (2013). Climate Change Politics. https://doi.org/10.1146/annurev-polisci-062011-154926.

Bhattacherjee, A., Clive Sanford. (2016). Influence Processes for Information Technology Acceptance: An Elaboration Likelihood Model1. *MIS Quarterly*, *30*(4), 805-825.

Biagini, B., Bierbaum, R., Stults, M., et al. (2014). A typology of adaptation actions: A global look at climate adaptation actions financed through the Global Environment Facility. *Global Environmental Change*, *25*(1), 97-108. https://doi.org/10.1016/j.gloenvcha.2014.01.003.

Biesbroek, G. R., Swart, R. J., Carter, T. R., et al. (2010). Europe adapts to climate change: Comparing National Adaptation Strategies. *Global Environmental Change*, *20*(3), 440-450. https://doi.org/10.1016/j.gloenvcha.2010.03.005.

Bollen, K. A. (1990). Overall fit in covariance structure models: Two types of sample size effects. *Psychological Bulletin*, *107*(2), 256-259.

Bord, R. J., O'Connor, R. E., Fisher, A. (2000). In what sense does the public need to understand global climate change? *Public Understanding of Science*, *9*(3), 205-218. https://doi.org/10.1088/0963-6625/9/3/301.

Boudet, H., Clarke, C., Bugden, D., et al. A. (2014). 'Fracking' controversy and communication: Using national survey data to understand public perceptions of hydraulic fracturing. *Energy Policy*, *65*, 57-67. https://doi.org/10.1016/j.enpol.2013.10.017.

Boykoff, M. T., Boykoff, J. M. (2007). Climate change and journalistic norms: A case-study of US mass-media coverage. *Geoforum*, *38*(6), 1190-1204. https://doi.org/10.1016/j.geoforum.2007.01.008.

Brossard, D., Shanahan, J., McComas, K. (2004). Are Issue-Cycles Culturally Constructed? A Comparison of French and American Coverage of Global Climate Change. *Mass Communication and Society*, *7*(3), 359-377.

Brulle, R. J., Carmichael, J., Jenkins, J. C. (2012). Shifting public opinion on climate change: an empirical assessment of factors influencing concern over climate change in the U.S., 2002-2010. *Climatic Change*, *114*(2), 169-188. https://doi.org/10.1007/s10584-012-0403-y.

Carter, L., Bélanger, F. (2005). The utilization of e-government services: citizen trust, innovation and acceptance factors. *Information Systems Journal*, *15*(1), 5-25. ttps://doi.org/10.1111/j.1365-2575.2005.00183.x.

Carter, N. (2013). Greening the mainstream: Party politics and the environment. *Environmental Politics*, *22*(1), 73-94. https://doi.org/10.1080/09644016.2013.755391.

Carvalho, A. (2007). Ideological cultures and media discourses on scientific knowledge: re-reading news on climate change. *Public Understanding of Science*, *16*(2), 223-243. https://doi.org/10.1177/0963662506066775.

Carvalho, A., Burgess, J. (2005). Cultural circuits of climate change in U.K. broadsheet newspapers, 1985-2003. *Risk Analysis*, *25*(6), 1457-1469. https://doi.org/10.1111/j.1539-6924.2005.00692.x.

Carvalho, A., van Wessel, M., Maeseele, P. (2017). Communication Practices and Political Engagement with Climate Change: A Research Agenda. *Environmental Communication*, *11*(1), 122-135. https://doi.org/10.1080/17524032.2016.1241815.

Chin, W. W. (1998). Commentary: Issues and Opinion on Structural Equation Modeling. *MIS Quarterly*, *22*(1), 1. https://doi.org/Editorial.

Corbett, J. B. (2004). Testing Public (Un)Certainty of Science: Media Representations of Global Warming. *Science Communication*, *26*(2), 129-151. https://doi.org/10.1177/1075547004270234.

Corner, A., Crompton, T., Davidson, S., et al. (2010). Communicating climate change to mass public audiences. *Climate Change*, (September), 14. Retrieved from http://psych.cf.ac.uk/ understandingrisk/docs/cccag.pdf.

Corner, A., Randall, A. (2011). Selling climate change? The limitations of social marketing as a strategy for climate change public engagement. *Global Environmental Change*, *21*(3), 1005-1014. https://doi.org/10.1016/ j.gloenvcha.2011.05.002.

Dawn. (n.d.). Climate change - Newspaper - DAWN.COM. Retrieved September 8, 2018, from https://www.dawn.com/news/1399723.

Dickinson, J. L., Crain, R., Yalowitz, S., et al. (2013). How Framing Climate Change Influences Citizen Scientists' Intentions to Do Something About It. *The Journal of Environmental Education*, *44*(3), 145-158. https://doi.org/10.1080/00958964.2012.742032.

Environment, S. I. (2012). Building bridges and changing minds: Insights from climate communication research and practice, 1-8.

Feldman, L., Maibach, E. W., Roser-Renouf, C., et al. (2012). Climate on Cable: The Nature and Impact of Global Warming Coverage on Fox News, CNN, and MSNBC. *The International Journal of Press/Politics*, *17*(1), 3-31. https://doi.org/10.1177/1940161211425410.

Hair, J. F., Black, W. C., Babin, B. J., et al. (2014). *Multivariate data analysis* (7th ed.). Upper Saddle River, N.J.: Pearson Education.

Hair, J. F., Black, W. C., Babin, B. J., et al. (2010). *Multivariate Data Analysis* (7th ed.).

Hart, P. S., Nisbet, E. C. (2012). Boomerang Effects in Science Communication. *Communication Research*, *39*(6), 701-723. https://doi.org/10.1177/0093650211416646.

Hong, S., Tam, K. Y., Hong, S. (2006). Understanding the Adoption of Multipurpose Information Appliances: The Case of Mobile Data Services, (January 2015). https://doi.org/10.1287/isre.1060.0088.

Hooper, D., Mullen, J., Hooper, D., et al. (2008). Structural Equation Modelling: Guidelines for Determining Model Fit. *The Electronic Journal of Business Research Methods*, *6*(1), 53-60.

Hu, L., Bentler, P. M. (1999). Cutoff criteria for fit indexes in covariance structure analysis: Conventional criteria versus new alternatives. *Structural Equation Modeling: A Multidisciplinary Journal*, *6*(1), 1-55.

Hulme, M. (2009). Why We Disagree About Climate Change. *ENDS Report*, *Carbon Yea*(June), 41-43. https://doi.org/10.1017/CBO9780511841200.

Ilyas, F. (n.d.). Parties urged to incorporate environmental issues in manifestos - Newspaper - DAWN.COM. Retrieved September 8, 2018, from https://www.dawn.com/news/1407884.

Jang, S. M. (2013). Framing responsibility in climate change discourse: Ethnocentric attribution bias, perceived causes, and policy attitudes. *Journal of Environmental Psychology*, *36*, 27-36. https://doi.org/10.1016/j.jenvp.2013.07.003.

Junaidi, I. (n.d.). Are politicians really serious about the climate? - Newspaper - DAWN.COM. Retrieved September 8, 2018, from https://www.dawn.com/news/1418778.

Kahan, D. M., Peters, E., Wittlin, M., et al. (2012). The polarizing impact of science literacy and numeracy on perceived climate change risks. *Nature Climate Change*, *2*(10), 732-735. https://doi.org/10.1038/nclimate1547.

Kahlor, L., Rosenthal, S. (2009). *If We Seek, Do We Learn? Science Communication* (Vol. 30). https://doi.org/ 10.1177/1075547008328798.

Leiserowitz, A. (2006). Climate change risk perception and policy preferences: The role of affect, imagery, and values. *Climatic Change*, *77*(1-2), 45-72. https://doi.org/10.1007/s10584-006-9059-9.

Lewandowsky, S., Gignac, G. E., Oberauer, K. (2013). The Role of Conspiracist Ideation and Worldviews in Predicting Rejection of Science. *PLoS ONE*, *8*(10). https://doi.org/10.1371/journal.pone.0075637.

Lorenzoni, I., Nicholson-Cole, S., Whitmarsh, L. (2007). Barriers perceived to engaging with climate change among the UK public and their policy implications. *Global Environmental Change*, *17*(3-4), 445-459. https://doi.org/10.1016/j.gloenvcha.2007.01.004.

Malaquias, R. F., Hwang, Y. (2016). An empirical study on trust in mobile banking: A developing country perspective. *Computers in Human Behavior*, *54*, 453-461. https://doi.org/10.1016/j.chb.2015.08.039.

Mccright, A. M., Dunlap, R. E. (2011). The Politicization Of Climate Change And Polarization In The American

Public's Views Of Global Warming, 2001-2010. *Sociological Quarterly*, *52*(2), 155-194. https://doi.org/10.1111/j.1533-8525.2011.01198.x.

Moser, S. C. (2010). Communicating climate change: history, challenges, processes and future directions. *WIREs Climate Change*, *1*, 31-53. https://doi.org/10.1002/wcc.011.

Nisbet, M. C. (2009). Communicating Climate Change: Why Frames Matter for Public Engagement. *Environment: Science and Policy for Sustainable Development*, *51*(2), 12-23. https://doi.org/10.3200/ENVT.51.2.12-23.

Nisbet, M. C., Scheufele, D. A. (2009). What's next for science communication? promising directions and lingering distractions. *American Journal of Botany*, *96*(10), 1767-1778. https://doi.org/10.3732/ajb.0900041.

O'Neill, S., Nicholson-Cole, S. (2009). 'Fear Won't Do It': Promoting Positive Engagement With Climate Change Through Visual and Iconic Representations. *Science Communication*, *30*(3), 355–379. https://doi.org/10.1177/1075547008329201.

Ockwell, D., Whitmarsh, L., Neill, S. O. (2009). Science Communication. https://doi.org/10.1177/1075547008328969.

Pavlou, P. A., Tan, Y. H., Gefen, D. (2003). The transitional role of institutional trust in online interorganizational relationships. *36th Annual Hawaii International Conference on System Sciences, 2003. Proceedings of the*, *0*(C), 1–10. https://doi.org/10.1109/ HICSS.2003.1174574.

Pike, B. D., Cara M. H. (2010). Climate Communications and Behavior Change: A Guide for Practitioners. *Climate Leadership Initiative*, *35*(1), 1–54. https://doi.org/10.1177/1075547012438465.

Qamar Uz Zaman Chaudhry. (2017). *Climate Change Profile of Pakistan. The Economic and Labour Relations Review* (Vol. 23). https://doi.org/10.1177/103530461202300301.

Qualitative, A. (n.d.). Methodology Series.

Ryan, D., Ramirez, A. (2016). Cdkn guide, (July).

Ryan, D. (2017). Politics and Climate Change: Exploring the Relationship Between Political Parties and Climate Issues in Latin America. *Ambiente & Sociedade*, *20*(3), 271-286. https://doi.org/10.1590/ 1809-4422asocex0007v2032017.

Salam, A. (n.d.). Pakistan is one of the world's leading victims of global warming. Retrieved September 8, 2018, from https://www.usatoday.com/story/news/world/2018/07/24/pakistan-one-worlds-leading-victims-global-warming/809509002/.

Sharif, A., Medvecky, F. (2018). Climate change news reporting in Pakistan: A qualitative analysis of environmental journalists and the barriers they face. *Journal of Science Communication*, *17*(1), 1-17. https://doi.org/10.22323/2.17010203.

Sönke, K., Eckstein, D., Dorsch, L., et al. (2015). Global climate risk index 2016: Who suffers most from Extreme weather events? Weather-related loss events in 2014 and 1995 to 2014. https://doi.org/978-3-943704-04-4.

Spence, A., Poortinga, W., Butler, C., et al. (2011). Perceptions of climate change and willingness to save energy related to flood experience. *Nature Climate Change*, *1*(1), 46-49. https://doi.org/10.1038/nclimate1059.

Sussex, W., Study, C. (n.d.). Climate Change Communication Strategy.

Ebrahim, T. Z. (n.d.). Does the environment matter in Pakistan's elections? - Pakistan - DAWN.COM. Retrieved September 8, 2018, from https://www.dawn.com/news/1419200.

Tauqeer Sheikh, A. (n.d.). Climate threats - Newspaper - DAWN.COM. Retrieved September 8, 2018, from https://www.dawn.com/news/1372877.

Vainio, A., Paloniemi, R. (2013). Does belief matter in climate change action? *Public Understanding of Science*, *22*(4), 382-395. https://doi.org/10.1177/0963662511410268.

Venkatesh, V. (2000). Determinants of perceived ease of use: Integrating perceived behavioral control, computer anxiety and enjoyment into the technology acceptance model. *Information Systems Research*, *11*(4), 342-365.

Weber, E. U., Stern, P. C. (2011). Public understanding of climate change in the United States. *American Psychologist*, *66*(4), 315-328. https://doi.org/10.1037/ a0023253.

Whitmarsh, L. (2009). What's in a name? Commonalities and differences in public understanding of 'climate change' and 'global warming.' *Public Understanding of Science*, *18*(4), 401-420. https://doi.org/10.1177/0963662506073088.

Public Participation and Government's Scientific Decision

——Taking the disputes caused by China's PX project as an example

Ma Jianquan, Qi Hailing

National Academy of Innovation Strategy, CAST, Beijing, China

Abstract: As technology is more and more popular in public's daily life, the scientific policy decision making become the burst point between the government and the public: if lack of public participation, government decision-making will be taken as against modern democratic and political aspirations, but once introduce public participation, the decision full of 'science and technology' is easy to be questioned and even boycotted by the 'ignorant' public. The PX project decision-making process in China is a typical case. PX is a chemical product with wide range of use, it has little environmental pollution and security incidents, but it causes general disputes in different areas of China. Based on these PX event cases in China since 2007, this paper will use the public understanding of science theory as the analytical framework, integrate of power legitimacy theory, risk society theory, trying to answer the following questions: first, how science become the cornerstone of the legitimacy of the government; second, why the authority of science is questioned by the public; third, how the public transform from the scientific passive recipients to scientific active participants.

1. Introduction

As technology is more and more popular in public's daily life, the scientific policy decision making become the burst point between the government and the public: if lack of public participation, government decision-making will be taken as against modern democratic and political aspirations, but once introduce public participation, the decision full of 'science and technology' is easy to be questioned and even boycotted by the 'ignorant' public. The PX project decision-making process in China is a typical case.

PX is a chemical product with wide range of use, it has little environmental pollution and security incidents, but it causes general disputes in different areas of China. Initially the government took the preparations underwater, exclude the public from the decision-making process, after several large-scale public opposition the government began to vigorously launch popular science campaigns of relevant knowledge, the public asked to participate in the decision-making process, but generally showed a desire to believe rumors rather than the government and scientists. From these PX events, we can see that the public on the one hand do not understand PX, on the other hand do not want to believe the interpretation of government and scientists, the public's understanding of PX is not simply based on science, but also based on their experience knowledge and relationship knowledge.

Based on these PX events cases in China since 2007, this paper will use the public understanding of science theory as the analytical framework, integrate of power legitimacy theory, risk society theory, trying to answer the following questions: first, how science become the cornerstone of the legitimacy of the government; second, why the authority of science is questioned by the public; third, how the public transform from the scientific passive recipients to scientific active participants.

2. Public Participation in Public Decision-making

1) Public Participation in Public Decision-making

In the period of traditional public administration, influenced by political and administrative dichotomy, bureaucratic theory, scientific management theory and other related theories, the government used to exclude public participation, even hostile to public participation. The government considers that the administrative field involves technical activities, and it should be 'completed by more professional administrative bureaucrats. Public participation in the administrative field will exert unnecessary intervention on administrative management and affect its efficiency and effectiveness'. (Goodnow, 1900)

With the development of the new public administration movement, scholars began to re-recognize the role and significance of public participation in the administrative field. The introduction of public participation in public decision-making has been enthusiastically and positively recognized by scholars. However, with the increasing practical experience of public participation, scholars began to realize that public participation is not a 'master key' to ensure public policy scientific and democratic. On the contrary, it may also have a negative impact on public administration.

Specifically, the critics' views are mainly focused on the following three points: first, the public cannot represent the public interest. 'Those who participate in public decision-making through participation often cannot represent the interests of the wider public. Participants often represent the needs of existing organized groups, expressing their special interests, not the general interests of the public. Their participation is often defensive and is used to protect the interests of existing groups.' (John, 1995) Second, public capacity is limited. Compared with experts, the public has a large gap in knowledge, ability, information, etc., which leads to doubts or even boycotts of public policies with scientific and technological content, which leads to the failure to implement public policies. Therefore, policy makers are not very confident in the public participation in public policy. Third, public participation leads to inefficient public administration. Although in general, in the public decision-making process, the more roles involved, the more people expect to participant, but the increase in public participants in public decision-making will inevitably lead to an increase in discussion making time and in the number of stakeholders that need coordination, thus affecting the efficiency of public administration.

Public participation is an inevitable requirement for the development of democratic politics, but public participation itself may also have a negative impact. Therefore, we must treat the issue of public participation dialectically. This study is to examine the behavioral logic of the government and the public in the process of public policy making.

2) Government and Public in Science and Technology Decision-making

In recent years, scientific and technological policy has received much attention. From space exploration, global climate governance, to the food safety and garbage disposal, it is necessary to make decisions based on science and technology. Science and technology decision-making has penetrated into our daily lives, but the government continues the traditional decision-making model—taking the public as ignorant in science and technology decision-making and rejecting public participation, which has caused strong public dissatisfaction. This led to a series of anti government science and technology decisions, such as opposition to the location of waste incineration, boycotts of PX projects, and opposition to the construction of nuclear power.

This paper takes the mass incidents triggered by the PX project as the research object, and attempts to explore the respective behavioral logics and the corresponding strategic choices of the government and the public in the decision-making of science and technology, to explain a series of issues such as why the government is promoting the PX project and how the government implemented the project, why the public will resist and oppose, and how to oppose government decision-making.

3) Where Does the Government's Power Come From?

The legitimacy of government power comes from the triple contract: democratic contract, political contract and service contract. According to the source of power, there must be two hidden meanings in every decision making: First, government power comes from the people, which means that one of the starting points of government decision-making must be for the public interest, or public interest should always run through the source and execute of power. Second, the execute of government power has strict procedure, that is, the authorization of power must conform to certain process, otherwise there may be acts of self-interest in the form of public opinion. That is to say, the legitimacy of a public policy comes from the contract, and also shown in the public interest and the legitimacy of the execute procedure. Only when public decision satisfy these two points, it is legal.

According to the three contract theory, we can find that in the decision-making of science and technology, the government use scientific knowledge to endorse the legality. The logic behind it is that the traditional authority of science, the expression of scientific public interest and the science can conform to the legitimacy of authority procedure.

First of all, scientific knowledge provides a source of explanation for the legitimacy of government decision-making. Scientific knowledge was developed rapidly after the Enlightenment and profoundly affected the current human living environment. From the perspective of modernity, science has become the only source of knowledge in modern society with a legitimate position, becoming the only spokesperson of 'truth', 'fact' and 'reality', and all 'non-scientific knowledge' will be rejected.

Second, science can realize the public interest. Science is the knowledge support of government science and technology decision-making, and it is important to human society. Since science promotes the progress of human society, science and technology decision-making relies on scientific knowledge to benefit the public, and obtained legitimacy.

Finally, science satisfies the inherent requirements of legal authority. In Weber's theory, the legal authority is based on the fairness procedure accepted by the people and is the leader or representative of the democratic election process. (Max, 1979) Modern administrative power is based on rational law and bureaucratic expertise. The obedience of the people is precisely because of the legal authority of the technocrats.

4) The Legalization Crisis of Science

In the decision-making of science and technology, there are actually two sets of contradictions: the first is the contradiction between professionalism and democracy. As the embodiment of science, the technocrats have the supreme right to speak, and the public cannot effectively control the administration because of lack of professionalism. Although the administration promote democracy with professionalism, but in turn, once there is professionalism, professional erosion of democracy will occur. (Jay et al., 1994)

The second is the contradiction between public interest and upward responsibility. Although the government should maintain and enhance the public interest, however, the government as a bureaucratic organization is responsible for the administrative organs of its superiors. This means that the starting point for the executives is not the public interest, nor whether the process of implementation is accountable to the public. This is precisely conflict to the spirit of public interest in the triple contract. Therefore, it is easy to alien the relationship between the state and the public, and the conflict between the government and the public often occurs.

As mentioned above, in the perspective of modernity, science has become the only source of knowledge in a modern society with a legitimate position. Science is authoritative, which is one of the reasons why the government borrows science to endorse the legitimacy of scientific and technological decision-making. However, since the end of World War Ⅱ, as science has penetrated into every aspect of people's lives, people have begun to see the other side of science, while science brings convenience to people, it also puts people into risk.

The immature results of scientific research have entered the society in the form of products, and the risks of science have entered the society. (Li, 2006) 'The high levels of urgency, risk and uncertainty caused by uncontrollable technological innovations have led to a loss of confidence in the trusted authority–science and public institutions to deal with tensions.' (Zhang et al., 2011) With the progress of the times, the legal authority of science has been slowly questioned, and science is not authoritative. Therefore, government science and technology decision-making endorsed by science has encountered a legalization crisis.

5) Public Understanding of Science: from Passive Acceptance to Active Participation

As scientific authority is challenged and public attitudes change, Western governments and scientific communities believe that public distrust of science is primarily based on their lack of understanding of science. According to this thought, they believe that in order to let the public re-support the development of science, the public should understand science, and the key issue in understanding science is to have scientific knowledge. As a result, in the 1980s, the government-led public understanding science movement was first established in the United Kingdom and the United States.

In the early public understanding science movement, the public is still considered to understand science, not science must better understand the human society. (Felt, 2003) Under such thinking, the public understanding science movement has not broken through the limitations of traditional science popularization. It still believes that the public is ignorant or that the public does not understand scientific facts, scientific theories, and scientific processes. The public is willing to understand science; while science is the representative of ration and is absolutely correct knowledge. In this process, there is a hypothesis that the public's attitude toward science is only caused by public ignorance, and there are no other factors, such as the public's value position, political considerations and so on. The public is only a passive recipient of scientific knowledge and an 'empty container' waiting to fill scientific knowledge.

However, will the public change their attitude towards science if they have more scientific knowledge? The results of subsequent empirical survey data show that, on the contrary, the more scientific information the public gets, the more reflective thinking they have on science. Public are paying more attention to the uncertainty and risk of scientific results, leading to a more nervous relationship between science and the public. (Xu et al., 2005)

Based on the reinterpretation of the 'Public Understanding Science', the 'Public Understanding of Science' movement has a major shift, the public is no longer considered ignorant, nor is it a passive recipient of scientific knowledge. The public's judgment and resistance to science can't be simply regarded as the expression of irrationality and lack of knowledge, it is the public's defense of their rights in the knowledge, interests and risks involved in the production and application of scientific knowledge, and also the process of democratic participation in social life. The public has changed from a passive recipient to an active participant. More importantly, the public is no longer an 'outsider' who has no voice in science and its applications, but an indispensable component that must listen to on the development of science, decision-making, and the application of science and technological achievements in society.

In summary, after half a century of development, 'Public Understanding Science' has demonstrated a major evolution in its paradigm: from the public being forced to accept scientific knowledge from scientists (one-way communication mode), to the dialogue between science and public (two-way interaction mode). In the early days, public understanding of science limited the relationship between science and the public to the field of communication, and concealed the political nature of the interaction between them. As a science in the political field, the public should not be excluded from science and technology affairs based on any knowledge reason. 'The public is not a bystander of science games, they should be participants, and express their own demands in the participation, modify the established "rules of the game".' The public understanding of science does not require the public to passively understand science, but to let the public take the initiative direct dialogue with science.

3. Case Analysis

1) Why Choose PX

The research method of this paper is case analysis. Why choose the PX project as a case? There are three reasons. First, the policy development process should be relatively open and transparent, or the administrative operation process is observable. The PX project is relatively open from planning to environmental impact assessment to later public participation. And relevant information can be obtained in government documents, official media and government reports. Second, stakeholders in the policy process (Government, people, enterprises, etc.) have complex conflicts. The government believes that the PX project is a good policy for the benefit of both the country and the people, and has done enough science popularization in advance, but the people not only 'don't appreciate', but succeeded in putting the policy on hold by walking on the streets. It can be seen that the PX incident fully demonstrates the tension and conflict between the government and the public. Third, the PX project has received extensive attention from the whole society, providing a basis for the acquisition of data and materials needed for research.

This paper has briefly summarized some recent px events, as shown in the following Table 1. As some scholars have concluded: no trouble, no solution; small trouble, part solution; big trouble, good solution. Whether the public make trouble is the key to determine whether industrial projects with environmental risks can continue.

As shown in Table 1, the PX project caused a series of environmental mass incidents. Most of the PX projects were forced to stop because of mass gatherings and protests. The common feature of these incidents is whether the project can continue to proceed smoothly is not directly determined by whether the project has qualifications for execution, whether it will bring environmental hazards, but whether the public agrees or not. It can be seen that with the development of China's economy, politics, society and culture, citizens pay more and more attention to their own rights and interests, and their environmental awareness is also increasing. They not only express their own ideas, wishes, and opinions, but also have more methods. Every project, policy implementation are under public attention and supervision.

Table 1 Different PX Cases in China.

Location	Time	Result
Xiamen	2007.03	Relocated to other city
Dalian	2011.08	Stop production immediately
Ningbo	2012.10	Stop pushing
Chengdu	2013.05	Built
Kunming	2013.05	Respect public's will
Jiujiang	2013.05	Built
Maoming	2014.03	Listening to public opinion before making a decision

2) Maoming PX Case

In the following, this paper will take Maoming PX case to analyse in detail.

On the first stage, the government secretly prepares for the PX project, public exclusion is beyond decision-making. The government's preparations have not been directly shown to the public, and even dare not mention the sensitive word of PX. The Maoming government information disclosure does not explicitly propose to build a PX project. The scope of the public announcement is narrow. Public didn't pay much attention on it.

On the second stage, the government vigorously carries out popular science, and the public demands to participate in the decision-making process. In view of Xiamen PX incident, Maoming government did not dare to directly publish. At the beginning of February 2014, the Maoming government invited the major media reporters to participate in the Spring Festival symposium. At the meeting, the organizers mentioned that in the future Maoming might build a PX project and consult with the reporters: How to guide the public opinion? What should government do if the public objected?

In order to cope with the thorny problem of the public 'every PX must be resisted', Maoming government has a very strict attitude, they only intensively publicizing related science knowledge and content of the PX. On March 17, the government organized media reporters to hold a closed-door meeting, announced the investment scale and site selection plan of the Maoming PX project, and also distributed the 'Aromatics Project Brochure' and 'Basic Situation of Maoming Aromatics Project' to the participants and played a video of correctly recognizes the PX project of Focus Interview. On the second day, Maoming local media began to publish PX information. The Maoming Daily began to forward articles related to the PX project. But these articles were only simple science popular articles, and did not mention any news of the Maoming PX project. At the same time, Maoming TV and Maoming Petrochemical Company TV repeatedly broadcast the 'Awareness of Maoming Aromatics (PX) Project' publicity video. Considering that some places are used to communicate in Cantonese, the Maoming government has also made a Cantonese-language video.

In order to enable public officials to better support the PX project, on the one hand, the Maoming government invited the academician of the Chinese Academy of Engineering and Professor Jin Yong from Tsinghua University to give a lecture on the 'Questions and Answers for Maoming PX (Aromatic Hydrocarbon) Project'. On the other hand, the signing of the letter of commitment to support PX is also in full swing in party and government organs, schools and petrochemical enterprises at all levels. But the practice of political missions not only did not gain the support of the public, but caused great pressure to the public. The suspicion of the PX project has continued, and more and more people have joined the online discussion. Most of these posts are opposing the construction of PX projects.

It can be seen that the intensive propaganda did not have the expected effect, but instead gave the public the illusion that the PX project is about to be launched. Therefore, the public is requiring to participate in decision-making and raise many questions to the government through internet. But at this time, the government did not respond positively to the public's doubts. On the one hand, the government's official Weibo is silent, not mentioning it. Relevant government officials do not accept any media interviews. On the other hand, in response to a large number of posts on the Internet about PX, the Maoming government took the same measures as local governments such as Xiamen, Dalian, and Chengdu—deleting the posts and blocking the news.

In the face of strong public opinion pressure and tense public relations, the Maoming government may have foreseen the people to protest against the PX project, so they invited local well-known netizens to participate in the PX project promotion meeting, trying to communicate with netizens rationally. But because the conference organizers are not ready to communicate with netizens sincerely, the present netizens were disappointed, which eventually aggravated the netizens' resentment. Subsequently, in the WeChat friends circle, Weibo, Post Bar and other multi-network channels, protest information spread on a large scale, and even the time and place of the protest has been arranged.

However, the Maoming government did not care about this information and did not have any early warning measures. On the morning of March 30, at the beginning, only a small number of citizens gathered in front of the city government to ask to stop the construction of the PX project, but no violence occurred. Later, people gathered more and more, even more than a thousand, and there were individual illegal snoring behaviors in the evening. As can be seen from the results of this conflict, the local government of Maoming had to declare that the government would not make decisions against the will of the public and would not launch the project without public's agreement.

3) Case Analysis of Maoming

Through the analysis of the three stages in the decision-making process of the Maoming PX project, we found that the Maoming government felt very helpless: they eagerly explained through various media that PX is very environmentally friendly and safe; let the experts come out and endorse; show a lot of foreign case; the newspaper also guides public opinion for the government. Why is the public still not appreciative? In the face of such popular science propaganda, why is the public always 'firm' that PX is poisonous?

This paper believes that the Maoming government knows that there is great resistance to the implementation of the PX project, but it still spares no effort to promote it. The reason is that the project meets the needs of political impulses. But the government's political impulses are not directly expressed, it will be achieved by seeking the legitimacy of the policy. The legality of policy includes two aspects: one is in accordance with legal procedures, and the other is accepted by the public. The former is formal legality and the latter is substantive legality. Therefore, the government adopted the dual discourse system of the public interest discourse system and the scientific discourse system, which is, 'I am good for you' and 'Don't be afraid' to achieve the legalization of the policy. Of course, among them, the government is still using different ways to promote PX science knowledge. However, these efforts by the government have not been recognized and supported by the public.

The government believes that 'the different attitudes of ordinary people and technocrats to science and technology policies are mainly due to the different amount of knowledge they have. According to this assumption, the more scientific and technical knowledge the general public has, the more rational they can be, the closer judgements and conclusions they can make compared to the technocrats.' As a result, a large number of popular science propaganda is carried out, and such practices will fall into a misunderstanding: the scientific and technological decisions that public oppose are often considered by technical officers to be technically no problem, but the ignorance of the public. However it is because the government only regards the PX project as a purely scientific issue, the government is simply talking about PX from a scientific perspective, but not the real problem that public are worried about.

But for the public, the public thinks that this is not just a scientific issue, it also involves profound social issues. Therefore, the public does not accept the government's constant science popularization, but instead shifts the topic to the government's credibility and regulatory issues. From the current scientific and technological development and foreign experience, the environmental problems brought by the PX project can achieve zero emissions and zero pollution. But even if PX's technology matures, the public still has too many doubts about China's PX project. This involves not only the problems in the field of science and technology, but also the distrust of the government.

The Maoming government has not shaken off the traditional popular science thinking mode, just emphasizing the role of scientific knowledge in decision-making, and believes that the transfer of knowledge is a simple one way communication mode, that is, the technocrats at the highest level communicate information to the public. This reflects the worship of decision makers on the authority of professional knowledge, while not recognizing the importance of the empirical knowledge of the public. In their eyes, knowledge from everyday life is not objective, scientific knowledge, it is subjective and self-interested, and there is no need to be included in decision-making.

But in fact, in many cases, the public is not ignorant. The public not only has the ability to obtain empirical knowledge about PX from life, but also can actively judge whether they want to believe the various rhetoric of the government. At the same time, as another knowledge system, empirical knowledge is more likely to become a public preconception than professional knowledge, because everyone can directly obtain empirical knowledge without the need for experts as an intermediary.

The empirical knowledge of the public mainly comes from the following aspects: First, the experience of living in the local area, which is the most basic judgment experience of public knowledge. Second, diverse information channels such as words from friends and family, or information from media and the Internet. Third, institutional participation. These three sources of information are mutually influential, and the first local life experience is the most important foundation, because no matter what information they get, the public must combine with their own experiences to make judgments, and the public no longer blindly believes in the government's rhetoric.

Through the above analysis, it can be seen that in the interaction between the government and the public, the public's empirical knowledge encounters the professional knowledge of the administrative department, and usually get such a result: no dialogue. One emphasizes empirical knowledge, one emphasizes professionalism, leads to a lack of consensus between the two sides. The local public has doubts based on their living experience in the local area. For example, if there is no poison, why do other cities people object? If there is no pollution, why other places choose places away from the residential area? In addition, previous life experiences have also hinted that the government will not do environmental protection and risk prevention and control very well. But for these doubts, the government has avoided talking about it. Instead, it uses the refutation of scientific data and other cites successful cases abroad to try to gain public trust through technical rationality. However, technical rationality is not omnipotent. It cannot answer the environmental, safety, health and other value issues that people care about. Therefore, when the public judges the value based on empirical knowledge and if the government cannot provide an effective response, this becomes the technical starting point for the public to resist the PX incident.

4. Advice on Enhancing Communication Between the Government and the Public on Science and Technology Decision-making

1) The Foundation of Trust Establishment: Information Disclosure and Active Communication

Active information disclosure and communication is more effective than strong advancement. Due to the inertia of the one-way decision-making thinking, local governments often pay less attention to the expression of public opinions during the decision-making stage, and even after the outbreak of mass incidents, they still focus their efforts on dealing with superiors and relevant regulatory authorities, while continuing to conceal information and avoid communication to the public. It can be seen from a series of PX incidents that this kind of strong working mode has begun to be challenged by various mass incidents. It is not only not conducive to the promotion of the project, but may cause the project to be stopped, damaging the credibility of the government. To change this situation, local governments should change the work ideas of relying on administrative resources to ensure that everything is done, take the initiative to disclose the details of the project, relevant advantages and disadvantages, understand the real needs, ideas and opinions of the public, and win public support for the project through building trust.

2) The Key to Effective Construction: Pre-Communication and Two-way Interaction

Compared with post-event communication, pre-communication is more conducive to establishing the social legitimacy of industrial projects. In many incidents, local governments organized representatives' visits, discussion and home-based announcements after event occurred, to make up for the lack of communication in the early stage, but did not actually change the public's attitude toward the project so that the project was eventually stopped. The lack of information disclosure and communication in the early stage makes the public's judgment on the project vulnerable to other information and easily make negative pre-forms judgments on the project. Coupled with the lack of government trust, the public is more likely to tend to view information disclosure and communication as formalism. Therefore, post-event communication is difficult to fundamentally change the public's impression of the project.

In addition, two-way communication is more effective than one-way information disclosure and knowledge dissemination. Local governments are often more likely to ignore public feedback, that is, when communicating, they tend to show the project advantages to the public in one direction, but ignore the idea of obtaining public opinions. In fact, the public's judgment on the projects is bound to be influenced by many factors such as the surrounding group's cognition and the way of government communication. The main appeals expressed by the public in mass incidents are often caused by various large and small interests that are difficult to directly explain. Only through two-way communication can the government understand these factors more comprehensively and promote the project more specifically. Therefore, two-way communication is more effective than one-way communication.

3) The Direction of System Improvement: Multi-participation and Full Supervision

Multi-participation helps to accurately target project stakeholders and their interests. In fact, on the one hand, the complexity of some project technologies and the uncertainty of social repercussions make it difficult for local governments to accurately target the project's audience; on the other hand, the lack of open and transparent research methods is likely to cause public suspicion and misunderstanding. Therefore, it is more helpful to gain the understanding and support of the public by actively absorbing the participation of multiple subjects. In addition, the participation of multiple subjects should not be limited to the expression of opinions and coordination of interests in the early stage of the project, but should also be extended to the supervision of the process of project construction, production and operation. The commitments made by many local governments in the previous communication lack the guarantee of a formal system. It is difficult for the public to supervise through the institutionalized way after the project is implementing and completed. Once the project has problems, it will cause public distrust. And the social legitimacy that has been constructed no longer exists. Therefore, the principle of information disclosure and multi-subject participation should be kept in the follow-up process of the project so that the public can participate in supervision at any time.

References

Frank, J. G. (1900). *Politics and administration: A study in government*. New York: Macmillan.

John, C. T. (1995). *Public participation in public decisions: New skills and strategies for public managers*. San Francisco: Jossey--Bass.

Max, W. (1979). *Economy and Society, 8*(2).

Jay, D. W., Guy, B. A. (1994). *Research in public administration: Reflections on theory and practice*. Sage Publications.

Li, R. (2006). *Risk, knowledge and public policy*. Tianjin: Tianjin People Publish.

Zhang, L., Wang, H. (2011). Scientific governance that moves toward hybrid forum—A survey of public participation in science. *Journal of Jiangsu University (Social Science Edition)*, 13(3), 16-20.

Felt, U. (2003). *OPUS-Optimising the Public Understanding of Science*.

Xu, Z., Mao, B. (2005). Science communication in risk society. *Studies in Science of Science*, 23(04), 9-13.

Thoughts on the Construction of Think Tanks in Association for Science and Technology System

Wang Yinqiu, Wang Shanshan, Zhang Danna

National Academy of Innovation Strategy, CAST, Beijing, China

Abstract: This paper mainly focuses on the thank tank operation of China Association for Science and Technology. In the first place, we investigate some international and local famous think tank, and review the literatures studying think tank. Then, the main work and achievements of China Association for Science and Technology and National Academy of Innovation Strategy in the process of building think tanks is concluded. In addition, the main advantages and characteristics of think tank belonging to Association for Science and Technology System is given and discussed a lot. Finally, the authors summarize some good experiences of building and improving the think tank of China Association for Science and Technology.

1. Introduction

Think tank is composed of experts from various disciplines to give advices and suggestions for decision-makers in economic, political, cultural, social, military, and diplomatic aspects. It also provides consulting research institutions with the best ideas, theories, methods and strategies. At present, think tank, as an integral part of national soft power, has played an important role in influencing governmental decision-making. In recent years, China has attached noticed the great importance to the development of think tank. In April 2013, General Secretary Xi Jinping first proposed the construction of 'new type of think tank with Chinese characteristics'. On January 20, 2015, the General Office of the CPC Central Committee and the General Office of the State Council jointly issued Opinions on Strengthening the Construction of New Type of Think Tanks with Chinese Characteristics, clearly stating that 'the new type of think tanks with Chinese characteristics mainly studies strategic issues and public policies, and is regarded as a non-profit research and consultation institution that serves the Party and the government in scientific and democratic decision-making'. 'The current science and technology innovation has increasingly supported and led the economic and social development'. Therefore, think tank has increasingly played a crucial part in the scientific and democratic decision-making for the Party and the government. On 30th May, 2016, General Secretary Xi Jinping clearly stated that 'it is necessary to speed up the establishment of scientific and technological decision-making mechanisms to support administrative decision-making through science and technology consultation, strengthen science and technology decision-making system, and build high-level science and technology think tank' at the National Science and Technology Innovation Conference, the Academician Conference of the two academies, and the Ninth National Congress of the China Association for Science and Technology[1]. The modern new type of think tank mainly studies the development trend of science and technology at home and abroad from the perspective of science and technology and its law, conducts scientific evaluation, and makes predictions so as to provide advice and suggestion on major science and technology innovation in economic and social development, as well as on national science and technology strategy, planning, layout, and policy-making.

Therefore, China Association for Science and Technology should undertake the important mission of building a new type of think tank for the Party and the nation[2,3], and meet the requirement of national scientific and technological, economic and social development. It will also benefit to the expansion and upgrading of China's science and technology cause. In addition, China Association for Science and Technology is served as the bridge and link between the Party and the government to connect with science and tech-

Corresponding author: Wang Yinqiu. wh6509@yahoo.com.

nology workers, and has unique organizational and academic advantages. It has the responsibility and obligation to build a science and technology innovation think tank for the Party and the government in scientific and technological fields.

2. The Status Quo of Think Tank Development at Home and Abroad

In response to various economic and technological problems brought about by globalization, many countries and institutions around the world make constant efforts to strengthen the construction of think tank. In foreign countries, especially in some western developed countries, the research on think tank started early. The earliest think tanks can be traced back to the period of World War Ⅱ. With the rapid development of the economy, national government decision-makings are increasingly dependent on think tanks. In recent years, the development of foreign think tanks has the following characteristics. The first is that the number of think tanks has increased rapidly. According to the Global Think Tank Evaluation Research Report, the number of global think tanks has grown significantly from 5,080 in 2007 to 6,846 in 2015[4]. The second is the intelligent positioning and clear development goal of think tanks. In the process of think tank development, think tanks are positioned in institutions that influence public policies and support government decision-makings with the goal of producing research results with high quality, independence, and influence. Thirdly, think tanks should have talents and system guarantee. A reasonable talent utilization and talent evaluation mechanism should be established. A wide range of talents can be gathered through hiring, contract and access system, as well as other forms. The procedure for think tanks' participation in decision-making should be legally clarified. Great efforts should be made to promote the construction of international network think tanks. The results of think tanks should be vigorously promoted through multiple channels and forms. Fourthly, the proportion of various think tank institutions should be reasonable. There are state-funded official think tanks, government-funded semi-official think tanks, and non-governmental think tanks. In the United States, for example, 75% of the existing think tanks are independent and separate non-governmental think tanks, and 15% ~ 20% are partially semi-independent and semi-official think tanks, which can better reflect the independence and objectivity of think tanks so as to produce convincing decision-making consultation results. Fifthly, the focus of think tank research has shifted from macro politics to micro people's livelihood. In recent years, studies have shown that foreign think tank research has shifted its focus to people's livelihood of science and technology innovation and health care from macro politics of government governance and risk management, which indicates that think tanks are serving the economic and social development[5–7].

Compared with the development of foreign think tanks, China's think tanks started late with little research on think tank in-depth theory. China's think tanks have developed rapidly thanks to the issue of Opinions on Strengthening the Construction of New Type of Think Tank with Chinese Characteristics and pilot work of think tanks. At the same time, the Party and the nation vigorously advocate the construction of science and technology innovation think tanks with Chinese characteristics based on China's reality. Therefore, science and technology think tanks have been growing significantly. Academy of Social Sciences, Party and administrative schools strive to become well-known think tanks with international influence. Chinese Academy of Sciences, Chinese Academy of Engineering, and China Association for Science and Technology play an important role in technological innovation, and strive to build high-end science and technology think tanks. Universities and colleges across the country and local institutes (associations) give full play to their own advantages and endeavor to build new type think tanks. Despite the great progress in China's science and technology innovation think tanks in recent years, there are a number of problems to be solved, such as lack of in-depth theoretical research compared with foreign think tanks, unreasonable talent utilization and evaluation mechanism, imperfect legal guarantee system, limited types of think tank institutions and lack of independence. Besides, think tank research focuses on primary theories featured by the concept and foreign study of think tanks. Therefore, there is a long way to go for China's think tanks, especially science and technology innovation think tanks[8,9].

3. Main Advantages and Characteristics of Think Tanks in Association for Science and Technology System

In the era of knowledge economy, think tanks play an integral part in knowledge innovation and dissemination, and power national and social progress and development. Based on the positioning of China Association for Science and Technology and its core tasks, think tank has obviously different characteristics and advantages from other think tanks existing in China, featured by organization, professional talents and resource sharing.

Firstly, the think tank in Association for Science and Technology boasts unique organizational advantages. A wide range of science and technology workers are mobilized to focus on the work for the Party and the government, conduct in-depth investigation and research, and actively give ideas and suggestions, which is beneficial to the development of Association for Science and Technology. As the organization for Chinese science and technology workers, China Association for Science and Technology is not a government department and not subject to the administrative restriction of the government, so it can give full play to its talents, various disciplines, extensive connection and unique position. Therefore, during the construction of think tank, China Association for Science and Technology has adopted the think tank model of 'small center, large periphery'. To be more specific, National Academy of Innovation Strategy is regarded as the center, relying on local Association for Science and Technology with location advantages and decision-making consultation basis, such as Chongqing. Then it conducts in-depth research on major difficulties and forward-looking issues of regional economic and social development, and optimizes industrial layout and collaboration so as to serve regional coordinated development and implement major national strategies. At the same time, CAST's affiliated societies work together to built a number of professional research institutes in the critical areas of national economic strategies and development, like information technology, materials technology, energy technology, intelligent manufacturing, life sciences, energy conservation and environmental protection, modern agriculture, health, military-civilian integration, resources and environment. It aims to make scientific predictions on the trend of science and technology development, propose the possible fields and directions of breakthroughs in the frontier of science and technology, and provide directional suggestions of reserving the basic scientific and technological knowledge for China's future core competitiveness[10, 11]. Meanwhile, China Association for Science and Technology has also established the nation's only survey site system with science and technology workers as the survey object. The whole system has 504 survey sites across the country, over 300 provincial survey sites and more than 100 prefecture-level survey sites, which can fully cover the institutions, regions and industries with a good number of science and technology workers. Besides, the system can report the thoughts, work, life and flow of science and technology workers, and related opinions and demands in real time.

Secondly, the think tank in Association for Science and Technology has the advantages of talents and disciplines. Association for Science and Technology is home to a majority of science and technology workers. In recent years, China's science and technology professionals have increased significantly. Statistics suggest that as of the end of 2014, the number of national science and technology workers in various professions has exceeded 81 million, and the number is expected to exceed 100 million by the end of 2017. Currently, China Association for Science and Technology has a total of 207 national societies, involving professional societies, associations and research associations in science, engineering, agriculture, medicine and inter-disciplines, covering most of natural sciences, engineering and technical fields and some integrated inter-disciplines. The national societies are regarded as the high-end talent pool. By taking its advantage, China Association for Science and Technology has built an expert network in science and technology, and has taken major projects as the link to attract a wide range of experts in science and technology, industry and government. Therefore, the Association has stable connection with high-end decision-making consultants and subject experts with solid profession, excellent academic level and strategic thinking. Based on various expert networks, especially in the field of science and technology, the think tank in Association for Science and Technology can give full play to talents and professions in each discipline centered on key technologies. It aims to make scientific predictions on the trend of science and technology development, propose the possible fields and directions of breakthroughs in the frontier of science and technology, make rational planning between technological breakthroughs and industrial development, guide related industries to develop in a new direction with refinement, depth and specialty, and finally provide directional suggestions of reserving the basic scientific and technological knowledge for China's future core competitiveness.

Thirdly, the think tank in Association for Science and Technology owns rich resources. Due to the role of China Association for Science and Technology in group organization and group work, it can cooperate with internationally famous science and technology innovation think tanks and share the results of think tanks. At the same time, China Association for Science and Technology has also taken the initiative to strengthen links with the National Science Foundation of the United States and the United Nations Educational, Scientific and Cultural Organization (UNESCO), actively communicated with foreign universities with international influence in such fields as innovative assessment, culture, talents, as well as scientific and technological data, and shared new ideas, methods, tools and data related to think tanks. Therefore, the abundant resources provide a strong guarantee for the construction of think tanks in Association for Science and Technology.

Finally, complete content and organizational editing before formatting. Please take note of the following items when proofreading spelling and grammar:

4. Main Experience in Building Association for Science and Technology System

Adhering to Opinions on Strengthening the Construction of New Type of Think Tank with Chinese Characteristics, Implementation Plan for Deepening Reform of Association for Science and Technology System, The 13th Five-Year Plan for the Development of China Association for Science and Technology, Opinions on Building High-Level Science and technology Innovation Think Tanks Issued by China Association for Science and Technology, and The 13th Five-Year Plan for the Construction of High-Level Science and technology Innovation Think Tanks by China Association for Science and Technology, the National Academy of Innovation Strategy, as the key part in building national think tanks by Association for Science and Technology, aims to establish national innovative think tanks and high-level science and technology innovation think tanks based on the development needs in national economy, society, science and technology, the Party's group organization and the reality of Association for Science and Technology System. There is some experience in the process of building think tanks.

Firstly, the think tank model of 'small center, large periphery' should be adopted. China Association for Science and Technology has been approved to establish National Academy of Innovation Strategy by State Commission Office for Public Sector Reform. It serves as a research institution, conducts strategic research centered on innovation evaluation, talents and environment, and analyzes survey statistics and literature information. Taking National Academy of Innovation Strategy as the core carrier, strategic platform and network hub, we have built the think tank model of 'small center, large periphery', and integrated national societies, local Associations for Science and Technology, universities and colleges, research institutes, enterprises and professional institutions. Therefore, the innovative think tank in Association for Science and Technology has been built featured by one nuclear and multiple poles, and network. At present, a number of professional and interdisciplinary research institutes conducting strategic research on major issues have been jointly established by National Academy of Innovation Strategy and China Information Technology Association, Intelligent Manufacturing Alliance of CAST Member Societies, and China Highway and Transportation Society. Besides, these research institutes rely on provincial and municipal Association for Science and Technology, such as Chongqing and Jiangsu, establish a number of regional strategic research bases, and carry out strategic research on major issues commonly seen in regions by adhering to major development strategies such as the 'Belt and Road Initiative' and the Yangtze River Economic Belt. A joint laboratory with distinctive professional features has been built by Tsinghua University, Inspur Group, Elsevier and enterprises.

Secondly, it conducts research on major issues related to science and technology and makes a third-party assessment[12, 13]. China Association for Science and Technology, especially National Academy of Innovation Strategy, has organized science and technology workers to conduct in-depth investigations and studies on key and difficult issues in economic and social development, as well as making extensive exchanges and discussions since the beginning of think tank research. Through in-depth research and discussion, we will promptly and acutely discover the problems in science and technology policies, plans and major engineering constructions, and the new development trend in the science, technology, economy and society. Therefore, we have made a great deal of important research results in think tank, which has attracted the attention of the central leadership. Some of the results have been added into the decision-making process and played an important role in solving related issues. Since 2015, China Association for Science and Technology has actively carried out third-party evaluation. It not only completed the evaluation of implementing ‘Mass Entrepreneurship and Innovation’ commissioned by the State Council, but also attracted wide attention among all sectors of the society and being appraised by leading comrades such as Premier Li Keqiang. Besides, it also made the assessment of the implementation of policies and measures for the ‘construction, use and management of grassroots public health care facilities’ entrusted by the State Council, which was highly praised by leading comrades such as Vice Premier Liu Yandong. In addition, it was commissioned by the National Science and Technology Reform Office, and it carried out the evaluation of policies implementation for ‘professional title system reform and income distribution incentive mechanism for high-level talents’, ‘863 Plan’s social impact during the 12th Five-Year period’, and the construction of key laboratories. The third-party assessment has played an increasingly leading role in the construction of national high-end think tanks by China Association for Science and Technology.

Thirdly, the scientific talent training mechanism and model has been established for think tanks. In order to improve the capacity of think tanks and build flexible employment mechanism, National Academy of Innovation Strategy has offered various research jobs, such as visiting researchers, visiting scholars and special experts, relying on societies and associations to attract high-level talents for think tanks. Great efforts has been made to the talent training for think tanks by assigning staff from China Association for Science and Technology, project cooperation, short-term training, and special training. These scientific measures for talent training will build a high-level and high-quality team for think tanks.

5. Summary

This paper briefly describes the main work and achievements of China Association for Science and Technology and National Academy of Innovation Strategy in the process of building think tanks. It also analyzes the main advantages and characteristics of think tank in Association for Science and Technology System, and summarizes the main experience of building Association for Science and Technology System with the purpose of promoting better development of think tanks.

Acknowledgments

The research results have been funded by ‘High-end Technology Innovation Think Tank Youth Project’ in China Association for Science and Technology (No. DXB-ZKQN-2016-005 and DXB-ZKQN-2017-010). We would like to express our heartfelt thanks.

References

[1] Xi Jinping. *Strive to build a powerful scientific and technological country in the world*[Z]. The speech on the National Science and Technology Innovation Conference, the Academician Conference of the two academies, and the Ninth National Congress of the China Association for Science and Technology, 2016.

[2] The General Office of the CPC Central Committee, the General Office of the State Council. *Opinions on strengthening the construction of new type of think tanks with chinese characteristics*[Z], 2015.

[3] The General Office of the CPC Central Committee. *Implementation plan for deepening the reform of the Association of Science and Technology*[Z], 2016.

[4] Jing Linbo, Wu min, Jiang Qingguo, et al. *Global think tank evaluation report*[J]. Social Sciences in China Review, 2016(1):90-124.

[5] Li Jianjun. *Research on global think tanks*[M]. People’s Publishing House, 2010.

[6] Ding Minglei, Chen Baoming. *Thoughts and suggestions on the construction of science and technology, innovation think-tank system with chinese characteristics*[J]. Science and Technology Management Research, 2016, 36(5):10-13.

[7] Zhang Zhiqiang, Su Na. *Trends and characteristics of global think tanks and suggestions for chinese think tanks construction*[J], 2016, 1(1):9-23.

[8] Zhu Hongbo, He Yu, Ma Yantao. *Construction of scientific and technological research think tank: status and countermeasure*[J]. Journal of Guizhou Institute of Socialism, 2016(3):54-64.

[9] Wang Chen, Wang Jianan. *Visualized comparative analysis on researches of think tanks at home and abroad: Comment on the new tendency of China's think tanks*[J]. *Think Tank:Theory & Practice*, 2016, 1(5): 18-27.

[10] China Association for Science and Technology. *The 13th Five-Year Plan for the development of China Association for Science and Technology*[Z], 2016.

[11] China Association for Science and Technology. *The 13th Five-Year Plan for the construction of science and technology, innovation high level think tank for China Association for Science and Technology*[Z], 2016.

[12] China Association for Science and Technology. *Main points of works on think tank of China Association for Science and Technology in 2016*[Z], 2016.

[13] China Association for Science and Technology, *Opinions on the construction of science and technology, innovation high level think tank for China Association for Science and Technology*[Z], 2016.

The Responsibility of Science Diplomacy for Society: The Story of the UK

Fu Zhenyu, Shi Lei

National Academy of Innovation Strategy, CAST, Beijing, China

Abstract: The function performed by science and diplomacy are well studied, even average citizens have recognised by the society as a whole. The intersection area of science and diplomacy, namely science diplomacy still lacks attention. Actually, science diplomacy could be an effective way people can utilise to transform scientific advancement to make ourselves better off. Within a country, science popularisation links scientific community with wider society. In international community, science diplomacy could finish a grander task, which is to tie scientific community in one country with other societies. Besides, science diplomacy could also help to facilitate scientists and technicians to shoulder their social responsibilities. This paper will first discuss what science diplomacy is, secondly, the author will review the development of science diplomacy in the UK. Then the current political and administrative-setting that regulate British science diplomacy activities will be analysed. Last but not least, the lessons regarding how to bridge the gap between science community and the rank and files drawn from British science diplomacy story will be demonstrated. Further research in this area will also be discussed.

Keywords: Science Diplomacy; The UK; Responsibility; Society

1. Science Diplomacy: A Very Short History

According to The Royal Society's report *New frontiers in science diplomacy Navigating the changing balance of power,* there is a long history of scientists supporting international co-operation (Koppelman, et al., 2010). Professor Lorna Casselton said, 'The Royal Society has a long history of using science to rise above military conflict and political and cultural differences. My post was instituted in 1723, nearly 60 years before the British Government appointed its first Secretary of State for Foreign Affairs' (Ibid, p1). His remarks pointed to Philip Zollman who became Foreign Secretary of the Royal Society in 1723. His role was to maintain regular correspondence with scientists overseas to ensure that the Society's Fellows remained up-to-date with the latest ideas and research findings. The significance of science diplomacy, however, was not recognised until the Second World War. It is after the use of the first atomic bomb that scientists actively engaged in activities that can bring peace to the world. Joseph Rotblat, who once joined the Manhattan Project but quitted after the war with Germany ended won Nobel Peace Prize 'for efforts to diminish the part played by nuclear arms in international affairs and, in the longer run, to eliminate such arms'[1]. During the cold war period, scientists in NATO, the USA and the Soviet Union performed important function in science diplomacy activities. They have contributed to a series of inter-governmental agreement which made the world safer.

After the end of the cold war, science diplomacy was not as critical as usual. The possibility of the world being destroyed in a nuclear war reduced to a large extent, science diplomacy's major theme during the cold war was to curb the irrationality of the political leaders, while nowhere can one find this irrationality after the end of the cold war. Recently, science diplomacy has discovered new problems to tackle with, consequently the new century has seen a revitalisation of science diplomacy. Dr Nina V. Fedoroff, Science and Technology Adviser to the Secretary of State and to the Administrator of the U.S. Agency for International Development pointed out that 'Science diplomacy is the use of scientific collaborations among nations to address the common problems facing 21st century humanity and to build constructive international partnerships. There are many ways that scientists

1 https://www.nobelprize.org/prizes/peace/1995/summary/, visited on 2018-9-1.

can contribute to this process' (Fedoroff, 2009). The UK established Science and Innovation Network in 2001, sending representatives all over the world, in the hope of servicing British scientists and their international partners. These activities marked a new change in science diplomacy from reduce conflicts to create friendship. This change echoes the changes appears in international community in the aftermath of the cold war. Science is no longer automatically linked with military and international politics, to do science diplomacy for science's sake or to do good to the whole world has gradually become the focal point of science diplomats. The Royal Society proposes that people's interest towards science diplomacy rises when a sharp change in international relations is spotted. Except for some multi-national organisations and sovereign states, a new and complicated international diplomatic network aimed at coping with grand scientific and technological problems is also being constructed. The biggest challenge facing relevant people is how to define and strengthen the role scientists could play in this ever-evolving governance and diplomacy network. Studies regarding this challenge are still in rudimentary stages. As is widely known, scientists and diplomats are two important groups of participants in science diplomacy activities, in what way could they make contribution to science diplomacy ought to be a key topic for studies concerned.

Generally speaking, Science diplomacy originated from scientists' intention of being informed their international counterparts' latest research outcomes and discoveries through diplomatic channel, thus leading to scientific community's common prosperity. Besides, science diplomacy is also useful in helping scientists setting their personal goal of research. In its burgeoning years, diplomats working for governments did not directly participate in science diplomacy. The operation of science diplomacy was mainly in a scientists' autonomous condition. After the two world wars, people all over the world reached a consensus on the hideous power of science and technology, which gave rise to governments' regulation on science diplomacy. Science and technology became a carrier for pushing the bilateral and multilateral communication. The recent development of science diplomacy can be attributed to the development of science and technology itself, but the new development in diplomacy area should get more credits. These developments include the diversification of the participating subjects and the newly designed mechanism for solving the problems that threaten the whole world. As time goes by, science has removed its veil of apathy, replaced with a more humanistic one, science is not the toy of the upper class any longer, its existence is almost everywhere. The function of science diplomacy is to accelerate and add an exotic flavour to this process.

2. Science Diplomacy: Concept

Different scholars have made different definition about science diplomacy. The most famous and influential one is given by The Royal Society, in a brochure: New Frontiers in Science Diplomacy, the Royal Society suggests that:

'Science diplomacy' is still a fluid concept, but can usefully be applied to the role of science, technology and innovation in three dimensions of policy:

informing foreign policy objectives with scientific advice (science in diplomacy);

facilitating international science cooperation (diplomacy for science);

using science cooperation to improve international relations between countries (science for diplomacy).

The first category stresses that issues contained with scientific elements entail scientific information and judgement to deal with, such as climate change, outer space utilisation, contagious disease, intellectual property rights, all those issues have scientific dimensions and no one country will be able to solve these problems on its own. The second category refers to use diplomacy as a tool to augment international scientific co-operation. Diplomacy can provide guarantee for scientific and technological co-operation. Many countries discuss critical scientific and technological problems in bilateral summit and sign governmental agreement that enhance international co-operation in science. This is not only a sign of mutual friendship, but can offer a comprehensive framework for co-operation among scientists in the two countries. Several big scientific projects are very costly and have high risks, such as International Thermonuclear Experimental Reactor, Large Hadron Collider, these projects require a long and reliable investment commitment made by relevant governments, which cannot be got without resort to science diplomacy. Science for diplomacy refers to improve bilateral or multilateral relations through science and technology co-operation. The mutual visiting activities started since 1972 between Chinese and American scientist's delegation enhance mutual understanding between Chinese and American science community, which paved the way for further development of Sino-America relationship. Not too long after establishing official diplomatic relationship in 1979, China and America signed governmental science and technology co-operation agreement. This agreement did not only promote the development of science and technology, but is very critical for the development of Sino-America relationship.

In short, this three-dimensional definition demonstrates a vivid picture of science diplomacy to the audiences. Actually, this is not a definition, it is a description about what science diplomacy is, what it contains. These three dimensions of science diplomacy are not incompatible, when we talk about a specific science diplomacy action, it may incorporate all the three dimensions mentioned above. To some extent, the Royal Society's definition is of significance in helping the laymen understand the connotation of science diplomacy. Meanwhile, it is also a superb analytical tool for researchers to probe into the realm of science diplomacy. It can be used to analyse can compare different science diplomacy activities, even the purpose of conducting science diplomacy might come out. Just as the Royal society proposed, science diplomacy is still a fluid concept, we need to look at other definitions.

United Nations has its own definition about science diplomacy, the difference is that it does not use the term science diplomacy, it uses science and technology diplomacy. According to a report publicised in 2003 by United Nations Conference on Trade and Development (UNCTAD), science and technology diplomacy means 'the provision of science and technology advice to multilateral negotiations and the implementation of the results of such negotiations at the national level. It therefore covers activities at the both international level and national level pursuant to international commitments' (UNCTAD,2003). This report is based on the view that 'advances in science and technology have become key drivers in international relations, and knowledge of trends in key fields is an essential prerequisite to effective international negotiations'(Ibid). This definition, especially the term being used by UN might be closer to Chinese context, as it attaches due importance to the technological elements. Science is the knowledge of what, while technology is the knowledge of how. In fact, science itself can hardly influence the world, except for some great discoveries that changed people's way of viewing the world and society, such as Newtonian theory of physics and Darwin's theory of evolution. Technology, however, can easily exert its influence on human society. Science diplomacy create opportunity for the dissemination of advance technology from its original birthplace to elsewhere, thereby enlarge technology's application area, it is the society that benefits from this process. Apart from that, the interactions of science and technology society from different countries have the potential of ignite the passion of the most gifted minds, scientists and technicians might be inspired so that they can produce more useful knowledge. This definition depicts the real science diplomacy process better, it is one of the best general definitions of science diplomacy.

Grunnet, an oversea student of Fudan University defined science diplomacy as the international communication and co-operation among scientists and scholars, it advances science and make different countries' people understand each other better, appreciate each other more simultaneously. The ultimate goal of conducting science diplomacy is to serve one's country's interest by changing other people's mind and action (Grunnet, 2010). This definition exposes the motivation of doing science diplomacy, namely to serve a specific country's interest. The aforementioned definitions are quite idealistic, while this definition is more realistic. The author proposed that both the hostile part and the friendly part are integral in science diplomacy, we should take both into consideration while studying science diplomacy.

The most commonly definition of science diplomacy used by Chinese scholars was given by Zhao Gang. According to Zhao Gang, science diplomacy's subjects include the supreme leader of sovereign states, diplomatic agencies, science and technology institutions, professional agencies (such as Chinese Academy of Sciences, National Natural Science Foundation of China) and enterprises, their aims are promoting the development of science and technology and realise the sustainable development of economy and society, the principle of science diplomacy is reciprocal and common development, these subjects adopt the form of negotiation, visit, international conferences establish research agencies with other countries, regions and international organisations so as to facilitate bilateral and multilateral communication and co-operation(Zhao, 2007; Zhao et al., 2008). Based on UNCTAD's definition, Zhu Yalan et al. make a similar definition about science diplomacy, it refers to sovereign states through their representatives (Department of diplomacy or department of science and technology, professional institutes or enterprises), under the principle of safeguarding national interests or serving national development strategy's guidance, aimed at science and technology advancement, economic and social development, organising science and technology co-operation and communication with other nations, regions or international organisations. It usually takes the form of visiting, negotiation and establishing research institutes (Zhu et al., 2016). Activities meet the above criteria can be attributed to science diplomacy. He also distinguishes three types of science diplomacy:

Table 1 Three types of sciecne diplomacy.

Type	Content
International science and technology co-operation and communication	Technology trade, international conference, research collaboration, personnel exchange, establish international science and technology co-operation base
International technical assistance	Technical service, technical training, personnel training, provide data, material, equipment, build up science and technology infrastructure
Technology control and sanction	Technology export control, Arms sale injunction, barrier on technology trade.

Besides, Zhao Gang also distinguishes science diplomacy and science co-operation, he is oppose the idea that regards the two concepts as the same. Science diplomacy is not direct and concrete international science co-operation; it utilises diplomatic approaches to promote international science co-operation. It makes relevant policies and conduct relevant international co-ordination that lead to better international science co-operation, it also paves the way for further international co-operation.

Though Zhao Gang's definition is being widely used, some other Chinese scholars still hold different views with him. Several scholars discuss the differences between science diplomacy and science co-operation too. Wang Baoqing proposed that science diplomacy is just one part of science co-operation, the latter has a wider scope than the former. Science diplomacy mainly adopts diplomatic approach in order to enhance inter-governmental relationship (Wang, 2009). The major actors of science diplomacy are governments, sovereign states and international organisations. Compared with science diplomacy, science co-operation's participants are not necessary to be officials or official organisations, individuals and academic institutions are key players as well. It seems that Wang holds that the following views, firstly, all science diplomacy activities belong to co-operations, secondly, all science diplomacy subjects must have official backgrounds. If these two situations can be met, Wang's conclusion is correct. But if one refers to the three types of science diplomacy pointed out by Zhao Gang, the first situation can only be viewed as a partial reflection of science diplomacy. If we take the fast development of track two diplomacy and people to people diplomacy, the second situation cannot depict a complete picture of science diplomacy either. As a result, science diplomacy and science co-operation have intersections, but their relationship is not one is the other's subset.

Fan Chunliang points out that one possible flaw of Chinese scholars' study in relation to science diplomacy is that they only cover the friendly and harmonious part of science diplomacy, the more hostile part is excluded (Fan, 2010). These researches tend to equate science diplomacy with science co-operation. This can be displayed by looking at the definitions of science diplomacy given by Chinese scholars, science and technology co-operation almost appears in every definition. Just as what has been mentioned, some scholars do have discussed the difference between science diplomacy and science co-operation, the discussion is not deep enough. These discussions concentrate on the differences in subjects and forms, not on the content itself. The discussants seem unconsciously turn a blind eye to other forms of science diplomacy. Diplomacy, as is widely acknowledged, is not always about co-operation or peace, negotiations on war and conflict are not rare, the same goes for science diplomacy. Based on the renewed understanding towards diplomacy, Fan Chunliang links science diplomacy with public diplomacy.

Traditional diplomacy is diplomacy between governments. Public diplomacy, however, is conducted by governments, private agencies and institutions, they wish to promote national interests and improve the country's international influence by change the view of targeted country's citizens towards their own country. Science produces public goods; therefore, it could play a key role in public policy. To conclude, the overemphasis on international science and technology co-operation might ignore the social responsibility that science diplomacy should assume. Focusing on science co-operation can benefit the development of science, but that's not enough, science diplomacy could do more. The science for diplomacy dimension can bridge the gap between scientific discoveries and social benefits. Apart from directly solving social problems, science's social function can be met by helping government achieve political goals, only by walking on two legs can science realise its potential of serving society.

Zhang Yiyan and Zhang Ning make a careful and in-depth analysis on the difference of science diplomacy and science co-operation. They are for the idea that science diplomacy has national interest and goal. Science co-operation could happen without the attendance of governments, it has only scientific or commercial considerations. Science co-operation aimed at contributing to the advancement of science and technology, it usually produces a win-win outcome. All the parties involved in science co-operation can harvest fruits. To the contrary, different parties engage in science diplomacy often have interest conflicts, the results of science diplomacy might be biased against some parties. (Zhang and Zhang, 2017) In conclusion, the driving forces behind science diplomacy and science co-operation are different, they are not on a par with each other. In some cases, they are possible to yield same results, but to equate the two is a distortion of reality. To some extent, science diplomacy is closer to society, as it always bears national interest in mind, which can be easily transferred into social development indicators. Science co-operation is a bit further away from its social responsibility, although it can be properly orchestrated, its outcomes could be used to augment social welfares.

By summarising the above definitions, the author finds that most definitions consist of the following component, the subject of science diplomacy, the major forms and content of science diplomacy, the purpose of conducting science diplomacy and the admirable outcomes of science diplomacy. The disputes against these elements of science diplomacy always appear. For the subjects of science diplomacy, some scholars prefer to restrain the subjects to official entities, other argues that non-governmental actors should be taken into consideration. When it comes to the purpose of science diplomacy, a group of scholars pay due attention to academic ones while another group of people focus on the political and social goals. This paper is not going to take part in this debate, and the author argue here that it is not very significant to try one's best just to create another fluid definition. The reason why a brief introduction about science diplomacy is given in this first two sections is that the author wishes to provide useful background information at the very beginning of this talk, so that the audiences can familiarise themselves to the theme of this paper. The question is that, for the convenience of further discussion, a definition of science diplomacy is supposed to be given here, no matter how fluid and unauthoritive it sounds. To the best of the author's knowledge, and based on the topic of this speech, science diplomacy would be defined as various forms of international activities conducted by states, inter-governmental organisations, enterprises, academic institutes and individuals, in the hope of realising specific national goals via the application of science and technology. As can be seen, both unofficial and official actors are included, national interest consideration is necessary, the outcome could be negative or positive. Pure international academic communication will not be regarded as typical science diplomacy activities. The three dimensions of science diplomacy will be automatically applied to all relevant actions, it could be a useful analytical tool.

3. British Science Diplomacy

As one of the most influential nations of the world, the UK has been making great contributions to mankind. In science, the UK is the home to a great many scientific giants, Isaac Newton Francis Bacon and Charles Darwin, to name but a few. They have left deep and profound impression to human society. In industry and technology, James Watt improved the quality and reliability of steam engine and thus release the energy of machines, thereafter, human enters a new epoch. Not only have these people's achievements changed the development trajectory of science and technology, but they have altered the course of human society as well. Besides, British politicians and diplomats have also played a pivotal role in history. The famous Splendid Isolation foreign policy which both guaranteed British involvement in European affairs and avoid directly establishing formal alliances with other European powers. British political influence is always sustained by its scientific and technological resources, at the same time, the prosperity of British science and technology relies on a stable international environment created by those sophisticated, if I may say so, politicians. It is safe to say that the contemporary world teems with British imprints As a result, a comprehensive study on the interactions of science and diplomacy in the UK will be of significance. This section looks at how British science diplomacy bridges the gap between British science community and society in the hope of finding some general laws.

1) A Brief History of British Science Diplomacy

Since the founding of the Royal Society, it had been paying special attention to correspondence with foreign scientists. As early as 1660s, the Royal Society established a committee specialised in dealing with its members' letters with their foreign counterparts. But during that period, the members of the Royal Society wrote and receive letters to scientists on European continent on behalf of themselves, rather than in the name of the Royal Society.

As time goes by, the quantity of these letters rise rapidly, as a result, the Royal Society establish an office that dealing with this kind of affairs. It is reported that the first foreign secretary is Philip Henry Zollman, he was appointed this post on April 11th, 1723. In fact, at time the full name for this post is called 'Assistant to the Secretaries for Foreign Correspondence', it is the modest antecedent of the Society's present-day Foreign Secretary (Massarella, 1992). During his interview, Zollman 'said he had knowledge in the following Languages, Latin, French, Dutch, German, Italian', which made him prefect candidate for this job. His German nationality, however, was a big stumbling block. Even Isaac Newton was against him at first. It is his language aptitude and strong recommendations from some distinguished government servants that secured him the post (Ibid). His role was to maintain regular correspondence with scientists overseas to ensure that the Society's Fellows remained up-to-date with the latest ideas and research findings (Koppelman, et al., 2010). This event is the earliest record of intended international science co-operation. But according to the definition of science diplomacy in this talk, strictly speaking, this is not science diplomacy as it did not have a clear national interest consideration. Actually, prior to the industrial revolution, the power of science and technology had not been recognised, so that they were not regarded as useful tools that can augment national interest. Full-fledged science diplomacy activities became usual after the end of the Second World War. Nevertheless, early years semi-science policy activities are still worthy to be studied. This study could help us understand the development trajectory of international science and the relationship between science and society.

Before the Second World War, the UK appointed its first foreign science representative, Sir Charles Galton Darwin(the grandson of Charles Darwin). During the Second World War, he spent much of the war years working on the Manhattan Project co-ordinating the American, British, and Canadian efforts[1]. Another famous British science diplomat is Joseph Needham, his Chinese name is more popular, Li Yuese. The The Needham Question, In Needham's words, 'Why did modern science, the mathematization of hypotheses about Nature, with all its implications for advanced technology, take its meteoric rise only in the West at the time of Galileo [but] had not developed in Chinese civilisation or Indian civilisation?' (Needham, 2004). In 1965, with a retired diplomat whom he first met in China, Needham established the Society for Anglo-Chinese Understanding, which for some years provided the only way for the British to visit the People's Republic of China[2]. Thess are two classic cases in which scientists deeply involved in politics. It can be seen that after the Second World War, the modern form of science diplomacy had become an important type of international communication channel.

During the cold war era, some great scientists

1 https://en.wikipedia.org/wiki/Charles_Galton_Darwin, visited on 2018/9/5.

2 https://en.wikipedia.org/wiki/Joseph_Needham, visited on 2018/9/5.

made a huge contribution to world peace. The Russell–Einstein Manifesto was issued in London on 9 July 1955 by Bertrand Russell in the midst of the Cold War. Both Albert Einstein and Bertrand Russell, are Nobel prize laureates and fellows of the Royal Society. The manifesto highlighted the dangers posed by nuclear weapons and called for world leaders to seek peaceful resolutions to international conflict[1]. Einstein died several days after he signed this manifesto. Thereafter, a conference in Pugwash was sponsored by a philanthropist Eaton. This conference became the first Pugwash Conferences on Science and World Affairs. Rotblat and the Pugwash Conference jointly won the Nobel Peace Prize in 1995 for their efforts on nuclear disarmament. This conference marked a milestone of science diplomacy, the greatest scientists began to walk out of the ivory tower and join the public to care about ordinary people's daily life. It is also a vivid reflection of the increasing influence of scientists could have on global issues. If science and technology had not developed to a extent that the destiny of mankind can be decided by a few people who has the power of using mass destruction weapons, if scientists had not played a pivotal role in the research and development of these weapons, they might not be able to have their suggestions accepted by political leaders. The Pugwash conference is a symbol that scientists started to actively fulfil their social obligation.

Unfortunately, the late 20th century witnessed a decline of science diplomacy, as the tension between the two super powers eased, and the possibility of having a nuclear war almost diminished. The revitalisation of science diplomacy was seen in the 21st century. This time, military plays only secondary role, civilian affairs become the core topic. This is because the trend that science and technology have been bringing remarkable change to ordinary people's life is becoming apparent. The UK did not lag behind in this wave.

In 2001, British government established Science and Innovation Network (SIN), this network has approximately 90 officers in over 30 countries and territories around the world building partnerships and collaborations on science and innovation. SIN officers work with the local science and innovation community in support of UK policy overseas, leading to mutual benefits to the UK and the host country[2]. Science and Innovation Network teams develop country-specific action plans and work to prosperity security influence and development. The Pugwash conference and Science and Innovation Network will be shed more light in the next section.

All in all, the UK has always been a pioneer in science diplomacy activities. It has the longest history of doing science diplomacy and wholeheartedly launches or joins some prestigious science diplomacy programs, British scientists used to hold a detached attitude towards social responsibility, this was due to the fact that science used to be a competition among aristocracies. Seldom did a common man have a chance to start an academic career. The development of science push itself rush towards the general public and tear down the walls that isolate science with society, science can no longer shirk its social responsibility. Science and the society have entered into a new epoch shoulder to shoulder, hand in hand.

2) Some Cases of British Science Diplomacy

(1) *Rotblat and Pugwash Conference on Science and World Affairs*

As what has been mentioned above, one of the most classic cases in which scientists assume their responsibility for society is the Russell–Einstein Manifesto and the ensuing Pugwash Conferences on Science and World Affairs called for in the manifesto. In this case, a group of scientists organise a conference to make their voice heard, their aim is to remind people the lurking danger of nuclear weapons, as this is a very professional field, ordinary people lack enough expertise to realise this. The most important organiser of Pugwash conference is Sir Joseph Rotblat, who shared, with the Pugwash Conferences, the 1995 Nobel Peace Prize 'for efforts to diminish the part played by nuclear arms in international affairs and, in the longer run, to eliminate such arms'[3]. Rotblat believed that scientists should always be concerned with the ethical consequences of their

1 https://pugwash.org/1955/07/09/statement-manifesto/ visited on 2018/9/6.

2 https://www.gov.uk/world/organisations/uk-science-and-innovation-network, visited on 2018/9/6.

3 https://www.nobelprize.org/prizes/peace/1995/summary/, visited on 2018/9/8.

work (Rotblat, 1999). This conference, of course, is an international conference, and Pugwash is a Canadian village, but as both Russell and Einstein are fellows of the Royal Society, and Sir Rotblat has British nationality, it is reasonable to conclude that the UK made the greatest contribution to make Pugwash Conference world famous. After the death of Cyrus Eaton, the prime sponsor of Pugwash Conference, International Student/Young Pugwash was established. This is an international organization that promotes awareness and action among students and young professionals in relation to ethical implications of science and technology policy, particularly matters of international security and weapons of mass destruction. As a matter of fact, Students have been invited as participants in Pugwash Conferences since the early 1970s. Following interest to create a forum specifically for students, Young Pugwash was established accordingly.

Not too long after the first Pugwash conference was held, did it begin to exert its influence in international community. Pugwash's first fifteen years coincided with the Berlin Crisis, the Cuban Missile Crisis, and the Vietnam War. Pugwash played a useful role in opening communication channels during a time of otherwise-strained official and unofficial relations. It provided background work to the Partial Test Ban Treaty (1963), the Non-Proliferation Treaty (1968), the Anti-Ballistic Missile Treaty (1972), the Biological Weapons Convention (1972), and the Chemical Weapons Convention (1993). It is not exaggeration to say Pugwash Conference and Rotblat made a huge contribution to international security.

In this case, science diplomacy performs the function of predicting the future earlier than common people. Scientists are the creator of mass destruction weapons, it is also the same group of people that foresaw the risks of these weapons. They tried their best to prevent the using of nuclear bombs and disseminate the knowledge concern the possible outcomes of putting nuclear bombs into use. They made the best use of their expertise and successfully swayed the opinion of politicians, who possessed the power of making decision about whether to use nuclear weapons or not. Scientist did not confine themselves to nuclear weapons only, they did not hesitate to utilise their reputation as long as it can help to reduce casualties. To save people's lives might be the loftiest responsibility one can imagine, Pugwash conference might be the most outstanding attempt to achieve that.

(2) *Science and Innovation Network (SIN)*

Although the UK is still a global leader in science and innovation, it faces various kinds of challenges. The UK government regards international collaboration is essential to maintaining the excellence of the UK's research base and the competitive advantage of British innovative businesses, for filling capability gaps and for ensuring value by leveraging international resources. Science diplomacy is one way to maintaining British science excellence (and reputation) and supporting innovation ensures the UK is a partner of choice, and helps UK companies with ambitions for rapid global growth[1].Actually, doing science diplomacy is the strategy the UK adopts to both boost its economy and science and improve its global influence and image by co-ordinating with foreign countries. The former can be viewed as take domestic social responsibility while the latter is a way to take international social responsibility.

Some examples will be introduced here. First, The UK has been at the forefront of global efforts to tackle antimicrobial resistance (AMR) and SIN has played a key role in coordinating international engagement to rally support. The £195m Fleming Fund aims to support improved AMR surveillance and response capability in low- and middle-income countries. SIN South Africa is collaborating with the Fund on a regional conference in November 2016 which will pave the way for funding. Apart from that, the UK led to a new, global approach to tackling dementia. SIN leveraged this momentum to engage over 100 of the next generation of dementia experts ('Future Leaders') in a series of global workshops that aimed to develop innovative ideas to address the global dementia challenge[2].

It is apparent that the UK is trying to kill two birds with one stone. It seeks methods to keep its prosperity and solve common problems simultaneously. If we refer to the three-dimension definition of science diplomacy made by the Royal Society, it belongs to science for diplomacy category. By promote the accomplishment of social responsibility, the UK get credits in its national image.

---

1 https://www.gov.uk/world/organisations/uk-science-and-innovation-network ,visited on 2018/9/8.

2 https://www.gov.uk/government/collections/uk-science-and-innovation-network-impact-stories, visited on 2018/9/8.

4. Discussion

In this section two interesting and important question will be raised. The first is about the relationship between science and social responsibility. It is useful to give science a brief definition before starting this discussion. As is widely known, in ancient Chinese classics, there was no such a word as Science, this word is borrowed from western language. In modern Chinese society, science is sometimes enshrined as if it can solve all the problems, not only scientific and technological problems, but also social problems, in some other occasions, however science is discredited as the cause of problems. In this discussion, science refers to the pursuit of knowledge about our world through empirical research and reasoning. In ancient Greek, science is one way to distinguish aristocracies with average citizens. Greek scientists or philosophers can finish their task of doing science and taking their social responsibility at the same time. In contemporary society, scientists might regard their research as their job or a scared mission, as for social responsibility, it is outside their research area. For some scientific luminaries, they concentrate on their research when they are young and productive, when they are not producing any significant useful knowledge any more, they turned their attention to improve social welfare. The Russell-Einstein Manifesto is a case in point. It is quite important for the society to design a way which can link scientists' work with social development. Force scientists to spend time taking social responsibility will squeeze out their time to conduct research. In the long run, it is the society that suffers lost. How to balance scientists' time spends in research and taking social responsibility is worth considering. Compared with ancient Greece, science and technology are increasingly being integrated into society, single-minded focusing on one's own research is not enough. Relevant platforms or channels that can establish connection between science and society are in need.

The above suggestion leads to the second discussing topic, is science diplomacy a potential candidate that can bind science and diplomacy together? In a globalised world, no country is an island, science diplomacy provides opportunities for scientists to directly get in touch with people from other societies. For science in diplomacy dimension, science is a excellent tool helping diplomats make more rational decision, better diplomatic decision is of significance for the avoidance of mutual misunderstanding, thereby contributing to a more harmonious international relations that will benefit every country. For science for diplomacy dimension, scientists usually take the task of improving national image by making contribution to local society, despite the political consideration involved, social welfare will be augmented accordingly. Last but not least, science diplomacy, if well orchestrated, can be helpful to domestic society too. Companies can take part in science diplomacy and get profit, this profit can be transferred to the improvement of employees' welfare. All in all, science diplomacy plays a key role in reduce the distance between science and society and probably will keep doing so in the future.

In conclusion, science has been changing people's understanding towards the world in a revolutionary way, it is taking more and more social responsibility now. There seem nothing can reverse this trend, it is predicted that the positive interaction between science and society will only be stronger than ever.

References

Fan, Chunliang. 2010. The new development of science diplomacy and the strategic countermeasures of China. *Strategy&Policy Decision Research.* 2010, 25(6): 621-627.

Grunnet, Rasmus Kaare. 2010. Science progress and soft power: How China can use bilateral science diplomacy. Fudan University Master's dissertation.

Massarella, Derek. 1992. Philip Henry Zollman, the Royal Society's first assistant secretary for foreign correspondence. *Notes and Records of the Royal Society of London*. Vol. 46, No. 2, pp. 219-234.

Fedoroff, N. 2009. Science diplomacy in the 21(st) century. *Cell,* Volume 136, Issue 1, 9 January 2009, Pages 9-11.

Koppelman, B., Day, N., Davison, N., et al. 2010. New frontiers in science diplomacy: Navigating the changing balance of power. Royal Society.

Needham, Joseph. 1969. The grand titration: Science and society in East and West.

Rotblat, Joseph. 1999. A hippocratic oath for scientists. *Science*. Vol. 286, Issue 5444, pp. 1475.

United Nations Conference on Trade and Development, 2003. Science and technology diplomacy concepts and elements of a work programme. United Nations: New York and Geneva.

Wang, Baoqing. 2009. The soft power of Chinese science and technology, *Quanqiu Keji Jingji Liaowang*. 24(11): 26-32.

Zhang, Yiyan, Zhang, Ning. 2017. Three-elements models of S&T diplomacy based on activity analysis. *Forum on Science and Technology in China*.2:171-177.

Zhao, Gang. 2007. The theory and practice of science diplomacy. Current Affairs Press: Beijing.

Zhao, Gang, Cheng, Jianrun. 2008. Analysis and Discussion on science and technology diplomacy of China. *Journal of Yunnan University of Finance and Economics*., 24(5): 10-13.

Zhu, Yalan, He, Kaihui, Huang, Suzhen. 2016. International talent cultivation to promote China's S&T diplomatic strength. *Global Science, Technology and Economy Outlook*.Vol. 31 No. 10: 62-67.

The Comparison of 'New Energy Laws' Researches from China and Other Countries

Zhao Jingya[1], Chu Jianxun[2]

[1] Department of Sci-Tech Communication and Policy,
University of Science and Technology of China, Hefei, China
[2] School of Public Affairs, University of Science and Technology of China, Hefei, China

Abstract: New energy has remarkable significance on relieving the pressure caused from energy crisis worldwide. Researching on 'new energy laws' is a vital aspect of using science communication to promote benefits for humans. This article compares and analyzes the research papers on the 'new energy laws' topic from China and other countries by 2017, using CNKI and Web of Science, these two datasets as data sources. The paper mainly uses bibliometrics method to explore the research status over the 'new energy laws' topic from amounts of relevant papers, institutions, funds and authors these different perspectives. Then using the CITESPACE software to analyze the development of research hotspots and paths on 'new energy laws' topic at home and abroad separately comes to some conclusions and suggestions. This paper reveals that both fields have some common and different research hotpots. However, they have significant difference in the research institutions. Moreover, China plays an important role in the researches on 'new energy laws' topic worldwide.

1. Introduction

New energy generally refers energy from resources which are naturally replenished on a human timescale, such as sunlight, wind, rain, tides, waves, and geothermal heat (Wikipedia, 2018). These new varieties of energy which are starting being exploited, developed and waiting for promotion are becoming important issues in social development around the world because they have great significance on easing the dilemma of resources shortcuts worldwide and improving the ecological environment. New energy laws provide basic frames and regulations for the progress of new energy and are in favor of it to be popularized and applied in order to satisfy human's needs.

New energy's researching in developed countries started early so they have some accumulation on the researches of new energy laws. Relevant studies about the same topic did not appear until the end of 20th century in China, but they are in high speed and there are abundant researches on the level of laws. This article uses Web of Science and CNKI as datasets, comparing the literatures of the new energy laws to reveal the dynamic changes and hotspots difference in this researching field in China and other countries.

2. Data Sources and Methods

1) Data Sources

Data in this research comes from: (1) CNKI dataset. This is a major Chinese dataset including research results which covers most subjects and has rich information resource. In the advanced searching function of CNKI, we used 'new energy' and 'laws' as subject words of searching. To assure the completeness of the literature in a year, we set the terminal time on December 31st, 2017. Through the preliminary searching, the earliest time of literature is March 1st, 2003. Hence, we set the searching period from January 1st, 2003 to December 31st, 2017. The searching results are 1189 literatures totally between 2003 and 2017. (2) Web of Science dataset. This dataset is the largest comprehensive literature database all over the world which is published by Thomson Reuters company, including major international research literatures. This article selected the core collection in Web of Science as data sources, using 'new energy' or 'new resource' and law or policy as searching subject words. The end time of searching is set in 2017. Finally, we got 792 literatures from 1975 to 2017 as analyzing data.

Funding project: Communication modeling and neural mechanism research of online social network emergency management based on interactive memory system (NSFC71573241)

Corresponding author: Chu Jianxun, Hefei, Anhui. chujx@ustc.edu.cn.

2) Methods

This research uses bibliometrical method to explore the development course of new energy laws thematic studies from time distribution, theme distribution, authors and institutions and hotspots these various perspectives. Bibliometrical uses statistical methods to analyze the literature information, which uncovers characteristics of the quantity and law of change of literatures (QIU, 2007).

3. Basic Information Analysis of Literature

By analyzing the search results from CNKI dataset and WOS dataset from literature quantity, research institutions, funding and authors these four aspects, we can find some similarities and differences on literatures' basic information.

1) Analysis of Literature Quantity

New energy laws researches in other countries started in 1975, while relevant researches in China began in 2003. In some overlapping years, both CNKI and WOS datasets have included objective literatures. Comparing the literatures quantity of these years, from 2003 to 2017, here is a comparison of new energy laws researches findings quantity history in China and other countries, as Fig. 1 shows.

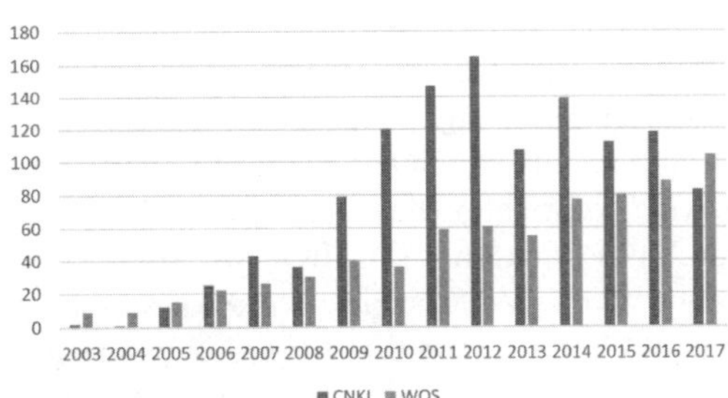

Fig. 1 Quantity history.

According to CNKI literature statistics in Fig. 1, Chinese new energy laws researches quantity is extremely low from 2003 to 2004, then stepped into the right direction in 2005. It kept rising until climbed the summit in 2012. There were 165 literatures in this year. After that, the quantity fluctuates slightly. In contrast to 2 articles in 2003, the numbers of research papers increased 82.5 times in nearly 10 years. After 2012, researchers' passion on this topic decreased a bit, but still on a high level. Average amounts of articles these years is more than 100. Thus, new energy laws researches in China in 21st century appear to be high-speed developing.

From statistics of WOS dataset in Fig. 1, new energy laws researches in other countries rise steadily. In 2017, the literature amounts reached the summit of 104. It increased 11.6 times compared with 9 articles in 2003. In general, researches of new energy laws in China and other countries are rising, but Chinese researches rise faster and have bigger number.

2) Research Institutions Distribution Analysis

By collecting the information of research institutions of searching results from two datasets, we list the top 10 institutions based on article amounts, as Table 1 shows. From CNKI relevant statistics, we find that research institutions on the new energy laws theme mainly are in the universities. More specifically, most in comprehensive universities and universities of politics and law. Top 7 universities of CNKI dataset in the Table 1 are all belong to these two kinds of universities. The articles from them take up 20% of the whole literatures. Therefore, new energy laws thematic researches in China place emphasis on law texts and are incline to humanistic and social studies.

According to WOS dataset situations, new energy laws researches also concentrate in universities, but there are also government institutions, such as Chinese Academy of Sciences and United States Department of Energy. Chinese universities play an important role in international researches and most of them are universities of science and engineering. Part of them closely relate to energy area, like north China electric power university and China university of petroleum. From above, we can know that international new energy laws are inclining to science and engineering studies, and Chinese universities behave positively in international environment.

3) Distributions of Sponsored Funding

Among the funding of CNKI articles, the number of articles sponsored by National Planning Office of Philosophy and Social Science is the highest, which is 69, far more than other funding institutions. As shown in Table 2, social science funding sponsor 71.9% of the articles that top 10 funding supported and have overwhelming advantage comparing to natural science funding. This trend reveals that Chinese researches on new energy laws mostly fall on social science aspect. From funding of WOS articles, Chinese funding institutions take a lead role in international new energy laws studies. Top 10 funding are all from China except for European union and National science foundation (USA). National

Table 1 Institutions of literature.

No.	Institution	Quantity	Percentage /%	Institution	Quantity	Percentage /%
	CNKI			WOS		
01	武汉大学	57	4.79	TSINGHUA UNIVERSITY	21	2.65
02	西南政法大学	38	3.19	NORTH CHINA ELECTRIC POWER UNIVERSITY	17	2.15
03	华东政法大学	36	3.02	BEIJING INSTITUTE OF TECHNOLOGY	15	1.89
04	吉林大学	34	2.85	CHINESE ACADEMY OF SCIENCES	15	1.89
05	华北电力大学	31	2.61	UNITED STATES DEPARTMENT OF ENERGY DOE	12	1.52
06	中国政法大学	24	2.02	UNIVERSITY OF CALIFORNIA BERKELEY	11	1.39
07	重庆大学	21	1.77	CHINA UNIVERSITY OF PETROLEUM	9	1.14
08	南京工业大学	21	1.77	UNIVERSIDADE DE LISBOA	8	1.01
09	华东理工大学	20	1.68	AALTO UNIVERSI	7	0.88
10	中国海洋大学	18	1.51	BEIJING JIAOTONG UNIVERSITY	7	0.88
Total		300	25.2		122	15.4

Table 2 Funding institution of literature.

No.	Funding Institution	Quantity	Percentage /%	Funding Institution	Quantity	Percentage/ %
	CNKI			WOS		
01	国家社会科学基金	69	5.80	NATIONAL NATURAL SCIENCE FOUNDATION OF CHINA	53	6.69
02	国家自然科学基金	11	0.93	FUNDAMENTAL RESEARCH FUNDS FOR THE CENTRAL UNIVERSITIES	9	1.14
03	山东省软科学研究计划	6	0.50	EUROPEAN UNION	8	1.01
04	江苏省教育厅人文社会科学研究基金	5	0.42	NATURAL SCIENCE FOUNDATION OF CHINA	5	0.63
05	湖南省教委科研基金	4	0.33	NATIONAL SCIENCE FOUNDATION	4	0.51
06	湖南省软科学研究计划	3	0.25	BEIJING NATURAL SCIENCE FOUNDATION OF CHINA	3	0.38
07	湖南省社会科学基金	3	0.25	CHINA POSTDOCTORAL SCIENCE FOUNDATION	3	0.38
08	广东省自然科学基金	2	0.17	CHINA SCHOLARSHIP COUNCIL	3	0.38
09	国家重点基础研究发展计划(973 计划)	2	0.17	PROGRAM FOR NEW CENTURY EXCELLENT TALENTS IN UNIVERSITY	3	0.38
10	中国博士后科学基金	2	0.17	TSINGHUA UNIVERSITY INITIATIVE SCIENTIFIC RESEARCH PROGRAM	3	0.38
Total		107	8.99		94	11.88

natural science foundation of China sponsored 53 articles, took up 60.6% of articles supported by top 10 funding institutions.

In contrast to China's situation, international researches are mostly sponsored by natural science funding, from both China and other countries. As can be seen, new energy laws researches in China and abroad have different research forces: China depends on social science funding while other countries mainly rely on natural science funding. Thus, their research content has different focus.

4) Analysis of Authors

Table 3 is a list of authors' research articles quantity. As shown in it, in the searching results of CNKI and WOS, the total number of top 10 authors' articles separately takes up 4.96% and 5.30% of the total quantity of the research papers. It illustrates that the research results of new energy laws topic are not concentrative enough. There are many authors while with limited results. This demonstrates that there are still no core authority authors group yet. It is well to be reminded that among top 10 authors of researches abroad, 7 of them are Chinese scholars, which implies that China is main research force of the international new energy laws researches.

Table 3 Authors of literature.

No.	Author	Quantity	Percentage/%	Author	Quantity	Percentage/%
	CNKI			SCI		
01	莫神星	11	0.92	ZHANG X	7	0.88
02	陈海嵩	9	0.75	LI Y	5	0.63
03	杨解军	8	0.67	LUND PD	5	0.63
04	岳树梅	7	0.59	WANG Y	5	0.63
05	杨泽伟	6	0.50	ZHOU Y	5	0.63
06	吕江	5	0.42	ANSARI N	3	0.38
07	刘超	4	0.33	DUIC N	3	0.38
08	于文轩	3	0.25	FANG YC	3	0.38
09	郭冬梅	3	0.25	HAO JM	3	0.38
10	张立锋	3	0.25	LIU YQ	3	0.38
Total		59	4.96		42	5.30

4. CNKI Articles Analysis

Import 1189 articles searched from CNKI dataset into the scientific knowledge mapping software CITESPACE, using its key words co-occurrence and timeline view functions to analyze the articles.

1) Key Words Analysis

After getting the key words co-occurrence network through CITESPACE software, we used GEPHI software to make the network more explicit, as Fig. 2 shows. Every round point in the Fig. 2 represents a keyword, whose color depth depends on the centrality. The point or the keyword has higher centrality, its color is deeper. It demonstrates that more links go through this point and it occupies a more important position in the network.

Then we used the network summary table function in CITESPACE software to get Table 4 which lists top 40 keywords based on their frequency of occurrence.

From these keywords we can find that new energy laws studies between 2003 and 2017 mainly focus on the following three aspects: firstly, using developed countries' new energy laws legislation experience for reference. Germany, USA, Japan are three words with high occurrence frequency in Table 4, because many researchers studied new energy laws in these developed countries who are early in building this module. They summarized the experience and lessons to provide suggestions for our country's new energy laws legislation which just started out. Secondly, economic aspects of new energy laws. Low carbon economy, subsidy these frequent key words reflect Chinese researchers pay great attention to the economic meaning of new energy laws. New energy has great significance on social economy's development. Traditional extensive economy with 'high input, high consumption, and high emission' characteristics does not conform to the trend of the times (Zhang, 2014). Developing low carbon new energy economy is an irresistible trend. Thirdly, new energy laws combine closely with industry sector. Key words like biomass energy, energy performance contracting and low-carbon technology represent this trend. Introducing new energy laws is an effective way to make publics realize the importance of developing new energy industry. Japanese government always put the laws-making on first place of new energy industry developing course, using laws as footstone and insurance for the new energy industry (Wang, 2013).

2) Analysis of Research Hot Spots Evolution

Using the Timeline view function of CITESPACE to demonstrate research hot spots' evolvement, as Fig. 3 shows. Based on the

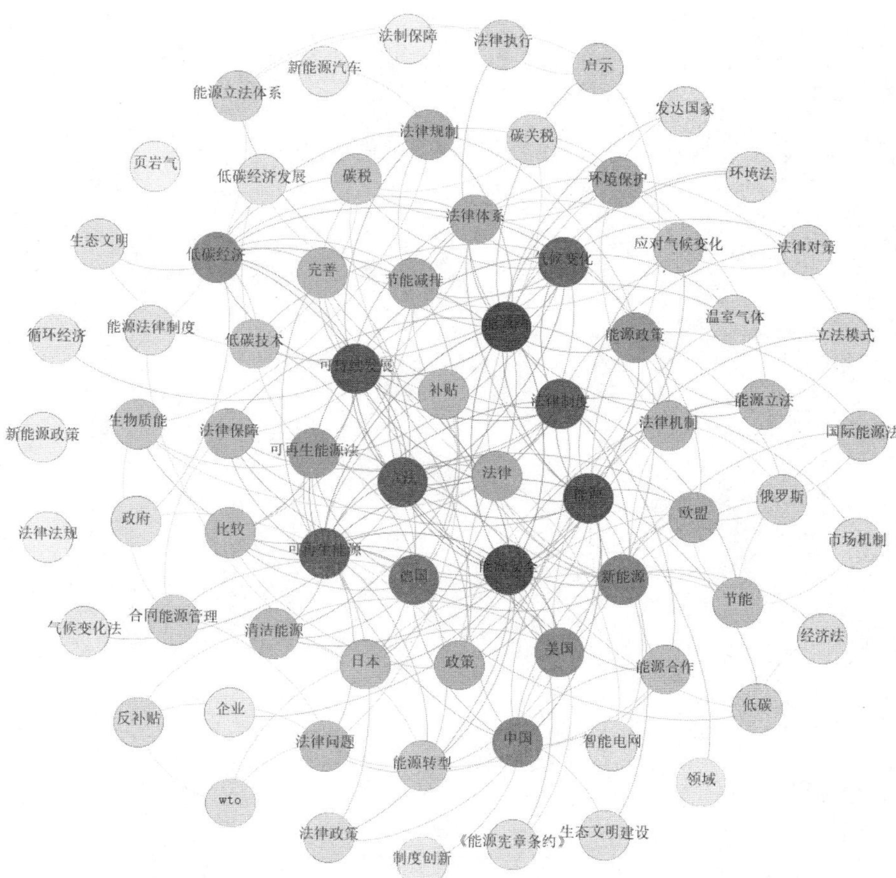

Fig. 2 Keywords network.

Table 4 Top 40 keywords in CNKI.

No.	Keyword	Frequency	No.	Keyword	Frequency
01	低碳经济	161	21	政策	14
02	可再生能源	99	22	法律机制	14
03	能源	89	23	能源政策	12
04	法律制度	73	24	日本	12
05	气候变化	66	25	补贴	12
06	能源法	64	26	节能	11
07	能源安全	60	27	低碳	11
08	新能源	49	28	环境保护	11
09	立法	47	29	欧盟	10
10	可持续发展	44	30	生物质能	9
11	节能减排	37	31	碳关税	9
12	法律规制	27	32	碳税	8
13	中国	23	33	启示	8
14	可再生能源法	22	34	合同能源管理	7
15	能源立法	21	35	法律政策	7
16	法律保障	17	36	低碳技术	7
17	能源合作	16	37	法律	7
18	德国	16	38	应对气候变化	6
19	美国	14	39	法律问题	6
20	法律体系	14	40	能源转型	6

Fig. 3 Timeline view of CNKI.

research hot spots summarized above, we can find that the heat of the key words changed with time. As early as 2004, there were research hotpots about renewable energy sources. From 2004 to 2007, many researches call for new energy legislation and providing legal protection. From 2007 to 2010, keywords like Japan, USA and inspiration reflected many researchers strived to borrow from new energy legislation experience of developed countries. Between 2010 and 2013, carbon tax, circular economy and subsidy these words gave expression to economization of this research. Until 2017, new energy vehicles these key words indicated that industrialization of new energy researches will become hot spots in recent years.

In general, new energy laws research in China mainly follow the paths below: From put forward the new energy legislation to explore the legislation experience of developed countries, then to researches content economization, finally to researches content industrialization. Chinese new energy laws researches emphasis transformed from theoretical to practical.

5. Analysis of International Researches from WOS

Put 792 articles searched from WOS dataset into CITESPACE software, using its key words co-occurrence network, citation co-occurrence network and timeline view functions to analyze these search results.

1) Key Words Analysis

From Table 5 and Fig. 4 we can see that international researches over new energy laws mainly concentrate on: firstly, technology and industry. Model, system, new energy vehicle, electric vehicle, subsidy, technology and industry these keywords new energy laws researches have close connections with new energy industry,

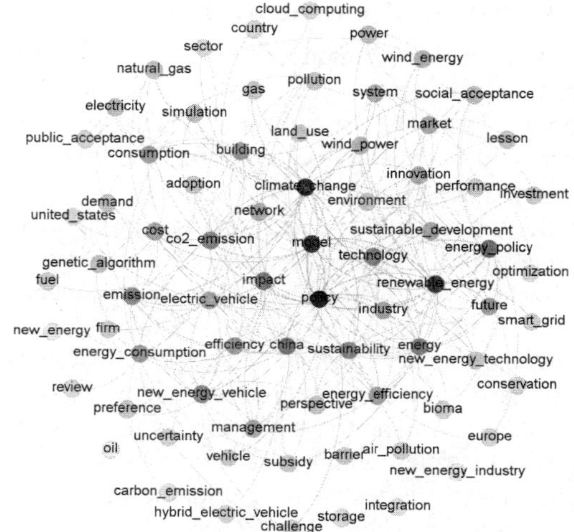

Fig. 4 Keywords network.

Table 5 Top 40 keywords in WOS.

No.	Keyword	Frequency	No.	Keyword	Frequency
01	policy 政策	75	21	wind energy 风能	10
02	renewable energy 可再生能源	52	22	energy consumption 能源消耗	10
03	model 模型	42	23	network 网络	10
04	china 中国	37	24	innovation 创新	10
05	system 系统	34	25	management 管理	10
06	energy 能源	33	26	demand 需求	9
07	climate change 气候变化	23	27	industry 产业	9
08	energy policy 能源政策	22	28	power 能量	9
09	new energy vehicle 新能源汽车	20	29	market 市场	8
10	electric vehicle 电动汽车	20	30	building 建筑	7
11	emission 排放	17	31	adoption 采用	6
12	energy efficiency 能源效率	17	32	perspective 观点	6
13	technology 技术	17	33	cost 花费	6
14	impact 影响	16	34	subsidy 补贴	4
15	efficiency 效率	12	35	sector 行业	4
16	sustainability 可持续性	12	36	vehicle 车辆	4
17	consumption 消耗	11	37	smart grid 智能电网	4
18	performance 表现	11	38	environment 环境	4
19	future 未来	11	39	country 国家	3
20	CO_2 emission 二氧化碳排放	11	40	fuel 燃料	3

especially the hot industry new energy vehicle. Secondly, these researches relate closely to market, which can be seen from market, demand, management and cost these key words. Thirdly, international researches pay great attention to the effect aspect. They keep a watchful eye on the thoughts of the public. From perspective, preference, public acceptance and social acceptance these key words we can conclude that international new energy laws researches put great effort on public's cognition and attitude towards new energy issues.

2) Analysis of High-cited Articles

From Table 6, among top 5 articles which are most frequently cited, the earliest one was published in 2009. The others were all published after 2010, especially flocked in 2013. As for their topics, other than the article An optimization model for renewable energy generation and its application in China: A perspective of maximum utilization of CONG RG, other four papers' topics are all new energy automobiles. From above can be seen that new energy vehicle are important subject of international new energy researches. Among the top 5 articles mentioned above, 4 articles' authors are from China. Their papers mainly discussed the development situation of Chinese new energy laws. These administrates that Chinese researchers are becoming the backbone of international new energy researches and China are important object of these studies.

3) Analysis of Research Hot Spots Evolution

Using the timeline view function of CITESPACE again, we can get process of international researches on new energy laws topic. Relevant research hot spots appeared in 2005, between 2007 and 2010, model, new energy technology, new energy vehicle and electric vehicle these key words show new energy laws researches focus on industry and technology aspects. Between 2010 and 2015, researches hot spots concentrate on market, which can be seen from investment, market, storage and demand these key words. From 2015 to 2017, the hot spots which concentrate on public's cognition, acceptance and attitudes arose in the research field, key words such as public acceptance, social acceptance and adoption express the trend.

Hence, new energy researches other countries started focusing on technology and industry aspects, then switching to market. For they emphasize the effect on the public. Moreover, these research hot spots are not independent completely to each other. They are overlapping with the change of time.

6. Conclusion

New energy laws researches have been lasting heat both in China and other countries because the topic has great significance with strategic meaning of developing new energy. International researches of new energy laws started earlier than China. Though researches in domestic began late, they progressed quickly. Relevant articles accumulation has exceeded international ones. The article compares and analyzes statistics of CNKI datasets and WOS datasets from quantity, authors and research hot spots these perspectives, then we can reach following conclusions of new energy laws researches in China and other countries.

Table 6 Citation information.

No.	Cited Frequency	Centrality	Year	Citation Information
01	9	0.07	2009	DIAMOND D, 2009, ENERG POLICY, V37, P972, DOI 10. 1016/J. ENPOL. 2008. 09. 094 *The impact of government incentives for hybrid-electric vehicles: Evidence from US states*
02	6	0.12	2013	GONG HM, 2013, MITIG ADAPT STRAT GL, V18, P207, DOI 10. 1007/S11027-012-9358-6 *New energy vehicles in China: policies, demonstration, and progress*
03	6	0.11	2014	HAO H, 2014, ENERG POLICY, V73, P722, DOI 10. 1016/ J. ENPOL. 2014. 05. 022 *China's electric vehicle subsidy scheme: Rationale and impacts*
04	6	0.03	2013	CONG RG, 2013, RENEW SUST ENERG REV, V17, P94, DOI 10. 1016/J. RSER. 2012. 09. 005 *An optimization model for renewable energy generation and its application in China: A perspective of maximum utilization*
05	4	0.01	2013	ZHANG X, 2013, ENERG POLICY, V61, P382, DOI 10. 1016/J. ENPOL. 2013. 06. 114 *The impact of government policy on preference for NEVs: The evidence from China*

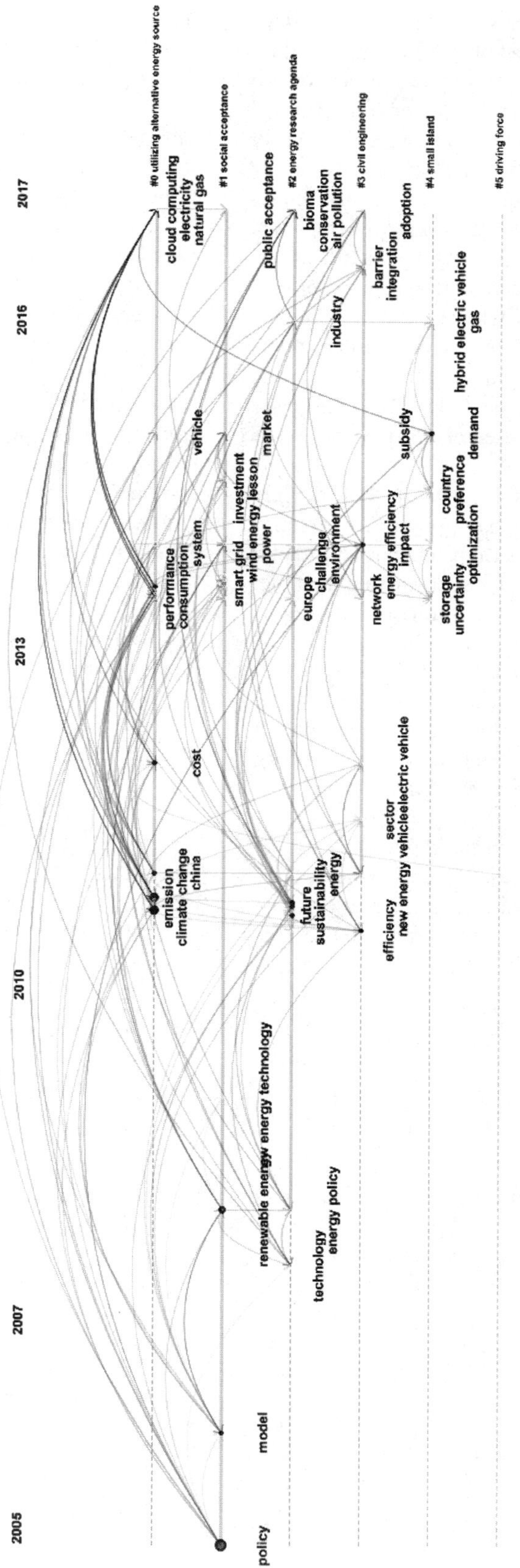

Fig. 5 Timeline view of WOS.

Firstly, the research's hot spots in two fields have both similarity and difference. This trend can be reflected from key words. In the new energy researching field, China and other countries both pay attention to industry aspects, especially on new energy vehicle this representative product. But China fall behind in this area. Chinese researches' emphasis on economy overlaps with international researches' hot spot on market. China has accumulated much legislation experience of western countries which is not included in international researches. In turn, international researches recently focus on public's cognition, acceptance and attitudes towards new energy stuff while Chinese researches have not noticed this aspect yet.

Secondly, research force over the topic of China and foreign countries have obvious distinction. By analyzing the research institutions and funding of searched results, it can conclude that new energy laws researches in China mainly rely on humanities and social science research force in universities. While international researches are mostly conducted by scholars from science and engineering subjects. This is also the reason why the portion of researches concerning technology in other countries is higher than in China.

Thirdly, China is an important field of as well as an important research object as for new energy laws researches. By analyzing research institutions, sponsored funding, prolific authors and high-cited articles of international research results, it can be concluded that China play a vital role in international researches of new energy laws. Among the top 10 institutions which published most research articles, Chinese universities and institutions take up 6 seats. Similarly, among top 10 authors who published most articles, 7 researchers are from China. China is also the fourth key word in the key words frequency list. Four articles of top 5 articles which are most frequently cited discuss Chinese new energy laws and new energy industry.

Acknowledgments

I would like to express my gratitude to all those who helped me during the writing of this article. My deepest gratitude goes first and foremost to Professor Chu Jianxun, my supervisor, for his constant encouragement and guidance. He has walked me through all the stages of the writing of this article. Last my thanks would go to my beloved parents for their loving considerations and confidence in me all through these years.

References

QIU, J. P. (2007). *Informetrics*. Wuhan, Wuhan University Press.

WANG, M. (2013). *Analysis of the evolution of Japanese New Energy industry*. Jilin University.

Wikiedia. (2018). *New Energy*. Available at https://e.wikipedia.org/wiki/New_energy.

ZHANG, H. L. (2014). *Research on the new energy development in China*. Jilin University.

Application, Evaluation and Implementation of PIIJ's Newly Launched STM Journals

Zhao Ji[1], Li Fang[2]

[1] National Academy of Innovation Strategy, CAST, echnology, Beijing, China
[2] China Association for Science and Technology, Beijing, China

Abstract: The Project for Enhancing International Impact of China STM Journals (PIIJ) was initiated in 2013 to support the launching of new English language STM journals in China, based upon their merits and in leading or preponderant disciplines. This paper presents certain procedures in PIIJ Category D, including application, selection, and implementation.

1. Background and Introduction

Science, Technology and Medicine (hereinafter referred to as STM) journals are the basic carriers to centralized recording and communication of scientific research results, an important means to discover and cultivate scientific and technological talents, an important symbol of the scientific and technological strength of a nation and play an important role in the Chinese national innovation system (http://www.cast.org.cn/n35081/n35488/16753578.html). In recent years, with the continuous increase of China's investment in science and technology, China's scientific research output has developed by leaps and bounds. The number of Chinese scientific papers in the SCI database has been ranked second in the world for consecutive seven years since 2008, following the USA, but the gap has been narrowed year by year[1]. However, most of the Chinese scientific papers that have been included are published in foreign journals. According to the statistics of the China Science and Technology Information Institute on the publication of Chinese scientific papers in 2013, the Chinese scholars included in the SCI published a total of 204,100 papers, of which only 23,200 were published in domestic journals, less than 11.4%. The reasons for this phenomenon are as follows: first, the need for international scientific and technological exchanges, and second, the number and service capacity of English STM journals in China are seriously inadequate. In terms of quantity, according to the relevant data of the State Administration of Press, Publication, Radio, Film and Television (hereinafter referred to as SAPPRFT) (http://www.gapp.gov.cn/govservice/1959.shtml), in 2013, there were 262 kinds of English STM journals in China, which was 5.3% of the 4,953 kinds of national STM journals in the same year. In terms of quality, the average value of the influence factor and total citation frequency of the 138 journals included in China's SCI was 1.042 and 1290 respectively, which were far lower than the international average of 2.208 and 5172[2], especially lacked of top English journals that are competitive in the international technology arena. The number of English STM journals in China is small, the structure is irrational, and the international influence is not big. It is difficult to meet the growing demand for international exchange of scientific and technological achievements in China.

In response to this situation, in 2013, the China Association for Science and Technology, the Ministry of Finance, the Ministry of Education, the SAPPRFT, the Chinese Academy of Sciences, the Chinese Academy of Engineering, began to implement 'Project for Enhancing International Influence of China STM Journals (PIIJ)' (hereinafter referred to as PIIJ) (http://www.cast.org.cn/n35081/n35668/n35758/n39435/n39450/15013509.html and http://www.cast.org.cn/ n17040442/n17041408/17266623.html), first, to select the excellent national English science and technology journals in category A, B and C, and to award them, to promote a number of important scientific fields in the field of STM journals into the forefront of the international rankings, which will be evaluated once every three years. The journals in Category A, B and C that are rated in three consecutive years will be annually funded with 2 million RMB,

1 million RMB and 500,000 RMB. Second, to launch a batch of high-level English science and technology journals, that can represent China's frontier disciplines, dominant disciplines, or fill gaps in domestic disciplines, which will be set up with category D in PIIJ and be evaluated once a year. For these, one-time funding of 500,000 RMB should be granted.

This paper takes the English journals of category D in PIIJ from 2013 to 2017 as object, focusing on the analysis of their application, evaluation, selection and implementation.

2. The Application and Evaluation of Category D Journals

Category D of PIIJ is nationwide open to organizations that intend to launch English scientific journals can participate in the application process. The materials are approved after formal evaluation. It is hereby called **the applicant journal**. The applicant journals shall be then elected by the procedures of on-site defense, expert evaluation and project publicity. It is called **the nominated journal**.

In the 2013-2017 five-year project, 269 applicant journals submitted the application one or more times, and the total number of applications was 369. The applicant journals can be divided into two categories: one is the journals that have obtained the International Standard Serial Number (hereinafter referred to as ISSN) and have been published overseas but have not obtained the China Uniform Serial Number (hereinafter referred to as CN). The other category is the proposed journals that have not yet been published.

From 2013 to 2017, a total of 70 journals were nominated. According to the PIIJ project implementation program, 10 kinds of start-up publications were supported by category D each year from 2013 to 2015, and the number of start-up publications that were supported was increased to 20 from 2016 to 2018. The application and nomination of category D in PIIJ are shown in Table 1.

Generally, in 2013, the number of applicant journals was the lowest at the beginning of the project. After that, new journals were involved every year, and the number of applicant journals has increased steadily. From 2014 to 2015, the competition was relatively fierce. From 2016 to 2017, the number of set-up publications supported increased from 10 to 20, and the proportion of nominated candidates remained above 20%.

The applicant journals were grouped and evaluated by discipline. In 2013, the applicant journals were few and divided into two groups. From 2014 to 2016, it was divided into three groups: medicine, engineering and others. In 2017, due to the increase in the number of applicant journals of engineering, it was divided into 4 groups (see Table 2 for details).

Table 1 Numbers of applicant & nominated journals for category D in PIIJ from 2013 to 2017.

Year	2013	2014	2015	2016	2017
Number of applicant journals	46	89	73	70	91
Number of journals applying for the 1st time	46	74	44	40	65
Number of nominated journals	10	10	10	20	20
Ratio of nominated journals and applicant journals/%	*22*	*11*	*14*	29	*22*

Table 2 Number of applicant & nominated journals and ratios for each group from 2013 to 2017.

Year	Group Number	1st Group	2nd Group	3rd Group	4th Group
2013	2	S & T (5/19，26%*)	Medicine (5/27，19%)		
2014	3	S & T (3/27，11%)	Medicine (3/34，9%)	Others (4/28，14%)	
2015	3	S & T (3/21，14%)	Medicine (4/31，13%)	Others (3/21，14%)	
2016	3	S & T (7/24，29%)	Medicine (6/23，26%)	Others (7/23，30%)	
2017	4	S & T - A (4/18，22%)	Medicine (6/27，22%)	Others (6/27，22%)	S & T - B (4/19，21%)

Note: 5/19，26% means this group has 19 applicants, 5 nominated, the ratio is 26%

From the number of journals nominated by different disciplines, the number of journals the medical and engineering was the largest, because the ratio allocation was based on the number of applicant journals. The number of journals in these two disciplines was large, so the number of nominated journals was also large.

3. Analysis of the Nominated Journals of Category D

According to the statistics of the three major supervisors of the nominated journals, there are 28 from the Ministry of Education, 20 from the Chinese Academy of Sciences, 17 from the Chinese Association of Science and Technology. The number of journals nominated by the Ministry of Education was the largest. Further analysis of the applicant journals, there are 73 from the Chinese Association of Science and Technology, 49 from the Ministry of Education, 35 from the Chinese Academy of Sciences. The number of the applicant governed by the Chinese Association of Science and Technology was the largest. From the rate of the nomination, 57% from the Ministry of Education, 57% from the Chinese Academy of Sciences, 23% from the Chinese Association of Science and Technology. Apparently, the rate of the Chinese Association for Science and Technology was significantly low.

In order to study the influence of nominated journals in the number of English science journals of various disciplines, this paper analyzes and compares the quantity distribution of 70 nominated journals and 282 other English science journals available in China as of May 2017. From the distribution by discipline, 1) zero breakthrough has been achieved in the field of safety science, which filled the gap in China's English journals; 2) in the interdisciplinary social sciences, environmental science, aerospace and chemistry, because the disciplines have developed rapidly, and the number of new journals has increased significantly; 3) There are new breakthroughs in China's strong disciplines, such as aviation, aerospace and space technology, materials science, energy science, and chemistry and chemical industry; 4) in the fields of medicine and engineering, there are many English journals and more new journals; 5) further analyzing from the engineering science in the appendix list, breakthroughs in cutting-edge disciplines such as data mining, network security, data intelligence and nanotechnologies. It can be said that category D project has better realized the original mission of the project, and the newly added journals have provided strong support for the development of the disciplines. However, how to reflect the project's support for basic, emerging, interdisciplinary and blank disciplines, it needs further study.

4. Implementation of Category D Project

The nominated journals, that are submitted to the establishment procedure in accordance with relevant Chinese policies and regulations, and granted with CN number by SAPPRFT for the publication, are called **the approved journals**. The approved journals, which are published in mainland China and abroad, are called **the published journals**.

For the nominated journals, the application materials shall be submitted to the SAPPRFT in accordance with relevant regulations, and a reply will be given on whether the materials will be accepted or not within 5 working days. If the materials meet the relevant regulations and are accepted, the approval of the publication will be approved within 60 working days from the date of acceptance. If the materials do not meet the relevant regulations and requirements, then they must be rectified within 20 working days, and the overdue materials will be out of date.

As of May 2017, 31 of the 70 nominated journals have been approved. Except the journal *Application of Natural Products*, which takes 32 months to be approved as the materials were replenished for many times to the SAPPRFT due to the qualifications of the publishing organization, other journals were approved within 24 months. Among the 39 unapproved journals, one in 2014 and two in 2015 were not granted with CN number formally by the SAPPRFT, and the materials were still being supplemented. The remaining 36 journals were nominated in 2016 or 2017 and are in the process of applying or preparing for application.

According to the implementation program of category D project, when a nominated journal becomes an approved journal, it can get a one-time support fund of 500,000 RMB. In view of the different time of each approved journal to come in force, and it may possibly be beyond the fiscal year, in order to strengthen the implementation of the project, if the nominated journal fails to obtain the approval of the publication within 24 months, it is recommended to consider canceling the fund.

As of May 2017, 27 of the 31 approved journals have been published. According to the publication situation, it can be divided into two categories: one was, journals that have obtained the ISSN number the application, and been published overseas before being approved, there were 11 kinds of such journals, their CN numbers were added to the online or paper version within 6 months, which achieved smooth transition to the publication; the other category was, new journals published after being approved, there were 11 kinds of such journals. And 9 of them were published within 12 months. Of the remaining four unpublished journals, three have been newly approved and are currently under preparation. *The Progress in Mathematics Research* has been delayed due to lack of manuscripts. Since the regulations on the administration of publishing and other relevant regulations clearly stipulate the time limit for the registration and publication of journals, it is recommended that the competent authorities strengthen the business guidance and supervision of the new journals to ensure that the journals are published according to the prescribed time. It is recommended that the office of PIIJ strengthens the follow-up management of the project. If the approved journals cannot be published within 12 months, corresponding responsibility measures should be taken to ensure the project performance.

After several years of hard work, some category D journals have shown to have a certain academic influence[3, 4] and have been included in the international mainstream search systems. According to the latest data released by Scopus in 2017, the journal *Photonics Research*, *Micron Nano Express*, *International Journal of Hepatobiliary and Pancreatic Diseases*, *Friction*, *World Journal of Pediatrics*, *Journal of Electronics* and *Environmental Accounting and Management*, these seven nominated journals were included, of which *Photonics Research* and *Micron Nano Express* entered the top 10% in their respective disciplines. According to the latest data released by JCR in 2017, three nominated journals, such as the journals *Micron Nano Express*, *Photonics Research* and *Chinese Science: Materials Science (English Edition)* were included and entered the Q1 area of their subject, with outstanding performance. The category A, B, and C of PIIJ are open to the national English science and technology journals in the whole China and grant funds to excellent journals. In the new round of selection in 2016, *Micron Nano Express* and *Friction* were nominated into category B, and the journal *World Journal of Pediatric* was nominated into category C. It can be seen that the nominated journals of category D have become an important supplement to the high-end English STM journals in China. To a certain extent, they have surpassed the level of domestic similar journals and moved closer to the top international journals.

5. Conclusions and Suggestions

This paper analyzes the four stages — the application, the nomination, the approval and the publication of category D of PIIJ. The implementation of category D project has been 5 years, in this period, 70 kinds of nominated journals have complemented the number of English STM journals in China, and its way of appearance optimizes the distribution in the disciplines of English scientific journals in China. At present, China's newspapers publication is in the stage of controlling the total amount and structural adjustment (http://old.chinacourt.org/public/detail.php?id=185680). The resources of the journal number are in short supply, so category D project has become a green channel for the publication of high- quality publishing resources, played an important role in expanding the scale, becoming excellent and making increments, and improving the supply of high-end English STM journals in China. It has also implemented the suggestions of the NPC deputies and CPPCC members on the decentralization of journals' approval authority.

The implementation of category D project has achieved phased results, but there is room for improvement and improvement in the implementation process. This paper suggests:

1) To continue implementing category D influence project and maintain an annual nomination of 20 new publications in the future. At the same time, optimize the evaluation system of category D project to ensure the quality of nomination, and to balance the development of China's journals of frontier disciplines, dominant disciplines, interdisciplinary disciplines and filling

blank disciplines (especially marginal emerging disciplines).

2) To strengthen the project guidance and process supervision of the selected journals and improve the implementation of the project; if the project tasks cannot be completed within a certain period of time, the qualifications for the selection will be cancelled.

3) To moderately expand the scope of the nomination and the funding of new publications. The successful implementation of category D project is a positive exploration of the reform of the publishing administrative evaluation and approval system and an enhancement of the scientific of the approval of the administrative evaluation. It is suggested to draw on the implementation experience of this project, and set up similar projects for Chinese-language science and technology journals, and decentralize the authority of scientific proofing to disciplinary professional academies (associations), to further promote the reform of the administrative evaluation and approval system for publications, and to promote the orderly transfer of government functions to science and technology associations.

References

[1] Pan Y T, Ma Z, Su C, et al. A Brief Report on Statistics and Analysis of Chinese Papers in 2015（in Chinese）. Chin J Sci Tech Period, 2017, 28(1): 58-67.

[2] Pan Y T, Ma Z, Su C, et al. A Brief Report on Statistics and Analysis of Chinese Papers in 2013(in Chinese). Chin J Sci Tech Preriod, 2015, 26(1): 73-81.

[3] Du Y W, Ning B. Accomplishments and Challenges — Review on English Language STM Journals of China in 2015，Sci-Tech & Publication, 2016, 35(2): 28-34.

[4] Ren S L. Review on English Language STM Journals of China in 2016，Sci-Tech & Publication, 2017, 36(2): 30-33.

The Service Invention System: Institutional Guarantee for Technological Innovation under New Conditions

Li Yi[1], Luo Hui[2], Wang Hongwei[3]

[1] National Academy of Innovation Strategy, CAST, Beijing, China
[2] China Association for Science and Technology, Beijing, China
[3] Institute of Quantitative & Technical Economics ,Chinese Academy of Social Sciences, Beijing, China

Abstract: The service invention system is an important part of the Intellectual Property System, which has become a major incentive to encourage scientific researchers to make innovations and an institutional guarantee to protect national scientific and technological competitiveness. As a new round of scientific and technological revolution accelerates, the brand-new characteristic, trend and change based on scientific research and rational analysis is about to cast a significant influence on the reform of service invention system. This paper attempts to study and analyze the influences on current national from three aspects, namely, the dual characters of disruptive technology, 'deadly' technical weakness, and motivations on the subjects of technological innovations; in the final part, the paper put forward some targeted policy recommendations on how to improve the weak links of service invention system.

1. Introduction

Under the interactive influence of new round of scientific and technological revolution and industrial transformation, the technological innovation has become the important promoting force to the reform of service invention system. Since the 18th CPC National Congress, the world scientific and technological innovation has gradually emerged the trends of social system transform, integration of subject intersection, indexed growth etc., the domestic economy has been entering the new normal of speed changing, structure optimization and dynamics conversion at the same time (Amirkolaii et al., 2017; Hsieh, 2009; Hwang et al., 2017; Saarikko, 2017). In this complex and grim environment, how to realize the reform direction of service invention system driven by technological innovation, how to promote the service invention system to adapt the transform under the guidance of new technology according to the new future, new trend and new change of the new round scientific and technological revolution and industrial reform, it becomes the key problem of invention reform at present stage in our country.

Corresponding author: Li Yi, No.3 Yuyuantan South Road, Haidian District, Beijing, 100038, China. liyi@cnais.org.cn.

2. A New Round of Technological Revolution and Industrial Transformation is Poised to Take off

1) The world enters an important period of structural reform, the service invention system becomes the key to guarantee and cultivate the new growing force

Currently, the global economy is showing some recovering tendency, however the economic growth still faces great uncertainty and many challenges. From the policy practice of various countries, though affected by the uncertain factors of macroeconomic imbalance, trade protectionism, financial risk agglomeration, etc., the service invention system has become the important system guarantee to cultivate the growing force and enhance the national scientific and technological competitiveness in US (Arora et al., 2016), Japan, and the developed countries represented by German. From the legal system experience of service invention system in US, Japan and German etc., we can get that the mature invention system is not only the legal base of development of high-tech industry and transformation of achievements with the Internet, new energy, intelligent manufacturing as the core, but also the effect is becoming more and more obvious in the aspects of attracting scientific and technological innovation talents, guiding innovation investment, shaping the global innovation ecosystem etc (Wu et al., 2015). As a key link of innovation value chain, the invention system has been deeply integrated into the global scientific and technological innovation system, the service invention is not only the inner problem of personal, company or scientific research institution, and it has become the basic requirement and system guarantee in the value chain of scientific research of 'Official, industrial, academic, research' and in the cooperation of transnational corporations, cross-regional R&D teams (Nordensvard et al., 2018).

2) Speed up the development of new technology, new industries and new forms of business, the service invention system has become the important mean of guiding interdisciplinary interaction

In the process of modernization, the needs of human life and development become the one of important motive powers which triggered the scientific revolution, technological revolution and industrial revolution (Van et al., 2018). As the accumulation, development and popularization of human knowledge, the technical innovation is developing to the direction of structure refinement, functional diversification, usage facilitation and cost-effective optimization (Cohen & Caner, 2016). Currently, the new round of scientific and technological revolution and industrial reform is speeding up obviously, vital scientific and technological innovation achievement, especially the rapid emergence, group breakthroughs, deep integration of subversive technology, it enhances the uncertainty of science and technology innovation and its influence fatherly (De et al., 2017; Gierej, 2017; Kiel et al., 2017; Krotov, 2017; Prybila, 2017). Under this trend, China's the achievements of service invention are accelerating (see Fig. 1 below). With the rise of a new round of scientific and technological revolution and industrial transformation, interdisciplinary interactive integration is becoming one of the important features of the current round of scientific and technological revolution (Hingley & Park, 2017). On one hand, the service invention system ensures the distribution of rights and interests among different scientific research subjects, provides the necessary legal guarantee for the cooperation between the subjects of scientific research of different disciplines; on the other hand, the service invention system ensures the interests of invention-team, scientific research organization and enterprises at the same time, it can also play a key role in guaranteeing the national scientific research ability and the economic development level.

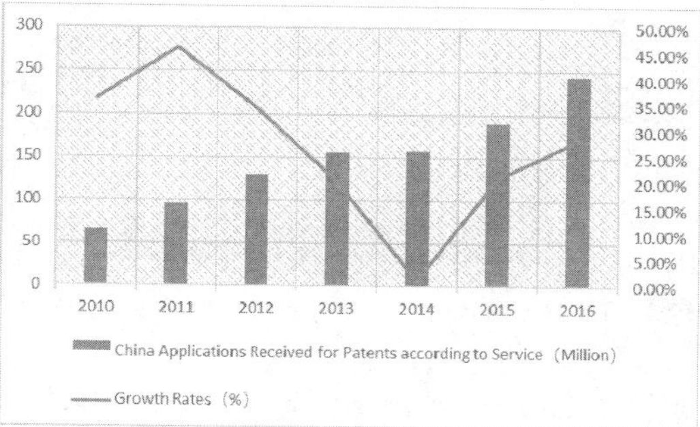

Fig.1 Number of patent applications for Chinese service invention.

3) The service invention system affirms the 'knowledge' ownership of property in the name of law, to stimulate the creative potential of researchers

It is different from the past scientific and technological revolution of 'personal activity' or 'government activity', diversification of input patterns has become an important feature of the present age. From stakeholders' perspective, the service invention system can not only protect the rights and interests of innovation input subjects in the name of law but also can encourage and stimulate the innovation enthusiasm of researchers, especially in the original innovative research (Zhang, 2015). Service invention system is a legal system to fix the property rights for benefits and results which obtained by the inventor in his organization and on duty (see Fig. 2). On one hand, the service invention system regulates the distribution of benefits of service invention by means of laws and regulations, corrects the unequal position of both sides in the legal relationship of on duty invention. On the other hand, through laws and regulations protect the interests of both sides too (Gu et al., 2012), improve the allocation efficiency of market innovation resources, improve the competitiveness of enterprises (Jou, 2018). The service invention system must not only need balance of relationship between employee-inventor's interest and profit of organization, innovation enthusiasm of employee-inventor and innovation competitiveness of organization, but also need to guarantee the national scientific and technological innovation strength by the construction of service invention system even the national comprehensive strength.

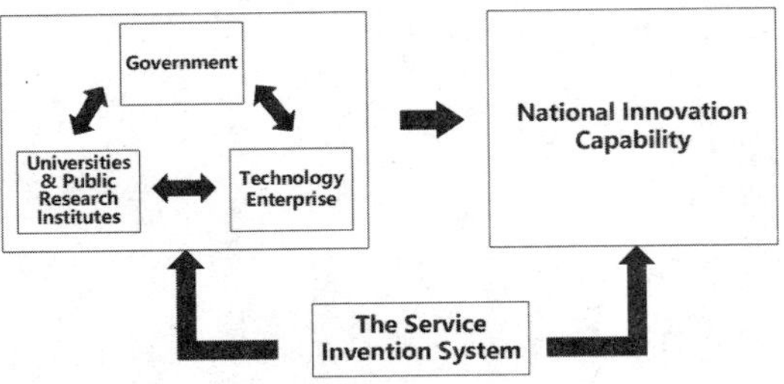

Fig.2 The mechanism of service invention system.

3. The Current Service Invention System Becomes an Important Weakness of Seizing the Opportunity of a New Round of Scientific and Technological Revolution

1) The dual character of disruptive technology will form 'destructive' impact to the current service invention system

From long period, disruptive technological innovation is fundamental dynamic of a systematic change in the service invention system. On one hand, most of the reason of great change of service invention system is the great inapplicability of the existing system to the application of new technology. With speeding up of economic globalization, the proliferation rate of subversive technology is greatly accelerated, however, the reform of the service invention system lags behind relatively, leads to the gradual decline of marginal benefits of institutional protection, makes it difficult for the current service invention system to match the huge transformative effect of the application of subversive technology. On the other hand, subversive technology will have a great impact on the original service invention system, also forces the service invention system to get rid of the institutional crisis through institutional innovation, and the institutional changes matched with the application of new technologies will be created, a new system to ensure benefits is formed, and then promote the evolution of subversive technology from single-point breakthrough to systemic explosion.

In recent decades, technological innovation and systems are increasingly becoming the dominant factors in maintaining economic prosperity and development. As the disruptive technology has the huge effect of disruptive innovation, its impact on the evolution of the short period of the economy will be more prominent, new technology spillover effect caused by FDI and ODI, made the technology exporting countries with the guarantee of perfecting the service invention system form the dimension reduction strike effect on the backward countries of the service invention system, the huge technological rent-seeking space will have a strong impact on the service invention system, new demand for disruptive technology innovation, new market becomes the important push hand of service invention system reform.

2) As the exposed 'cutthroat' technique weakness of ZTE incident, reflects the structural defects of the present service invention system

Firstly, under the current service invention system, the pattern of 'strong capital and weak Labor', gives the organization the right to decide on the invention right. As the special character of the result of the service invention, a number of departments have made laws and regulations from different perspectives to monitor and protect the achievement of service invention. Therefore, the current service invention system is consist by *law of contract, patent law, Science and technology progress law, Promoting the transformation of scientific and technological achievements law* and *Detailed rules for the implementation of the Patent Law etc.*, differences in the level of effectiveness and inconsistency of criteria provide different opinions in dispute resolution for service invention, the absence of the rules of the referee buried the hidden danger, and shackles the transformation of scientific and technological achievements and releasing of new dynamic energy. For example, the corporation headquarter is the central enterprise, the institute is an enterprise, it is inconsistent provisions on examination and approval authority for the management of state-owned assets according to the *Measures on Supervision and Administration of transactions of State-owned assets of Enterprises (decree No. 32)* made by SASAC (State-owned Assets Supervision and Administration Commission) and *Interim Measures for the management of state assets of central institutions* made by the Ministry of Finance, it caused the problem of that want to transform but dare not to transform, enterprise dares not to make decision. For the provisions of the *Law on the Promotion of Transformation of Scientific and technological achievements* on incentives and remuneration for breaking through the total amount of wages, there is a discrepancy between the specific operational level and the existing provisions, the rewards cannot be realized.

Secondly, it must be perfected that the maintenance and appreciation of state-owned assets and the service invention system in promoting the transformation of scientific and technological achievements. When the state-owned enterprises and institutions manage the achievements of the service invention according to the regulations, the state-owned assets of organization owned by the whole people are owned by the state, however, the loss of state-owned assets involved in the process of the transformation of achievements has become the core problem that restrains the inventors' enthusiasm and conversion rate. Although the State Council has introduced a special exemption policy for pricing scientific and technological achievements, the logical premise of the existing exemption mechanism is that there is an 'exemption' after the existing 'responsibility'. So, heads of many state-owned enterprises, universities and research institutions are still worrying. On the other hand, many supporting policies do not solve the core issues, it is difficult to realize the equity rewards. As the property of state-owned assets of service scientific and technical payoffs, state-owned shares naturally formed after valuation, the rewards to the employee-inventors will be realized by the state equity trading system. So, some provinces and cities are exploring the reform of mixed ownership of service invention, but the documents issued by local governments to promote the transformation of scientific and technological achievements into new policies obviously exceed the 'ceiling' superior law and the superior department management method.

Thirdly, it is needed to strengthen the rationality of the design of invention report system. The current focus of controversy on invention reporting system is turned from 'The necessity of establishing invention reporting system' to 'rationality of the design of invention report system'. As the relevant laws and regulations do not provide the necessary legal basis for the invention reporting system, in the process of management of service invention report, each organization has the problem of non-standard operation to varying degrees, it increases the legal risk of disputes between enterprises and inventors over service inventions, increases the possibility of a dispute.

3) The current service invention system does not inspire the original innovation dynamic of technological innovation subject completely

Firstly, the duty inventor's interest protection is insufficient. Guided by the principle of 'employer first' legislation, our current service invention system is more focused on the protection of the interests of the organization. The pattern of 'strong Capital and weak Labor' endows the enterprise with the right to make decision of invention interest, it directly results in the passive situation of the inventor of the enterprise. In the private enterprise, the intellectual property system established by most small and medium-sized private enterprises informally, although business owners have improved their understanding of the importance of the system of service inventions to a greater extent than in the past, the phenomenon of infringing the rights and interests of inventors occurs from time to time due to the internal implementation rules, the procedure and management chain is not perfect.Take the employee-inventor's signature right as an example, there is a phenomenon that some private enterprises still have hundreds of patents a year, person in charge of the enterprise has signed them. On one hand, due to the high cost of protecting rights through judicial channels, difficult in proving infringement, weak awareness of business owners and employee-inventors, many inventors are often unwilling and afraid to defend their rights. On the other hand, construction of service invention system in some enterprises is not standard, enterprise scientific research and technical personnel don't sign or stamp on the test drawings, related process documents and program documents, and it caused the problems of infringing upon the right of signature of technical personnel in enterprises.

Secondly, the rewarding system of service invention system and its supporting policy design are needed to be improved. Enterprises and patent inventors are always facing the problem of contradiction between reward system and organization management and insufficient reward amount and unclear distribution of both parties' rights and interests. The system and method according to the commercial practicability of the service invention, the intellectual contribution of the inventor is determined to compensate and motivate the inventor is still lack of operability. And in the process of service invention transformation, the related fiscal and taxation policies are not perfect. Take the tax policy as an example, the questions of the scope of personal income tax and the amount of payment in the award of service invention become the new focus of service inventions award. Currently, organization pays to an individual in the form of a scientific research incentive or year-end bonus which is transferred from the service patent income, and the final tax rate can be as high as 45%, it seriously dampens the creative initiative of the employee-inventor.

4. Filling up the Weakness of the Service Invention System Needs Comprehensive Precision Policy Power

1) Explore the new system model guided by the priority of the rights and interests of the 'Employee-inventors'

Change the long-standing legislation principle of 'employer first', try to explore the reforming of the service invention system which takes into account the rights of the inventors and the interests of the organization at the same time. Under the principle of promoting scientific and technological progress, encouraging organization and inventors to agree, by contract, on the attribution of patent rights and the distribution of benefits for service inventions, two sides agreed on an equal footing, propose the organization to pay attention to protect the inventor's interest, so then stimulate the scientific research personnel to promote the effectiveness and applicability of service invention and creation. Speed up the revision and perfection of laws, regulations and standards related to the service invention system, make the service invention system bring the function to the supervision and promotion of technological innovation. Establish bonus pool for the transformation of the country's sustained and steady growth in the achievements of service inventions, focus on supporting enterprises that can undertake major technology research and development and demonstration applications, awards for scientific research institutions that have achieved the invention results of the core key technical positions.

2) Perfect the legal system of service invention, provide effective system guarantee for the development of new technology

Strengthen research on legal, ethical and social issues related to service invention in key scientific and technological fields, perfect the framework of laws, regulations and ethics of service invention system to ensure the sustainable and healthy development of new technology. Around the detailed fields of autopilot, service robot, etc, speed up the study and improvement of the legal applicability of the system of service invention. Research on behavioral science and ethics in the field of artificial intelligence, gene editing and other emerging technologies, establish multi-level structure and frame of ethics and morality. Research and formulate codes of ethics and codes of conduct for new technology product developers. Study and perfect the policy system of education, medical treatment, insurance, social assistance and so on, which can adapt to the application of new technology. Deepen the international co-operation on international rules for the service invention system in the field of new technology and standard making etc.

3) Play the roles of market and enterprise efficiently, encourage the enterprises to take breakthrough attempt to transform the achievements of service invention actively

As the uncertainty of the development of science and technology innovation, innovations of many policies and systems, break out in constant trial and error. In the process of developing new technologies, must strengthen the guarantee function of the service invention system, while promoting enterprises to undertake major national science and technology projects, actively organize conditional enterprises to actively expand the transformation system of service invention achievements. At the same time, the government must play an important role on planning guidance, policy support, safety prevention, market supervision, environmental construction etc, by the features of new technologies, new formats, and new models, encourage enterprises to actively carry out the breakthrough attempt and exploration of the achievements of service invention, local governments should sum up successful experiences and organize promotion in time.

4) Speed up the construction of the training system of service invention system, improve the whole cognition and application level of the service invention system in the whole society

Speed up on the research on the impact of new technology on the service invention system, establish the training system of service invention system to meet the needs of new economic development, Support institutions of colleges and universities, professional schools and socialized training institutions to establish service systems for service inventions, raise the consciousness of the employed personnel about the service invention system. Encourage enterprises and institutions to provide staff with service invention system training. Strengthen the training and guidance of the staff and workers on the legal policies related to the of service invention system, ensuring the employee-inventors' legitimate rights and interests.

References

Amirkolaii K N, Baboli A, Shahzad M K, et al. Demand forecasting for irregular demands in business aircraft spare parts supply Chains by using artificial intelligence (AI)[J]. IFAC-PapersOnLine, 2017, 50(1): 15221-15226.

Arora A, Cohen W M, Walsh J P. The acquisition and commercialization of invention in American manufacturing: Incidence and impact[J]. Research Policy, 2016, 45(6): 1113-1128.

Cohen S K, Caner T. Converting inventions into breakthrough innovations: The role of exploitation and alliance network knowledge heterogeneity[J]. Journal of Engineering and Technology Management, 2016, 40: 29-44.

De Marco A, Scellato G, Ughetto E, et al. Global markets for technology: Evidence from patent transactions[J]. Research Policy, 2017, 46(9): 1644-1654.

Gierej S. The framework of business model in the context of Industrial Internet of Things[J]. Procedia Engineering, 2017, 182: 206-212.

Gu D, Yin H, Yang G. Research on the driving mode of scientific and technological achievements transformation in colleges and universities[J]. Science & Technology Progress and Policy, 2012. (in chinese)

Hingley P, Park W G. Do business cycles affect patenting? Evidence from European Patent Office filings[J]. Technological Forecasting and Social Change, 2017, 116: 76-86.

Hsieh K L. Applying an expert system into constructing customer's value expansion and prediction model based on AI techniques in leisure industry[J]. Expert Systems with Applications, 2009, 36(2): 2864-2872.

Hwang J, Choi M, Lee T, et al. Energy prosumer business model using blockchain system to ensure transparency and safety[J]. Energy Procedia, 2017, 141(37): 194-198.

Jou J B. R&D investment and patent renewal decisions[J]. The Quarterly Review of Economics and Finance, 2018.

Kiel D, Arnold C, Voigt K I. The influence of the Industrial Internet of Things on business models of established manufacturing companies—A business level perspective[J]. Technovation, 2017, 68: 4-19.

Krotov V. The Internet of Things and new business opportunities[J]. Business Horizons, 2017, 60(6): 831- 841.

Nordensvard J, Zhou Y, Zhang X. Innovation core, innovation semi-periphery and technology transfer: The case of wind energy patents[J]. Energy Policy, 2018, 120: 213-227.

Prybila C, Schulte S, Hochreiner C, et al. Runtime verification for business processes utilizing the bitcoin blockchain[J]. Future Generation Computer Systems, 2017.

Saarikko T, Westergren U H, Blomquist T. The Internet of Things: Are you ready for what's coming?[J]. Business Horizons, 2017, 60(5): 667-676.

Van Roy V, Vértesy D, Vivarelli M. Technology and employment: Mass unemployment or job creation? Empirical evidence from European patenting firms[J]. Research Policy, 2018.

Wu Y, Welch E W, Huang W L. Commercialization of university inventions: Individual and institutional factors affecting licensing of university patents[J]. Technovation, 2015, 36: 12-25.

Zhang Y. Thought on some basic problems on service invention bill [C]// Patent Law Research. 2015. (in chinese)

The Technology Innovation System for Military Enterprises under the Background of Civil-military Integration

Tang Saili, Li Zhancheng, Yu Xia, Liu Meng, Dai Kun

Research and Development Center, China Academy of Launch Vehicle Technology, Beijing, China

Abstract: In the past decades, the civil-military integration has been promoted to the national strategic level, which is beneficial from a great achievement by long-term coordination development between economic construction and national defense construction of our party and country. At present, our civil-military integration is in the transition stage from initial integration to deep integration, problems still remain in elements, fields and effectiveness. This paper illustrates contents and component elements of civil-military dual-use technical innovation system including institutional mechanism, organization, talent team, technological foundation. And then analyses the path of innovation and application from technical identification stage, technology research and development stage, trial production stage, product application stage. Furthermore, the paper analyzes the ways of technology innovation of military enterprises from the perspective of technology stock and technology increment. Finally, the research presents a hypothesis of construction of civil-military dual-use technical innovation system for military enterprises, which can provide references for improvement of the conversion efficiency and promotion of depth development of civil-military integration.

Keywords: Civil-military Integration; Technology Innovation; System Construction; Management Efficiency

1. Introduction

In the reports of the 19th National Congress of the Communist Party of China, President Xi Jinping who deployed great strategy about forming a deep development pattern of civil-military integration, and building an integrated national strategic system and capacity, thrice mentioned the development of civil-military integration from different perspectives including persisting and developing the basic strategy of socialism with Chinese characteristics in the new era, firmly implementation of the national strategy for building a moderately prosperous society, and adhering to the road of strengthening the army with Chinese characteristics. Thus, civil-military integration was formally incorporated into the national strategic system and was written into the party constitution. In order to respond positively to government orders, domestic military enterprises have integrated civil-military integration into the emphasis of comprehensive deepening reform. In the practical proceed of promoting market-oriented transformation, based on the constantly thinking and exploring the development mode of combining civil-military integrative development with technology innovation, the paper proposed a dual use technology innovation system, with a view to carry forward civil-military integrative development of domestic military enterprises in depth.

2. The Connotation of the Dual Use Technology Innovation System

1) Related Concepts

In 1994, United States Congress presented explicitly that dual-use technology included scientific research, technology and products that can meet both military and non-military needs. The dual-use scientific research referred to the scientific and technology achievements that can be applied to both military industry and commercial production. The dual-use technology referred to the technology that can be applied to both military production and civilian manufacturing technique. The dual-use products were products that could be purchased by both military and commercial customers. In 1997, the United States Department of Defense redefined the term 'dual-use technology' in the National Defense Authorization Act which could satisfy both military application and sufficient commercial

Corresponding author: Tang Saili, No. 1 South Dahongmen Road, Fengtai District, Beijing, 100076. tangsaili@126.com.

value and support feasible production base. And for the first time, the act put forward the native dual-use technology plan (Peng et al., 2017).

Based on the definition and interpretation of dual-use technology in developed countries, the paper defined dual-use technology as the existing high-tech which could be applied to both military and civil purposes, or the technology which had good military use, significant commercial value and industrial development potential to be developed.

According to the related concepts of technology innovation system and the characteristics and practice of technology innovation in military enterprises, this paper proposed that the dual-use technology innovation system was guided by the market demand of dual-use technology, centered on the research and development of dual-use technology, focused on promotion and application, which innovated administrative system and operation mechanism, and would help to achieve a system that each element of dual-use technology innovation were coordinative and efficiently interactive.

2) Elements of the Dual-use Technology Innovation System

Generally speaking, there were four main elements of the dual-use technology innovation system (Fig. 1). The first element was the system and mechanism, mainly including management system, innovation institution, equipment sharing mechanism, information communicative and interactive mechanism, incentive mechanism, etc. (Dong et al., 2015). The second element was the organization that about dual-use technology management, research and development, promotion and other institutions. The third element was the talent team that suitable for the dual-use technology development. The fourth element was the technical basis, mainly referred to the existing technology and capacity.

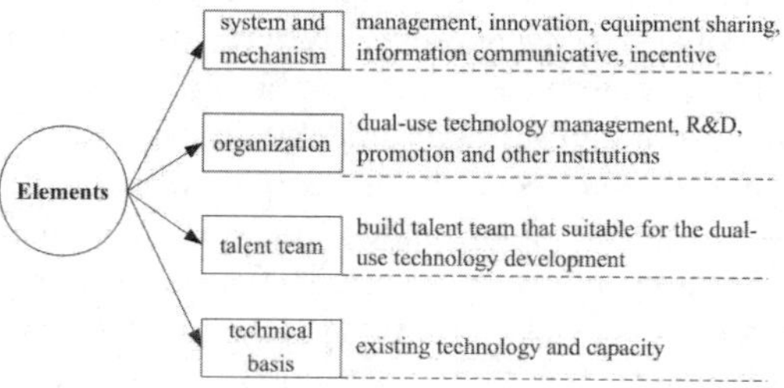

Fig. 1 Elements of the dual-use technology innovation system.

3) Innovation Method of the Dual-use Technology

The main innovation method of the dual-use technology included the existing military technology extended to the civilian field, the existing civil technology extended to the military field, and technological development met the needs of both military and civilian fields at the beginning.

In the field of technology stock, the enterprises transferring the dual-use technology could tap the applicative potential of the technology in another field through the secondary development of their original technology. They could also obtain the existing dual-use technology achievements from home and abroad through the introduction of the dual-use technology, but this process was usually accompanied by the digestion absorption and re-innovation of technology.

In the field of technology increment, the enterprises transferring the dual-use technology could carry out independent research and development according to the demand of the dual-use technology market, or obtain the results of dual-use technology through the cooperation of enterprises and institutions within and between fields, and focusing on future dual-use performance from the beginning of technology development (An Liu and Jia, 2016).

4) Path of the dual-use technology innovation and application

According to the literature on the path of the dual-use technology innovation and application at home and abroad, the thesis summarized the following four stages (as shown in Fig. 2).

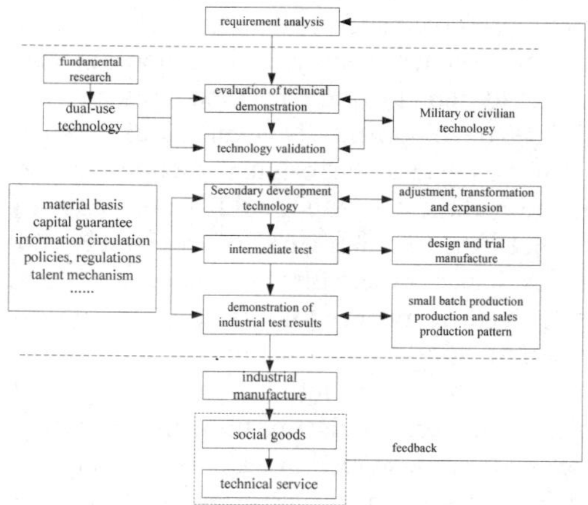

Fig. 2 Path of the dual-use technology innovation and application.

(1) *Technological identification phase*

First of all, the implementation of the dual-use technological achievements should clarify the goals and motivations of their social needs with the demands as traction and market as orientation, and attach importance to the uniform principle of technological advancement, practicability and applicability. Afterwards, market forecast for the promotion of the dual-use technology should be done, in order to evaluate the possibility of the popularization and application of the technical achievements and the possible military benefits, economic benefits and the social benefits after the popularization and application. At the same time, this was also an important basis for evaluating the effectiveness of the dual-use technology.

(2) *Technological development stage*

The generation of the dual-use technology in this stage referred to the broad sense, which included mature dual-use technology in military or civilian fields with potential for popularization and application, as well as the dual-use technology that considered both military and civilian needs in research and development. The evaluation and experimental development of the dual-use technological demonstration in this stage was not only the continuation of the previous stage, but also the premise and foundation of the latter stage.

(3) *Trial production stage*

This stage was an intermediate link in the whole process of the dual-use technology promotion, and it was also a core link. In addition to a few the dual-use technology could be directly promoted and applied, the promotion of many military technology, civil technology and dual-use technology often involved the secondary technical development. This link could adjust, reform or expand the original technology, so that the technical results could better adapt to and meet the requirements of new areas for technical indicators, performance and other aspects, and played a major role in the successful promotion of the dual-use technology results. The intermediate test link was the redesign and trial-manufacture of the existing or secondary developed dual-use technology, so as to lay the foundation for the next industrial test. Industrial test and demonstration included small batch production, sales and production finalization, which further promoted the popularization and application of dual-use technology achievements. The three links were interrelated and inseparable. At the same time, each link needed to be effectively guaranteed in terms of material basis, fund raising, information circulation, policies and regulations, personnel mechanism and so on, so as to jointly promote the successful promotion of the dual-use technology achievements.

(4) *Product application stage*

This stage was the industrialized manufacture and application of the dual-use technology achievements, which formed products, processes or technology for military or civilian use. Industrialized manufacture and application included commercialization and industrialization of the dual-use technology: commercialization was to achieve normal production scale, at the same time products were really sold in the market; industrialization referred to the production can form a larger group of manufacturers, or even the formation of new industries or industries, and ultimately achieve popularizing and applying of the dual-use technology results. Finally, through the information feedback mechanism and according to the social demand and market forecast in the first stage, the paper evaluated whether the promotion of the dual-use technology had achieved the expected military, economic and social benefits, so as to analyze and evaluate the success of the promotion of dual-use technology.

3. Dual-use Technology Innovation System for Military Enterprises

In order to promote the innovation and development of the dual-use technology in military enterprises, the construction of the dual-use technology innovation system in military enterprises should take the R&D of the dual-use technology as the core, the demand as the traction, the integration of social superior resources, and the improvement of various systems and mechanisms.

Referring to the development experience of the dual-use technology system of foreign military enterprises and combining with the practice of civil-military integration of domestic military enterprises, the thesis put forward the '341' development mode of the dual-use technology innovation system, that is, forming 'three subsystems', perfecting 'four mechanisms' and optimizing 'one platform'. This mode would provide solid technical support for the development of military and civilian integration of the military enterprises.

1) Three Subsystems

(1) *Organization subsystem*

The military enterprises should establish the dual-use technology management departments to improve the interaction between two-level unit and three-level unit, improve the research institutions of dual-use products in two-level unit, and promote the group company's dual-use technology innovation and development.

(2) *Technology subsystem*

Starting from the market demand, this paper established a demand response mechanism and developed the dual-use technology suitable for the market demand by integrating civil technology, joint development, independent research and development, transference military technology to civilian market and other technology innovation paths.

(3) *Promotion subsystem*

The dual-use technology could achieve market-oriented and industrialized, through national high-tech park, technology trading platform, technology incubator, financing platform.

2) Improve the 'Four Mechanisms'

(1) *Achievement evaluation and feedback mechanism*

The military enterprises could establish the evaluation and compensation mechanism for the transformation of the dual-use technology achievements, and form a mutually beneficial and complementary model in the military and civil fields.

(2) *Financial support and incentive mechanism*

A special support fund for the dual-use technology should be established to encourage the integrated development of the dual-use technology, and a corresponding incentive system should be established to encourage units and individuals that promoted the development of the dual-use technology.

(3) *Bidirectional flow mechanism for talents*

The bidirectional flow mechanism for talents should be built in order to effectively integrate and utilize human resources, and stimulate the innovation vitality of the whole military enterprise.

(4) *Service guarantee mechanism*

The military enterprises could unify the interface standards of the dual-use technology, adapt to the requirements of product scale and industrialization development under the market conditions actively respond to competitive market, establish full-time promotion team of the dual-use technology, and expand the application channels of dual-use technology.

3) Optimize 'One Platform'

The dual-use technology exchange platform needed be built or optimized if the platform had established. The military enterprises should further strengthen top-level planning, look ahead to the future, aim at the market, upgrade and improve the exchange platform, strengthen daily management, expand the functions of the platform, update the information of the technical requirements and advantages of the enterprise in time, and better serve promotion and transformation of the dual-use technology.

4. Suggestions

1) Strengthen Ideological Guidance, Change Ideas, and Consolidate the Consensus and Action of Civil-military Integration

The long-term military research and development task had brought heavy pressure to ensure the success of the military enterprises, which had not fully released the vitality of dual-use technology innovation. In the face of the new normal and new situation of civil-military integration development needs, military enterprises should deeply understand the importance of building a dual-use technology innovation system, achieve ideological consensus on 'equal emphasis on both the military and civilian need' and turn the dual-use technology innovation system into 'a game of chess'.

2) Take Measures to Accelerate the Construction of the Dual-use Technology System

Under the unified deployment, military enterprises should set up dual-use technology management and promotion departments, that should not only sort out the technology they have mastered, understand their own capabilities and advantages, but also find out the core technology suitable for the dual-use. The new technology that guides the future development should start from the top level design and develop aiming at the dual-use need. At the same time, the military enterprises could make market demand as orientation, improve the corresponding safeguard mechanism, incentive mechanism, two-way personnel flow mechanism, evaluation and feedback mechanism; explore the application for dual-use technology projects suitable for research and protection conditions, in the process of military capacity-building; establish and improve the dual-use technology information exchange platform, accelerate the construction of the dual-use technology system.

3) Optimize the Link of Investment Decision-Making and Operation Authorization Management to Improve Management Efficiency

The military enterprises should further optimize the investment decision-making process, improve the market response speed, adjust the management authorization at all levels, and adapt to the market operation law. Because of the different market industries, the different actual stages of industrial development, and the different investment decision-making processes, the demand for business authorization varies from unit to unit. Therefore, the military enterprises should actively explore and study decision-making mechanism fitting the market, differential business authorization mechanism and improve management efficiency in accordance with the actual situation of their civil-military integration. In this way the military enterprises could create a good environment for the industry to develop according to market rules.

5. Conclusion

In the new era, the establishment of the dual-use technology innovation system is an inevitable requirement for the military enterprises to comply with the reform and development of state-owned enterprises, and an important measure to implement the spirit of the Nineteenth National Congress. The military enterprises should thoroughly implement the policies of the Party and state, combine with the practice of civil- military integration in various enterprises, formulate a dual-use technology development line that meets the requirements of their own development, construct a compatible and win-win development model, and gradually form a civil-military integration development situation with market demand- oriented, wholesome system and mechanism and benignant interaction between the military and the civilian, so as to provide powerful technical support for accelerating the comprehensive and deepening reform of the military enterprises, promoting the formation of a deep development pattern of civil-military integration, fulfilling the development goals of the 13th Five-Year Plan and realizing the dream of a powerful country.

References

Peng Zhongwen, Liu Tao, Zhang Shuangjie (2017). Research on the construction of synergic innovation system of military and civilian integration science and technology industry——based on international comparative perspective. Science & Technology Progress and Policy , 34(11), 102-107.

Dong Xiaohui, Qi Yi, Zhang Weichao (2015). On the characteristics and model of the transformation of dual-use scientific achievement from innovation-driven. Science & Technology Progress and Policy, 32(21), 135-139.

An Mengchang, Liu Xingang, Jia Yi (2016). China's dual-use technological achievements transformation system. Dual Use Technologies & Products, 15:12-14.

Collaborative Innovation Mechanism Construction in Social Governance of China in the New Era

Gao Yuanying[1,2], Chen Rui[1]

[1] National Academy of Innovation Strategy, CAST, Beijing, China
[2] Institute of Science and Technology Policy and Management Science, CAS, Beijing, China

Abstract: As socialism with Chinese characteristics enters new era, there exit a series of new situation and new problems to be settled in social governance innovation. In order to maximize people's fundamental interests, constructing a multi-center governance body and adopting various means of governance as the main direction of social governance collaborative innovation has become an inevitable requirement of social governance in China. At present, the factors that restrict the collaborative innovation of social governance mainly include the laggard of management concept, the imperfect governance mode and structure and the imperfect benefit distribution mechanism. Mechanism of public participation, interest integration, evaluation and supervision need to be better constructed to promote the collaborative innovation in social governance.

1. Introduction

With China comes into the new era, social management is faced with a series of new problems and new situation. Traditional social management model has been unable to adapt to the growing complexity of social transformation, and a series of innovation need to be adapted to the new requirements of the new era[1]. Since the 18th national congress of the communist party of China (CPC), the CPC central committee, with Xi Jinping at its core, has repeatedly proposed to strengthen and innovate social management and innovate the social governance system. However, as a large and complex system, the innovation of social governance cannot be led by the government alone. It must be realized through collaborative innovation. As an effective form of innovation, both theory and practice of collaborative innovation has been extensively explored. From the viewpoint of collaborative innovation to analyze social governance, the features of social governance will be summarized, the limit innovation factors will be found out, what's more, the cooperative innovation mechanism to promote social governance will be built up.

2. Characteristics of Social Collaborative Governance Innovation

First, the aim of social collaborative governance is to maximize the fundamental interests of the people. Different from the traditional management to safeguard the interests of the ruling class[2], the core of social collaborative governance is to safeguard the public interests. The relationship between the government and citizens is no longer a simple management relationship, on the contrary, the government should be able to take the initiative to provide citizens with more high-quality public services and create a good environment for development. The relationship between citizens, organizations and the government is no longer simply public service recipients and public services providers, instead citizens also should take the initiative to create more public interest so as to realize the virtuous circle of maximizing public interest. All governance bodies should work together to create social interests, solve social problems and coordinate social relations so that the fruits of social governance will be shared by all.

Second, the body of social collaborative governance changes from single to diversified and multi-center. The government is no longer a single governance body, instead a diversified multi-center is an important feature of social collaborative governance. Government as the core of the traditional single-center management already can not adapt to the development of modern society, the disadvantages are growing and variety kinds of social contradictions are increasing. In addition to the government, citizens and all kinds of social organizations have more and more obvious complementary roles and advantages in social governance. For citizens, are the basic elements of the social structure, they can directly sense the social situation and social information and they are the broadest beneficiary of the fruits of social governance. For social organizations, with their continuous development and improvement, they will play unique roles in social governance as a bridge between government and citizens[3]. Therefore, on the basis of giving full play to the leading role of the government, the sufficient and necessary conditions have been met to form a multi-center and diversified social governance body.

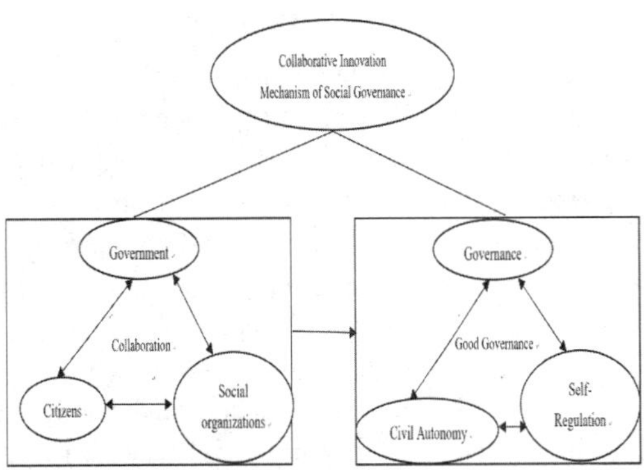

Fig. 1 Frame of social collaborative governance innovation system.

Third, the governance means and modes of social collaborative governance change to flexibility and diversification. The original compulsory means, such as administrative means and legal means, will play a limited role and will be supplemented by flexible means guided by service concepts, and market-oriented means will play an increasingly important role[4]. In addition, some moral constraint means are also more and more necessary. Governance by all is the most important means. Compared with the traditional means, collaborative governance is a combination of top-down and top-down governance that promotes each other. The government delegated power from top to bottom to citizens and social organizations, and citizens and social organizations participated in social governance from bottom to top in various ways, thus forming a two-way channel of social governance. With the advent of the information age and the new media era, the network governance has been formed, which further facilitates the participation of citizens and social organizations. For example, microblog of government affairs and WeChat of government affairs both greatly facilitate the participation and communication of citizens and promote the collaborative innovation of social governance.

3. Factors Restricting the Collaborative Innovation of Social Governance

1) Restrictions of Traditional Management Concept

The ideology of official standard and elitism still exists. Some leaders are still accustomed to traditional controls, treating citizens and social organizations as objects of control, and trying to bring citizens to their knees with tough top-down executive orders. Accustomed to a rigid hierarchy of power, they are subservient and unwilling to listen to the voices of the grassroots. Elitist philosophy thinks that the main body of social management should be a handful of social elites, only those who are qualified professional skills of social elite to manage society, such as a politician or a leader[5]. The presence of these ideas will make the government ignore the citizens and social organizations in social governance, and even opposite to the masses. In this way, the social foundation of collaborative innovation cannot be formed, which hinders the formation of collaborative governance innovation.

2) Restrictions of Social Governance Modes

Some government agencies are not good at or even reject the use of the Internet to open government affairs and interact with citizens. As more and more government departments have launched their own Weibo and WeChat platform, through these platforms they release information, interpret policies and directly deal with corresponding business, but there are still some department did not adopt the new way, some opened an account but rarely publish contents. This shows that some managers have not accepted these new governance methods fundamentally and have not correctly realized the importance of these new methods in the new era.

3) Restrictions of Social Governance Structure

In order to optimize the governance structure, the diversified subjects should cooperate with each other and form synergy. However, in China, the governance structure is far from perfect, which is mainly reflected in the fact that the pattern and mechanism of the joint efforts of the whole people have not yet been formed [6]. Apart from the reasons of the concept of government, first, the development of social organizations in China is not yet perfect, and the participation of citizens in social governance is not enough. Second, there is no enough synergy or corresponding institutional arrangement between the government, citizens and social organizations so that the effect of collaboration is far from being achieved.

4) Restrictions of Uncoordinated Benefit Distribution

The new normal is an important feature of China's new era. Under the new normal, the slowing down of the economic growth and the adjustment of the structure could lead to unemployment rising and social welfare reduction, then there may be more and more social contradictions exposed. These contradictions potentially restrict the social governance collaboration. The more contradictions are exposed, the more likely the conflicts are to erupt and the less likely the collaborative innovation will be. Therefore, only by coordinating the interest relations and establishing the interest mechanism shared by the whole people can further promote the collaborative innovation of social governance at a deep level and better safeguard the results of social governance.

4. Construction of the Collaborative Innovation Mechanism of China's Social Governance in the New Era

1) Establish the Concept of Collaborative Governance

Managers should set up the concept of service and cooperation, and realize that their role is not only controller, manager or stabilizer, but also server and coordinator. To establish the concept of service means to start from serving the people wholeheartedly and safeguarding the fundamental interests of the largest number of people, attach importance to the construction and development of people's livelihood and social security, and make full use of various resources to provide sound and high-quality public services to the society. To establish the concept of cooperation is to insist on building and sharing in the process of social governance[7]. Managers should abandon the concept of bureaucratic and elitist governance, fully recognize the enormous energy and important role of citizens and social organizations in social governance, respect and safeguard their rights, and actively create favorable conditions for public participation. In addition, in the process of collaborative innovation of social governance, managers should have an inclusive mentality. In order to realize collaborative innovation, there must be many different voices. Managers should learn to draw nutrition and strength from the masses and try to resolve all kinds of contradictions arising from the public.

2) Perfect the Public Participation Mechanism

We need to build a multi-center collaborative governance mechanism for public participation. On the one hand, in order to ensure long-term social stability, it is necessary for the government to continue to perform its supervisory functions and continue to play a leading role. On the other hand, it is necessary to cultivate and improve all kinds of social organizations, fully mobilize the enthusiasm of citizens, build diversified and compound governance bodies, stimulate social vitality, optimize the governance structure, and improve governance efficiency. At the same time, all governance bodies should make use of network technology to widen the participation channels[8]. The government should make full use of online platforms and new media forms to expand exchanges with the public, promote openness in government affairs and encourage citizens to participate. The public should pay timely attention to the new media forms of governments, actively participate in politics through the Internet, and fully exercise their rights. In addition, a response mechanism should be established. For multiple governance bodies, the premise of collaboration is response, and the lack of response mechanism is one of the main factors that influence the participation enthusiasm of all governance bodies. Therefore, it is necessary to establish a stable response mechanism on the Internet platform, respond to the content of public participation in a timely manner, and deal with Internet violence or online rumors in a timely manner, so that a long-term network interaction mechanism can be established.

3) Perfect the Interest Synergy Mechanism

The government should enhance its capacity to provide public services and encourage social organizations and institutions to join in the provision of public services and the creation of public interests. At the same time, accelerate reform of the income distribution system, further establish and improve the mechanism for coordinating interests, set standards for the equalization of public services, and work to narrow the development gap between regions. What's more, guide the public to correctly and reasonably express their own interests and properly handle the interest conflicts between different stakeholders.

4) Perfect the Monitoring and Early Warning Mechanism

It is necessary to establish an effective monitoring and early warning mechanism for social governance to predict and prevent possible problems and risks in the process of collaborative innovation in advance, and improve the emergency management ability of the governance bodies. At the same time, it is need to improve the evaluation and supervision mechanism of social governance and timely evaluate and supervise the governance effect of collaborative innovation. The government should build a reasonable evaluation index system, carry out timely management performance evaluation, and timely adjust unreasonable governance models and methods. A supervision mechanism should also be set up among all governance bodies to test the effect of collaborative innovation. In particular, the public should strengthen the supervision of government power, prevent the abuse of power, and ensure the effective combination of governance among all subjects.

5) Encourage the Autonomy of Citizens and Communities

As the basic unit of society and the basic carrier of social governance, the improvement of the ability of citizens and communities to govern themselves can make a qualitative leap in social governance. The government should actively encourage and correctly guide citizens' self-government, encourage citizens and communities to innovate means and behaviors in self-government, and form a good atmosphere of collaborative innovation. For citizens, they should change their behavior of relying on and attaching importance to the government, instead, they should establish a sense of ownership, actively explore ways and means of social governance, actively create public interests and participate in social governance. At the same time, community management needs to be further improved, and the content of community construction needs to be further enriched, so as to give full play to the combination of community autonomy and public services. Strengthen the sense of identity of community residents, resolve some problems and contradictions within the community, and work together to enhance the interests of community residents.

5. Conclusion

In a word, it is an inevitable trend to build a long-term collaborative innovation mechanism in social governance. On the basis of correctly understanding the necessity and connotation of collaborative innovation in social governance, it is still an urgent problem to take concrete actions to overcome the limitations of existing concepts and institutional mechanisms and to truly push forward collaborative innovation. The government, citizens and organizations need to work together and make positive innovations to deal with the problems and contradictions in China's social governance in the new era and achieve long-term social stability.

References

[1] Watts, Duncan J., Small Worlds: The Dynamics of Networks between Order and Randomness, Princeton, N.J.: Princeton University Press, 1999.

[2] Newman, M.E.J., The Structure and Function of Complex Networks, SIAM Review, vol.45,2003, pp. 167-256.

[3] Luis A. N. A, Brian U. Complex Systems—A New Paradigm for the Integrative Study of Management, Physical, and Technological Systems, Management Science, vol.53, 2007, pp.1033-1035.

[4] Filip A., John S., Node Centrality in Weighted Networks: Generalizing Degree and Shortest Paths, Social Networks, vol.32, 2010, pp.245-251.

[5] Fan R.G., Synergetic Innovation in Social Governance in a Complex Network Structural Paradigm, Social Science in China, vol.04,2014, pp.98-120.

[6] Gu L.M., Thinking of Internet Participation and Governance Innovation, Chinese Public Administration, vol.301, 2010, pp.11-14.

[7] Ansell C, Torfing J, How Does Collaborative Governance Scale? Policy & Politics, vol.43, 2015, pp.315-329.

[8] Bryson JM, Crosby BC, Stone MM, Designing and Implementing Cross-Sector Collaborations: Needed and Challenging, Public Administration Review, vol.75, 2015, pp.647-663.

Study on Salaries of Engineering Staff in China

Yang Guang[1], Wang Yanni[2]

[1] National Academy of Innovation Strategy, CAST, Beijing, China
[2] Beijing Wuzi University, Beijing, China

Abstract: Engineering staffs play a vital role in promotion of industry innovation and international competitiveness. Salary is one of the most important motivators and closely relates to engineering staff's performance. Based on the data from the 4th National S&T Workers Condition Survey and interviews with engineering staff and HR, this paper describes salary status of the Chinese engineering staff, and tries to analyze deep-seated reasons behind some problems. The paper makes a useful attempt in understanding current situation of Chinese engineering staff.

Keywords: Engineering Staff; Salary; Incentive; China

1. Introduction

In the new age of innovation-driven development, the vital role of engineering staff remains considerable, in particular, technology innovation practitioners and impellors play core roles in the application and recreation of advanced science and technology[1,2]. As indicated by some scholars in some study, engineering staff are playing vital roles in improving enterprises' innovation ability and enhancing technical level of industries[3].

The countries all over the world make much account of the basic working and living conditions of science and engineering personnel; through the investigation and analysis, specific organizations make overall understanding on the working/living conditions and satisfaction degree of science and engineering personal and provide the references for policies made by the country. Among the information above, salary condition and satisfaction are important items highlighted in the investigation, because salary affects the engineering staff' enthusiasm to devote themselves to work and to play their core roles. For instance, US National Science Foundation (NSF) conducts a survey on the bachelors or masters in science, engineering or health fields every two years, and knows about and predicts the trends of education, employment opportunity and salary of graduates. Meanwhile, China Association for Science and Technology (CAST) also makes a sampling survey on science and technology practitioners every five years, and learns about their conditions in such aspects as employment mode, scientific research environment, living condition, flowing trend and thinking. Now how is the salary condition of the Chinese engineering staff? Are they competitive? How about their satisfaction for work? However, a few research data are disclosed for the description and analyses on such aspects, and some researchers have made some surveys and researches on only the salary condition of S&T workers. For instance, Wu Xianhua, et al. makes the research on the salary condition of the S&T workers in Changzhou and Xuzhou in Jiangsu Province[4]; Chen Tao make the research on the salary incentive effect of the S&T workers in Jiangsu Province[5], These researches are not made specially against the group of engineering staff and the surveys are conducted with limited sample size and area coverage, so the research results are difficult to reflect the nationwide circumstances of the engineering staff.

Based on the data from the 4th National S&T Workers Condition Survey, the research project team makes further survey on the salary condition of the science and technology practitioners and takes in-depth interviews for learning the salary conditions in some enterprises and institutions. In support of these primary data and existing secondary data, this article is focused on the analysis on the salary construction, influencing factors and satisfaction condition of the research object 'engineering staff' and raising the policy suggestions against the problems found. Salary comes with broad and narrow definitions: broad salary mainly refers to the sum of monetary incomes, specific services and welfare obtained by a party under an employee relationship, which includes basic salary, performance salary, long-term incentive salary, welfare and service[6]. Narrow salary refers to the compensation paid to worker in form of money, also called as salaries [7]. This article is to analyze the income condition and satisfaction degree of the Chinese engineering staff in narrow salary angle.

2. Salary Conditions of the Chinese Engineering Staff

1) The salary level of the engineering staff is improving gradually, but not ideal

With progressive development of the national economy, the income level of the Chinese engineering staff improves continuously. In 2017, the average annual income (including salary income and additional income) of engineering staff is RMB 41,993[8], about twice of that in 2007, with an average annually increase rate of 12.8%. As shown in the correlation analysis (Table 1 and Table 2), incomes of engineering staff are positively correlated to the education background and job title: the higher the education background and job title, the higher the gross income is. Such phenomenon is more obvious especially in the education background distribution: among the group with an income of RMB 150,000, engineering staff with junior college education account for only 7.3 %, ones with master degree account for 16.8 %, and ones with doctor degree account for 29.1%. This means that education background and job title play vital roles in salary obtaining of engineering staff.

Table 1 Relationship of engineering staff income with education background.

		Education	Gross Annual Income
Education	Pearson correlation coefficient	1	0.106*
	Sig. (two-tailed)		0.000
	Number of cases	14087	14087
Gross Annual Income	Pearson correlation coefficient	0.106*	1
	Sig. (two-tailed)	0.000	
	Number of cases	14087	14510

* At level 0.01 (two-tailed), with notable correlation.

Table 2 Relationship of engineering staff income with job title.

		Gross Annual Income	Job Title
Gross Annual Income	Pearson correlation coefficient	1	0.143*
	Sig. (two-tailed)		0.000
	Number of cases	14510	14510
Job Title	Pearson correlation coefficient	0.143*	1
	Sig. (two-tailed)	0.000	
	Number of cases	14510	14510

* At level 0.01 (two-tailed), with notable correlation.

According to surveys, the annual income of engineering staff has been growing continually; the average growth rate is 12.8% in 2007–2017. However, compared with the average salary growth rate (19.0% [9]) of employees in urban units in 2007—2016, the income growth rate of engineering staff is not high enough. And compared with the income (96440 USD [10]) of American engineering staff in 2016, the difference is still relatively large. In addition, compared with all S&T workers, engineering staffs who want to change jobs account for higher rate (more than 40%). Among the engineering staff who want to change jobs, nearly 50% are due to poor income, which is beyond the average of all S&T workers. More than 40% of the engineering staffs consider economic income as the largest source of stress, which is far beyond the average of all S&T workers. This fully illustrates that the salary situation of Chinese engineering staff is not ideal. Poor economic income may affect the family life and working condition of engineering staff, inhibit their work motivation and innovation enthusiasm, and also lead to the job change of engineering staff, force them to pursue and to be engaged in some jobs not related to their professions but with higher income, which will have negative impacts on improvement of enterprise innovation capability and sustainable innovation of industrial technology.

2) Diversified Salary Structure Is Reflected Apparently in Engineering Staff

In 2006, China's public ownership institutions carried out the Fourth Reform of Wage System (also known as Reform of Performance Salary). Employees' salaries included basic salary, post salary and performance salary, which highlighted the great influence of employees' actual contribution on their salaries. Especially in recent years, the Chinese government has specially issued a number of national policies, which require that the actual contribution of S&T workers should be highlighted in the salary distribution, and a more diversified salary system should be implemented. According to surveys, the diversified salary structure, which includes fixed salary system, annual salary system, dual-structure salary structure (basic salary + performance salary), ternary-structure salary system (basic salary + post salary + performance salary), negotiated salary system, salary + equity/option + bonus, is reflected apparently reflected in engineering staff. Most of the engineering staff' salaries belong to fixed salary system, dual-structure salary system or ternary-structure salary system; engineering staff in only a few entities have implemented annual salary system, and few have followed negotiated salary system or salary + equity/option + bonus. This shows that the diversified salary structure of Chinese engineering staff has made great progress, but there are still many people enjoying fixed salary system, annual salary system and other salary systems with great incentive have been implemented rarely, having a long way to achieve a perfect salary structure for Chinese engineering staff.

3) Most Engineering Staff Consider Their Own Salary Levels as Acceptable

Salary satisfaction refers to the positive or negative emotional attitude level of an individual to its remuneration[11]. The salary satisfaction is an important factor in determining employee's behavior, which plays an important role in the working enthusiasm and efficiency of the employees. According to this survey, the salary satisfaction of engineering staff is not low. Nearly 30% of engineering staff are very satisfied or relatively satisfied with their own salary level, and over 50% of engineering staff consider their own salary level as acceptable. Only less than 30% of engineering staff are not very satisfied or very dissatisfied with their own salary level. This shows that the salary level of Chinese engineering staff has achieved their expectation to a certain extent. Although relatively few are very satisfied and relatively satisfied, they can generally accept their own salary level, and only a few of them are not satisfied with their own salary level. At the same time, in the interview, most engineering staff also expressed that the current salary level was acceptable, but not very satisfying.

4) Differences in Salary Satisfaction of Different Engineering Staff

In this paper, a single factor analysis of variance (ANOVA) is carried out for different groups of engineering staff. The results are as follows: engineering staff with different titles have significant differences in salary satisfaction (p=0.000<0.05, see Table 3), and it is found in the post-inspection that in addition to the intermediate title and deputy senior title, every other two different titles have significant differences (p<0.05), and the salary satisfaction is improved with the title accordingly. The differences in salary satisfaction of engineering staff with different educational backgrounds are not significant (p=0.323>0.05, see Table 4). Engineering staff under different salary structures have significant differences in salary satisfaction (p=0.000<0.05, see Table 5), and it is found in the inspection that the engineering staff under annual salary system, salary + equity/option + bonus and negotiated salary system have relatively high salary satisfaction, and no significant difference in salary satisfaction; the engineering staff under fixed salary system, dual-structure salary system and ternary-structure salary system have lower satisfaction; especially engineering staff under fixed salary system have the lowest satisfaction, and have significant differences compared with those under other salary systems. This phenomenon proves two facts: first, salary structures, such as annual salary system, salary + equity/option + bonus and negotiated salary system, have their own advantages that are more flexible and can reflect the actual contribution of employees, thus are welcomed by engineering staff; fixed salary system has great disadvantage that cannot reflect the actual performance and contribution of the employees, thus receives the lowest satisfaction. Second, it further highlights existing problems in salary structure of engineering staff in our country, although dual-structure salary system and ternary-structure salary system implemented belong to performance salary system, salary distribution is not carried out according to the actual contribution of engineering staff, which leads to not very high satisfaction of the employees.

Table 3 Inspection on differences in salary satisfaction of engineering staff with different titles.

	Square Sum	Degree of Freedom	Mean Square	F	Significance
Among Groups	75.983	4	18.996	21.004	0.000
Within Group	13117.158	14504	0.904		
In Total	13193.141	14508			

Table 4 Inspection on differences in salary satisfaction of engineering staff with different educational backgrounds.

	Square Sum	Degree of Freedom	Mean Square	F	Significance
Among Groups	3.139	3	1.046	1.162	0.323
Within Group	12681.666	14083	0.9		
In Total	12684.805	14086			

Table 5 Inspection on differences in salary satisfaction of engineering staff with different salary structures.

	Square Sum	Degree of Freedom	Mean Square	F	Significance
Among Groups	99.319	6	16.553	18.331	0.000
Within Group	13089.726	14496	0.903		
In Total	13189.045	14502			

5) The salary conditions of young engineering staff should be paid more attention

At a status of most vigorous creativity, young S&T workers are the most energetic fresh troops in the science and technology team. Currently, the proportion of young S&T workers' acceptance of systematic education is relatively high, with high overall academic degree, so the work and innovation capability of young S&T workers should be given the largest incentive and excavation. However, the satisfaction survey on salary reveals that, the satisfaction degree of engineering staff below 40 years old for their own salary is far below that of those above 40 years old, and over 40% of the engineering staff below 40 years old consider that their own income is at the middle and low level. During the interview, although engineering staff reflect the practical problems of high living pressure, but the feelings of young engineering staff are stronger: they are faced with housing problem, marriage problem, children problem and parents support problem, so they are concerned with economic salary and their dissatisfaction of their economic situation result in their eager to seek for better way out, which influences their professional loyalty to their current job and reversely affects their innovation enthusiasm.

3. Reason Analysis of Existing Problems in Salary Incentive of Chinese Engineering Staff

1) Diversified salary structure has not been practically implemented

The ultimate objective of diversified salary is to promote the enthusiasm and initiative of the employees and improve their work performance. Although the investigation discloses that most engineering staff realize diversified performance related payment system, but also that the salary of most engineering staff are relatively largely influenced by years of working, post, title and leadership approval, only the salary of a minority of engineering staff is influenced by actual performance and project undertaking quantity. This leads to a relatively large gap from their expected salary influence factors. This demonstrates that diversified salary system has not been truly implemented in organizations and enterprises, which has not truly reflected the salary system oriented to actual contribution. Particularly in public ownership organization, this phenomenon is more obvious. During analysis, Zhao Min[12] points out that the egalitarian thought of scientific research institution is deeply rooted and the assessment system of salary distribution does not reflect the actual contribution of employees. Wang Na[13] holds that in the salary of scientific research institution, post salary derails from the actual post work, doesn't adjust incentive action via different salary system of post salary, and lacks internal fairness. Interviews with some state-owned enterprises also reveal that the performance salary system they implement for engineering staff is actually fixed salary system with insufficient incentive. The salary distribution of Chinese most public ownership units substantially still adopts the traditional allocation system of institutions, so salary distribution is still mainly based on such traditional elements as seniority, academic degree, while the role of such elements as knowledge, management, labor and technology is insufficient, and then the motivating performance salary is relatively small in the whole salary system, difficult to truly reflect the difference of actual contribution, the internal unfairness of salary allocation weakens the incentive action of performance salary of institutions [14].

2) Total quantity control restrains the exertion of salary incentive mechanism

In China, most public ownership organizations adopt control system gross salary, i.e. the annual gross salary of every organization is established, including basic salary, post salary and performance salary. If an organization improves basic salary and post payroll scale, the performance salary share will be compressed, which on the long term, will gradually weaken its incentive effect. Therefore, the implementation of total amount control of salary will result in the ceiling effect of incentive effect. Shen Yixuan[15] points out that, the total amount control of finance is an external difficult point of salary allocation system and the personnel caliber policy and financial caliber policy are in deviation status and policy objective and policy measures are in discordance, resulting more contradiction and harassment of salary allocation system implementation. Yi Wei[16] points out that as the dynamic adjustment mechanism of gross salary of public institutions has not been established, as long as the total amount is established, it is hard to be readjusted for some time, which, to a certain extent, influences the effect of public institutions to implement salary incentives. Among the investigated S&T workers, over 70% engineering staff work in public ownership organizations, so their salary incentive mechanism is similarly at the mercy of gross salary system and hard to exert incentive action of performance salary

References

[1] Zhang Haishui. Actual State, Problem and Countermeasures of Engineering staff Human Resource [J]. China Journal of Commerce, 2013, (5): 177-178.

[2] Huang Yuanxi. Research on Current Status and Distribution of Engineering Staff in China [J]. Overview of China Technology and Economy, 2017, (2): 42-48.

[3] Meng Fan, Xie Lianrui, et al. Empirical Analysis on Occupational Capability and Performance of Engineering staff [J]. Technology and Economy, 25(3): 86-90.

[4] Wu Xianhua, Guo Ji, et al. Demonstration Survey and Policy Advices on Salary incentive Status of Scientific and Technological Personnel - Xuzhou, Yangzhou and Changzhou in Jiangsu Province [J]. Scientific Research Management, 2011, (3): 77-90.

[5] Chen Tao. Research on Salary incentive Effect of Scientific and Technological Personnel in Enterprises [J]. Scienology and S&T Management, 2010, (7): 163-169.

[6] Carolyn Milkovich, Jerrr. M. Newman. Compensation (Edition 6), translated by Dong Keyong, et al. [M]. China Renmin University Press, Beijing, 2002.

[7] Xie Xuanzheng. Analysis on Salary Satisfaction of Human Resource Managerial Personnel in Enterprises [D]. Jinan University, 2009.

[8] Project Group of National S&T Workers Condition Survey. Report of the 2rd National S&T Workers Condition Survey[R]. China Science and Technology Press, 2010.

[9] National Bureau of Statistics of the People's Republic of China. China Statistical Yearbook (2017) [M]. China Statistics Press, 2017.

[10] Source: American employment income data up to May 2016 released by US Bureau of Labor Statistics on March 2017.

[11] Yang Tao, Tao Rong, Li Guohong. Analysis on Salary Satisfaction of Physicians in Shanghai Public Hospitals (Based on Investigation of 4 Tertiary Hospitals) [J]. Journal of Shanghai Jiaotong University (Medicine), 2016, 36 (6): 896-900.

[12] Zhao Min. Optimized Research on Performance Assessment on Knowledge Employees in Scientific Research Institutions, with the Example of GL [D], Northwest University, 2016.

[13] Wang Na. Study on Employee Salary Management System of Scientific Research Institutions [J]. Human Resource Management, 2014 (12): 92.

[14] Zhang Yingzi. Study on the Reform Issues of Salary System of Scientific Research Institution in China [D], Jilin Finance and Economics University, 2015.

[15] Shen Yixuan. Study on Salary Allocation System of Scientific Research Institution in Beijing [J]. Modern Business, 2013, (23).

[16] Yi Wei. Reflection on Sichuan's Agricultural Scientific Research Institutions Implementing Performance Salary Reform [J]. Sichuan Agricultural Science and Technology, 2016, (2): 8-10.

Research on the Legal Regulation of Preventing Academic Misconduct in China: Current Situation and Problems

Xia Ting

National Academy of Innovation Strategy, CAST, Beijing, China

Abstract: This article sorts out the legal regulation of China's prevention and treatment of academic misconduct from five levels: national laws, administrative regulations, departmental rules, group conventions, and administrative measures for academic misconduct in colleges and universities. Based on the above findings: the current legislative standards for prevention and control of academic misconduct are low, lack of basic legal documents; many advocacy rules, lack of processing rules, and lack of operability; academic misconduct has exceeded the current regulation range. In view of the above problems, this paper puts forward suggestions on prevention and treatment of academic misconduct from the perspective of legal regulation.

Keywords: Academic Misconduct; Law; Administrative Regulations; Departmental Rules

The experience of countries all over the world proves that it is necessary to rely on the law to prevent academic misconduct. Many countries have also formed a relatively complete legal policy system in the prevention and treatment of academic misconduct. In recent years, China has attached great importance to the prevention of academic misconduct and has continuously increased its punishment for academic misconduct. To explore the existing laws and policies for the prevention and treatment of academic misconduct in China, and to study the problems, it plays an important role in preventing and controlling academic misconduct in China, building an academic credit system, optimizing the academic environment, and improving academic ecology.

1. Definition of Academic Misconduct

The academic circles have a lot of names for academic misconduct, such as scientific research, academic corruption, and academic misconduct. At present, there is no unified definition of academic misconduct at home and abroad.

In 1992, the definition of academic misconduct was given by a group of 22 scientists consisting of the National Academy of Sciences, the National Academy of Engineering, and the National Academy of Medical Sciences: fabrication, tampering, or plagiarism in the process of applying for a project or implementing the results of a research report. That is, misconduct is mainly restricted to 'forgery, falsification, plagiarism' (FFP).[1] *Opinions of Chinese Academy of Sciences on Strengthening the Construction Code of Conduct for Scientific Research* defines scientific misconduct as various fabrications, falsifications, plagiarism and other violations of the recognized morality of scientific community in research and academic fields; other acts that violate social morality in the process of abusing and defrauding scientific research resources. As mentioned in *Measures for Handling of Scientific Misconduct in the Implementation of the National Science and Technology Plan (Trial)*, scientific misconduct refers to the violation of the scientific research code of conduct recognized by the scientific community, include(1) Providing false information on the title, resume and research basis of research personnel; (2) copying and plagiarizing other scientific research results; (3) data falsification; (4) in the study of human body, violation of informed consent, privacy protection and other provisions;(5) violation of laboratory animal protection regulations; (6) other scientific research misconduct. *Measures for Preventing and Handling Academic Misconduct in Colleges and Universities* clarifies six kinds of academic misconduct such as plagiarism, plagiarism, encroachment on the academic achievements of others, inappropriate authorship, academic paper selling or ghostwriting.

In this study, academic misconduct is defined as the behavior that violates the scientific community's code of conduct, falsification, plagiarism or other violations of the public code of conduct in the course of academic research. According to the existing research at home and abroad, the main academic misconducts are roughly divided into the following four categories: plagiarism, forgery, falsification and others. 'Others' mainly include inappropriate authorship, multiple submissions, multiple publications.

2. The Status Quo of Legal Regulation Against Academic Misconduct

According to the legal order of validity, combined with the actual situation, this part mainly elaborates the existing laws on prevention and treatment of academic misconduct in China from five levels: national laws, administrative regulations, local regulations and departmental rules, university management methods, and group conventions.

1) National Law

The national laws related to academic misconduct mainly include *Law of the People's Republic of China on Progress of Science and Technology, Copyright Law of the People's Republic of China, Patent Law of the People's Republic of China, Criminal Law of the People's Republic of China, and Law of the People's Republic of China on Teachers. Law of the People's Republic of China on Progress of Science and Technology* is the basic law in the field of science and technology in China. In article 55, it clearly states that scientific and technical personnel should carry forward the scientific spirit, abide by academic norms, abide by professional ethics, be honest and trustworthy, and shall not falsify in scientific and technological activities. Article 71 stipulates that researchers cheat National Science and technology awards should bear the legal consequences of withdrawing rewards and recovering bonuses according to law, the recommended unit or individual providing false data and materials to assist in defrauding the national scientific award should bear the consequences of suspending or canceling the qualification of the recommendation. In Article 47 of the *Copyright Law of the People's Republic of China*, it clearly stipulates that infringements that plagiarize the works of others, in accordance with the circumstances, should bear civil liabilities such as stopping the infringement, eliminating the effects, apologizing, and compensating for losses.

Table 1 National laws for preventing academic misconduct (sorted by time of promulgation).

Num	Legal Name	Enactment Department	Implementation Time
1	Law of the People's Republic of China on Progress of Science and Technology	the NPC Standing Committee	2008
2	Patent Law of the People's Republic of China	the NPC Standing Committee	2009
3	Copyright Law of the People's Republic of China	the NPC Standing Committee	2012
4	Criminal Law of the People's Republic of China	the NPC Standing Committee	2015

2) Administrative Regulations

Administrative regulations are the general term for the political, economic, educational, scientific, cultural, foreign affairs and other regulations formulated by the State Council for lead and manage all administrative tasks in the country, in accordance with the Constitution and laws, and *Rules for Enacting Administrative Regulations*. Administrative regulations are formulated by the State Council. The administrative regulations related to academic misconduct mainly include *Regulations on State Science and Technology Prizes*, *Regulations of the National Natural Science Foundation of China*, and *Outline of the National Intellectual Property Strategy*.

In articles 21 to 23 of *Regulations on State Science and Technology Prizes* clearly stipulate academic misconduct such as plagiarism, infringement of others' discoveries, inventions or other scientific and technological achievements, provide false data and materials, and play favoritisms and commit irregularities in the process of peer review, should receive appropriate punishment. In the third part of *Outline of the National Intellectual Property Strategy*, clearly pointing out that in the whole society, we should promote the moral values which being proud of innovation, being ashamed of plagiarism, being proud of honesty and trustworthiness, and being ashamed of fabrication and deception, cultivate an intellectual property culture that respects knowledge, advocates innovation, and integrity and law-abiding. In the sixth part of *Regulations of the National Natural Science Foundation of China*, it stipulates different forms of academic misconduct from applicants, participants, project leaders, participants, support units, review experts, fund management staff in the entire scientific research activities, then undertake different legal responsibilities according to different situations.

Table 2 Administrative regulations or documents for preventing academic misconduct (sorted by time of promulgation).

Num	Legal Name	Enactment Department	Implementation Time
1	Regulations on State Science and Technology Prizes	The State Council	2003
2	Regulations of the National Natural Science Foundation of China	The State Council	2007
3	Outline of the National Intellectual Property Strategy	The State Council	2008

3) Departmental Regulations

In this paper the departmental regulations refer to the normative documents formulated by the constituent departments and department directly under the State Council within their terms of reference. All regulatory documents contain requirements for academic norms or treatment for dealing with academic misconduct.

These departmental regulations can be divided into three categories, one is specific to the treatment of academic misconduct, such as *Academic Standards for Philosophy and Social Science Research in Colleges and Universities (Trial)*, *Treatment on violation of scientific ethical behavior about member of China Academy of Engineering*, etc. The second category is the guiding opinions for strengthening academic norms, such as the *Chinese Academy of Sciences' Opinion on Strengthening the Norms of Scientific research*, *Suggestions on Strengthening the Construction of Scientific Research Integrity in China*, and *Opinions on Further Regulating Scientific Research Behaviors in Colleges and Universities*; The third category is the guidance or management methods related to academic misconduct in the rules on fund management and project management, such as *Notice of the Ministry of Finance and the Ministry of Science and Technology on the Measures for the Administration of Special Funds for the National Science and Technology Support Program*, and *Notice of the Ministry of Finance and the Ministry of Science and Technology on the Measures for the Administration of Special Funds for the National Key Basic Research and Development Program*, and *Opinions on Further Strengthening the Management of Scientific Research Projects in Colleges and Universities*.

In article 3 of *Opinions on the Code of Conduct for Science and Technology Workers* , it clearly stipulates that in scientific and technological research activities, it is not allowed to fabricate, tamper with, or piece together the research results or experimental data for the conclusion, it also does not allow for opportunistic and out of context to make research conclusions that are inconsistent with objective facts. *Opinions on the Treatment of Academic Misconduct in Philosophy and Social Science in Colleges and Universities* stipulates that the following academic misconduct must be dealt with seriously: (1) plagiarism, embezzlement of others' academic achievements; (2) tampering with others' academic achievements (3) Forging and tampering with data, literature, and fabricating facts; (4) forging citations; (5) signing on the academic achievements of others; (6) improperly using the signature of others without permission of others; (7) other academic misconduct. Article 7 of *The Ministry of Education's Opinions on Further Strengthening and Improving the Construction of Teachers' Morality* stipulates that it make a stand against to misconduct in the scientific research work, such as falsification, plagiarism, which violates academic norms and encroaches on the labor results of others.

4) Group Convention

In 1999, the 231 National Society of China Association of Science and Technology signed *National Society Science and Technology Journal Ethics Convention* in Beijing. The scientific journals of the National Society promised to abide by the regulations and supervise each other. *National Society Science and Technology Journal Ethics Convention* includes a total of seven articles. In the prevention and treatment of academic misconduct, the Convention proposes that those who have academic misconduct will be given a written warning, refuse to publish their papers, inform his unit or public exposure according to different situations. For serious violations of the

Table 3 Departmental regulations for preventing academic misconduct (sorted by time of promulgation).

Num	Legal Name	Enactment Department	Implementation Time
1	Opinions on the Code of Conduct for Science and Technology Workers	MOST、MOE、CAS、CAE、CAST	1999
2	Standards of Ethical Conduct in Scientific Research for CAS Members	CAS	2001
3	Opinions on Strengthening Academic Ethics Construction	MOE	2002
4	Academic Standards for Philosophy and Social Science Research in Colleges and Universities (Trial)	MOE	2004
5	The Ministry of Education's Opinions on Further Strengthening and Improving the Construction of Teachers' Morality	MOE	2005
6	Measures taken by Supervision Committee of the National Natural Science Foundation of China on the Misconduct in the Funding of Science Funds (Trial)	NSFC	2005
7	Opinions on Establishing Socialist Concept of Honor and Disgrace and Further Strengthening Academic Morality Construction	MOE	2006
8	Notice of the Ministry of Finance and the Ministry of Science and Technology on the Measures for the Administration of Special Funds for the National Science and Technology Support Program	MOF、MOST	2006
9	Notice of the Ministry of Finance and the Ministry of Science and Technology on the Measures for the Administration of Special Funds for the National Key Basic Research and Development Program	MOF、MOST	2006
10	Measures for the Handling of Scientific Misconduct in the Implementation of National Science and Technology Plan (Trial)	MOST	2006
11	Declaration on the Concept of Science	CAS	2007
12	Chinese Academy of Sciences' Opinion on Strengthening the Norms of Scientific research	CAS	2007
13	Detailed rules for the Implementation of the State Science and Technology Awards Ordinance	MOST	2008
14	Opinions on the Treatment of Academic Misconduct in Philosophy and Social Science in Colleges and Universities	MOE	2009
15	Notice of the Ministry of Education on Seriously Handling Academic Misconduct in Colleges and Universities	MOE	2009
16	Suggestions on Strengthening the Construction of Scientific Research Integrity in China	MOST、MOE、MOF、MOHRSS、MOHC、The General Reserve Department of PLA、CAS、CAE、NSFC、CAST	2009
17	The Academic Degrees Committee of the State Council on Strengthening the Construction of Academic Ethics and Academic Norms in the Degree Awarding Work	ADCSC	2010
18	Implementation Opinions on Strengthening and Improving the Construction of Academic Atmosphere in Colleges and Universities	MOE	2011
19	Treatment on false behavior of Academic degree's dissertation	MOE	2012
20	Scientific code of ethics of Chinese Academy of Engineering academicians	CAS	2012
21	Opinions on Further Regulating Scientific Research Behaviors in Colleges and Universities	MOE	2012
22	Opinions on Further Strengthening the Management of Scientific Research Projects in Colleges and Universities	MOE	2012

(Continued)

Num	Legal Name	Enactment Department	Implementation Time
23	Opinions of the Ministry of Education and the Ministry of Finance on Strengthening the Management of Scientific Research Funds in Colleges and Universities	MOE、MOF	2012
24	Five requirements for publishing academic papers	CAST、MOE、MOST、NHFPC、CAS、CAE、NSFC	2015
25	Measures for the Prevention and Treatment of Academic Misconduct in Colleges and Universities	MOE	2016
26	Treatment on violation of scientific ethical behavior about member of China Academy of Engineering	CAE	2017

professional ethics of scientists and technicians with bad circumstances and extremely bad impacts, the relevant departments will be referred to deal with them seriously. It may also be referred to the Committee on the Ethics and Rights of Scientists and Technologists of China Association of Science and Technology for necessary investigation.

In 2007, China Association of Science and Technology formulated *Scientific Ethics of Scientific and Technological Workers (Trial)* in accordance with relevant national laws and regulations.

In 2014, 29 universities including Peking University and Tsinghua University held the '985 Project' University Graduate Research Integrity Seminar and released the first *Research Integrity Convention of Postgraduate in China.* It mainly consists of five parts. The Convention advocates a good academic atmosphere, continuously strengthens its academic moral cultivation, and jointly creates an academic atmosphere that is energetic, rigorous, pragmatic, open and inclusive. Refused to make quick success and resist the impetuous wind.

Table 4 Group Conventions on Prevention of Academic Misconduct (sorted by time of promulgation).

Num	Legal Name	Enactment Department	Implementation Time
1	National Society Science and Technology Journal Ethics Convention	the 231 National Society of CAST	1999
2	Scientific Ethics of Scientific and Technological Workers (Trial)	CAST	2007
3	Research Integrity Convention of Postgraduate in China	29 universities	2014

5) Administrative Measures for Preventing Academic Misconduct in Colleges and Universities

With the proliferation of academic misconduct, universities have also released their own administrative measures or academic norms to prevent academic misconduct. *Basic Academic Norms of Graduate Students in Peking University* stipulate that if there are serious violations of academic norms during the period of the graduate school, as soon as passes through the verification, the relevant rewards, diplomas and degree certificates obtained at that time will be revoked. *Measures for Prevention and Treatment of Academic Misconduct in Tsinghua University* has detailed provisions on the acceptance, investigation, identification, handling, review, prevention and supervision of academic misconduct. According to *Handling Methods for Postgraduates' Violation of Academic Ethics in Jilin University*, the degree office is responsible for accepting reports and complaints about violations of academic norms by graduate students. I*nterim Measures for the Prevention and Treatment of Academic Misconduct in Postgraduate Thesis in Jinan University* stipulates that degree evaluation committee, degree evaluation subcommittee, subject group, thesis defense committee, postgraduate instructor are responsible for review of academic dissertations, identification and treatment of academic misconduct in accordance with their respective limits of authority and the requirements of degree application review.

3. Problems

1) Lack of Basic Legal Documents

In our country, according to the order of

Table 5 Management measures for preventing academic misconduct in colleges and universities (sorted by time of promulgation).

Num	Legal Name	Enactment Department	Implementation Time
1	Basic Academic Norms of Graduate Students in Peking University	Peking University	2007
2	Graduate Academic Norms in ZheJiang University	Zhejiang University	2008
3	Interim Measures of Nankai University for Handling Academic Misconduct	Nankai University	2009
4	Scientific Research Code of Conduct and Style of Study Construction in Nanjing University (Trial)	Nanjin University	2009
5	Interim Measures for the Treatment of Academic Misconduct in Postgraduate Thesis of Renmin University of China	Renmin University of China	2010
6	Measures for the Treatment of Academic Misconduct in China Ocean University	Chinese Marine University	2010
7	Academic Conduct Standards and Management Measures in Xi'an Jiaotong University (Discussion Draft)	Xi'an Jiao Tong University	2010
8	Science and Technology Academic Ethics and Academic Misconduct Regulations in Huazhong University of Science and Technology (Trial)	Huazhong University of Science and Technology	2010
9	Measures for the Treatment of Academic Misconduct in Degree (Graduation) Papers in Sichuan University (Trial)	Sichuan University	2010
10	Interim Measures for the Prevention and Treatment of Academic Misconduct in Postgraduate Thesis in Jinan University	Jinan University	2011
11	Handling Methods for Postgraduates' Violation of Academic Ethics in Jilin University	Jilin University	2013
12	Academic Ethics Code of Conduct in China Agricultural University	China Agriculture University	2014
13	Methods for Detection and Treatment of Academic Misconduct on Postgraduate Thesis in Northwest University	Northwest University	2014
14	Implementation Opinions on Investigation and Handling of Suspected Academic Misconduct in Southeast University (Revised Draft)	Southeast University	2014
15	Measures for the Treatment of Academic Misconduct in Sun Yat-sen University	Sun Yat-sen University	2014
16	Methods for Identifying and Handling Academic Misconducts of Postgraduates in Beijing Normal University	Beijing Normal University	2015
17	Measures for Dealing with Academic Misconduct in Chongqing University (Trial)	Chongqing University	2016
18	Detailed Rules for Investigation of Academic Misconduct in Wuhan University	WuHan University	2016
19	Measures for Prevention and Treatment of Academic Misconduct in Tsinghua University	Tsinghua University	2017

Note: This table is incomplete statistics.

effectiveness of the law, it can be divided into constitution, law, administrative regulations, departmental regulations and local regulations. Judging from the current situation of legal regulation of academic misconduct, the first is the legislative level for preventing and controlling academic misconduct is relatively low. The level of legislation is relatively low, many rules and regulations belong to the departmental regulations. Prevention of academic misconduct lacks a basic legal document. The existing laws are not specific to the basic law for the prevention and treatment of academic misconduct. Secondly, *Patent Law of the People's Republic of China* and *Copyright Law of the People's Republic of China* mainly responsible for adjusting intellectual property disputes between equal subjects. Second, there are cross-coincidences in the rules and regulations of various departments, and the rights between departmental laws are unclear and the scope of adjustment is ambiguous. Third, for serious academic misconduct, the main punitive measures are administrative sanctions, not criminal liability, so the crime cost of serious academic misconduct is very low in China, which provides an opportunity for the breeding of academic misconduct. For example, *Measures for the Prevention and Treatment of Academic Misconduct in Colleges and Universities* clearly stipulates, according to the nature of the behavior and the severity of the circumstances, the person responsible for academic misconduct will be dealt with as follows: (1) circulate a notice of criticism; (2) to terminate or revoke the relevant scientific research projects, and to cancel the application qualification within a certain period of time; (3) to revoke academic awards or honorary titles; (4) to dismiss; (5) other treatments by laws, regulations and rules. From the perspective of legal significance, these are still administrative penalties, not civil penalties and criminal penalties.

2) Existing legal Regulations Lack Implementation Rules

Many existing laws and regulations are advocacy-based content, lacking implementation rules. From a legal perspective, the administrative punishment for people who have academic misconduct in China can't find the corresponding legal basis. Administrative legislation is still a blank area in dealing with academic misconduct.[2] For example, in article 4 'the Supervision of academic misconduct' of *Scientific Ethics of Scientific and Technological Workers (Trial)*, there is no clear handling of the rules on the treatment of academic misconduct, so it is difficult to implement specific penalties from the realistic level.

3) Academic Misconduct Has Exceeded the Scope of Current Regulation

With the rapid development of network technology, the channels of information acquisition are more and more diversified, the acquisition speed is greatly improved; the digitalization of journal publishing is becoming more and more common, and the degree of intensification is getting higher and higher; the phenomenon of academic paper selling or ghostwriting is common; the academic paper selling or ghostwriting company begins commercial operation, all of which make academic misconduct more concealed and complicated, and it is increasingly difficult to identify academic misconduct. According to public network information, a peer reviewer found an author explain a new scientific discovery in the paper, he immediately revealed the relevant information to his own student, and at the same time put forward some review comments to delay review process. Then his student falsifies some data, and put pictures taken by others into his own paper, and rushed to publish the paper in a small magazine with the same academic views. [3] This kind of academic misconduct is hidden and not easy to identify, and it is beyond the scope of current legal regulations.

4. Suggestions

1) Clarify the Criteria for Defining Academic Misconduct

There are many kinds of academic behaviors, and the specific problems and environments are characterized by diversity and complexity, so it is difficult to exhaust all the academic misconduct. [4] Existing legal regulations should be based on different types of academic activities and characteristics, combined with the current phenomenon of academic misconduct, to clarify the constituent elements and main characteristic of academic misconduct. By explain clearly the basic standards of academic misconduct, it can provide clear judgment criteria for dealing with academic misconduct in practice, and is more conducive to accurately identifying and punishing academic misconduct, and providing a sufficient legal basis for dealing with academic misconduct.

2) Strengthening Legislation on Academic Norm

Academic misconduct may result in civil liability, administrative liability and criminal responsibility. At present, the punishment of academic misconduct in China's legal regulation is mainly administrative responsibility, and the punishment of administrative responsibility is generally inside the unit and department where the person is punished. It is even a non-public punishment. First, it is difficult to play a warning role in the academic community. Second, for the person being punished, there is no financial compensation or mental compensation and other criminal penalties. The processing result of each unit is not exactly the same. So the cost of conducting academic misconduct is very low. Therefore, this paper suggests strengthening academic norms at the legal level, and at the same time gradually improve the regulation of academic misconduct in criminal law and civil law, so the punishment for academic misconduct have a legal basis.

3) Strengthening the Construction of Supervision System for Academic Misconduct

The lack of supervision can be said to be an important reason for the spread of academic misconduct in China. This paper believes that the construction of the supervision system can be strengthened from three levels, first, from the academic subject level, strengthening the author's supervisory responsibility for the paper is an important part of curbing academic misconduct from the source. In many cases, the actual author of one paper is only 1-2 people, but the co-author(s) are as high as 6-8 people. Second, from the level of administrative management, a research integrity office should be established, the members of the office should have both scholars, legal expert, administrator who have dealt with academic misconduct, the office of research integrity should establish a supervisory reporting mechanism, and improve the investigation procedure for academic misconduct, to deal with academic misconduct in accordance with the law. Third, from the level of supervision by public opinion, timely publish the investigation and treatment results of academic misconduct, give full play to the supervision role of the media and the network, and create a fair and positive academic environment.

References

[1] Study Style Construction Committee of Science and Technology Committee of the Ministry of Education. Guide to academic norms of science and technology in colleges and universities[M]. China Renmin University Press, 2010.

[2] Xu Heping, Yuan Yuli. Discussion on Administrative Legal Regulation of Academic Misconduct[J]. Academics in China， 2011(10): 48-56.

[3] Jin Dongyan. Resisting hidden academic misconduct [EB/OL]. (2011.08.22) [2018-07-17] http://blog.sciencenet.cn/home.php?mod=space&uid=216627&do=blog&id=478089.

[4] Zhao Tingting, Feng Lei. The structure and characteristics of American academic misconduct governance system[J]. University Education Science, 2016(05): 93-101.

Analysis of Chinese Unicorn Phenomenon and Suggestions for Innovation Policy

Chen Qianqian

National Academy of Innovation Strategy, CAST, Beijing, China

Abstract: This article mainly describes the current hot topic – the unicorn phenomenon. The unicorn phenomenon plays a very important role in the country's economic development and innovation. This paper summarizes the status and characteristics of the unicorn phenomenon and unicorn enterprises in China firstly, and then compares and analyzes with the American unicorn enterprises. At last we sort out the four relationships that make good use of the unicorn phenomenon, and finally puts forward suggestions.

The innovation group represented by the unicorn enterprise has led a new round of technological revolution and has become an important indicator to demonstrate a country's ability of innovate. China is the second largest gathering area of unicorn enterprises in the world. Especially since the release of the innovation-driven strategy, the unicorn company has entered an accelerated expansion period in a good policy environment. However, unicorn enterprises in China also have problems such as low valuation, insufficient capacity for sustainable development, lack of technological innovation, etc. In order to thoroughly implement the innovation-driven development strategy, and use government and market functions better, we study the unicorn phenomenon to provide the basis for policy development.

1. Introduction

General Secretary Xi Jinping pointed out at the Davos Forum that a new round of industrial transformation is moving forward at an exponential rate. According to the research, in recent years, unicorn enterprises have concentrated in emerging technologies such as big data and mobile internet. The scale of income that conforms to the trend of technology evolution and market demand tends to grow by order of magnitude. This reflects that the industrial transformation has not been ready to go, but it's already popping up.

For a long time, innovation ecology of China has been dominated by the government and the research institutes, all belong to the administrative plan. After the innovation-driven development strategy was put forward, innovation ecology of China has undergone positive changes. It is not difficult to find five changes from the unicorn phenomenon: (1) The superposition effect of comprehensive innovation begins to appear. The concept of comprehensive innovation is deeply rooted in the hearts of the people, and the company has achieved leapfrog development through the policy dividend. (2) High-growth companies play a key role in promoting disruptive innovation and industrial change. The unicorn enterprise has become an engine to promote industrial transformation. At the same time, the infiltration development of the digital economy, the sharing economy, and the intelligent economy has become an unstoppable force. (3) Market-oriented innovative resource allocation mechanisms are emerging. Venture capital plays an increasingly important role in the innovation system. The market determines the technical direction, route selection, and factor price allocation. (4) The enterprise-led innovation ecosystem has emerged. In particular, the internet platform companies play a key role in China's new innovation ecosystem, and its influence on capital, talents, technology and innovation system cannot be underestimated. (5) Key areas are forming an ecological environment that is inclusive and innovative. The entrepreneurial services, agglomeration of innovative resources, perfect market mechanism, and strong entrepreneurial culture of Beijing, Shanghai, Shenzhen and Hangzhou make these cities become the source of innovation with global influence gradually.

In addition, the unicorn phenomenon is a

Block C, Science and Technology Hall, Haidian District, Beijing. chenqkc@126.com.

microscope that financially supports innovation. The unicorn company has a short establishment time and high market valuation, which is sought after by investors. This reflects the fact that high-growth enterprises often adapt to the new laws of industrial transformation and market demand. The unicorn company are creating new productivity. They improve the production efficiency and competitiveness, and attract the support of capital much easier. Discovering and cultivating unicorns and enhancing their demonstration will help guide resources to areas innovation and competitiveness and promote the rebalancing of finance and the real economy. Studying the phenomenon of unicorns has positive significance innovation and development of China.

2. The Characteristics of Unicorn Enterprises of China

Unicorns represent the high-growth enterprise in the capital market. The concept of the unicorn was originally proposed by Aileen Lee, the founder of the seed round fund Cowboy Ventures in 2013. Unicorns refer to start-ups that are fast-growing and highly regarded by investors. The standard is about 10 years of entrepreneurship, and the company's valuation is over $1 billion. Among them, companies with a valuation of more than 10 billion US dollars are called 'super unicorns'. The global unicorn is mainly concentrated in the emerging field represented by information technology. However, there is no authoritative unicorn statistics in the world. The valuation of enterprise is not based on the unified evaluation of the current net assets and capital return rate, but the judgment of one or more investment institutions based on the growth of the enterprise.

According to the *Report of China Unicorn Enterprise Development in 2017* released by the Torch Center of the Ministry of Science and Technology, there are 164 unicorn enterprises in China with a total valuation of $628.4 billion. In general, China's unicorn enterprises have 'Four concentration':

1) Founding Time Concentration

Most of the unicorn enterprises in China were founded after the 18th National Congress of the Communist Party of China. The unicorn enterprises established after 2012 accounted for 56%, of which the largest number is in 2014, there are 31. 29 in 2012, and 25 in 2015.

2) Regional Concentration

Unicorn enterprises have appeared in 16 cities of China. Among them, Beijing, Shanghai, Shenzhen and Hangzhou rely on a good innovation and entrepreneurial ecology become the most concentrated area of unicorn enterprises, accounting for 88% of the total.

3) Field Concentration

More than 80% of China's unicorn enterprises are distributed in the internet information service industry. The four industries of e-commerce, Internet finance, culture and entertainment, and transportation have accounted for 56%. From the valuation scale, 28% comes from internet finance and 14% comes from e-commerce. From the global perspective, the top seven unicorn companies in India are engaged in internet information services, this situation is similar to that in China. However, American unicorn companies are widely distributed in 73 sub-sectors, including software, artificial intelligence, biotechnology, commercial aerospace and other technological innovation-driven areas.

4) Platform Concentration

Most unicorns are incubated or invested by large platform companies. Unicorn companies that have been incubated or invested by large internet companies such as Alibaba, Baidu, Tencent, Xiaomi, and Jingdong account for the majority.

3. Comparison of Chinese and American Unicorn Enterprises

As of the first half of 2018, according to data from US consulting firm CB Insights, there are 125 US unicorn companies with a total valuation of $435.8 billion, involving 26 industries including transportation, corporate services, financial technology, and big data. These unicorn companies are mainly distributed in 23 cities including Silicon Valley, San Francisco, New York and Los Angeles. China and the United States have great differences in the development of unicorn enterprises. They are more obvious in terms of the number and valuation, industry distribution, and urban distribution.

1) China and the United States are Different in the Unicorn Industry

From the perspective of the top ten industries of unicorn companies, there is a clear difference between China and the United States. In China, the number one unicorn company is the financial technology, and the United States is transportation. Among the top ten industries of American unicorn companies, there are corporate services, big data, aviation technology and social networking. These five industries have not entered the top ten in China. Similarly, among the top ten industries in china, cloud services, cultural entertainment, logistics, new energy vehicles and new media have not entered the top ten in the United States. This shows that China and the United States have significant differences in key industries for innovation and entrepreneurship.

2) Some Emerging Areas and Traditional Areas in China are Blank

In several emerging and traditional areas, unicorn companies have emerged in the United States, while they are blank in China. In the fields of aviation technology, building technology, 3D printing, VR/AR, clean technology, etc., there are unicorn companies with a valuation of more than 1 billion US dollars in American. The above-mentioned industries, there are have no unicorn company with a valuation of more than 1 billion US dollars in China. In some traditional fields, such as clothing accessories, fitness and other fields, the United States also has unicorn companies with a valuation of more than \$1 billion. But traditional industry in China have unicorn companies rarely. In addition, there is a unicorn company JUUL Labs with a valuation of up to 15 billion dollars in the US e-cigarette market. There is no similar unicorn company in China in this field.

3) Unicorn Companies in China and the United States are Different in Object Orientation

From the perspective of the distribution of unicorn enterprises in China and the United States, Chinese unicorn companies focus on user-oriented and business model innovation. For example, the financial technology field mainly focuses on P2P loan platforms. The industries like transportation, e-commerce, culture and entertainment, new energy vehicles and new media are focused on users. American unicorn companies focus on enterprises and technological innovation. For example, the unicorn companies of corporate services sector in the US totaled valuation of \$61.3 billion, second only to the transportation sector. The corporate service sector have the most unicorn companies, the number is 21. There are only five unicorn companies in China's corporate services sector, with a total valuation of only \$6.9 billion.

4. Thoughts and Suggestions

At present, China's unicorns have three blinds. First one, blind valuation. With the help of big internet companies and venture capital giants, the valuations of related companies have grown too fast, and some have deviated from their actual value and brought unsustainable risks. Secondly, blind worship. The rapid development of the unicorn enterprise has created the myth of internet-making riches, causing worship blindly. Many entrepreneurs have built up model innovations of internet business, more emulating and more failing. There are even some local governments take the number of unicorn companies as the measure of local innovation performance. The third is blind expansion. Most capital that unicorn companies receive is not used in the research and development of core technologies, but in the crazy 'burning money' that quickly takes advantage of the scale of the internet. These areas are often policy-sensitive. If there are changes in regulatory policies or broken financial chains, business is easy to close.

Take the unicorn fever dialectically, four pairs of relations need to grasp.

(1) The relationship between business model innovation and original technology innovation. In the past, the independent innovation emphase on original innovation, integrated innovation, introduction, digestion, absorption, and innovation, all focused on technological innovation. However, in the context of the digital economy, major changes have taken place in the innovation ecosystem and innovation system. The 'technology + model' innovation represented by the unicorn enterprise has attracted more and more attention. The new format and new model reflect strong vitality. The national revitalization must hold the original innovation and the enterprise's development must firmly grasp the core competitiveness of technological innovation.

Business model innovations that are independent of technological innovation like passive waters and unrooted trees.

(2) The relationship between the new business model and the real economy. Currently, unicorn companies are mainly concentrated in the internet field. Internet-based cross-border integration is advanced productivity, gathering a lot of resources. There are even opinions that the internet format model has attracted resources that should be invested in manufacturing, jeopardizing the foundation of the real economy. However, dialectical analysis is not difficult to find that the rise from the internet to 'internet +' and to the integration of the internet and manufacturing is an irreversible historical trend. The survey found that Suzhou GCL Solar introduced Alibaba Cloud's big data application. In just half a year, PV production improved by one percentage point, and annual production cost savings of over 100 million RMB. Practice has proved that the real economy and the internet are parallel relationship and should complement each other. The internet service is also a real economy, and it can fully blend with the traditional real economy. The new mode and new business are inseparable from the foundation of manufacturing, service industry and agriculture. The realization of integrated development is the only way to revitalize the real economy.

(3) The relationship between the pursuing valuation and long-term healthy development. The development of venture capital in China has accelerated and its scale has been among the best in the world. But there are also many startups and venture capital firms obsessed with the concept of speculation, raising the valuation in order to seeking greater returns when exiting. The study believes that emerging industries are inseparable from venture capital, but if entrepreneurs are excessively pursuing high valuation financing and ignoring the value of technology and products, it is inverted. Innovative enterprise growth is like long-distance race. It needs to be deeply cultivated around the core technology, strengthen the body and continuously improve the development quality and brand. Capital investment is more like 'milk powder' or 'transfusion', which cannot be without, and cannot be over-reliant, otherwise it is prone to bubbles, and even jeopardize economic development.

(4) The relationship of market-led and government-led. Whether the success of platform companies such as BAT or the growth of hundreds of unicorns, it reflects the development of China's market-oriented technology innovation system. However, behind the 'unicorn fever' is lack of original innovation, the similarity of business models, and lack of stamina for sustainable development. These problems reflect that rely solely on market orientation is not enough even in the field of fully competitive emerging industries. A healthy and sustainable innovation ecosystem requires the participation of government, research institutions, universities, enterprises, and financial institutions. The government must increase investment in strategic areas and market failures. The rapid development of the unicorn reflects the acceleration of the digital economy transformation. However, the ultimate success of the transition depends on the degree of institutional innovation and technological innovation at the national level. The government's planning and guiding role cannot be neglected.

According to the above analysis, the suggestions are following.

(1) Innovation drive can no longer fall into the scale orientation, should pay more attention to strategic value and work hard on technological innovation and institutional innovation. The innovation-driven development strategy is essentially an economic transformation that relies on innovation to improve the quality and efficiency and cannot take the old path of factor-driven. It is necessary to avoid the formation of new scale expansion and low-level repetitive patterns in emerging fields and it is not possible to say succeed or defeat by the number of unicorns simply. In the next step, it is necessary to rationally divide the government and market boundaries. The government's innovation investment should aim at the new round of scientific and technological revolution and dominant power of industrial change and more focus on strategic innovation. It is necessary to promote institutional innovation with greater efforts, create a more suitable environment, release more resources, and transform the scale advantages of internet users and big data resources into real strategic advantages and form overall competitiveness.

(2) Supporting the development of emerging

fields cannot be limited to venture capital investment. It is necessary to exert the role of government funds to complement the short-comings and work hard to guide the resources investment strategy. Most of the emerging industries are in the field of full competition. It is must play a decisive role in the allocation of resources in the market. In recent years, governments at all levels have reduced support for direct subsidies in emerging areas, and have injected to guide funds, hoping to indirectly cultivate emerging industries through the development of venture capital. At present, the profit-seeking characteristics of capital have influenced the main investment of venture capital. Some areas overheated while strategic areas invest insufficient. In this context, the policy objectives cannot be limited to supporting venture capital itself and should strengthen the guidance of industrial investment to play a fill the short board effect. In the next step, we can consider formulating a special policy performance evaluation method for the national emerging industry venture capital guiding fund, improve the policy transmission mechanism, and guide more venture capital investment in technological innovation fields such as genetic engineering, artificial intelligence, Nano materials, biomedicine, etc. Innovative companies investing in the initial and seed periods. On the other hand, it is necessary to enhance government investment in key areas and strategic links and accelerate the implementation of major systemic projects in the '13th Five-Year Plan'. Government resources should be leaner in 'no-man's land' and 'strategic highland' to prevent government background industry funds from drifting along and becoming a booster for investment bubbles.

(3) Entrepreneurial culture can't be quick and should pay more attention to social values and human development, and work hard to the source of innovation and stimulate the spirit of exploration. The innovation of mass entrepreneurship is not to guide everyone to create wealth, to become rich overnight, but to build a fair stage and superior conditions for more people to realize their values. In the next step, we must guide entrepreneurial culture to return to the source of innovation and introduce more hard measures in the areas of open sharing of data and technology resources, integration of artisan spirit and entrepreneurial culture, and synergy between institutions and dual-innovation. Guide entrepreneurs to carry out technological innovations based on real problems, social needs, and life needs. To create a tolerant entrepreneurial environment, encourage and support more entrepreneurs to explore freely in cutting-edge areas.

(4) 'Internet +' cannot only to build up new business model, but also to promote the integration of the internet into the real economy, improve quality and efficiency. To grasp the direction of public opinion, we cannot simply understand innovation as internetization, nor can we rely on business model innovation for internet development. On the one hand, it is necessary to strengthen technological breakthroughs in basic fields such as integrated circuits, network security, operating systems, and fifth-generation mobile communications, and expand the technological advantages of internet development. On the other hand, we must make great efforts to promote the integration of the internet and the real economy. Guide internet companies and manufacturing enterprises to carry out in-depth cooperation. Rely on internet technology to enhance the value chain of the real economy and expand the industrial chain. Let more unicorn companies integrate into the real economy to improve quality and efficiency

(5) The development of unicorn enterprises cannot rely solely on the scale advantage of the domestic market. It is necessary to emphasize more global perspectives and international development. On the one hand, it is necessary to strengthen cooperation between government and enterprises, jointly improve the domestic application market environment, and shape more 'Chinese programs' and 'Chinese brands' of internet applications. It is necessary to create conditions for more internet platform companies and unicorn enterprises to undertake the strategic research of the country and improve their innovation ability and competitiveness. On the other hand, we must also create conditions to support these superior enterprises to go global. Combined with the implementation of the 'One Belt, One Road' strategy, it will provide a greater platform, integrate resources and open up markets for China's e-commerce, internet finance and sharing economy to 'go global', form an international influence and a new global competitive advantage.

References

Wu Beibei, Wang Shengnan (2017). Research on the explosive growth of enterprises in the new economy. *Frontier theory.* 10: 23-24.

Wang Lijun, Li Ziying, Kang Zhengguang and Zhang Hua (2017). Thoughts on Developing Unicorn Enterprises in Jiangsu Province. *Jiangsu Science and Technology Information.* 36: 46-47.

Liu Zhuoqun (2018). Research on Risk Management of Unicorn Enterprises in China. *Knowledge economy.* 13: 41-43.

Studies on Culture of Science under the Perspective of Big Data

Zhang Haodong, Wu Hong

National Academy of Innovation Strategy, CAST, Beijing, China

At 10:30-12:00 on September 16, the 21st Session 'Studies on Culture of Science from Perspective of Big Data' of 2018 Science & You International Symposium and Summit on Culture of Science was held in conference room 208 of the National Convention Center. The moderator was Wu Hong from the National Academy of Innovation Strategy and the minutes taker was Zhang Haodong.

In this session, academic exchange was conducted on the theme of 'Studies on Culture of Science from Perspective of Big Data', studying and discussing the topics such as solution of practical problems with big data of science and technology, cyberspace and new economic transformation, governance of computer technology society, application of big data in studies of science of science, big data research in standard development and text mining in scientific publications & patents.

1. Help Enterprises to Innovate and Develop with Big Data of Science and Technology

Zhu Lijun, the executive deputy director & researcher of the National Engineering Center of Institute of Scientific and Technical Information under the Ministry of Science and Technology, presented the report titled 'Help Enterprises to Innovate and Develop with Big Data of Science and Technology'. The report is divided into three major parts: problems confronted by technology innovation, analysis of cases in electric vehicle industry, and core processing technique for big data of science and technology.

According to Zhu Lijun, there are five stages in the life cycle of a technology-based enterprise, namely seed stage, start-up stage, growth stage, stable stage and transformation stage, and every stage has different need for innovation. The main problems in the innovation network with government-industry-university-research-user synergy are: enterprises haven't become the subject of technology innovation; the industry-university-research combination is loose and the continuously stable cooperation around industrial technology innovation chain is insufficient; the innovation resources are scattered and repeated, with unbalanced layout; and enterprises, especially small- and medium-sized enterprises, are short of comprehensive effective support services.

He introduced the relevant studies on the electric vehicle technology forecast and decision support system. By collecting data from the traditional industry chain and the Internet, a one-stop support service platform for electric vehicle technology decision and plan management is formed and a new model of information service for management of science and technology, namely 'demonstration, decision, management and innovation with data', is developed. For instance, via big data analysis, the competitive and cooperative relationship of the listed companies in the electric vehicle field can be studied.

With the core technology of big data processing, the literature database of theses, patents and technology reports is linked and integrated with scientific data, project data, personnel data, achievement data, Internet information and enterprise data; the information resources are restructured, linked and mined deeply; index models can be established; computing tools can be developed; services can be provided for information decision; and platforms of basic services and value added services, for example, for finding technology, experts or projects, can be built.

2. Cyberspace Innovation Trend and New Economic Transformation

Shen Yushi, an expert among the 13th group of the 'Thousand Talents Program' and the vice president & (Cloud) CTO of 21Vianet Group, presented the report titled 'Cyberspace Innovation Trend and New Economic Transformation'. He believed that the development of science and technology has entered a new revolution era. Scientific research is evolving more quickly and deeply in various scales including microscopic and cosmoscopic scale. Physical sciences keep on expanding towards macroscopic, microscopic and extreme conditions, entering the age of big science and comprehensively expanding the space of world perceived by human. Cutting-edge technologies are taking on the trend of breakthrough at multiple points; information, biological, new energy and new materials become the most important directions leading industry changes; and a chain revolution with mutual support and common development of multiple technology clusters is coming into being. The development of science is showing a trend of high integration based on high differentiation more and more; the integration of human, machine and material is accelerated; and more emerging research areas and broad innovation spaces are being exploited. The revolution in scientific research, innovation and industrial paradigm is changing the traditional scientific organization mode, innovation competition mode and industry existence mode. Technological innovation becomes the core power for economic growth; it is triggering huge changes in economic system, catalyzing new industrial structure, power structure, production pattern and consumption pattern and realizing new leap of social productive forces. The allocation and flow of innovative elements worldwide is accelerated, the fight-alone era has past and the global innovation is taking on a new layout. Technological innovation is showing more complexity and uncertainty and may bring unprecedented challenges to the economic & social development or even the global governance.

Cyberspace becomes the second space for human survival and the fifth domain independent of land, ocean, air and space for human — the information domain (network domain). Communication technology, information technology and artificial intelligence technology are further integrated. Informatization becomes socialized and molecular. Internet is not all of cyberspace, but only its first wave, initiated by the Silicon Valley and led by the USA, with a basically established international pattern. The second wave of cyberspace is led and guided jointly by a group of mathematicians, algorithm scientists and IT elite, covering cloud computing, big data, AI, global blockchain, super internet, crypto-currency and digital economy. The next generation of network information system is characterized by transition from personal data to personal data center; from digital living to digital life; from interconnection to interlink; and from inorganic world to organic world. The ultimate thinking and consciousness will last forever.

Create a new generation of industrial public blockchain via Token + real economy, to promote the integration development of digital currency and real economy. In particular, blockchain represents the underlying protocol and the future information infrastructure. Token represents the ticket and language for the computation of man, machine and material. In the future, if you want the service from AI or a robot, you must pay tokens, because it doesn't recognize legal tender but recognize digital currency.

3. Technology Society and Culture of Science

Ma Huimin, the vice chairman & secretary general of China Society of Image and Graphics and an associate professor of the Department of Electronic Engineering, Tsinghua University, presented the report titled 'Technology Society and Culture of Science'. China Society of Image and Graphics was founded in 1990. The 7th board of directors is in office from 2016 to 2020, of which the chairman is academician Tan Tieniu. The society sticks to a 'member-based, democratic and academics-first' philosophy.

1) Technology society management culture

According to the Plan for Implementation of Deepening Reform of Science & Technology Association System and the work arrangement of the Party group of China Association for Science and Technology for deepening reform, in terms of the culture of directors and the academic leadership of the board, the election of directors follows the principle of combining academic level with service awareness and the technology leading personnel are over 1/3 of the total number of directors; the size of the board is moderate, with 157 persons, convenient for discussion and decision and capable to carry out work comprehensively; affairs are handled strictly according to rules, everyone performs his/her duties, the attendance rate of the standing council is over 91% and the attendance rate of the board is over 70%. The scope of work of the chairman is clear, and a work structure with leadership under chairman, guidance & supervision under secretary general and specific implementation under the working committee is formed. The 24 special committees carry out academic activities actively, covering various research directions of the image and graphics field. The existing special committees will be reformed, new special committees will be created and the management of special committees will be standardized. Rules and regulations are the safeguard for the culture of technology society. The society has developed and revised 12 rules, regulations and management measures including the Regulations for Management of CSIG Special Committees, to guarantee the standardized and efficient operation of the society. On November 10, 2016, the society set up a functional Party committee at the level of standing council, to the play the leading role of Party members, increasing the cohesion of the Party organization and the board of directors significantly.

2) Service culture of technology society

According to the needs of different member groups, the society has constructed an academic exchange system integrating large academic annual meetings, high-end cutting-edge conferences, forums and saloons; for social needs, it holds academic promotion activities such as advanced disciplines lectures, China visits and technology forums; holds science popularization activities such as 'Walk into Primary School Classroom'; builds a platform for scientific payoff sharing of the society, and publishes the journals such as the Journal of Image and Graphics and the Communication of China Society of Image and Graphics; strengthens academician recommendation, young talent recommendation and rewarding, allowing the society to be a channel for innovation talents to stand out; carries out evaluation of scientific and technological achievements, to evaluate the scientificalness, advancement and feasibility of technological achievements; holds exhibitions of science and technology such as Vision China, to serve technology enterprises; develops and implements the plan for member development and service, to make member services accurate and diversify the service modes.

3) Reflection on technology society and culture of science

In the construction of the society, we are exploring to build a technology society meeting the law of development of technology society with academic influence, member cohesion, social credibility and independent development capability, develop sound and vital systems for academic exchange and member honor and become a founder and promoter of culture of science.

4. Studies of Science of Science under Academic Big Data

Jia Tao, a selected candidate of the national 'Thousand Young Talents Program' and a professor of Southwest University presented the report titled 'Studies of Science of Science under Academic Big Data'.

He believed that for the application of big data, three aspects need to be considered: Complexity, Data availability and Practical need. Seen from the emerging research area 'science of science', science is a complicated system; the quantity of scientific articles takes on an exponential increase, doubled every nine years worldwide; there have been many indexes used to quantify and evaluate science activities, such as SCI, CCF-A, Impact factor and H-index, to understand the regular pattern of science.

According to the study of Quantifying Patterns of Research-interest Evolution, in the publications of Web of Science in 1990–2016, there were over one million articles and over ten million citations every year. The study showed that the international citations of Chinese articles were decreasing gradually in 2010-2016. The number of domestic citations of Chinese articles was very large. As the representativeness of Chinese research work was insufficient, the citations by international counterparts can't match with the enormous quantity of domestic citations of international counterparts. Therefore, for Chinese articles, the quantity of citations may not be a good measuring standard, and the quantity of international citations may be a better way.

5. Big Data Research in Standard Development of China

Gan Keqin, the director of the Information Technology Department, National Library of Standards, China National Institute of Standardization, presented the report titled 'Big Data Research in Standard Development of China'.

Standards are the basic system for the modernization of governance system and governance capability and are playing an increasingly significant role in facilitating business and economic exchange, supporting industry development, promoting progress of science and technology and regulating social governance. In the Circular of the State Council on Issuing the Executive Summary for Promotion of Big Data Development, it is indicated that big data have become a new power driving economic transformation and development, a new opportunity for rebuilding the competitive edge of the country and a new approach to promote the political governance capability. Standards are an important component of national quality and technical base and a manifestation of the soft power of a country, region or unit, representing the discourse power, governance capability and technical & quality level.

1) Characteristics of standard development in China

In 2016, there were 6009 drafting units of national standards. From 2001 to 2016, the number of drafting units of national standards increased by 12.2% annually. In 2015, there were 8592 drafting units of industry standards. From 2001 to 2015, the number of drafting units of industry standards increased by 11.8% annually. About 14000 drafting units were involved in both national standards and industry standards, and about 26000 drafting units were only involved in national standards. In terms of the concentration of standard development contribution indexes, in 2015, the concentration of drafting units of industry standards was relatively high, with the top 5% drafting units contributing 45%; in 2016, the concentration of drafting units of national standards was relatively high, with the top 5% drafting units contributing 38.85%. In terms of standard development in different regions, in 2016, the national standard indexes of Beijing, Guangdong Province, Zhejiang Province, Jiangsu Province, Shanghai and Shandong Province accounted for 67.9% of the national total. Among them, Beijing made the largest contribution. The leading role of research institutes in standard development is significant. In 2016, research institutes occupied 20.3% of the drafting units of national standards and led the development of 43.7% of national standards. The top ten drafting units of national standards were all research institutes or enterprises transformed from research institutes. The contribution of enterprises to the development of national standards and industry standards is taking on a rising trend, and the contribution of research institutes to the development of national standards and industry standards is taking on a declining trend.

2) National standard development contribution index

In 2016, the national standard index was 5786.9, the quantity of national standards issued was 2435 and the number of drafting units of national standards was 6009. Compared with 2015, the quantity of national standards issued increased by 7.1%, the national standard index decreased by 10.0% and the number of drafting units decreased by 11.5%. From 2001 to 2016, the annual growth rate of the quantity of national standards issued was 5.5%, the annual growth rate of national standard index was 8.8% and the annual growth rate of drafting units of national standards was 12.2%. From 2001 to 2016, the degree of participation in standard development increased significantly, and the number of drafting units per standard increased from 1.8 in 2001 to 6.5 in 2016, by 261%. In 2016, the concentration of national standard development was high, with the top 5% of drafting units (300) contributing 38.85%. In 2016, the drafting units of national standards were mainly concentrated in Beijing and coastal regions, including Guangdong Province, Zhejiang Province, Jiangsu Province, Shanghai and Shandong Province. Among them, the national standard index of Beijing was the highest, accounting for 27%. Enterprises were the main forces in national standard development. In 2016, the national standard index and the number of drafting units of enterprises both exceeded 50%. The leading role of research institutes in national standard development was clear: accounting for 20.3% of drafting units of national standards and leading the development of 43.7% of national standards. In 2016, the national standard indexes of enterprises nationwide were mainly distributed in: Zhejiang Province, Guangdong Province, Beijing, Jiangsu Province, Shanghai and Shandong Province. From 2001 to 2016, the national standard indexes of enterprises nationwide were mainly distributed in: Zhejiang Province, Beijing, Jiangsu Province, Guangdong Province, Shanghai and Shandong Province.

3) Industry standard development contribution index

In 2015, the industry standard index was 11088.2, the quantity of industry standards issued was 4599, and the number of drafting units of industry standards was 8705. Compared with the annual mean value from 2001 to 2015, the industry index increased by 61.2%, the quantity of industry standards issued increased by 31.9% and the number of drafting units increased by 86.3% in 2015. Compared with 2010, the last year of the '11th Five-year Plan', the industry standard index increased by 47.2%, the quantity of industry standards issued increased by 19.9% and the number of drafting units increased by 85.3% in 2015, the last year of the '12th Five-year Plan'. From 2001 to 2015, the annual average of industry standard index was 6879.2, the annual average of the quantity of industry standards issued was 3485.9 and the annual average of the number of drafting units of industry standards was 4672.7. From 2001 to 2015, the annual growth rate of the industry standard index was 10.3%, the annual growth rate of the quantity of industry standards issued was 7.4% and the annual growth rate of drafting units of industry standards was 11.9%. From 2001 to 2015, the degree of participation in industry standard development increased significantly, and the number of drafting units per standard increased from 2.1 in 2001 gradually to 4.4 in 2015, by 110%.

6. An Approach for Topic Linkages Between Science and Technology

Xu Shuo, a professor of Beijing University of Technology, presented the report titled 'An Approach for Topic Linkages between Science and Technology'. He discussed the studies on text mining of scientific publications and patents.

Scientific publications and patents are generally considered as an important index for basic scientific research and technological development. Intuitively, there should be some hidden interaction and exclusive relationship between scientific publications and patents. Linear models show that science is driving technology. The Two-branched Model (Rip, 1992) studied the technological development, pilot process and feedback. The researches of (Narin & Noma, 1985; Brooks, 1994) showed that technology is closely related to science, and they can be regarded as a pair of dancers or two strands of DNA. Furthermore, now high technology has almost become nothing different from science. Narin and his colleagues started the research on the connection between publications and patents. Such connection can help the understanding of technology opportunity discovery (Albert, 2016; Lee et al., 2011) and can help the understanding of university-industry-government interaction (Leydesdoref & Meyer, 2007; Tian 2015). Relevant research works also include: Non-Patent Reference (NPR) (Narin et al., 1997; Narin & Olivastro, 1992, 1998; Meyer, 2000, 2001, 2002), Citations of patents in scholarly articles (Glanzel & Mey, 2003), The correspondence tables between patent classes and scientific disciplines (Bassecouolard & Zitt, 2004), etc.

Academic publications aim at communicating scientific discovery to relevant research circles and the public, and patents are designed to prevent other manufacturers from commercializing their technical process or equipment. In order to guarantee the quality, peer review is generally used to filter, improve and plan the scientific publications. The review of patents is only driven by legal requirements, focusing on whether there is anything overlapping with existing patent documents and other public materials.

Professor Xu Shuo introduced his research framework in Topic Linkages, which originated from his article 'A CRF-based System for Recognizing Chemical Entity Mentions (CEMs) in Biomedical Literature' published in the Journal of Cheminformatics. Professor Xu compared the applications of CCorrLDA2 and CorrLDA2 model in scientific papers and patents, and introduced the studies in Entity Mentions, Word Tokens Clustering, Topic Similarity and Topic Linkage. He gave the research data of Topic Linkage of CCorrLDA2, CorrLDA2 and LDA model in Dataset.

According to Professor Xu, the topic linkages he is studying are similar to hyper-links. That is to say, the topic of one resource can be linked to multiple topics from other resources, and the topic links are asymmetrical. This study will be helpful to the topic navigation, technology frontier detection and business opportunity identification among different information resources. However, now there is no public reference dataset available for topic linkage yet and it is not easy to assess the results of topic linkage.